Ecohydrology

Ecohydrology

Editor:

Quan Cui, Ph.D.

ArclerPress

www.arclerpress.com

Ecohydrology

Editor: Quan Cui, Ph.D.

© 2017 Arcler Press LLC
708 3rd Avenue, 6th Floor
New York
NY 10017
United States of America
www.arclerpress.com

ISBN: 978-1-68094-483-9

Library of Congress Control Number: 2016951354

About the Editor

Quan Cui, Ph.D.

Quan Cui obtained her PhD from Beijing Normal University in 2015. She worked as environmental impact assessment engineer for years. Aiming at linking science to policy, her study interest spans from ecosystem modelling to water resources management. She published peer-reviewed articles about using machine learning models in inflow forecasting, CO2 management and watershed ecosystem health assessment.

Contents

List of Contributors

Junguo Liu
School of Nature Conservation, Beijing Forestry University, Beijing, 100083, China

Giri Kattel
School of Nature Conservation, Beijing Forestry University, Beijing, 100083, China
Self-Sustaining Regions Research and Innovation Initiative, Collaborative Research Network (CRN), Federation University Australia, Mt. Helen Campus, Victoria 3350, Australia

Hans Peter H. Arp
Norwegian Geotechnical Institute, Sognsveien 72, NO-0855, Oslo,Norway

Hong Yang
CEES, Department of Biosciences, University of Oslo, Blindern, Oslo 0316, Norway

Justin Huntington
Desert Research Institute, Reno, Nevada, United States
Western Regional Climate Center, Reno, Nevada, United States

Kenneth McGwire
Desert Research Institute, Reno, Nevada, United States

Charles Morton
Desert Research Institute, Reno, Nevada, United States

Keirith Snyder
USDA Agricultural Research Service, Reno, Nevada, United States

Sarah Peterson
Bureau of Land Management, Reno, Nevada, United States

Tyler Erickson
Google, Inc., Mountain View, California, United States

Richard Niswonger
U.S. Geological Survey, Menlo Park, California, United States

Rosemary Carroll
Desert Research Institute, Reno, Nevada, United States

Guy Smith
Desert Research Institute, Reno, Nevada, United States

Richard Allen
University of Idaho, Kimberly, Idaho, United States

B. Grizzetti
European Commission Joint Research Centre (JRC), I-21027 (VA), Italy

D. Lanzanova
European Commission Joint Research Centre (JRC), I-21027 (VA), Italy

C. Liquete
European Commission Joint Research Centre (JRC), I-21027 (VA), Italy

A. Reynaud
European Commission Joint Research Centre (JRC), I-21027 (VA), Italy

A.C. Cardoso
European Commission Joint Research Centre (JRC), I-21027 (VA), Italy

Ouko M. Chrisphine
Jomo Kenyatta University of Agriculture and Technology, Department of Geomatic Engineering and Geospatial Information, Nairobi, Kenya

Odhiambo A. Maryanne
Jomo Kenyatta University of Agriculture and Technology, Department of Geomatic Engineering and Geospatial Information, Nairobi, Kenya

Boitt K. Mark
Jomo Kenyatta University of Agriculture and Technology, Department of Geomatic Engineering and Geospatial Information, Nairobi, Kenya

Lavinia Suberg
National Oceanography Centre Southampton (NOCS), European Way, Southampton SO14 3ZH, UK

Russell B. Wynn
National Oceanography Centre Southampton (NOCS), European Way, Southampton SO14 3ZH, UK

Jeroen van der Kooij
Centre for Environment, Fisheries and Aquaculture Science (CEFAS), Pakefield Road, Lowestoft, Suffolk NR33 0HT, UK

Liam Fernand
Centre for Environment, Fisheries and Aquaculture Science (CEFAS), Pakefield Road, Lowestoft, Suffolk NR33 0HT, UK

Sophie Fielding
British Antarctic Survey, High Cross, Madingley Road, Cambridge CB3 0ET, UK

Damien Guihen
British Antarctic Survey, High Cross, Madingley Road, Cambridge CB3 0ET, UK

Douglas Gillespie
Scottish Oceans Institute, East Sands, University of St Andrews, St Andrews, Fife, KY16 8LB, UK

Mark Johnson
Scottish Oceans Institute, East Sands, University of St Andrews, St Andrews, Fife, KY16 8LB, UK

Kalliopi C. Gkikopoulou
Scottish Oceans Institute, East Sands, University of St Andrews, St Andrews, Fife, KY16 8LB, UK

Ian J. Allan
Norwegian Institute of Water Research, Oslo Centre Interdisciplinary Environmental & Social Research, Gaustadalléen 21, NO-0349 Oslo, Norway

Branislav Vrana
Masaryk University, Research Centre for Toxic Compounds in the Environment (RECETOX), Kamenice 753/5, 62500 Brno, Czech Republic

Peter I. Miller
Remote Sensing Group, Plymouth Marine Laboratory, Prospect Place, Plymouth PL1 3DH, UK

David Smeed
National Oceanography Centre Southampton (NOCS), European Way, Southampton SO14 3ZH, UK

Alice R. Jones
National Oceanography Centre Southampton (NOCS), European Way, Southampton SO14 3ZH, UK
The Environment Institute & School of Ocean and Earth Sciences, University of Adelaide, South Australia 5005, Australia

Michael B. Gracz
Conservation Biology Program, University of Minnesota, Saint Paul, MN 55108, United States
Department of Earth Sciences, University of Minnesota, Minneapolis, MN 55455, United States

Mary F. Moffett
Mid-Continent Ecology Division, United States Environmental Protection Agency, 6201 Congdon Blvd., Duluth, MN 55804, United States

Donald I. Siegel
Department of Earth Sciences, 204 Heroy Geologic Laboratory, Syracuse University, Syracuse, NY 13244, United States

Paul H. Glaser
Department of Earth Sciences, University of Minnesota, Minneapolis, MN 55455, United States

C. Lehr
Leibniz Centre for Agricultural Landscape Research (ZALF), Institute of Landscape Hydrology, Müncheberg, Germany
University of Potsdam, Institute for Earth and Environmental Sciences, Potsdam, Germany

F. Pöschke
Leibniz-Institute of Freshwater Ecology and Inland Fisheries (IGB), Department Ecohydrology, Berlin, Germany

J. Lewandowski
Leibniz-Institute of Freshwater Ecology and Inland Fisheries (IGB), Department Ecohydrology, Berlin, Germany
Humboldt University Berlin, Geography Department, Berlin, Germany

G. Lischeid
Leibniz Centre for Agricultural Landscape Research (ZALF), Institute of Landscape Hydrology, Müncheberg, Germany
University of Potsdam, Institute for Earth and Environmental Sciences, Potsdam, Germany

Keith Beven
Lancaster Environment Centre, Lancaster University, Lancaster, UK
Department of Earth Sciences, Geocentrum, Uppsala University, Sweden
Centre for Analysis of Time Series, London School of Economics, London, UK

Magdalena K. Grove
School of Environment and Technology, University of Brighton, Cockcroft Building, Lewes Road, Brighton BN2 4GJ, UK
Aquaread Ltd., Bridge House, Northdown Industrial Estate, Broadstairs, Kent, CT10 3JP, UK

Gary S. Bilotta
School of Environment and Technology, University of Brighton, Cockcroft Building, Lewes Road, Brighton BN2 4GJ, UK

Robert R. Woockman
Department of Civil and Environmental Engineering, The University of Tennessee, 325 John D Tickle Engineering Building, Knoxville, TN 37996, USA

John S. Schwartz
Department of Civil and Environmental Engineering, The University of Tennessee, 325 John D Tickle Engineering Building, Knoxville, TN 37996, USA

J.W. Bloodworth
School of Energy, Environment and Agrifood, Cranfield University, Cranfield, Bedfordshire MK43 0AL, United Kingdom
Scottish Water, Castle House, 6 Castle Drive, Dunfermline, Fife KY11 8GG, United Kingdom

I.P. Holman
School of Energy, Environment and Agrifood, Cranfield University, Cranfield, Bedfordshire MK43 0AL, United Kingdom

P.J. Burgess
School of Energy, Environment and Agrifood, Cranfield University, Cranfield, Bedfordshire MK43 0AL, United Kingdom

S. Gillman
Scottish Water, Castle House, 6 Castle Drive, Dunfermline, Fife KY11 8GG, United Kingdom

Z. Frogbrook
Scottish Water, Castle House, 6 Castle Drive, Dunfermline, Fife KY11 8GG, United Kingdom

P. Brown
Scottish Water, Castle House, 6 Castle Drive, Dunfermline, Fife KY11 8GG, United Kingdom

Clare Duncan
Institute of Zoology, Zoological Society of London, Outer Circle, Regent's Park, London NW1 4RY, UK
UCL Department of Geography, University College London, Gower Street, London WC1E 6BT, UK

Jurgenne H. Primavera
Conservation Programmes, Zoological Society of London, Outer Circle, Regent's Park, London NW1 4RY, UK
Zoological Society of London-Philippines, 43-E Burgos Street, Barangay Magdalo, La Paz, 5000 Iloilo City, Philippines

Nathalie Pettorelli
Institute of Zoology, Zoological Society of London, Outer Circle, Regent's Park, London NW1 4RY, UK

Julian R. Thompson
UCL Department of Geography, University College London, Gower Street, London WC1E 6BT, UK

Rona Joy A. Loma
Conservation Programmes, Zoological Society of London, Outer Circle, Regent's Park, London NW1 4RY, UK
Zoological Society of London-Philippines, 43-E Burgos Street, Barangay Magdalo, La Paz, 5000 Iloilo City, Philippines

Heather J. Koldewey
Conservation Programmes, Zoological Society of London, Outer Circle, Regent's Park, London NW1 4RY, UK
Centre for Ecology and Conservation, University of Exeter, Penryn, Cornwall TR10 9EZ, UK

Steven E. Orchard
Sustainability Research Institute (SRI), School of Earth and Environment, University of Leeds, Leeds, LS2 9JT, UK

Lindsay C. Stringer
Sustainability Research Institute (SRI), School of Earth and Environment, University of Leeds, Leeds, LS2 9JT, UK

Claire H. Quinn
Sustainability Research Institute (SRI), School of Earth and Environment, University of Leeds, Leeds, LS2 9JT, UK

Fei Tian
State Key Laboratory of Urban and Regional Ecology, Research Center for Eco-Environmental Sciences, Chinese Academy of Sciences, 18 Shuangqing Road, Haidian District, Beijing 100085, China
CSIRO Land and Water Flagship, Canberra, ACT 2601, Australia

Yi He Lü
State Key Laboratory of Urban and Regional Ecology, Research Center for Eco-Environmental Sciences, Chinese Academy of Sciences, 18 Shuangqing Road, Haidian District, Beijing 100085, China

Bo Jie Fu
State Key Laboratory of Urban and Regional Ecology, Research Center for Eco-Environmental Sciences, Chinese Academy of Sciences, 18 Shuangqing Road, Haidian District, Beijing 100085, China

Yong Hui Yang
Key Laboratory of Agricultural Water Resources, Center for Agricultural Resources Research, Institute of Genetics and Developmental Biology, Chinese Academy of Sciences, 286 Huaizhong Road, Shijiazhuang 050021, China

GuoYu Qiu
Key Laboratory for Urban Habitat Environment Science and Technology, School of Environment and Energy, Peking University, Shenzhen 518055, China

Chuanfu Zang
Forestry College of Inner Mongolia Agricultural University, Daxue East Road, Saihan District, Huhehaote, 010018, China

Lu Zhang
CSIRO Land and Water Flagship, Canberra, ACT 2601, Australia

Kim de Mutsert
Department of Environmental Science and Policy, George Mason University, 4400 University Drive, Fairfax, VA 22030, USA

Jeroen Steenbeek
Ecopath International Initiative, C/ Sant Eduard 22-A-2-1, Sant Cugat del Valles SP 08172, Barcelona, Spain

Kristy Lewis
Department of Environmental Science and Policy, George Mason University, 4400 University Drive, Fairfax, VA 22030, USA

Joe Buszowski
Ecopath International Initiative, C/ Sant Eduard 22-A-2-1, Sant Cugat del Valles SP 08172, Barcelona, Spain

James H. Cowan Jr.
Department of Oceanography, Louisiana State University, 1002-Y Energy, Coast & Environment Building, Baton Rouge, LA 70803, USA

Villy Christensen
Fisheries Centre, University of British Columbia, Vancouver, B.C., Canada V6T 1Z4

J. Forster
School of Environmental Sciences, University of East Anglia, Norwich NR4 7TJ, UK
School of Marine Science and Technology, Ridley Building 2, Newcastle University, NE1 7RU UK

I.R. Lake
School of Environmental Sciences, University of East Anglia, Norwich NR4 7TJ, UK

A.R. Watkinson
School of Environmental Sciences, University of East Anglia, Norwich NR4 7TJ, UK

J.A. Gill
School of Biological Sciences, University of East Anglia, Norwich NR4 7TJ, UK

D. Lancaster
University of Victoria, David Turpin Building, B243, Victoria BC V8W 2Y2, Canada
School of Environmental Studies, University of Victoria, Victoria, BC, Canada

P. Dearden
University of Victoria, David Turpin Building, B243, Victoria BC V8W 2Y2, Canada
Geography Department, University of Victoria, Victoria, BC, Canada

N.C. Ban
University of Victoria, David Turpin Building, B243, Victoria BC V8W 2Y2, Canada
School of Environmental Studies, University of Victoria, Victoria, BC, Canada

Jieqiong Su
State Key Laboratory of Water Environment Simulation, School of Environment, Beijing Normal University, Beijing 100875, China
Key Laboratory for Water and Sediment Sciences of Ministry of Education, School of Environment, Beijing Normal University, Beijing 100875, China
Chinese Academy for Environmental Planning, Ministry of Environmental Protection, Beijing 100012, China

Xuan Wang
State Key Laboratory of Water Environment Simulation, School of Environment, Beijing Normal University, Beijing 100875, China
Key Laboratory for Water and Sediment Sciences of Ministry of Education, School of Environment, Beijing Normal University, Beijing 100875, China

Shouyan Zhao
Management Office of Miyun Reservoir, Beijing 101512, China

Bin Chen
State Key Laboratory of Water Environment Simulation, School of Environment, Beijing Normal University, Beijing 100875, China

Chunhui Li
Key Laboratory for Water and Sediment Sciences of Ministry of Education, School of Environment, Beijing Normal University, Beijing 100875, China

Zhifeng Yang
State Key Laboratory of Water Environment Simulation, School of Environment, Beijing Normal University, Beijing 100875, China
Key Laboratory for Water and Sediment Sciences of Ministry of Education, School of Environment, Beijing Normal University, Beijing 100875, China

Lee E. Brown
School of Geography, University of Leeds, Woodhouse Lane, Leeds LS2 9JT, UK

Sheila M. Palmer
School of Geography, University of Leeds, Woodhouse Lane, Leeds LS2 9JT, UK

Kerrylyn Johnston
School of Geography, University of Leeds, Woodhouse Lane, Leeds LS2 9JT, UK

Joseph Holden
School of Geography, University of Leeds, Woodhouse Lane, Leeds LS2 9JT, UK

Frans H.M. van de Ven
Deltares, 3508 AL Utrecht, The Netherlands
Delft University of Technology,Dept. Water Management, Stevinweg 1, 2628 CN Delft, The Netherlands

Robbert P.H. Snep
Alterra Wageningen University & Research, 6700 Wageningen, The Netherlands

Stijn Koole
Bosch Slabbers landscape architects, 1e Sweelinckstraat 30, 2517 GD Den Haag, The Netherlands

Reinder Brolsma
Deltares, 3508 AL Utrecht, The Netherlands

Rutger van der Brugge
Deltares, 3508 AL Utrecht, The Netherlands

Joop Spijker
Alterra Wageningen University & Research, 6700 Wageningen, The Netherlands

Toine Vergroesen
Deltares, 3508 AL Utrecht, The Netherlands

Daniel Leitner
Computational Science Center, University of Vienna, Austria

Félicien Meunier
Earth and Life Institute, Environmental Sciences, Université catholique de Louvain, Louvain-la-Neuve, Belgium

Gernot Bodner
Division of Agronomy, BOKU-University of Natural Resources and Life Sciences, Vienna, Austria

Mathieu Javaux
Earth and Life Institute, Environmental Sciences, Université catholique de Louvain, Louvain-la-Neuve, Belgium
Forschungszentrum Jülich GmbH, Agrosphere (IBG-3), Jülich,Germany

Andrea Schnepf
Forschungszentrum Jülich GmbH, Agrosphere (IBG-3), Jülich,Germany

Preface

Ecohydrology deals with the interacts between ecological and hydrological processes which may happen in the water body or ecosystem. The scale of ecohydrology studies may up to water regime. This is an interdisciplinary science which may cross hydrology, ecology, environmental science and management. From studying the fundamentals of ecohydrology process, ecosystem management suggestions can be drawn for decision-makers. Water is our most precious natural resource. The availability and quality of fresh water not only impact human health and wellbeing, but also the functioning of essential ecosystems, including rivers, wetlands, lakes and coastal ecosystems. Most available fresh water is present as groundwater. Exchanges between groundwater and surface water occur throughout the landscape and support a plethora of key ecosystem services. This multidisciplinary text in ecohydrology is dedicated to advancing the understanding of the fluxes and transformations of nutrient elements and metals at the groundwater-surface water interface, and assessing their consequences for the health and functioning of aquatic ecosystems.

This book covers the freshwater ecosystem management, the assessment of water ecosystem services, the assessment of hydrological impacts on ecosystems, the studies on peatland, the studies of relationship between ecohydrology science and climate change and the management and assessment of pollutants. The models of water ecosystem resilient are included as well.

Ecohydrology aims to advance ecohydrology as the study of the interplay between ecological and hydrological processes from molecular to river basin scales, and to promote its implementation as an integrative management tool to harmonize societal needs with biosphere potential. An essential component of ecohydrology is a rigorous understanding of hydrobiology, from ecosystem properties, dynamics and functions to modelling of abiotic and biotic interactions in relation to their hydrological determinants at the basin scale. Ecohydrology is an interdisciplinary area that draws on both ecology and hydrology, and is related to watershed management. It places particular emphasis on interactions and associated feedbacks in both space and time between ecological systems and the hydrological cycle. Ecohydrology is especially important in arid and semiarid systems where the feedbacks between ecology and hydrology can be particularly tight. Research in ecolohydrology emphasizes considering "green" water associated with evapotranspiration as well as "blue" water such as streamflow and groundwater considered more traditionally in hydrology and watershed management. The information contained in this book will be of valuable for students, researchers as well as practitioners.

Editor
Quan Cui, Ph.D.

INTRODUCTION

Ecohydrology is an interdisciplinary field studying the interactions between water and ecosystems. These interactions may take place within water bodies, such as rivers and lakes, or on land, in forests, deserts, and other terrestrial ecosystems. Areas of research in ecohydrology include transpiration and plant water use, adaption of organisms to their water environment, influence of vegetation on stream flow and function, and feedbacks between ecological processes and the hydrological cycle.

The developmental conditions required for comprehensive, integrative and interdisciplinary scientific research is „maturity" of the empirical disciplines that participate in the integration process. The progress that took place in ecological sciences in the last years of the 20th century, allowed for major advancements of knowledge. A level was attained that permitted an attempt to integrate ecological sciences with the more advanced scientific fields to great extend expresses by physics and mathematics hydrology.

Key concepts

The hydrologic cycle describes the continuous movement of water on, above, and below the surface on the earth. This flow is altered by ecosystems at numerous points. Transpiration from plants provides the majority of flow of water to the atmosphere. Water is influenced by vegetative cover as it flows over the land surface, while river channels can be shaped by the vegetation within them. Ecohydrologists study both terrestrial and aquatic systems. In terrestrial ecosystems (such as forests, deserts, and savannas), the interactions among vegetation, the land surface, the vadose zone, and the groundwater are the main focus. In aquatic ecosystems (such as rivers, streams, lakes, and wetlands), emphasis is placed on how water chemistry, geomorphology, and hydrology affect their structure and function.

Principles

The principles of Ecohydrology are expressed in three sequential components:

- Hydrological: The quantification of the hydrological cycle of a basin, should be a template for functional integration of hydrological and biological processes.

- Ecological: The integrated processes at river basin scale can be steered in such a way as to enhance the basin's carrying capacity and its ecosystem services.
- Ecological engineering: The regulation of hydrological and ecological processes, based on an integrative system approach, is thus a new tool for Integrated Water Basin Management.

Their expression as testable hypotheses may be seen as:

- H1: Hydrological processes generally regulate biota
- H2: Biota can be shaped as a tool to regulate hydrological processes
- H3: These two types of regulations (H1&H2) can be integrated with hydro-technical infrastructure to achieve sustainable water and ecosystem services

Ecohydrology principles:

1. **Hydrological** (framework): integrating of understanding water and biota interplay at a catchments scale (GIS) for identification of threats and opportunities for sustainable water, economy and society. It covers such aspects as:
    - **scale:** mesocycle water circulation in a basin - the template for quantification of ecological processes;
    - **dynamics:** water and temperature - the driving forces for terrestrial and freshwater ecosystems;
    - **hierarchy of factors:** abiotic factors to hydrology dominate, however, when these are stable and predictable, biotic interactions begin to manifest themselves
2. **Ecological** (target): understanding of the evolutionarily established resilience and resistance of ecosystems to stress and patterns of ecological succession to increase their carrying capacity against human impacts.
3. **Ecological Engeneering** (methodology): using ecosystem properties as a management tool for restoration of biodiversity, improve water quality and enhancement of ecosystem services for society by:
    - **Dual regulation**: of hydrology by shaping biota and, vice versa, regulation of biota by altering hydrology;
    - **Integration:** at the basin scale various types of regulations should be integrated towards achieving synergy to stabilize and improve the quality of freshwater resources at a basin scale
    - **Harmonization** of ecohydrological measures with necessary hydrotechnical solutions (e.g., sewage treatment plants, levees in urbanized areas, etc.)

Aims and Scope

Ecohydrology & Hydrobiology is an international journal that aims to advance ecohydrology as the study of the interplay between ecological and hydrological processes from molecular to river basin scales, and to promote its implementation as an integrative management tool to harmonize societal needs with biosphere potential.

An essential component of ecohydrology is a rigorous understanding of hydrobiology, from ecosystem properties, dynamics and functions to modelling of abiotic and biotic interactions in relation to their hydrological determinants at the basin scale.

The journal will publish high-quality science aimed at increasing the understanding of water–biota interplay in all types of ecosystems and exploring how this knowledge can be used to regulate hydrological and biogeochemical processes to enhance the carrying capacity especially of anthropogenically modified ecosystems. Research addressing one or several of the following issues will be of prime interest to the journal if it:

- Leads to new insights into the interactions of water, nutrient and pollutant cycles with biotic components of ecosystems;
- Employs this understanding for developing ecosystem biotechnologies and system solutions;
- Identifies possibilities to integrate engineering infrastructure with ecosystem biotechnologies into system solutions at the catchment scale;
- Translates transdisciplinary knowledge into decision-support tools;
- Targets biodiversity, quality and quantity of water resources, ecosystem processes and functioning (including environmental flows), ecosystem services and/or ecosystem resilience as necessary components for enhanced carrying capacity;
- Is policy-oriented and, within the scope of the journal, addresses the priorities and objectives of international initiatives such as the Millennium Development Goals, UNESCO's International Hydrological Programme, Water Framework Directive, etc.

Ecohydrology Assumptions and Principles

Up to the time when the ecohydrology concept was defined, hydrologists considered aquatic biota mostly as an indicative system for monitoring while hydro biologists considered hydrological processes as a disturbance factor.

Key assumptions of EH

- Regulation of hydrology by shaping biota and, vice versa, regulation of biota by altering hydrology.
- Integration - at the basin scale various types of regulations (E <-> H) act in a synergistic way to improve and stabilize the quality of water resources.
- Harmonization of ecohydrological measures with necessary hydrotechnical solutions (e.g., dams, sewage treatment plants, levees at urbanized areas, etc.)

Following these assumptions the concept of ecohydrology is based on three principles.

Technical approach

The importance of the effort to develop the ecohydrology approach increased with the publication of the paper by Meybeck in which he justifies the name of Anthropogenic for the present era. Based on an in-depth analysis of published studies, he demonstrated that the modification of aquatic systems by human pressures (e.g., flood regulation, fragmentation, sedimentation imbalance, salinization, contamination, eutrophication, etc.) has increased to a level that no longer can be considered as being controlled by only natural processes (climate, relief, vegetation, limnology), thus defining a new era that we have already entered.

The decline in water quality and biodiversity, observed at the global scale in both developed and developing countries, has provided evidence that the traditional „mechanistic" approach focused on elimination of threats, such as point source pollution and flood control, is crucial but not sufficient. This is because purely technical control, without understanding and considering biotic dynamics, constitutes a more trial and error approach to water management than the implementation of a policy toward sustainable water use. While elements of this approach remain valid and viable, a technical solution alone is clearly insufficient for the sustainable use of the world's water resources. To guarantee the sustainability of freshwater resource use, it is necessary not only to reduce or eliminate the discharge of pollutants, but also to extend the number of potential tools to manage the degradation of ecological processes in landscapes. Such a more efficient approach must be based on an understanding of the temporal and spatial patterns of catchment scale water dynamics.

Ecohydrology - Creating Opportunities

Human survival and the preservation of biodiversity on Earth are dependent on our ability to maintain the integrity of ecological processes. Therefore, one of the fundamental tenets for the sustainable development of water resources is the maintenance of a homeostatic equilibrium within an ecosystem. At the present level of human impacts on ecosystems, it is necessary to increase the opportunities for ecosystems. It can be achieved by increasing the absorbing capacity of ecosystems against human impacts that continue to increase. Ecohydrology as an approach provides tools to achieve this goal by defining new approaches to freshwater protection, restoration and management.

Ecohydrology as an Integrative Approach

The formulation of the ecohydrology concept defined in UNESCO IHP V was to a large extent a logical consequence of the progress of river ecology. The awareness of a need for integration of hydrology and ecology appears in the hydrobiology and hydrology scientific papers of the 1970`s. However, only in the 1990`s did independent research directed to the interactions between the hydrosphere and biosphere become a subject of research for scientists in various fields. This created a basis for the holistic approach to understanding interactions between ecological and hydrological processes at a catchment scale and directed at the development of practical approaches for sustainable watershed management.

Among others, the broad scope of the research covered the following aspects:

- The relationship between vegetation, soil and water based on an understanding of the physiological properties of plants was presented by Baird & Wilby.
- Considerable progress was made in understanding the role of vegetation in water cycling processes in a landscape through research by Rodriguez-Iturbe and that done within the IGBP BAHC programme.
- The multidimensional role of the buffering by ecotone zones between land and water have been well defined within the framework of the UNESCO MAB Programme.

Vegetation and water stress

A fundamental concept in ecohydrology is that plant physiology is directly linked to water availability. Where there is ample water, as in rainforests, plant growth is more dependent on nutrient availability. However, in semi-arid areas, like African savannas, vegetation type and distribution relate directly to the amount of water that plants can extract from the soil. When insufficient soil water is available, a water-stressed condition occurs. Plants under water stress decrease both their transpiration and photosynthesis through a number of responses, including closing their stomata. This decrease in the canopy water flux and carbon dioxide flux can influence surrounding climate and weather. Insufficient soil moisture produces stress in plants, and water availability is one of the two most important factors (temperature being the other) that determine species distribution. High winds, low atmospheric relative humidity, low carbon dioxide, high temperature, and high irradiance all exacerbate soil moisture insufficiency. Soil moisture availability is also reduced at low soil temperature. One of the earliest responses to insufficient moisture supply is a reduction in turgor pressure; cell expansion and growth are immediately inhibited, and unsuberized shoots soon wilt.

The concept of water deficit, as developed by Stocker in the 1920s, is a useful index of the balance in the plant between uptake and loss of water. Slight water deficits are normal and do not impair the functioning of the plant, while greater deficits disrupt normal plant processes. An increase in moisture stress in the rooting medium as small as 5 atmospheres affects growth, transpiration, and internal water balance in seedlings, much more so in Norway spruce than in birch, aspen, or Scots pine. The decrease in net assimilation rate is greater in the spruce than in the other species, and, of those species, only the spruce shows no increase in water use efficiency as the soil becomes drier. The two conifers show larger differences in water potential between leaf and substrate than do the hardwoods. Transpiration rate decrease less in Norway spruce than in the other three species as soil water stress increases up to 5 atmospheres in controlled environments. In field conditions, Norway spruce needles lose three times as much water from the fully turgid state as do birch and aspen leaves, and twice as much as Scots pine, before apparent closure of stomata (although there is some difficulty in determining the exact point of closure). Assimilation may therefore continue longer in spruce than in pine when plant water stresses are high, though spruce will probably be the first to "run out of water".

Temporal and spatial considerations

Ecohydrological theory also places importance on considerations of temporal (time) and spatial (space) relationships. Hydrology, in particular the timing of precipitation events, can be a critical factor in the way an ecosystem evolves over time. For instance, Mediterranean landscapes experience dry summers and wet winters. If the vegetation has a summer growing season, it often experiences water stress, even though the total precipitation throughout the year may be moderate. Ecosystems in these regions have typically evolved to support high water demand grasses in the winter, when water availability is high, and drought-adapted trees in the summer, when it is low.

Ecohydrology also concerns itself with the hydrological factors behind the spatial distribution of plants. The optimal spacing and spatial organization of plants is at least partially determined by water availability. In ecosystems with low soil moisture, trees are typically located further apart than they would be in well-watered areas.

Hydrologic Cycle

The Earth as we know it is made up of primarily water, considered the most precious of our natural resources. Snow storms, rain showers and thunderstorms play just a small part of what is known as the hydrologic cycle, or water cycle. The hydrologic cycle is made up of water in all three of its forms: solid, liquid and gas. The hydrosphere is the area which contains all the water in the atmosphere. Every cycle occurs in five different stages.

The hydrologic cycle starts in the evaporation stage. All water on the ground — including lakes, rivers and oceans — is at the mercy of the sun. Sunlight aids in the process to raise the water temperature and eventually change it from a liquid to a gas. As a gas, the water vapor rises into the atmosphere and starts to condense. During the condensation stage, the water vapor condenses on particles in the atmosphere and forms clouds. A smaller version of this process is dew on the grass in the morning.

From condensation to precipitation, the water can come in a variety of forms. Even though this is known as the precipitation stage, hail is one of the forms which can fall. The other likely scenarios are snowfall and rain showers. Once clouds form, upper atmosphere winds spreads the clouds across the entire globe until the clouds cannot hold all the moisture. Weather systems over lakes in the winter can actually stall out and lead to a longer more intense system as it sits over a larger body of water which speeds up the hydrologic cycle.

The infiltration and runoff stages are the final two parts of the hydrologic cycle. Both stages occur simultaneously. When the snow and rain hit the ground, it begins to seep into the ground during the infiltration process. During a flood, the ground has already absorbed the amount of water it can hold and begins to accumulate on the ground. The runoff stage goes from the ground, either above or below ground and flows into rivers and streams. The streams and rivers flow back into lakes and oceans which lead back to the start of the process, evaporation.

Hydrometeorology

Hydrometeorology is a scientific field of study that incorporates aspects of both meteorology and hydrology. Meteorology is an interdisciplinary field of study specializing in atmospheric science, specifically as it relates to weather patterns. Hydrology, in the simplest terms, is the study of water on earth. Hydrometeorology, by comparison, studies water as it relates to earth's lower atmosphere and interacts with earth's surface. Hydrometeorology is chiefly concerned with the transfer of energy between atmospheric water and the earth's surface. In other words, when atmospheric water impacts landmasses, hydrometeorologists are keenly interested. Any time atmospheric water changes from gaseous to liquid or solid form, a hydrometeorological process has occurred.

At any given time, most of earth's water is found in oceans, lakes, and rivers. It is important to remember, however, that atmospheric water vapor goes through a continual cycle. Even though atmospheric water is a small percentage of the earth's water, a large amount of water moves through the water vaporization process on a daily basis. Water vaporizes and falls to earth at astounding rates, leading to intense movement of the atmosphere. On a daily basis, water moves from the oceans to the atmosphere and back again.

Floods, hurricanes, thunderstorms, and other water-rich weather all represent an intense transfer of energy. Such water-based weather patterns are predicted and studied using the latest hydrometeorological methods. Using Doppler radar and other devices, hydrometeorologists study the atmosphere, looking for weather patterns indicative of known atmospheric water disturbances. Emphasis is also placed on careful study of precipitation. By studying a region's precipitation levels, hydrometeorologists are better able to predict the effect atmospheric water will have on the nearby populations.

In addition to studying precipitation cycles and water movement, hydrometeorology also studies static bodies of atmospheric water. Atmospheric water vapor that has not fallen to earth poses a specific curiosity for hydrometeorologists. Scientists expect regular cycling of water vapor, and lapses in this cycle are a phenomenon that can lead to drought.

Hydrometeorology not only studies phenomena, it seeks proactive solutions to climate problems. During a drought, atmospheric water content is locked in the clouds, leading to decreased rain, sleet, and snow. Hydrometeorology studies those clouds and also looks for solutions to unforeseen lapses in the atmospheric water cycle. In some instances, hydrological engineers can use hydrometeorological data determine when and if it is appropriate to "seed" stubborn clouds. Cloud seeding is a process whereby an airplane drops silver iodide and frozen carbon dioxide into a cloud, thereby forcing the gaseous cloud to liquefy and fall to the ground as rain.

CHAPTER 1

THRESHOLD-BASED MANAGEMENT OF FRESHWATER ECOSYSTEMS

Junguo Liu, Giri Kattel, Hans Peter H. Arp, Hong Yang

Keywords: Ecological threshold, Ecosystem management, Ecosystem services, Water management, Climate change

INTRODUCTION

Freshwater ecosystems, defined as aquatic systems with average salinities less than 0.5 parts per thousand (Moss, 2009), provide a diverse range of essential services such as food products, clear water, waste recycling, nutrient cycling, carbon storage, as well as cultural and recreational amenities. However, freshwater ecosystems are changing rapidly due to factors like urbanization and climate change. Over the past 50 years, economic and population growth have resulted in more rapid changes in the structure and function of freshwater ecosystems than in any other comparable time period of human history (MEA, 2005). Populations of freshwater species in North America, Europe, Australia, and New Zealand are estimated to have declined on an average by 50% between 1970 and 2000 (MEA, 2005). Freshwater ecosystems in urban areas are among the most affected (Kozlowski and Bondallaz, 2013), in that urbanization has led to dramatic changes in freshwater ecosystems throughout the globe (Alberti et al., 2007 and Kozlowski and Bondallaz, 2013).

Adding to the ongoing burdens of intensive agriculture, industrialization and urbanization, climate change is an additional serious threat to freshwater ecosystems and biodiversity worldwide. Climate change can alter freshwater ecosystems via various direct and indirect mechanisms (Chu et al., 2005 and Vörösmarty et al., 2010). Rising temperature and changing precipitation directly influence shifts in habitats and seasons, and also the physiological adaptation and phenology of freshwater species, thereby altering food web structure and ecosystem dynamics (Doak and Morris, 2010 and Walther et al., 2002). Climate change can also indirectly affect freshwater ecosystems via geomorphological alterations of lake and river systems, changes in nutrient and ionic loads (leading towards alteration of photosynthetic rates, eutrophication, acidification, salinization) as well as enhancing the impacts of prevalent diseases, chemical pollutants, biological invasions, and changes in predation and competition among species (IPCC, 2007). There have been a large number of discussions on the challenges and solutions facing human interventions to freshwater ecosystems (e.g., Chen et al., 2014), considering experimental and theoretical ecological thresholds (e.g., Horan et al., 2011), and how to best implement risk management of aquatic ecosystems (e.g., Chen et al., 2011 and Chen et al., 2013). However, it remains rare to explicitly discuss the freshwater ecological thresholds associated with climate change. There is an urgent need for integrating the scientific understanding of the diverse, complex and interrelated impacts of climate change on the thresholds of freshwater ecosystems. Here we review freshwater ecosystem thresholds in the context of climate change, and suggest the need for collaborative efforts across scientists, decision makers and stakeholders at all levels.

Citation: Junguo Liu, Giri Kattel, Hans Peter H. Arp, Hong Yang, Threshold-based management of freshwater ecosystems, Ecological Modelling, Volume 318, 24 December 2015, Pages 265-274, ISSN 0304-3800, http://dx.doi.org/10.1016/j.ecolmodel.2014.09.010.

WHAT ARE ECOLOGICAL THRESHOLDS?

There are several definitions for the term "ecological threshold". Most of these definitions commonly emphasize the non-linearity of ecological or biological responses to pressures caused by human interventions or natural processes. As defined by Groffman et al. (2006), an ecological threshold is the point, or "tipping point", at which there is an abrupt change in an ecosystem quality, property or phenomenon, or where small changes in environmental drivers can lead to dramatic changes to an ecosystem. Thresholds and their associated stability towards different environmental drivers can be conceptualized within a coupled socioeconomic–ecological system (Horan et al., 2011). Any restoration of losses in ecosystem services after a threshold is crossed could be difficult or costly (Groffman et al., 2006).

Identifying ecological thresholds related to climate change is complex, as various climate-change control variables (e.g., atmospheric and surface water temperature) can be related differently to changes in the ecosystem service responses (e.g., fish supply, volume of potable water) (Rockström et al., 2009).

Fig. 1 presents a general schematic diagram on how to view the interactions between climate change control variables and ecosystem service response variables using a threshold-based approach. Ecosystem services can be linearly dependent, non-linearly dependent, or independent of the climate control variables. When a shift in a climate control variable causes a substantial depletion in a dependent ecosystem service, a "threshold", point F_1, may be crossed, after which the drop in this service can become depleted rapidly with slight increases in the climate control variable. Often it is difficult to predict when a threshold occurs, or even ascertain the relationship between the climate change variable and the ecosystem service. Thus, a range can be assigned as a "zone of uncertainty". The "ecosystem boundary", point F_0 in Fig. 1 refers to the lowest value of a climate change control variable (compared to the present value) in which the probability of passing a "threshold" becomes higher than an assigned margin of safety (Rockström et al., 2009). Sometimes, ecosystems exhibit resilience (Kattel et al., 2013). After the thresholds are crossed, ecosystem services can recover either due to the climatic and other environmental variable reversing to an earlier level or some delayed feedback or buffering mechanism. This restoration is usually not instantaneous. It often requires the control-variable to decrease to a level substantially below the threshold, which is a phenomenon, referred to as ecosystem hysteresis (Scheffer et al., 2001). However, if ecosystem thresholds are crossed and restored periodically (e.g., from forest fires), this behavior would be "cyclical". However, if the change in a climatically sensitive and controlled variable becomes extreme, a "point of no return" can be reached, indicated as point F_2 in Fig. 1, i.e., a 'catastrophic' transition where the ecosystem can no longer return to its original state (e.g., Kattel et al., 2013). The condition between F_1 and F_2 can have profound implications for 'early warning signals' for ecosystem management (Wang et al., 2012). However, this is often unnoticed (Scheffer et al., 2001 and Folke et al., 2004).

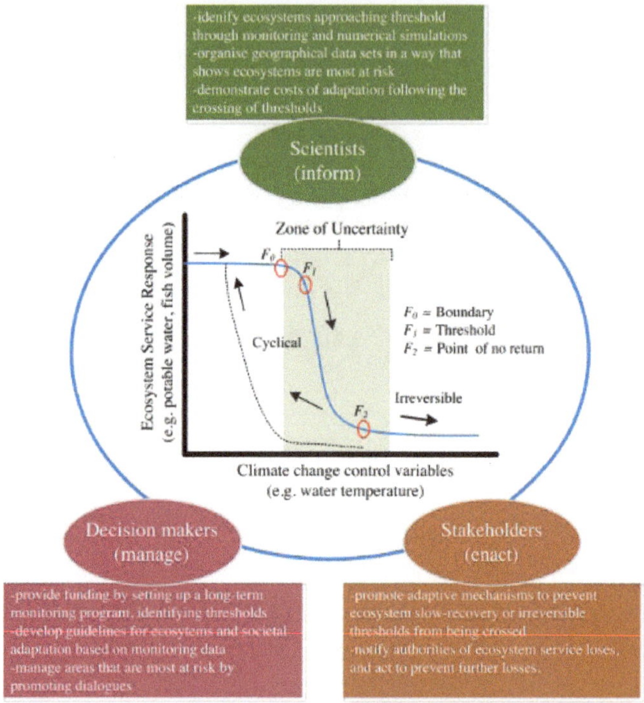

Figure 1: Conceptualization of threshold-based ecosystem change. Climate change control variables are on the *x*-axis and the response variables on ecosystem services are on the *y*-axis. The climate change variables could be one or any combination of parameters that are directly or indirectly affected by climate change (such as temperature, number of invasive species, nutrient concentration). The ecosystem service could refer to any required by the local population (such as potable water, fish population). Conceptualizing ecosystem change in this way could serve as a basis for communication in ecosystem management, where scientists inform of susceptibility to threshold changes, decision makers manage appropriately, and stakeholders get involved through enacting mitigation measures to prevent loss of services.

A key goal of ecosystem management is to avoid thresholds with slow recovery or points of no return. Though difficult for scientists to anticipate, boundaries and thresholds are highly useful for mobilizing policy makers or stakeholders. For instance, the use of 350 ppm CO_2 being a safe level to avoid a tipping point for global ecosystem change has been successful for mobilizing stakeholders and policy efforts (Hansen et al., 2008), as has been the prediction that a warming of 1–2.5 °C above pre-industrial levels could be a threshold value where substantial loss of Arctic summer ice or the Greenland ice sheet could occur (IPCC, 2007).

EFFECTS OF CLIMATE CHANGE AND THRESHOLD TENDENCIES IN FRESHWATER ECOSYSTEMS

The most practical thresholds for use by policy makers and stakeholders are those based on measurable control variables that have a straightforward link to an ecosystem response (Table 1). The easier the climate control variable is to quantify, as well as its effect on ecosystem services, the more practical it is to implement policy and effective monitoring measures. However, in practice direct links between climate change control variables and ecosystem responses are difficult to establish. Anthropogenic climate change is quantified by several parameters that could be used as control variables for defining the thresholds of freshwater ecosystems, e.g., atmospheric and surface water temperature, the length and onset of seasons, and precipitation (Table 1). However, in most cases, multiple control variables acting together must be considered. For example, riverine ecosystems are strongly and simultaneously influenced by variables like streamflow, erosion rates, temperature, and concentrations of micro-pollutants; though how to account for all the variables in a threshold type model is not always straightforward (Groffman et al., 2006). Other control variables that are indirectly influenced by climate change as well as other anthropogenic processes include water volume, salinity, dissolved oxygen (DO) and pH. All of these parameters, in isolation or in combination, could influence ecosystem services. In Section 4, we review the state-of-the-art understanding of how shifts towards ecological thresholds for freshwater ecosystems can be influenced directly or indirectly by climate change. A schematic overview of this presentation is provided in Fig. 2.

Table 1: Example freshwater ecosystem monitoring approaches with possible scientific, policy and stakeholder involvement to assess or prevent possible thresholds and loss of socio-economic services

Environmental parameter monitoring	Potential affected threshold response monitoring[a]	Potentially affected socio-economic services monitoring[a]
Climate: surface to bottom water temperature, evapotranspiration rates, precipitation, humidity season onset and length	Population of algae and plants, migration of fish and other species, habitat range	Changes in local economy due to change in products (prices of food, water, plants, real estate), as well as quality of life (health, opinion pools, recreational land area)
Hydrology: water level, water volume, water resident time	Population of algae and plants, areas of excess flooding, drowning, loss of water habitat	
Light: UV radiation, visible water depth	Population of algae and plants (planktonic and benthic)	
Oxygen: dissolved oxygen (DO) levels, lake stratification.	Phytoplankton blooms, aerobic benthic species, vertically migrating fish species, oxygen sensitive species,	
Geomorphology: erosion rates, sedimentation, siltation	Riparian fish habitat, formation of anoxic zones	
Water quality: pH, micropollutants, dissolved organic carbon (DOC), nutrients including total nitrogen (TN), total phosphorus (TP) and chlorophyll a (chl a)	Loss of sensitive species, change in species dynamics,population of algae and plants	
Species diversity: presence of invasive species, changes in foodweb structure, biodiversity losses	Loss of locally important species	

[a]The threshold like responses and socio-economic service may be affected by a combination of climate and local anthropogenic impact parameters (e.g., pollution levels, water over usage, urbanization).

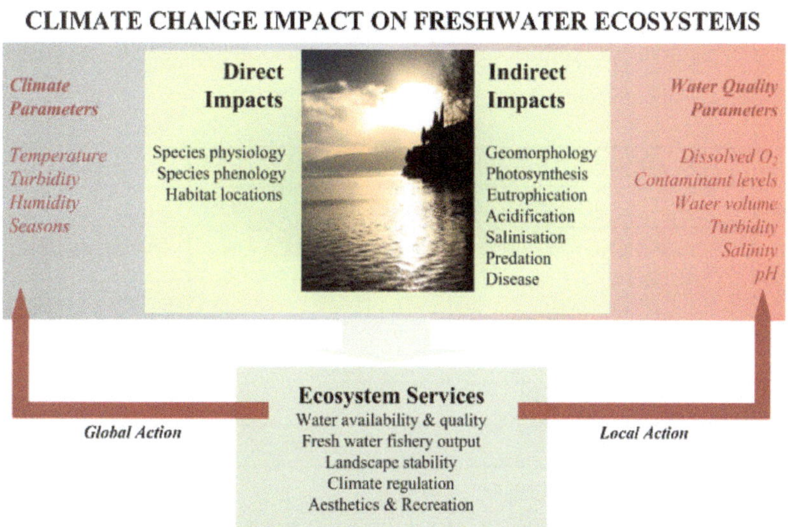

Figure 2: Overview of ecosystem services that are directly and indirectly impacted by climate change and local anthropogenic impacts. The figure is an illustration of the complex, cyclical nature of how the use of ecosystem services can, through direct and indirect mechanisms, affect those same ecosystem services. Appropriate global action to reduce harmful climate change impacts and local action to reduce harmful disturbances in water quality can help mitigate losses of ecosystem services.

THRESHOLDS ASSOCIATED WITH DIRECT EFFECTS OF CLIMATE CHANGE

Over the past decades, freshwater ecosystems have exhibited direct responses to changes in temperature, precipitation, and atmospheric humidity (Hader et al., 2007), particularly in regard to the phenology and physiology of freshwater species, as well as their habitats. These responses become more pronounced with an increasing global surface temperature (Schippers et al., 2004).

Phenological thresholds

Climate change has directly influenced the phenology of aquatic biota. Changes in air temperature influence water temperature and the timing and duration of the stratified period, while changes in precipitation can influence water residence time. These changes in turn influence the seasonal dynamics of the biota present (Peeters et al., 2007). Phytoplankton phenology in lentic systems is directly affected by an increased water temperature, changes in the onset and the duration of thermal stratification, and earlier ice break up (Meis et al., 2009). For instance, phytoplankton blooms in the US lakes and Taihu Lake, the third largest lake in China, have begun to occur earlier due to the warmer spring (Qin et al., 2010 and Winder and Schindler, 2004). In addition, high air temperatures and low wind speeds have directly led to an early onset of the spring phytoplankton bloom in Upper Lake Constance of the western European Alps (Peeters et al., 2007).

Changes in temperature and seasons influence multiple levels of biological organization, including foraging behavior, phenology and metabolic rates, potentially leading to profound shifts in the stoichiometry of elemental fluxes between consumers and resources at the base of the food web (Woodward et al., 2010). A rise in surface-water temperature and a regional decrease in wind velocity in central Africa's Lake Tanganyika, which provides 25–40% of the animal protein for surrounding populations, has lead to a reduction in primary productivity by ~20%, implying a ~30% decrease in fish yields (O'Reilly et al., 2003). Arctic lake and river systems, however, experiencing the opposite effect, as longer growing seasons have led to increased primary productivity (Smol et al., 2005). Such differences in phenology and life cycle cues can induce strong variance in community composition (Burgmer et al., 2007), trophic levels and fish production, ultimately changing ecosystem services (Ficke et al., 2007).

Physiological thresholds

Some species that are exposed to warmer temperatures for longer periods will show biochemical variations at the cellular level. Heatwaves, which are likely to increase in severity due to the changing climate (Schar et al., 2004), may influence freshwater gastropods and mussels by affecting the release of the stress-associated enzymes (e.g., protein kinases), altering metabolic activity and causing hypoxia in these animals (Anestis et al., 2010). Physiological processes in aquatic organisms, including rates of oxygen uptake, movement, feeding, developmental rate, and immune function are strongly driven by temperature (Helmuth, 2009). Warmer surface temperatures of freshwater lakes can lead directly (through water currents) and indirectly (through affecting plant and algae growth as discussed above) to the alteration of thermal stratification in temperate lakes, which can lead to an increased dissolved organic carbon (DOC) concentrations in the hypolimnion, limiting the DO level for vertically migrating fish species, potentially causing hypoxia (Hader et al., 2007). A mismatch between the demand for oxygen and the capacity of oxygen supply to the tissues can deviate the physiological adaptation of aquatic species and restrict tolerance of these animals to a range of climatic extremes (Pörtner and Knust, 2007). Decreased oxygen in lakes also decreases the processing of organic matter, recycling of nutrients, and the microbial breakdown of pollutants (Carpenter et al., 2011).

To place this in the context of a freshwater ecosystem threshold, a previous modeling study suggests that a doubling of CO_2 concentrations would cause earlier onset of lake stratification and a 1–7 °C increase in summer epilimnetic temperature, consequently limiting the vertical migration of zooplankton in addition to physiological and respiration failures in cold water fish species (De Stasio et al., 1996). This example and other examples of freshwater ecosystem thresholds related to climate change are presented in Table 2.

Table 2: Thresholds for freshwater ecosystem regime shift

Regime-shift	Parameter	Threshold	Location	References
Physiological related				
Early lake stratification, limiting vertical migration and causing physiological failure in cold fish	Atmospheric CO2 concentrations	A doubling of CO2 concentrationfrom the 1980s level	North temperate lakes	De Stasio et al. (1996)
Habitat related				
Reduction in fish habitats by 50%	Summer mean temperature	2–6°C	US lakes	Smol et al. (2005)
Shift of small bass and yellow perch habitat 500km north	Mean air temperature	4°C	US lakes and rivers	Shuter and Post (1990)

Substantial loss of ice sheet	Temperature	1–2.5°C increase above pre-industrial levels	Arctic, Antarctic, Greenland and other mountain areas	IPCC (2007)
Water quality related				
Shift in diatom community	Nitrogen deposition loads	1.4kgNha −1y−1	Lakes around the eastern Sierra Nevada and the Greater Yellowstone National Park, US	Saros et al. (2011)
Shift in biotic assemblages, water quality and carbon budget	Dissolved organic carbon	91% increase	22 upland lakes and streams, UK	Evans et al. (2005)
From clear to turbid state 1)	Orthophosphate phosphorus concentration in the water	10–12μg/L	26 temperate lakes, Northern Italy	Chiaudani and Vighi (1974)
From clear to turbid state	Total phosphorus in water	150μg/L	Lake Velume, Netherland	
From turbid to clear state	Total phosphorus in water	100μg/L	Lake Velume, Netherland	
From clear to intermediate state 2)	Total phosphorus in water	61μg/L	Gehu Lake, China	Tao et al. (2012)
From intermediate to turbid state 3)	Total phosphorus in water	115μg/L	Gehu Lake, China	Tao et al. (2012)
From clear- to turbid-state	Total phosphorus in water	70–100mgm−3	46 small and medium lakes along the mid-lower Yangtze River	Wang (2007)
From turbid- to clear-state	Total phosphorus in water	20–30mgm−3	46 small and medium lakes along the mid-lower Yangtze River	Wang (2007)
From turbid to clear state 4)	Phosphorus loading	0.3gm−2y−1	Lake Mendota, Wisconsin, USA	Carpenter and Lathrop (2008)
From clear to turbid state	Dynamic linear model standard deviation (DLM SD) of phosphorus in soil, water and sediment	Around 1gm−2	Lake Mendota, Wisconsin, USA	Carpenter and Brock (2006)

Note: Although different words are used for the states of freshwater ecosystems, we used the terms clear and turbid states (widely accepted now); the original words for the states were listed: (1) from meso- to eutrophic; 2) from grass-stable to grass-algae intermediate state; (3) from grass-algae intermediate to algae-stable state; (4) from persistent eutrophy to oligotrophy.

Habitat-associated thresholds

Climate change also directly influences fish populations, as changes in climate lead to changes in the hydrological regimes and habitat niches that endemic fish have evolved to cope with (Ficke et al., 2007). Global warming has been attributed with decreasing the thermal niche and habitat availability for cold water fish, such as galaxid and salmonid fishes (Carpenter et al., 2011). A 3 °C change in mean annual temperature corresponds to a shift in approximately 300–400 km latitudinal and 500 m altitudinal isotherms (Smol et al., 2005). One study estimates that global warming could shift natural habitat ranges by approximately 6.1 km per decade towards the poles (Parmesan and Yohe, 2003). This change, though subtle, is enough to impact fish communities and ecosystem adaptation in the higher latitudes. To place habitat changes in the context of climate change related thresholds, a coupled CO_2–climate model suggests that an increase in 2–6 °C summer mean temperature will reduce 50% of the suitable habitats for cold water fish in the US (Eaton and Scheller, 1996). A further 4 °C increase in mean air temperature will extend the habitat of small bass mouth and yellow perch as far as 500 km towards the north (Shuter and Post, 1990).

THRESHOLDS ASSOCIATED WITH INDIRECT EFFECTS OF CLIMATE CHANGE

Climate change can affect freshwater ecosystems indirectly via a range of positive and negative feedbacks. These indirect effects may be gradual and subtle in the short term but could be catastrophic in the long term, leading to slowly recovering or irreversible threshold type losses. Changes in geomorphology, water depth, flow regimes, siltation and sediment flux in lake and river systems can occur as a result of indirect effects of climate change (Lake et al., 2000). Biotic and abiotic interactions with climate can further lead to unprecedented ecological conditions, including the alteration of photosynthetic rates, eutrophication, acidification and salinization (Schindler, 2001 and Woodward et al., 2010). Widespread invasion from exotic flora and fauna, and changes in predation and competition among species are also exacerbated by indirect interactions between climates and human activities (Baron et al., 2002).

Below, we take the trophic state threshold as an example of indirect effects of climate change. Other thresholds related to indirect effects are described in Supplementary information, and they include morphological, photosynthetic, acidification, salinity, pollutant thresholds, disease, biological invasion, as well as predation and competition-associated thresholds.

Decreased precipitation and lower water flows imply smaller dilution volumes and thus, higher concentrations of nutrients downstream from point pollution source, particularly untreated sewage in developing countries. On the other hand, the increased occurrence of storm events can cause an overflow and discharge of nutrient into receiving rivers or lakes (Whitehead et al., 2009). Severe storms can entrain large amount of deeper and anoxic sediments, particularly when the water level is low. As an example, a storm in 1996 that occurred at Lake Võrtsjärv, Estonia, resulted in a greater release of sediments on one stormy day compared to the total nutrient inflows over 1 y (Nõges and Kisand, 1999). The effects of wind-mixing on the vertical transfer of nutrients from sediment are more important in small lakes than the large lakes. The relationship between North Atlantic Oscillation Winter Index (NAOw) and phosphate has been observed in some European lakes (e.g., George et al., 2004 and Weyhenmeyer, 2004). In temperate Europe, positive values of NAOw imply milder winter, higher water levels and lower spring concentrations of phosphate, while negative values indicate drier winters, low water levels and larger amounts of phosphate released from sediment. In cold areas, the extension of the ice-free period can cause marked increases in phosphorus concentration, as has been observed in the hypolimnion and epilimnion in Lake Erken, Sweden (Pettersson et al., 2010). This additional phosphorus is released annually from the sediment zone to the water column as a result of increased temperature and low oxygen concentration in the bottom water (Malmaeus and Rydin, 2006). Increased temperature and precipitation intensity and changes in winter precipitation are expected to enhance phosphorus loading in the temperate freshwater lakes and reduce the loading in Mediterranean lakes (Jeppesen et al., 2009).

Nutrient loads are predicted to increase under climate change in many areas such as Denmark and UK (Andersen et al., 2006 and Whitehead et al., 2006). Higher temperature increases the release of nitrogen from soil, and lower stream flows decrease the dilution capacity of rivers (Whitehead et al., 2006). However, the impact on eutrophication is not straightforward due to the complex interaction between nutrient, light, temperature, residence time and flow conditions (Jeppesen et al., 2005 and Yang and Flower, 2012). Increased surface water temperature can indirectly lead to stronger stratification and regeneration of water column nutrients, which can intensify eutrophication (Rabalais et al., 2009). Climate change has exacerbated eutrophication through nutrient dynamics of the north temperate lakes due to a longer growing season, depending on epilimnion and mineral mixing (Smol et al., 2005). The analyses of 103 Chinese lakes indicated that the mean precipitation was one of the main predictors of eutrophication (Liu et al., 2010). A 41-year data series (1968–2008) from Blelham Tarn UK, where eutrophication has reduced hypolimnetic DO, indicates negative effects that are likely to be exacerbated by changes in climate and the thermal structure of lakes (Foley et al., 2012). The interactive effects of future eutrophication and climate change on harmful cyanobacterial blooms are difficult to understand, though much of the current knowledge suggests that climate change will likely enhance the magnitude and frequency of these events (O'Neil et al., 2012). Recently eutrophication in lacustrine environments has been adopted as a way to define ecological and societal thresholds in space and time (Carpenter et al., 1999). Experimental management programs have been replicated among several lakes based on these thresholds (Carpenter et al., 1999).

The phosphorus thresholds for the regime shift between clear and turbid states have been suggested for different lakes (Table 2). Compared to the primary drivers of eutrophication such as nutrients, the role of climate change to modulate eutrophication thresholds is not well understood. However, increasing temperature, changing precipitation and atmospheric circulation can indirectly influence the nutrient dynamics, and increase the likelihood of exceeding thresholds of nutrients in freshwater ecosystems (Battarbee, 2000).

IDENTIFYING THRESHOLD RESPONSES FOR MANAGING FRESHWATER ECOSYSTEMS

Identifying and predicting threshold responses in ecological systems is a challenging, though useful task. Thresholds related to lake eutrophication causing regime shift is a well studied area (Scheffer et al., 2001) and one that can be used to understand thresholds from other climate-change control variables. Excessive nutrient levels, particularly phosphorus, have been widely accepted as the leading factor of eutrophication (Schindler, 1974 and Carpenter et al., 1999). The orthophosphate phosphorus concentration 10–12 $\mu g\ L^{-1}$ in water was suggested as the threshold from clear to turbid states in the lakes of Northern Italy (Chiaudani and Vighi, 1974) (Table 2). Higher concentrations were also found to be thresholds for the regime shift in Dutch and Chinese lakes (Tao et al., 2012). Although the phosphorus concentration in water can be controlled to some extent by lowering nutrients emissions, such as by controlling fertilizer usage; internal sediment release of phosphorus in addition to temporal changes in external catchment input can cause thresholds to be exceeded e.g., the level of 0.3 $g\ m^{-2}\ y^{-1}$ external phosphorus input in Lake Mendota, Wisconsin for threshold crossing benchmark (Carpenter and Lathrop, 2008). With many dynamic influential factors, changing non-linearly with time, the use of Bayesian forecasting through Dynamic Linear Models (DLM), which allows influential parameters to have a prior distribution (West and Harrison, 1989), is regarded as a more reasonable approach for modeling changes in release of phosphorus from the lake sediment (Cottingham et al., 2000), than steady linear models. DLM standard deviation (DLM SD) of phosphorus load in soil, water and sediment was also suggested as a way to anticipate a threshold of lake regime shift (Carpenter and Brock, 2006).

Thresholds from other climate-control variables, similar to those caused by phosphorus induced eutrophication are not only related to external driving forces, but are also mediated by internal changes in the ecological regimes (Walker and Meyers, 2004). The direct and indirect impacts listed above do not occur in isolation but simultaneously, and are often indicated by a change in several quantifiable parameters (Fig. 2). Often these changes are inconsistent and may behave non-linearly across and within ecosystems. Even a minor disturbance may move a system to a new regime, for instance, a gradual increase in nutrients can transform an oligotrophic lake into an eutrophic one (Limburg et al., 2002). Approaches to identifying the thresholds of fresh water ecosystems should involve the use of continuous monitoring, paleolimnological data and simulations, and thereby account for changes in the present, past and future.

Implementing effective monitoring programs

Environmental policies designed to maintain freshwater quality are often dependent on identifying "threshold" dose–response relationships (Groffman et al., 2006). Effective monitoring is useful for tracking ongoing ecological system shifts that are driven by external environmental forces or internally-mediated drivers, and thus for predicting threshold type responses, and devising priorities and practical strategies for biological conservation (Doak and Morris, 2010 and Lindenmayer and Likens, 2009). For example, a time series analysis of DOC in 22 upland lakes and streams of the UK obtained during a 15 y monitoring program indicated a 91% increase in DOC concentration has triggered a shift in biotic assemblages, water quality and the carbon budget in the region (Evans et al., 2005). A 20-year monitoring study of water conductivity of shallow lakes across the Arctic region (in combination with paleolimnological data) has provided a description of thresholds related to wetland permanency as a result of climate warming, in which evaporation/precipitation ratios are a suitable climate-control variable (Smol and Douglas, 2007). For all monitoring programs, it is essential to establish an archive of environmental data, such as the Vannmiljø system used in Norway (http://vannmiljo.miljodirektoratet.no/) for analyzing trends and predicting threshold responses.

Key variables, such as precipitation, temperature, total nitrogen (TN), total phosphorus (TP) and chlorophyll *a* (chl*a*) should be continuously monitored based on field sampling and laboratory measurement. In addition, advancement of technologies including remote sensing will improve techniques of monitoring environmental forces over a regional scale. A few examples of further monitoring approaches to obtain relevant data are presented in Table 1.

Reconstructing historical data

Palaeolimnological techniques can help identify thresholds of ecosystem change. For example, critical loads of nitrogen deposition were determined for alpine lake ecosystems in the Western US using fossil diatom assemblages in lake sediment cores (Saros et al., 2011). A transfer function technique, where calibration training sets are constructed to show relationships between individual species and one or more quantifiable (climate change related) parameters, followed by a reconstruction of changes in ecosystem functions and services based on a biological data, is useful for identifying ecological regime shifts caused by environmental perturbations (Battarbee, 2000). As an example, sub-fossil assemblages of diatoms in lakes around the eastern Sierra Nevada and the Greater Yellowstone National Park suggests that 1.4 kg Nitrogen ha $^{-1}$ y^{-1} can cause a regime shift in the diatom community (Saros et al., 2011). The advancement of dating techniques and the use of multiple species approaches for lake sediments can further help improve climate reconstruction and define climate thresholds of freshwater ecosystems (Battarbee, 2000), but it is still necessary to disentangle the relative role of all influential factors using a combination of both limnological and paleolimnological data (Battarbee et al., 2012). It also needs to be pointed out that the historically caused perturbations result primarily from internal shifts and climate change, whereas current human induced perturbation (e.g., land use, changed nutrient budgets) is now superimposed on internal shifts and climate-related changes, making it difficult to use past changes as direct analogs of current and future thresholds.

Developing appropriate numerical tools and models

Precision in ecosystem projections will increase when quantitative estimates of thresholds are backed by reliably measured data (Woodward et al., 2010). Scenario-based models can provide decision-makers a range of possibilities, as well as integrate natural and social science based strategies for adaptation (IPCC, 2007). The Intergovernmental Panel on Climate Change (IPCC) has implemented a variety of scenario-based climate-change models, including models associated with historical freshwater ecosystems (IPCC, 2007). However, models to describe thresholds that influence individual freshwater ecosystems have not yet been developed, due to the uncertainty in future climate projections as well as scale issues, as small lakes cannot be adequately represented using coarse resolution climate models (Groffman et al., 2006). Some attempts have been made for improved modeling of the ecological thresholds such as the trophic cascade model for game fish (Carpenter and Brock, 2004) and the water quality model for DO concentrations in North American lakes (Stefan et al., 1993). Advancement of computer software and numerical modeling will continue to improve the knowledge of threshold based ecological management strategies.

Several types of models, e.g., statistical, rule-based, the previously mentioned DLM models, and process-based models, have been used to guide ecosystem management under global change. Process-based models are based on a scientific understanding of relevant ecological processes, and provide a valuable framework to include responses to the changed environmental conditions (Cuddington et al., 2013). The current understanding of the role of thresholds is often based upon conceptual models of how ecosystems work (Dennison et al., 2007). Process-based threshold models can be specifically designed to anticipate ecological consequences of human activities on freshwater ecosystems, and should play a key role in setting conservation targets by water resources managers.

Testing climatological hypotheses

Laboratory or field-based experimental designs have the potential to characterize ecological thresholds (Adler et al., 2009). For instance, a laboratory experiment on burrowing mayfly, *Hexagenialimbata*, from the Lower Mobile River, Alabama, suggests that the nymphs of this animal can survive at elevated salinity levels (6.3%) only for temperature below 18 °C, however, when the salinity level is reduced to 2.4% they can survive at higher temperatures up to 28 °C, which was further supported by field observations (Chadwick and Feminella, 2001). Similarly, an experiment which reduced the pH of a small lake from 6.1 to 4.7 was found to cause a substantial decline in species richness of many freshwater taxa, reducing the ecosystem's adaptive capacity (Hogsden et al., 2009). These studies assist with understanding which species within ecosystems are the most vulnerable to climate change, and how the disappearance of these species will influence the broader ecosystem services.

Relating threshold-associated costs to ecological and societal adaptation

Ecosystem service changes also influence the societal structure of the population dependent on those services. Loss of ecosystem services, including water quality, water quantity, fishery resources, and recreational amenities, can lead to the crossing additional thresholds associated with social systems and unacceptable societal costs. Integrating key components of societal development can promote ecological resilience, including information management, cultural integrity, technological development, institutional responsibility, and education (Falkenmark, 2003). Consideration of a probabilistic framework with a range of components can offer policy and management responses to emerging crisis from freshwater ecosystem changes worldwide (Vörösmarty et al., 2010). Specifying the thresholds related to time, location, species, disturbance and scale, and understanding their economic and societal implications can help set up regulations, improve adaptive capacity of freshwater ecosystems, and facilitate societal adaptation (Naiman and Turner, 2000). The socio-economic costs of crossing ecological thresholds in freshwaters need to be placed clearly into the context of threshold-based management program (Table 1). This involves a careful cost-benefit analysis of the services affected before and after crossing thresholds.

Certain threats to freshwater ecosystems may be addressed through investments and advances in infrastructure, science and technology (Vörösmarty et al., 2010). Though this may be mainly a question of political will in richer countries, poorer countries often lack the resources to invest in freshwater infrastructure (UNFCCC, 2007). It is estimated that the United States currently would have to put in $60 billion dollars for the restoration of its ecosystems. In regions like the Middle East and North Africa, water pollution and over exploitation of water is expected to incur environmental damage costs equivalent to 2.1–7.4% of their gross domestic product(GDP). Similarly, the cost of water related crises in China was 2.3% of China's GDP in 2003.

Assigning an economic value to the crossing of ecological thresholds is an effective way of communicating the societal and economical costs to decision makers and stakeholders (Balmford et al., 2002). For example, the use of artificial drainage in South Ontario marshes for the purpose of increasing agricultural productivity was found to be less beneficial economically than not draining (van Vuuren and Roy, 1993). The cost-benefit analysis suggested that the naturally intact marsh ecosystems would cost less (3700 USD ha^{-1}) than the artificially drained marshes (8800 USD ha^{-1}). Thus the use of scenario-based numerical models of freshwater systems coupled with economical models is strongly encouraged, as it can help implement threshold-based management approaches of freshwater ecosystems effectively, and thereby underscore the services that intact ecosystems can provide to society (de Groot et al., 2002). Risk analysis should be conveyed together with threshold analysis, in which models are used to predict the risk and benefits (environmental, economical, societal) from thresholds being crossed under different climate change scenarios, as well as the risk and benefits from potential mitigation efforts.

THRESHOLD-BASED MANAGEMENT IN FRESHWATER ECOSYSTEMS

The goal of ecosystem management is to restore or preserve some main attributes of an ecosystem that are desirable for humans (Mayer and Rietkerk, 2004). Due to the inherent complexity of ecosystems, this can only be achieved through a framework (e.g., Fig. 1) involving long-term collaborations between scientists, policy makers and stakeholders. The main goal of such collaboration would be to set up a warning and prevention system from approaching thresholds (near F_0 or F_1), as well as mitigation strategies to prevent such thresholds being reached. In this effort, scientists with the assistance of local stakeholders would monitor relevant environmental parameters and potential threshold responses (examples are presented in Table 1). Decision makers would set up a program to encourage such monitoring efforts, as well as be prepared in advance to enforce appropriate mitigation strategies as needed (e.g., lowering fish quotas, capping contaminated areas, preventing excess nutrients from entering the local area). Mitigation strategies would ideally have to be planned far well in advance based on previously conducted simulations and risk analysis. Clear communication with stakeholders regarding risks from crossing thresholds is critical, particularly in cases where long term and short term interests collide (e.g., over fishing, urbanization). Further, stakeholders should be encouraged to notify authorities in case a rapid loss in ecosystem services occurs (potentially indicating a threshold F_1 being met)(Fig. 1), to get involved in monitoring programs (e.g., species counting, environmental parameter logging), as well as be presented with opportunities to act in a way to mitigate risks from boundaries F_0being reached.

Disagreement may often occur amongst actors on practices, remediation and logistic issues. Scientists being too cautious about the position of F_0, may risk losing trust by decision makers and stakeholders, particularly if no change in a response variable is evident after F_0 is crossed (Fig. 1). On the other hand, if scientists are under cautious, predicting an F_0 that is close to F_2, then trust in scientists would also be lost alongside the ecosystem service.

Ecological systems are not simple to understand, neither are threshold responses. After a threshold is crossed, the situation may become restored by various feedback loops (Fig 1). The prudent option is, however, to assume that this does not happen (Briske et al., 2010). Avoiding disagreements amongst actors is essential when ecosystem is at a risk to cross thresholds. Disagreements are resolved through regular interactions by organizing public involvement (e.g., through fairs, festivals, workshops, panel discussions, news media), as well as research projects and societal awareness programs implemented amongst ecologists, modelers, economists, sociologists, resource managers, stakeholders and decision makers, activists and reporters (Carpenter et al., 2009 and Lindenmayer and Likens, 2009). Bringing these actors in a common, formerly established platform can be a successful approach for freshwater ecosystems globally (Carpenter et al., 2009 and MEA, 2005). This reflects an adaptive management approach, where solutions to the problems are proposed and implemented, and the management strategies are constantly reviewed over the course of ecological response to climate change (Williams, 2011). Due to multiple spatial and temporal scales of ecosystems, Côté and Darling (2010) argued that the ecosystem management to control local anthropogenic disturbance, for example nutrient input, and to reverse the ecosystem degradation will inadvertently lower resilience to climate disturbance. Ongoing research on freshwater ecosystem management is needed, especially in regards to integrating rare or extreme events in freshwater ecology (e.g., see Fuentes et al., 2006 and Denny et al., 2009), and evaluating evidence and uncertainty in threshold concepts and models (Katharine and Richard, 2009). Establishing programs that seek to identify or prevent thresholds from being crossed, and which have the involvement of the scientific, policy and stakeholder communities, are a way to effectively manage threats of local impacts and global climate change on freshwater ecosystem services.

ACKNOWLEDGEMENTS

This study was supported by the International S&T Cooperation Program from the Ministry of Science and Technology of China (No. 2012DFA91530), Special Fund for Forestry Scientific Research in the Public Interest (No. 201204204), Projects of International Cooperation and Exchanges NSFC (41161140353, 91325302), the 1st Youth Excellent Talents Program of the Organization Department of the Central Committee of the CPC, the Fundamental Research Funds for the Central Universities (TD-JC-2013-2), and the NGI GBV Fund (H.P.H. Arp). We also thank Potsdam Institutes of Climate Change Impacts (PIK) in Germany for supporting J. Liu's visits, and the University of Leeds for providing J. Liu a Cheney Senior Fellowship. We thank Steve Carpenter from the University of Wisconsin and Elena Bennett from McGill University for sharing discussions through the preparation of the manuscript.

REFERENCES

1. Adler, P.B., Leiker, J., Levine, J.M., 2009. Direct and indirect effects of climate change on a prairie plant community. PLos One 4, e6887.
2. Alberti, M., Booth, D., Hill, K., Coburn, B., Avolio, C., Coe, S., Spirandelli, D., 2007. The impact of urban patterns on aquatic ecosystems: an empirical analysis in Puget lowland sub-basins. Landsc. Urban Plann. 80 (4), 345–361.
3. Andersen, H.E., Kronvang, B., Larsen, S.E., Hoffmann, C.C., Jensen, T.S., Rasmussen, E. K., 2006. Climate-change impacts on hydrology and nutrients in a Danish lowland river basin. Sci. Total Environ. 365, 223–237.
4. Anestis, A., Portner, H.O., Michaelidis, B., 2010. Anaerobic metabolic patterns related to stress responses in hypoxia exposed mussels Mytilus galloprovincialis. J. Exp. Mar. Biol. Ecol. 394, 123–133.
5. Balmford, A., Bruner, A., Cooper, P., Costanza, R., Farber, S., Green, R.E., Jenkins, M., Jefferiss, P., Jessamy, V., Madden, J., Munro, K., Myers, N., Naeem, S., Paavola, J., Rayment, M., Rosendo, S., Roughgarden, J., Trumper, K., Turner, R.K., 2002. Economic reasons for conserving wild nature. Science 297, 950–953.
6. Baron, J.S., Poff, N.L., Angermeier, P.L., Dahm, C.N., Gleick, P.H., Hairston, N.G., Jackson, R.B., Johnston, C.A., Richter, B.D., Steinman, A.D., 2002. Meeting ecological and societal needs for freshwater. Ecol. Appl. 12, 1247–1260.
7. Battarbee, R.W., Anderson, N.J., Bennion, H., Simpson, G.L., 2012. Combining limnological and palaeolimnological data to disentangle the effects of nutrient pollution and climate change on lake ecosystems: problems and potential. Freshwater Biol. 57, 2091–2106.
8. Battarbee, R.W., 2000. Palaeolimnological approaches to climate change, with special regard to the biological record. Quat. Sci. Rev. 19, 107–124.
9. Briske, D.D., Washington-Allen, R.A., Johnson, C.R., Lockwood, J.A., Lockwood, D.R., Shugart, T.K.S., 2010. Catastrophic thresholds: a synthesis of concepts, perspectives and applications. Ecol. Soc. 15, 37.
10. Burgmer, J., Hillebrand, H., Pfenninger, M., 2007. Effects of climate-driven temperature changes on the diversity of freshwater macroinvertebrates. Oecologia 151, 93–103.
11. Côté, I.M., Darling, E.S., 2010. Rethinking ecosystem resilience in the face of climate change. PLoS Biol. 8.
12. Carpenter, S.R., Brock, W.A., 2004. Spatial complexity, resilience, and policy diversity: fishing on lake-rich landscapes. Ecol. Soc. 9.
13. Carpenter, S.R., Brock, W.A., 2006. Rising variance: a leading indicator of ecological transition. Ecol. Lett. 9, 308–315.
14. Carpenter, S.R., Lathrop, R.C., 2008. Probabilistic estimate of a threshold for eutrophication. Ecosystems 11, 601–613.
15. Carpenter, S., Ludwig, D., Brock, W., 1999. Management of eutrophication for lakes subject to potentially irreversible change. Ecol. Appl. 9, 751–771.
16. Carpenter, S.R., Mooney, H.A., Agard, J., Capistrano, D., DeFries, R.S., Diaz, S., Dietz, T., Duraiappah, A.K., Oteng-Yeboah, A., Pereira, H.M., Perrings, C., Reid, W.V., Sarukhan, J., Scholes, R.J., Whyte, A., 2009. Science for managing ecosystem services: beyond the millennium ecosystem assessment. Proc. Natl. Acad. Sci. U. S. A. 106, 1305–1312.

17. Carpenter, S.R., Stanley, E.H., Vander Zanden, M.J., 2011. State of the world's freshwater ecosystems: physical, chemical, and biological changes. Annu. Rev. Environ. Resour. 36, 75–99.

18. Chadwick, M.A., Feminella, J.W., 2001. Influence of salinity and temperature on the growth and production of a freshwater Mayfly in the lower mobile river, Alabama. Limnol. Oceanogr. 46, 532–542.

19. Chen, S.Q., Fath, B.D., Chen, B., 2011. Information-based network environ analysis: a system perspective for ecological risk assessment. Ecol. Indic. 11 (6), 1664–1672.

20. Chen, S.Q., Chen, B., Fath, B.D., 2013. Ecological risk assessment on the system scale: a review of state-of-the-art models and future perspectives. Ecol. Modell. 250, 25–33.

21. Chen, S.Q., Chen, B., Fath, B.D., 2014. Urban ecosystem modeling and global change: potential for rational urban management and emissions mitigation. Environ. Pollut. 190, 139–149.

22. Chiaudani, G., Vighi, M., 1974. The N:P ratio and tests with Selenastrum to predict eutrophication in lakes. Water Res. 8, 1063–1069.

23. Chu, C., Mandrak, N.E., Minns, C.K., 2005. Potential impacts of climate change on the distributions of several common and rare freshwater fishes in Canada. Divers. Distrib. 11, 299–310.

24. Cottingham, K.L., Rusak, J.A., Leavitt, P.R., 2000. Increased ecosystem variability and reduced predictability following fertilisation: evidence from palaeolimnology. Ecol. Lett. 3, 340–348.

25. Cuddington, K., Fortin, M.-J., Gerber, L.R., Hastings, A., Liebhold, A., O'Connor, M., Ray, C., 2013. Process-based models are required to manage ecological systems in a changing world. Ecosphere 4 (2), art20.

26. De Stasio, J.B.T., Hill, D.K., Kleinhans, J.M., Nibbelink, N.P., Magnuson, J.J., 1996. Potential effects of global climate change on small north-temperate lakes: physics, fish, and plankton. Limnol. Oceanogr. 41, 1136–1149.

27. Dennison, W.C., Lookingbill, T.R., Carruthers, T.J.B., Hawkey, J.M., Carter, S.L., 2007. An eye-opening approach to developing and communicating integrated environmental assessments. Front. Ecol. Environ. 5 (6), 1540–9295.

28. Denny, M.W., Hunt, L.J.H., Miller, L.J., Harley, C.D.G., 2009. On the prediction of extreme ecological events. Ecol. Monogr. 79, 397–421.

29. Doak, D.F., Morris, W.F., 2010. Demographic compensation and tipping points in climate-induced range shifts. Nature 467, 959–962.

30. Eaton, J.G., Scheller, R.M., 1996. Effects of climate warming on fish thermal habitat in streams of the United States. Limnol. Oceanogr. 41, 1109–1115.

31. Evans, C.D., Monteith, D.T., Cooper, D.M., 2005. Long-term increases in surface water dissolved organic carbon: observations, possible causes and environmental impacts. Environ. Pollut. 137, 55–71.

32. Falkenmark, M., 2003. Freshwater as shared between society and ecosystems: from divided approaches to integrated challenges. Philos. Trans. R. Soc. London Ser. B Biol. Sci. 358, 2037–2049.

33. Ficke, A.D., Myrick, C.A., Hansen, L.J., 2007. Potential impacts of global climate change on freshwater fisheries. Rev. Fish Biol. Fish. 17, 581–613.

34. Foley, B., Jones, I.D., Maberly, S.C., Rippey, B., 2012. Long-term changes in oxygen depletion in a small temperate lake: effects of climate change and eutrophication. Freshwater Biol. 57, 278–289.

35. Folke, C., Carpenter, S., Walker, B., Scheffer, M., Elmqvist, T., Gunderson, L., Holling, C. S., 2004. Regime shifts, resilience, and biodiversity in ecosystem management. Ann. Rev. Ecol. Evol. Syst. 35, 557–581.

36. Fuentes, M., Kittle, T.G.G., Nychka, D., 2006. Sensitivity of ecological models to their climate drivers: statistical ensembles for forcing. Ecol. Appl. 16, 99–116.

37. George, D., Maberly, S., Hewitt, D., 2004. The influence of the North Atlantic sscillation on the physical: chemical and biological characteristics of four lakes in the English Lake district. Freshwater Biol. 49, 760–774.

38. Groffman, P., Baron, J., Blett, T., Gold, A., Goodman, I., Gunderson, L., Levinson, B., Palmer, M., Paerl, H., Peterson, G., Poff, N., Rejeski, D., Reynolds, J., Turner, M., Weathers, K., Wiens, J., 2006. Ecological thresholds: the key to successful environmental management or an important concept with no practical application? Ecosystems 9, 1–13.

39. Hader, D.P., Kumar, H.D., Smith, R.C., Worrest, R.C., 2007. Effects of solar UV radiation on aquatic ecosystems and interactions with climate change. Photochem. Photobiol. Sci. 6, 267–285.

40. Hansen, J., Sato, M., Kharecha, P., Beerling, D., Berner, R., Masson-Delmotte, V., Pagani, M., Raymo, M., Royer, D.L., James, C., 2008. Target atmospheric CO2: where should humanity aim? Open Atmos. Sci. 2, 217–231.

41. Helmuth, B., 2009. From cells to coastlines: how can we use physiology to forecast the impacts of climate change? J. Exp. Biol. 212, 753–760.

42. Hogsden, K.L., Xenopoulos, M.A., Rusak, J.A., 2009. Asymmetrical food web responses in trophic-level richness, biomass, and function following lake acidification. Aquat. Ecol. 43, 591–606.

43. Horan, R.D., Fenichel, E.P., Drury, K.L., Lodge, D.M., 2011. Managing ecological thresholds in coupled environmental-human systems. Proc. Natl. Acad. Sci. 108 (18), 7333–7338.

44. IPCC, 2007. Climate Change 2007: The Physical Science Basis. Cambridge University, Cambridge pp. 1–18.

45. Jeppesen, E., Søndergaard, M., Jensen, J.P., Havens, K.E., Anneville, O., Carvalho, L., Coveney, M.F., Deneke, R., Dokulil, M.T., Foy, B., Gerdeaux, D., Hampton, S.E., Hilt, S., Kangur, K., Köhler, J., Lammens, E.H.H.R., Lauridsen, T.L., Manca, M., Miracle, M.R., Moss, B., Nõges, P., Persson, G., Phillips, G., Portielje, R., Romo, S., Schelske, C.L., Straile, D., Tatrai, I., Willén, E., Winder, M., 2005. Lake responses to reduced nutrient

loading – an analysis of contemporary long-term data from 35 case studies. Freshwater Biol. 50, 1747–1771.

46. Jeppesen, E., Kronvang, B., Meerhoff, M., Søndergaard, M., Hansen, K.M., Andersen, H.E., Lauridsen, T.L., Liboriussen, L., Beklioglu, M., Ozen, A., Olesen, J.E., 2009. Climate change effects on runoff, catchment phosphorus loading and lake ecological state, and potential adaptations. J. Environ. Qual. 38, 1930–1941.

47. Katharine, N.S., Richard, J.H., 2009. Threshold models in restoration and conservation: a developing framework. Trends Ecol. Evol. 24 (5), 271–279.

48. Kattel, G.R., Elkadi, H., Meikle, H., 2013. Developing a complementary framework for urban ecology. Urban For. Urban Greening 12, 498–508.

49. Kozlowski, G., Bondallaz, L., 2013. Urban aquatic ecosystems: habitat loss and depletion of native macrophyte diversity during the 20th century in four Swiss cities. Urban Ecosyst. 16 (3), 543–551.

50. Lake, P.S., Palmer, M.A., Biro, P., Cole, J., Covich, A.P., Dahm, C., Gibert, J., Goedkoop, W., Martens, K., Verhoeven, J.O.S., 2000. Global change and the biodiversity of freshwater ecosystems: impacts on linkages between above-sediment and sediment biota. BioScience 50, 1099–1107.

51. Limburg, K.E., O'Neill, R.V., Costanza, R., Farber, S., 2002. Complex systems and valuation. Ecol. Econ. 41, 409–420.

52. Lindenmayer, D.B., Likens, G.E., 2009. Adaptive monitoring: a new paradigm for long-term research and monitoring. Trends Ecol. Evol. 24, 482–486.

53. Liu, W., Zhang, Q., Liu, G., 2010. Lake eutrophication associated with geographic location, lake morphology and climate in China. Hydrobiologia 644, 289–299.

54. MEA, 2005. Ecosystems and Human Well-beings: Wetlands and Water Synthesis. World Resources Institute, Washington, DC. Malmaeus, J., Rydin, E., 2006. A time-dynamic phosphorus model for the profundal sediments of Lake Erken, Sweden. Aquat. Sci. Res. Across Boundaries 68, 16–27.

55. Mayer, A.L., Rietkerk, M., 2004. The dynamic regime concept for ecosystem management and restoration. BioScience 54, 1013–1020.

56. Meis, S., Thackeray, S.J., Jones, I.D., 2009. Effects of recent climate change on phytoplankton phenology in a temperate lake. Freshwater Biol. 54, 1888–1898.

57. Moss, B.R., 2009. Ecology of fresh waters: man and medium, past to future. Blackwell, Oxford. Nõges, P., Kisand, A., 1999. Forms and mobility of sediment phosphorus in shallow eutrophic Lake Võrtsjärv (Estonia). Int. Rev. Hydrobiol. 84, 255–270.

58. Naiman, R.J., Turner, M.G., 2000. A future perspective on North America's freshwater ecosystems. Ecol. Appl. 10, 958–970. O'Neil, J.M., Davis, T.W., Burford, M.A., Gobler, C.J., 2012. The rise of harmful cyanobacteria blooms: the potential roles of eutrophication and climate change. Harmful Algae 14, 313–334.

59. O'Reilly, C.M., Alin, S.R., Plisnier, P.D., Cohen, A.S., McKee, B.A., 2003. Climate change decreases aquatic ecosystem productivity of Lake Tanganyika, Africa. Nature 424, 766–768.

60. Pörtner, H.O., Knust, R., 2007. Climate change affects marine fishes through the oxygen limitation of thermal tolerance. Science 315, 95–97.

61. Parmesan, C., Yohe, G., 2003. A globally coherent fingerprint of climate change impacts across natural systems. Nature 421, 37–42.

62. Peeters, F., Straile, D., Lorke, A., Livingstone, D.M., 2007. Earlier onset of the spring phytoplankton bloom in lakes of the temperate zone in a warmer climate. Global Change Biol. 13, 1898–1909.

63. Pettersson, K., George, G., Nõges, P., Nõges, T., Blenckner, T., 2010. The impact of the changing climate on the supply and re-cycling of phosphorus. In: Glen, G. (Ed.), The Impact of Climate Change on European Lakes. Springer, Dordrecht, Heidelberg, London, New York.

64. Qin, B., Zhu, G., Gao, G., Zhang, Y., Li, W., Paerl, H.W., Carmichael, W.W., 2010. A drinking water crisis in Lake Taihu, China: linkage to climatic variability and lake management. Environ. Manage. 45, 105–112.

65. Rabalais, N.N., Turner, R.E., Diaz, R.J., Justic, D., 2009. Global change and eutrophication of coastal waters. Ices J. Mar. Sci. 66, 1528–1537.

66. Rockström, J., Steffen, W., Noone, K., Persson, Å., Chapin, F.S., Lambin, E.F., Lenton, T. M., Scheffer, M., Folke, C., Schellnhuber, H.J., Nykvist, B., De Wit, C.A., Hughes, T., Van Der Leeuw, S., Rodhe, H., Sörlin, S., Snyder, P.K., Costanza, R., Svedin, U., Falkenmark, M., Karlberg, L., Corell, R.W., Fabry, V.J., Hansen, J., Walker, B., Liverman, D., Richardson, K., Crutzen, P., Foley, J.A., 2009. A safe operating space for humanity. Nature 461, 472–475.

67. Saros, J.E., Clow, D.W., Blett, T., Wolfe, A.P., 2011. Critical nitrogen deposition loads in high-elevation lakes of the western US inferred from paleolimnological records. Water Air Soil Pollut. 216, 193–202.

68. Schar, C., Vidale, P.L., Luthi, D., Frei, C., Haberli, C., Liniger, M.A., Appenzeller, C., 2004. The role of increasing temperature variability in European summer heatwaves. Nature 427, 332–336.

69. Scheffer, M., Carpenter, S., Foley, J.A., Folke, C., Walker, B., 2001. Catastrophic shifts in ecosystems. Nature 413, 591–596.

70. Schindler, D.W., 1974. Eutrophication and recovery in experimental lakes: implications for lake management. Science 184, 897–899.

71. Schindler, D.W., 2001. The cumulative effects of climate warming and other human stresses on Canadian freshwaters in the new millennium. Can. J. Fish. Aquat. Sci. 58, 18–29.

72. Schippers, P., Lurling, M., Scheffer, M., 2004. Increase of atmospheric CO2 promotes phytoplankton productivity. Ecol. Lett. 7, 446–451.

73. Shuter, B.J., Post, J.R., 1990. Climate, population viability, and the zoogeography of temperate fishes. Trans. Am. Fish. Soc. 119, 314–336.

74. Smol, J.P., Douglas, M.S.V., 2007. Crossing the final ecologicalthreshold in high Arctic ponds. Proc. Natl. Acad. Sci. U. S. A. 104, 12395–12397.

75. Smol, J.P., Wolfe, A.P., Birks, H.J.B., Douglas, M.S.V., Jones, V.J., Korhola, A., Pienitz, R., Ruhland, K., Sorvari, S., Antoniades, D., Brooks, S.J., Fallu, M.A., Hughes, M., Keatley, B.E., Laing, T.E., Michelutti, N., Nazarova, L., Nyman, M., Paterson, A.M., Perren, B., Quinlan, R., Rautio, M., Saulnier-Talbot, E., Siitonen, S., Solovieva, N., Weckstrom, J., 2005. Climate-driven regime shifts in the biological communities of Arctic lakes. Proc. Natl. Acad. Sci. U. S. A. 102, 4397–4402.

76. Stefan, H.G., Hondzo, M., Fang, X., 1993. Lake water quality modeling for projected future climate scenarios. J. Environ. Qual. 22, 417–431.

77. Tao, H., Pan, J.Z., Shen, Y.L., Li, W.C., Huang, F., Zhao, H.G., 2012. Phosphorus threshold for the shift between grass-and algae-stable states in Dahong Harbor of Gehu Lake. Chin. J. Appl. Ecol. 23, 264–270.

78. UNFCCC, 2007. Climate Change: Impacts, Vulnerabilities and Adaptation in Developing Countries. United Nations Framework Convention on Climate Change, Bonn, Germany 64 pp..

79. Vörösmarty, C.J., McIntyre, P.B., Gessner, M.O., Dudgeon, D., Prusevich, A., Green, P., Glidden, S., Bunn, S.E., Sullivan, C.A., Liermann, C.R., Davies, P.M., 2010. Global threats to human water security and river biodiversity. Nature 467, 555–561.

80. Walker, B., Meyers, J.A., 2004. Thresholds in ecological and social-ecological systems: a developing database. Ecol. Soc. 9, 3.

81. Walther, G.R., Post, E., Convey, P., Menzel, A., Parmesan, C., Beebee, T.J., Fromentin, J. M., Hoegh-Guldberg, O., Bairlein, F., 2002. Ecological responses to recent climate change. Nature 416, 389–395.

82. Wang, R., Dearing, J.A., Langdon, P.G., Zhang, E., Yang, X., Dakos, V., Scheffer, M., 2012. Flickering givers early warning signals of a critical transition to a eutrophic lake state. Nature 492, 419–422.

83. Wang, H.J., 2007. Predictive Limnological Researches on Small- to Medium-sized Lakes Along the Mid-lower Yangtze River. Institute of Hydrobiology, Chinese Academy of Sciences, Wuhan, China.

84. West, M., Harrison, P., 1989. Bayesian Forecasting and Dynamic Models. SpringerVerlag, New York.

85. Weyhenmeyer, G.A., 2004. Synchrony in relationships between the North Atlantic oscillation and water chemistry among Sweden's largest lakes. Limnol. Oceanogr. 49, 1191–1201.

86. Whitehead, P.G., Wilby, R.L., Butterfield, D., Wade, A.J., 2006. Impacts of climate change on in-stream nitrogen in a lowland chalk stream: an appraisal of adaptation strategies. Sci. Total Environ. 365, 260–273.

87. Whitehead, P., Wilby, R., Battarbee, R., Kernan, M., Wade, A.J., 2009. A review of the potential impacts of climate change on surface water quality. Hydrol. Sci. J. 54, 101–123.

88. Williams, B.K., 2011. Adaptive management of natural resources—framework and issues. J. Environ. Manage. 92, 1346–1353.

89. Winder, M., Schindler, D.E., 2004. Climatic effects on the phenology of lake processes. Global Change Biol. 10, 1844–1856.

90. Woodward, G., Perkins, D.M., Brown, L.E., 2010. Climate change and freshwater ecosystems: impacts across multiple levels of organization. Philos. Trans. R. Soc. B Biol. Sci. 365, 2093–2106.

91. Yang, H., Flower, R.J., 2012. Effects of light and substrate on the benthic diatoms in an oligotrophic lake: a comparison between natural and artificial substrates. J. Phycol. 48, 1166–1177.

92. de Groot, R.S., Wilson, M.A., Boumans, R.M.J., 2002. A typology for the classification, description and valuation of ecosystem functions, goods and services. Ecol. Econ. 41, 393.

93. van Vuuren, W., Roy, P., 1993. Private and social returns from wetland preservation versus those from wetland conversion to agriculture. Ecol. Econ. 8, 289–305.

CHAPTER 2

ASSESSING THE ROLE OF CLIMATE AND RESOURCE MANAGEMENT ON GROUNDWATER DEPENDENT ECOSYSTEM

Justin Huntington, Kenneth McGwire, Charles Morton, Keirith Snyder, Sarah Peterson, Tyler Erickson, Richard Niswonger, Rosemary Carroll, Guy Smith, Richard Allen

Keywords: Landsat, NDVI, Cross-sensor calibration, Groundwater dependent ecosystems, Riparian restoration, Groundwater pumping, Phreatophytes, Complementary relationship, Evaporative demand

INTRODUCTION

Groundwater dependent ecosystems (GDEs) rely on near-surface groundwater. They provide critical habitat for many sensitive species in arid and semi-arid environments and include phreatophyte shrub lands, meadows, spring areas, and riparian zones. Improved understanding of how climate and anthropogenic impacts affect the spatiotemporal variability of GDEs is needed to increase the effectiveness of ecosystem assessments, monitoring, adaptive management frameworks, and designations of protected areas (e.g. sage-grouse habitat). However, the lack of observations provides a constraint on the utilization of data for decision-making and scientific research. Long-term remote sensing observations from the Landsat archive have repeatedly demonstrated value for ecosystem monitoring, and are increasingly being used to evaluate GDE changes relative to changing climate, drought, groundwater pumping, and agricultural disturbance (Elmore et al., 2003, Groeneveld, 2008, Yang et al., 2011, Pritchett and Manning, 2012, Nguyen et al., 2014 and Homer et al., 2015). The longevity and continuity of measurements from sensors in the lineage of Landsat's Thematic Mapper (TM), Enhanced Thematic Mapper Plus (ETM +), and Operational Land Imager (OLI), provide important baseline and current conditions that would not otherwise be attainable.

With free access to the Landsat archive, resource managers can now rely on long time series of Landsat derived vegetation index information in order to evaluate important factors affecting vegetation vigor within GDEs, including natural background variability due to climate, weather, land surface-atmospheric feedbacks, and anthropogenic factors such as land use, restoration, and groundwater pumping. However, long time series analysis with Landsat must deal with number of challenges, including data storage, computational efficiency, and changes in sensor bandwidths over time (i.e. TM, ETM +, OLI). The first two issues can be easily managed with parallelized cloud computing within Google's Earth Engine (EE), a powerful new cloud computing and environmental monitoring platform. The third issue of changing sensor bandwidths and accurate interpretation of Landsat vegetation index time series, requires quantification of how changing spectral response functions between sensors interact with the spectral variability of dynamic vegetation communities (Li et al., 2013). Effective quantification of GDE changes, and the credibility of derived resource management decisions, in part, depends on the compatibility of vegetation indices derived from different sensors, and the ability to isolate the effects of climatic and natural hydrologic variability on vegetation vigor from anthropogenic effects of land and water management.

Citation: Justin Huntington, Kenneth McGwire, Charles Morton, Keirith Snyder, Sarah Peterson, Tyler Erickson, Richard Niswonger, Rosemary Carroll, Guy Smith, Richard Allen, Assessing the role of climate and resource management on groundwater dependent ecosystem, Remote Sensing of Environment, Available online 20 July 2016, ISSN 0034-4257, http://dx.doi.org/10.1016/j.rse.2016.07.004.

OBJECTIVE

The objective of this work is to 1) use the Landsat 8 under-fly dataset to quantify differences in spectral reflectance and vegetation indices between Landsat 7 ETM + and Landsat 8 OLI for a range of vegetation communities in arid and semiarid regions of the southwestern United States, and 2) demonstrate the value of 30-year historical vegetation index datasets derived from sensors of the Landsat TM, ETM +, and OLI lineage for assessing GDEs in a variety of land and water management contexts. The study approach relies on sensor cross-calibration, cloud computing of Landsat and meteorological data, and statistical evaluation of vegetation index time series relative to annual precipitation and evaporative demand, restoration, and changing groundwater levels. Landsat 7 and 8 images from a brief under-fly test period are used to develop correction factors that account for differences between ETM + and OLI, and to assess how discrepancies may be correlated with the typical spectral reflectance curves of dominant vegetation communities within the Great Basin, USA. Two forms of atmospheric correction are used to evaluate inter-sensor variability of spectral indices: 1) the at-surface reflectance products of Landsat climate data records available through the USGS Earth Resources Observation and Science (EROS) Center Science Processing Architecture (ESPA; USGS, 2015a), and 2) the Tasumi et al. (2008) atmospheric correction algorithm that utilizes meteorological datasets within EE. Time series of corrected Normalized Difference Vegetation Index (NDVI) are analyzed for six GDE study areas within the Great Basin to demonstrate the utility of the Landsat archive for local-scale GDE assessments.

STUDY AREAS AND BACKGROUND

Specific study areas were chosen to represent a range of GDEs and environmental conditions important for three scenarios: baseline monitoring of vegetation and climate, riparian restoration, and groundwater level changes (Fig. 1 and Table 1).

Figure 1: Study areas within the Great Basin and location of the contemporaneous Landsat 7/8 image data. The Landsat 8 under-fly dataset was acquired on WRS-2 path 38 from rows 31–38, which spans the Great Salt Lake to Mexico. The brown path represents Landsat 7 with SLC off, and the green path represents Landsat 8.

Table 1: Study area names, locations, Landsat path and row, and vegetation and groundwater information

Study area	Location (lat, long)	Altitude (m)	WRS-2 path/row	Vegetation class	Water table (m below land surface)	Plant types
Spring Valley, NV	38.8010, −114.4760	1758	39/33	Alkali shrub	2–10	Greasewood, sagebrush, saltgrass
Indian Valley, NV	38.8010, −117.5000	2247	41/33	Meadow	0.5–2.5	Greasewood, sagebrush, saltgrass, meadow grass
Maggie Creek, NV	40.8860, −116.1870	1599	41/32	Riparian	0–2	Willow, sedge, rush
Susie Creek, NV	40.8350, −116.0242	1564	41/32	Riparian	0–2	Willow, sedge, rush
Fish Lake Valley, CA/NV	37.8036,-118.0667	1476	41/34	Alkali shrub	4–10	Greasewood, saltgrass
Snake Valley, UT/NV	38.7556, −114.0296	1662	39/33	Wetland	0–3	Cattail, greasewood, sagebrush, saltgrass

Baseline assessment of vegetation and climate

Spring Valley (Fig. 2a) is located in eastern Nevada, and is of interest to federal, state, and local water resource and land managers due to the potential for groundwater development. State of Nevada groundwater permit terms, along with stipulated agreements by local and federal agencies, require detailed hydrologic and biological monitoring associated with groundwater development (NSEO (The Office of the State Engineer of the State of Neva, Burns and Drici, 2011 and BWG (Biological Work Group), 2009). The alkali shrub phreatophyte area analyzed in this study is located in the southern portion of Spring Valley, and is a primary groundwater discharge area down gradient from proposed pumping wells. Depth to groundwater within the study area ranges from 2 to 10 m below land surface (Moreo et al., 2007). Phreatophyte shrubs obtain their water requirement from surface water, groundwater, or both, through root systems that range from shallow to 15 m depth (Robinson, 1958, Glancy and Rush, 1968 and Dawson and Pate, 1996). While phreatophytes within the study area consume groundwater, they primarily rely on shallow soil water derived from precipitation, and typically only consume harder to access groundwater during summer and early fall when shallow soil moisture levels are low (Dawson and Pate, 1996 and Chimner and Cooper, 2004), thereby making summer phreatophyte vegetation vigor (i.e. NDVI) a function of interannual precipitation, soil moisture, and shallow groundwater level variations.

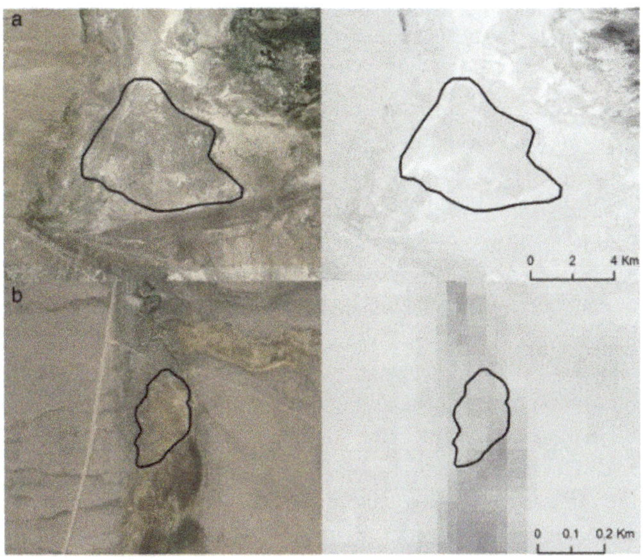

Figure 2: Spring Valley alkali shrub (a) and Indian Valley meadow (b) natural variability study areas (black polygons). 2006 National Airborne Imagery Program (NAIP) imagery (left) and April–October 2014 Landsat 8 median NDVI (right). NDVI values linearly scaled and range from 0.0 (white) to 0.8 (black).

Indian Valley (Fig. 2b) is located in central Nevada and is of interest to many wildlife and land managers due to the presence of the greater sage-grouse (*Centrocercus urophasianus*), a threatened species that has been petitioned for formal protection under the Endangered Species Act (DOI, 2015). Indian Valley has been identified as priority habitat due to its high sage-grouse lek (the male's mating arena) count, remote location, and relatively undisturbed phreatophyte meadow and shrub areas (BLM, 2012). An active area of cross-disciplinary research on sage-grouse is focused on the use of climate and Landsat archives to better understanding how climate, vegetation vigor, and sage-grouse habitat co-vary in time and space, and to identify which areas are resistant to prolonged drought (Aldridge and Boyce, 2007, Homer et al., 2015 and Donnelly et al., 2016).

Riparian restoration

Maggie Creek (Fig. 3a) and Susie Creek (Fig. 3b), both located in north-central Nevada, are tributaries to the Humboldt River that support Lahontan cutthroat trout (*Oncoryhnchus clarki henshawi*). This species has been federally listed under the Endangered Species Act due to its sensitivity to changes in land and water use, prolonged drought, and changing climate (Williams et al., 2015). Vegetation communities within riparian zones of Maggie and Susie Creeks are typical of the Great Basin and include both obligate and facultative herbaceous and woody phreatophyte species. The Bureau of Land Management (BLM) collaborated with federal and local government agencies, non-profit organizations, and local mining and livestock companies to implement comprehensive watershed and riparian restoration efforts beginning in the early 1990s. Restoration efforts included fencing, culvert replacement, prescriptive livestock grazing, and development of livestock water sources away from stream areas (Elliott et al., 2004). These efforts led to a recolonization of beaver and advanced restoration through increased shallow groundwater, stream bank storage, and habitat resilience (Williams et al., 2015).

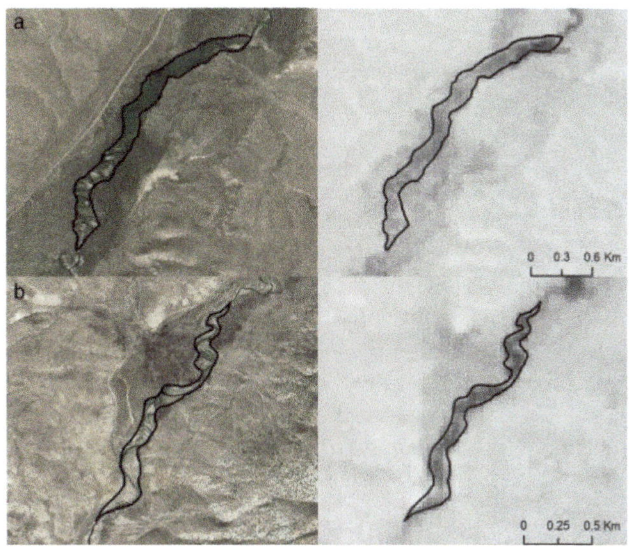

Figure 3: Maggie Creek (a) and Susie Creek (b) riparian zone study areas (black polygons). 2006 NAIP imagery (left) and April–October 2014 Landsat 8 median NDVI (right). NDVI values linearly scaled and range from 0.0 (white) to 0.8 (black).

Groundwater level changes

In the Great Basin, insufficient surface water storage requires growers to irrigate with groundwater. Groundwater pumping for irrigation often times results in lowering of the groundwater table (i.e. phreatic surface), leading to reduced phreatophyte evapotranspiration (ET) and vegetation vigor (Bredehoeft et al., 1982, Bredehoeft, 2002, Nichols, 2000, Elmore et al., 2003, Elmore et al., 2006, Naumburg et al., 2005, Cooper et al., 2006, Patten et al., 2008 and Groeneveld, 2008). Shallow groundwater and spring areas within Fish Lake Valley, Nevada (Fig. 4a) and Snake Valley, Utah, (Fig. 4b) respectively, were chosen to illustrate multi-year declines of shallow groundwater levels and vegetation vigor associated with groundwater pumping. These areas support obligate and facultative herbaceous and woody phreatophyte species, and associated fauna. The spring study area in Snake Valley, Needle Point Spring, is the subject of litigation between federal agencies and local irrigators since the spring ceased to flow in 2001 (Halford, 2015 and Childress and Smith, 2015).

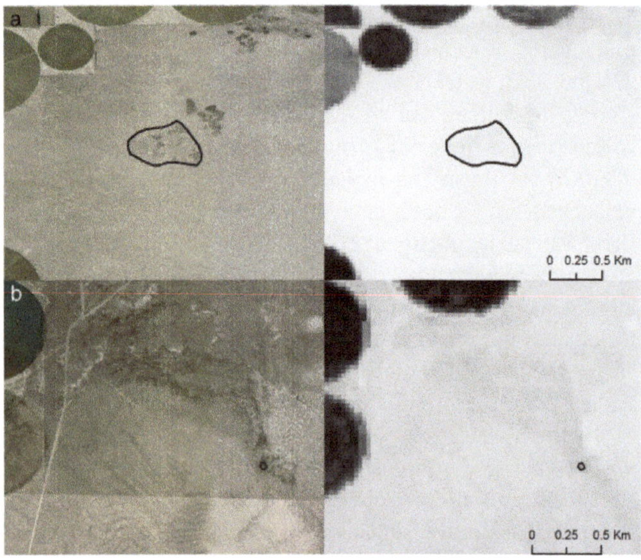

Figure 4: Fish Lake Valley shallow groundwater (a) and Snake Valley Needle Point spring (b) groundwater change study areas (black polygons). 2006 NAIP imagery (left) and April–October 2014 Landsat 8 median NDVI (right). NDVI values linearly scaled and range from 0.0 (white) to 0.8 (black).

METHODS

Landsat data processing for each study area was performed through the development of Python computer programs that use the EE application programming interface (API) to access Landsat and hourly meteorological data archives. The processing workflow implements simple automated cloud masking (Irish et al., 2006), integration of hourly meteorological data for atmospheric correction and at-surface reflectance estimation (Tasumi et al., 2008), and computation of NDVI from 1985 to the present. Programming within the EE API allowed efficient scene selection and processing of 403 Landsat images collected over the six study sites (average of 67 images per study site from 1985 to 2014).

Sensor cross-calibration

The bandwidths of spectral channels on Landsat satellites have changed over time. Fig. 5 illustrates the spectral response functions of red (TM: 630–690 nm, ETM +: 630–690 nm, OLI: 640–670 nm) and near infrared (NIR; TM: 760–900 nm, ETM +: 770–900 nm, OLI: 850–880 nm) bands of Landsats 5, 7, and 8 along with representative spectra from Clark et al. (2007) of five plant species that are found in the study area (fir, juniper, piñon, sagebrush, and rabbitbrush). ETM + response functions are seen to be relatively similar to the original TM, while the bandwidths of the OLI sensor are narrowed, particularly for the NIR band. Steven et al. (2003) suggest that there is a 2% increase in NDVI from TM to ETM +, but their result appears to have been biased upwards by questionable outliers that perhaps should have been removed (Steven et al., 2003, Fig. 3). Martínez-Beltrán et al. (2009) found increases in NDVI from TM to ETM + to be < 1%, and Vogelmann et al. (2001) support the interchangeability of the two sensors, so this analysis made no attempt to compensate between TM and ETM +. However, the more dramatic difference of OLI spectral bands has been identified as requiring attention (Li et al., 2013 and Roy et al., 2016).

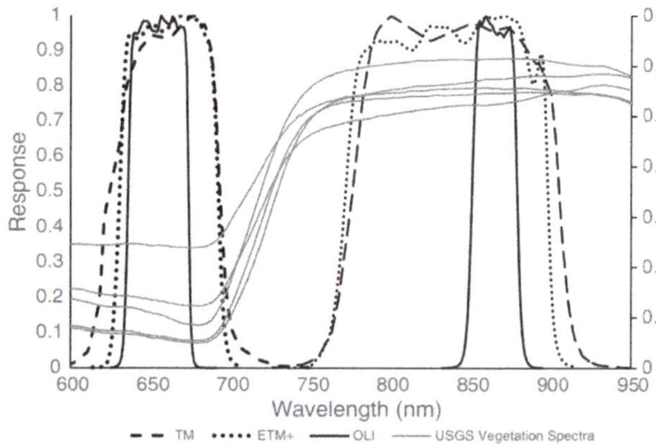

Figure 5: Spectral response functions of red and near infrared bands for TM (630–690 nm, 760–900 nm), ETM + (630–690 nm, 850–880 nm), and OLI along with plant type spectra (gray lines) for fir, juniper, piñon, sagebrush, and rabbit brush from a USGS spectral database (ENVI software library file usgs_min.sli).

The effect of changes in the OLI spectral bands for vegetation in the Great Basin was assessed using images from OLI and ETM + that were acquired within 7 min of each other on March 29, 2013 during the "under-fly" testing of the Landsat 8 system (Roy et al., 2014 and USGS (U.S. Geological Survey), 2015b). This under-fly dataset was acquired on WRS-2 path 38 from rows 31–38, which spanned the Great Salt Lake to Mexico (Fig. 1). Only rows 33–37 were used in this analysis, as the others had too much cloud cover or haze to be useful. Atmospherically corrected OLI and ETM + images were intersected with up to 54 natural vegetation classes of the LANDFIRE vegetation map (USGS, 2010) to determine how OLI must be rescaled to match ETM +, and to examine the degree to which differences might be dependent on vegetation class type. Vegetation-dependent changes in NDVI are detected if different classes exhibit differing spectral reflectance distributions within the narrower OLI bands as compared to the broader ETM + bands.

Two methods of radiometric and atmospheric corrections were tested with the Landsat imagery, 1) the Landsat surface reflectance climate data record provided through the ESPA on-demand interface of the USGS EROS Data Center (Masek et al., 2006 and USGS (U.S. Geological Survey), 2015a), and 2) a method following Tasumi et al. (2008) and Trezza and Allen (2013). The ESPA surface reflectance product uses the USGS Landsat Ecosystem Disturbance Adaptive Processing System (LEDAPS) software for Thematic Mapper imagery from Landsats 5 and 7, and the preliminary USGS L8SR (Landsat 8 Surface Reflectance) software for OLI imagery (USGS, 2015a). The Tasumi/Trezza method was implemented within EE with spatially distributed, near-surface hourly vapor pressure from the North American Land Data Assimilation System (NLDAS; Mitchell et al., 2004) to estimate precipitable water and atmospheric transmittance (Tasumi et al., 2008). Spatially distributed atmospheric pressure was estimated from the 30 m National Elevation Dataset (NED) following ASCE-EWRI (2005).

While the under-flight of Landsat 8 was nearly contemporaneous with Landsat 7, the two paths were not identical (Fig. 1) - the ETM + scan line corrector mirror created data gaps, and clouds moved between acquisitions. As such, a mask of the most consistent regions between the two sensors was required for inter-sensor calibration. Clouds and cloud shadows in both images were manually masked out with a buffer distance of approximately 1 km to ensure that gradients and cloud cover and haze were removed. Areas with a NIR reflectance of < 0.05 were masked out to remove water bodies and areas of deep shadow that affect NDVI. Pixels on either side of boundaries between classes in the LANDFIRE map were removed to reduce the influence of map misregistration, as well as misregistration between the two satellite overpasses in areas of high relief that arose from the offset in overpass position. LANDFIRE classes with masked areas containing < 1000 pixels were removed from the analysis to ensure that a stable mean value was calculated for each land cover class. The mean red, NIR, and NDVI for each vegetation class was calculated, and OLI mean values were regressed against ETM + to determine the appropriate transformation to make NDVI from OLI consistent with the prior sensors.

Data preparation for GDE assessments

Boundaries were defined around the six study areas (Fig. 2, Fig. 3 and Fig. 4) and used to spatially and temporally average summer NDVI derived from cloud-free Landsat images that were closest in time to the middle of June, July, and August from 1985 to 2014. For riparian restoration study sites, average summer NDVI was computed from July–August Landsat images so that the potential for standing water to decrease NDVI was minimized. Based on results of the sensor cross-calibration (detailed in the Results section), NDVI was computed using the Tasumi/Trezza at-surface reflectance method. NDVI was chosen over other indices (i.e. Soil Adjusted Vegetation Index or Modified Soil Adjusted Vegetation Index) since NDVI has the breadth and history of usage that is not matched by other indices, is a popular standard measure, does not require soil parameter calibration, and has been shown to outperform other indices for quantifying sparse vegetation cover in arid environments (McGwire et al., 1999 and Wu, 2014). To detect the signal of interannual groundwater availability within GDEs, it was important to focus on periods when shallow soils were dry, and when spectral variability due to the happenstance timing of individual precipitation events was minimized. Precipitation over the study areas is generally at its minimum in June for the more southern and eastern study areas, and in July and August for the northern study areas. Images acquired during the summer months of June–August were selected to limit the NDVI signal of shallow rooted grasses and forbs, and maximize the phreatophyte NDVI signal derived from groundwater.

Summer streamflow, spring flow, and shallow groundwater levels in the Great Basin are highly correlated to annual precipitation due to the fact that the majority of annual precipitation falls during winter as snow, and spring snowmelt provides the majority of groundwater recharge that is later discharged during summer and fall (Huntington and Niswonger, 2012 and McEvoy et al., 2012). Water-year (October–September) precipitation (PPT) for each study site was estimated from a hybrid 4 km spatial resolution gridded weather dataset, GRIDMET (Abatzoglou, 2013), based on the Parameter Regression on Independent Slopes Model (PRISM) (Daly et al., 1994) and NLDAS (Mitchell et al., 2004). Evaporative demand (ET_0) was estimated using the standardized Penman-Monteith equation, which is a function of daily solar radiation, air temperature, humidity, and wind speed (ASCE-EWRI, 2005). Daily GRIDMET data were accessed within EE to estimate water-year PPT and ET_0 totals at each study site.

Groundwater level data from the Arlemont Ranch well located within the Fish Lake Valley study area (Fig. 4a), was acquired from the Nevada Division of Water Resources water level database (NDWR, 2015). Two water level readings per year were typically reported, therefore an average of the two readings was computed for each year. Groundwater level and nearby agricultural pumping data for the Needle Point Spring study area (Fig. 4b) was acquired from Summers (2001) and Halford (2015). Annual average groundwater levels and pumping rates were compared with summer NDVI and water-year PPT and ET_0 to assess historical vegetation response to climate and changing groundwater levels.

RESULTS

Sensor cross-calibration

Average LANDFIRE vegetation class red, NIR, and NDVI values derived from USGS ESPA products and the Tasumi/Trezza method are illustrated in Fig. 6. The Tasumi/Trezza method shows a markedly higher degree of correspondence in NDVI between ETM + and OLI observations. The greater degree of scatter observed in the ESPA plots is likely due to the mixing of atmospheric correction methods between sensors, and limitations the current L8SR software has for areas of high topography (USGS, 2015a). Due to the close correspondence in NDVI between ETM + and OLI observations, results presented in the following sections were derived solely from the Tasumi/Trezza method.

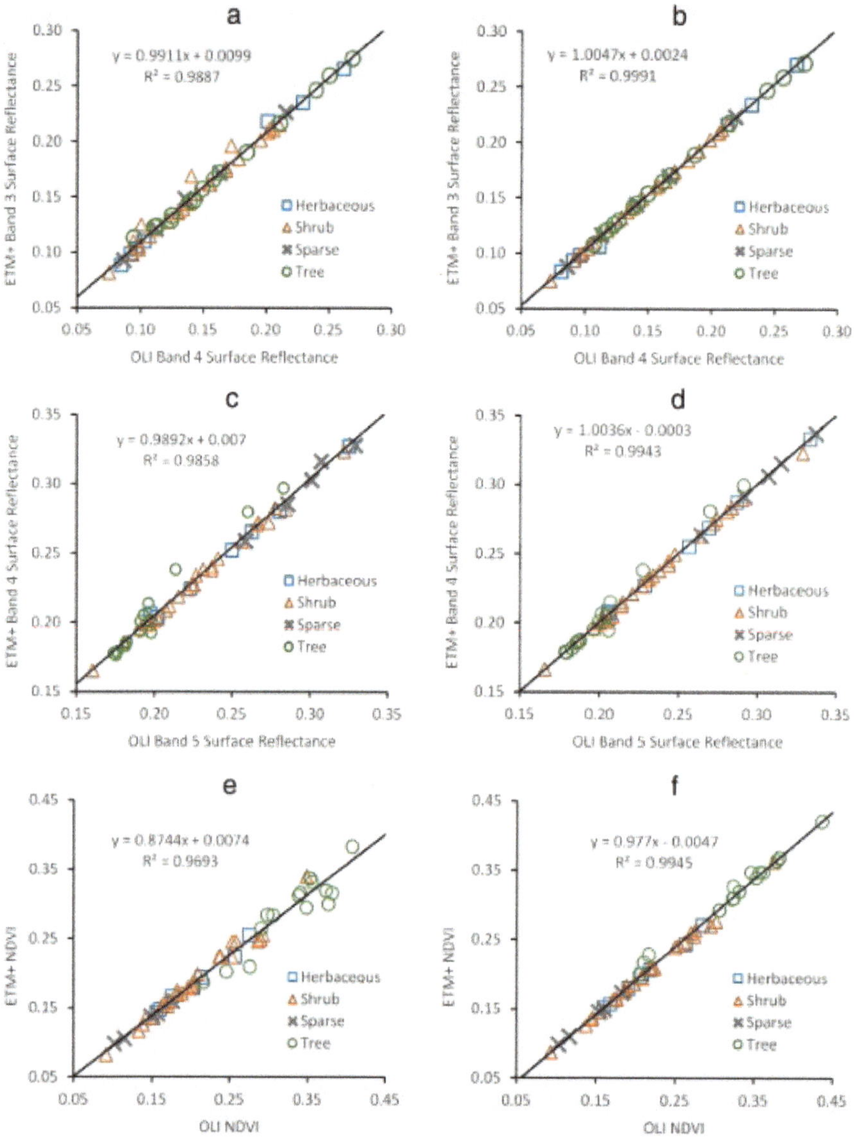

Figure 6: Mean values of ETM + versus OLI in LANDFIRE vegetation classes for red (a, b), NIR (c, d), and NDVI (e, f) using USGS ESPA Landsat surface reflectance products (a, c, e) and Tasumi/Trezza at-surface reflectance (b, d, f). Symbols indicate the generalized life form attribute of each vegetation class.

As might be expected by the greater difference in bandwidths, scatter in Fig. 6 is more pronounced in the NIR than in the red band. While regression residuals for NDVI were relatively small (maximum absolute value < 0.02), the four most under-predicted classes in Fig. 6 were all mixed conifer classes, while the two most over-predicted classes contained Gambel oak *(Quercus gambelii)*. This suggests that some vegetation communities did have a detectable difference in spectral reflectance between the ETM + and OLI bandwidths. However, there were seven other conifer classes that were quite close to the regression line, so a generalization about conifers could not be made. None of the vegetation classes that appear as outliers (Fig. 6) are present in the study areas. The Tasumi/Trezza NDVI regression in Fig. 6 was applied in the EE workflow to compute OLI NDVI for each of the six study areas.

Baseline assessment of vegetation and climate

Interannual variations of vegetation vigor and climate for Spring Valley and Indian Valley study areas are illustrated by plotting summer NDVI with annual PPT and ET_0 (Fig. 7 and Fig. 8). Results show that NDVI and PPT time series co-vary and have similar long-term trends (Fig. 7). While it is evident that NDVI co-varies with annual PPT, statistical correlation between the two variables is only moderate, with R^2 values of 0.49 and 0.41 for Spring Valley and Indian Valley, respectively. Factors limiting the strength of the relationship between annual PPT and summer average NDVI could include: 1) antecedent soil moisture conditions, 2) sparse leaf area of Spring Valley setting a lower limit on detectability; 3) influence of background soil reflectance; 4) presence of shallow groundwater stabilizing minimum vegetation vigor and NDVI, and 5) wide range in NDVI during summer months as illustrated by maximum and minimum values shown in Fig. 7. The deeply rooted phreatophytes of Spring Valley appear to be buffered against years with extremely low PPT more than the shallow rooted meadow grass in Indian Valley (i.e. multi-year drought periods of 1989 to 1994).

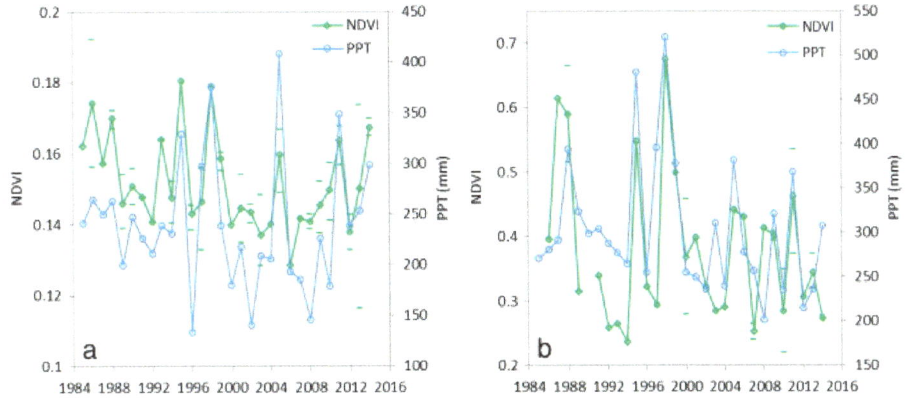

Figure 7: Spatially averaged summer NDVI and annual PPT time series for Spring Valley (a) and Indian Valley (b) phreatophyte shrub and meadow vegetation study areas. NDVI tics represent June–August maximum and minimum NDVI values. Years with missing maximum and minimum NDVI tics indicate that only one image was available.

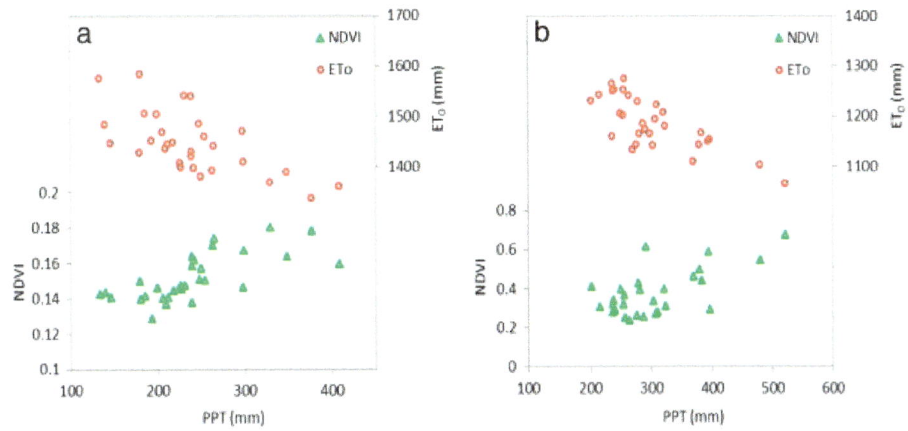

Figure 8: Spring Valley (a) and Indian Valley (b) summer NDVI, annual ET_0, and PPT, showing a complementary relationship between NDVI and ET_0. As NDVI (i.e. proxy for ET) generally increases with PPT, ET_0 decreases due to near surface atmospheric feedbacks.

Annual vegetation and near-surface atmospheric and climatic feedbacks are evaluated by plotting summer average NDVI and annual ET_0 on the y-axes, with annual PPT on the x-axis (Fig. 8). Illustrating PPT and ET_0 with vegetation and land surface states and fluxes is useful for evaluating the well-known complementary relationship between actual ET and ET_0 in arid environments (Brutsaert and Stricker, 1979, Hobbins et al., 2004, Huntington et al., 2011 and Jaksa et al., 2013). The use of summer vegetation indices as proxies for phreatophyte annual ET has been well established within the Great Basin (Nichols, 2000, Groeneveld et al., 2007, Smith et al., 2007, Beamer et al., 2013 and Garcia et al., 2014). Fig. 8 illustrates that as annual PPT increases, NDVI increases and ET_0 decreases. Conversely, as PPT decreases, so does NDVI, while ET_0 increases.

These results are consistent with land surface energy balance theory. When water is limited, and available energy is fairly uniform in space, energy that would have been used for ET is instead used in the production of sensible heat flux, thereby increasing air temperature, vapor pressure deficit, and ultimately ET_0 (Brutsaert and Stricker, 1979). This drying scenario is well illustrated in Fig. 8, where ET_0 increases as PPT and NDVI decrease. The complementary relationship between actual ET and ET_0 has been shown to be prevalent at multiple time scales in Spring Valley and other phreatophyte shrub areas in Nevada using measured energy balance data (Huntington et al., 2011 and Beamer et al., 2013). These results clearly highlight the complementary relationship between vegetation vigor, ET_0, and PPT, and are consistent with similar remotely sensed vegetation index and complementary relationship studies (Goward et al., 1994, Szilagyi, 2002 and Mo et al., 2014). The long time history of Landsat allows for baseline variability of vegetation vigor and complementary relationships between vegetation and climate to be established at local to regional scales.

Riparian restoration

The ecological benefits from watershed restoration of Maggie Creek and Susie Creek have been previously well documented (Elliott et al., 2004 and Williams et al., 2015). However, interannual Landsat time series of vegetation vigor offer a new perspective for evaluating how riparian areas within these restored watersheds have responded to multiple drought cycles over several decades. Watershed and riparian area restoration efforts within the Maggie and Susie Creek drainages began in the early 1990s, and since that time riparian area NDVI has markedly increased (Fig. 9). Maggie and Susie Creek NDVI trends from 1985 to 2014 are statistically significant at the 95% confidence level using the Mann-Kendall trend test (Helsel and Hirsch, 1992), with *p*-values of 1.4E − 06 and 4.1E − 04, respectively. July–August maximum and minimum NDVI trends are also statistically significant for both sites (*p*-values of 1.8E − 04 and 4.1E − 05 for Maggie Creek, and 1.7E − 03 and 3.1E − 03 for Susie Creek, respectively). A more direct metric of vegetation change since restoration is the percent increase in July–August average NDVI from pre- to post-restoration periods (1985–1989 and 1990–2014). Maggie Creek NDVI increased by 54% from 0.288 to 0.444, and Susie Creek NDVI increased by 67% from 0.236 to 0.396. Average annual PPT increase from pre- to post-restoration was 13% and 14% for Maggie and Susie Creek, respectively, so only part of the post-restoration NDVI increase can be attributed to increased PPT and streamflow given that annual PPT and annual and summer streamflow are highly correlated in this region (Berger, 2000, Prudic et al., 2006 and McEvoy et al., 2012). An important attribute of successful riparian restoration is reduced drought stress (Stromberg, 2001 and Shafroth et al., 2002). Reduced drought stress since restoration is evident during droughts of the late 1990s and early 2000s, 2007 and 2008, and 2012 and 2013, by the increase of July–August average, maximum, and minimum NDVI values illustrated in Fig. 9.

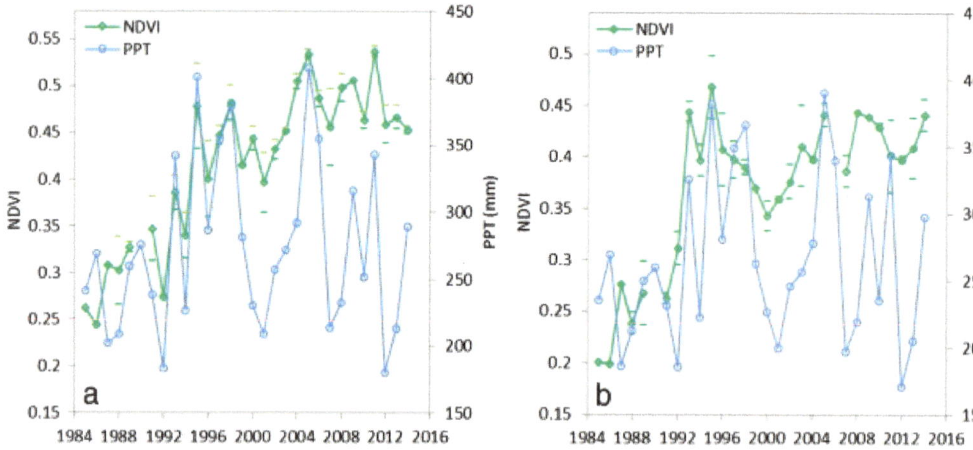

Figure 9: Spatially averaged July–August NDVI and annual PPT time series for Maggie Creek (a) and Susie Creek (b) riparian restoration sites. NDVI tics represent July–August maximum and minimum NDVI values.

Groundwater level changes

Annual PPT, groundwater level, and corresponding summer NDVI changes for the Fish Lake study area are illustrated in Fig. 10. Groundwater levels have steadily decreased since 1979 due to groundwater pumping for irrigated agriculture adjacent to the study area (Fig. 4). Depth to groundwater was approximately 5 m in 1985, and has declined at a rate of ~ 0.167 m/year to reach a depth of 10 m in 2014. NDVI has also declined during the period of groundwater level decline, with intermittent NDVI increases that correspond to anomalously high annual PPT. The trend of June–August average NDVI from 1985 to 2014 is statistically significant at the 95% confidence level, with a p-value of 4.2E − 06. Summer maximum and minimum NDVI trends from 1985 to 2014 are also statistically significant, with p-values of 6.9E − 04 and 8.7E − 05, respectively.

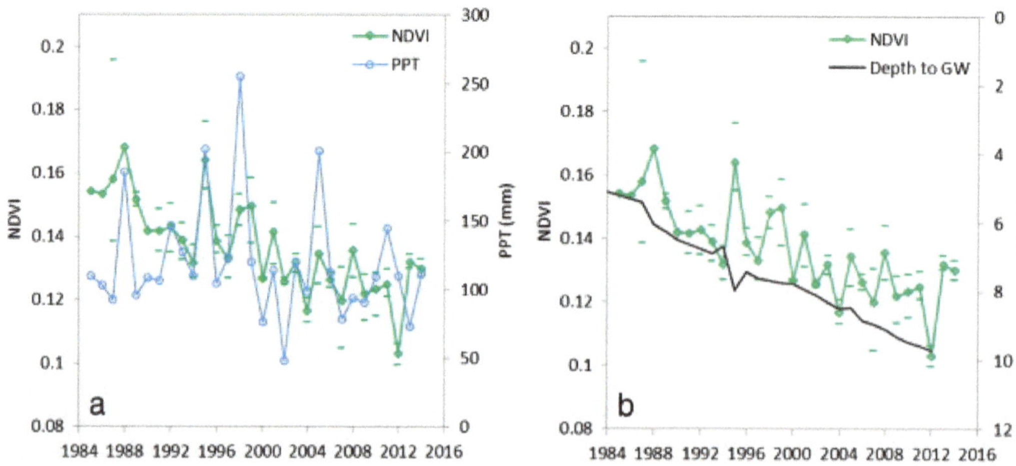

Figure 10: Spatially averaged summer NDVI and annual PPT time series (a) and spatially averaged summer NDVI and average depth to groundwater (b) for the Fish Lake Valley study area. NDVI time series illustrate a general decline as depth to groundwater increases. NDVI tics represent June–August maximum and minimum NDVI values.

Fig. 11 and Fig. 12 illustrate spatially averaged summer NDVI paired with annual PPT, depth to groundwater, groundwater pumping, and ET_0 for the Needle Point Spring study area, where it is observed that NDVI begins to decline following the year 1999, and is coincident with the onset of pumping and drought. According to Summers (2001), Needle Point spring ceased flowing in 2001. NDVI has continued to decline since 2001, with NDVI increases that coincide with years of anomalously high annual PPT. The downward trend in NDVI from 1985 to 2014 is statistically significant at the 95% confidence level (p-value of 3.4E − 04), and there is a slight positive trend in PPT during this same period. Summer maximum and minimum NDVI trends are also statistically significant, with *p*-values of 3.7E − 03 for both maximum and minimum NDVI. Average summer NDVI for pre- and post-groundwater pumping periods (1986–1999 and 2000–2014) was 0.37 and 0.29, respectively, which equates to a 22% decrease. Average annual PPT was 239 and 221 mm for pre- and post-groundwater pumping periods, an 8% decrease. These results suggest that the post-pumping NDVI decline (2000 to 2014) could be partly attributed to lower annual PPT. However, given the supporting evidence of measured and simulated groundwater level declines (Summers, 2001 and Halford, 2015), long-term NDVI declines are most likely related to groundwater pumping adjacent to the spring area (Fig. 12a). Reduced maximum, minimum, and average summer NDVI peaks during post-pumping further support this hypothesis. For example, average NDVI values coincident with anomalously high PPT years of 2005 and 2011 never reach the pre-pumping average NDVI of 0.37 (Fig. 11b). Additionally, the relationship between PPT and NDVI markedly changed during the post-pumping period (Fig. 12b).

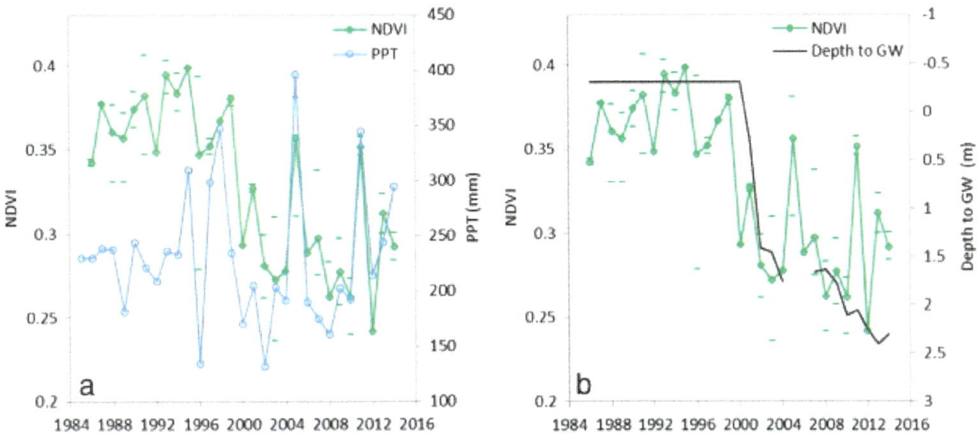

Figure 11: Spatially averaged summer NDVI and annual PPT time series (a) and spatially averaged summer NDVI and average depth to groundwater (b) for the Needle Point Spring study area. NDVI tics represent June–August maximum and minimum NDVI values.

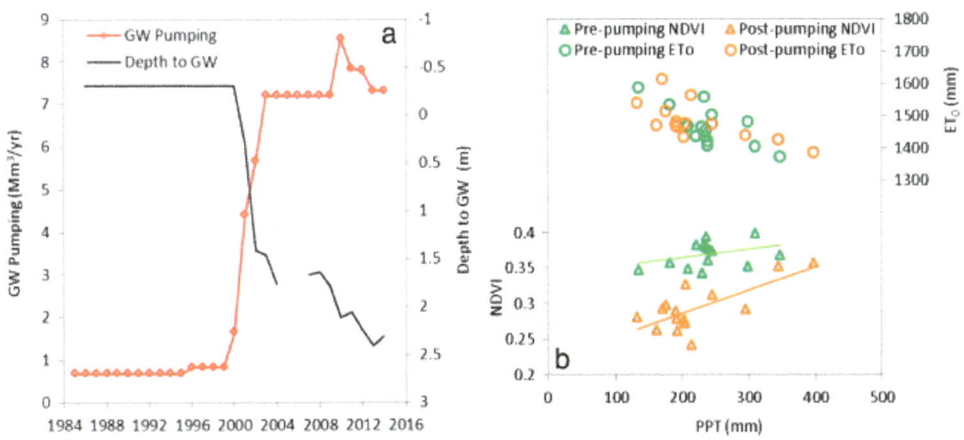

Figure 12: Groundwater pumping and depth to groundwater for the Needle Point Spring study area (a), and annual PPT, ET_0, and summer NDVI, for pre and post-pumping periods (b).

DISCUSSION

Isolating the impacts of natural climatic and hydrologic variability on vegetation vigor, from of the impacts of anthropogenic land and water management is challenging, yet important for identifying cause and effect relationships. Having the ability to readily evaluate lengthy and paired time series of climate and Landsat derived vegetation vigor, offers land and water managers new and valuable information for long-term assessment of GDEs. Such lengthy Landsat derived time series require effective sensor cross-calibration so that cause and effect are not misinterpreted (Roy et al., 2016). Fig. 6demonstrates the importance of using a consistent atmospheric correction method across sensor systems. The magnitude of OLI/ETM + NDVI regression residuals using the ESPA products was double that of the Tasumi/Trezza method.

While the use of differing atmospheric correction algorithms from ESPA may be dictated by operational data processing flows, consistent implementation of the Tasumi/Trezza method using EE cloud computing allowed rapid on-demand processing and delivery to the desktop computer with consistent image data products. The near-contemporaneous collection of images during the Landsat 8 under-fly testing period, and our averaging of NDVI within boundary-masked areas of different vegetation classes, essentially eliminated the effects spatio-temporal misregistration on the regressions of Fig. 6. The high level of correspondence between Tasumi/Trezza-corrected NDVI across a wide variety of natural vegetation communities suggests that the dramatic change in NIR bandwidths from ETM + to OLI is well corrected by linear regression and should not generally interfere with time series analyses. These findings suggest that vegetation index analyses originating from different Landsat sensors can be made easily compatible, allowing seamless analysis over the entire archive. However, the consistency of species composition among the most positive and the most negative regression residuals indicates that land-cover specific adjustments may be, in some cases, necessary.

While time series analyses of Landsat data over spatially limited study areas are often powerful and enlightening, cloud storage computing capabilities (i.e., co-located storage and parallel processing of Landsat, meteorology, and climate datasets with geospatial analysis capabilities) provided by EE make it possible to easily and rapidly spatially up-scale. A computation that would take a single computer minutes to hours can be performed in a matter of seconds. This allows for exploration of multiple interannual temporal and spatial signals over large regions. Large scale spatiotemporal analyses, previously precluded due to massive storage and computational volumes, are for the first time, practical due to geospatially-enabled cloud computing systems. However, regardless of newly available computing power, a limitation with respect to seasonal and interannual time series analysis is the limited number of cloud free Landsat images available for time periods of interest (e.g. 8 or 16 day returns for calculation of one to three month summer maximum, minimum, and average NDVI). Integration of newly available Sentinel 2 imagery with Landsat processing workflow will increase the probability of cloud free images available for seasonal and interannual time series analysis.

Future research that combines both time series and spatial analyses will greatly enhance our ability to rapidly evaluate the magnitude, spatial extent, and cause of GDE changes related to climate, land, and water management (Kennedy et al., 2014). Sophisticated tools for Landsat spatial time series analysis that have traditionally been focused on forest disturbance and recovery such as LandTrendr and TimeSync (Kennedy et al., 2010 and Cohen et al., 2010), could prove extremely useful for rapid GDE monitoring within EE. Currently, simple products such as Landsat scale NDVI anomalies (i.e. per pixel difference from long-term average) can be rapidly generated within EE to explore and discover relative change, and be combined with climate, meteorological, and field data. For example, Fig. 13 illustrates the Needle Point Spring June–August 2014 NDVI anomaly (i.e. difference) relative to the 1985 to 2014 average using Landsat 5, 7, and 8 top-of-atmosphere reflectance image collections within EE. While Fig. 13 clearly illustrates that Needle Point Spring has anomalously low NDVI during summer of 2014, it also shows anomalously low NDVI within groundwater discharge areas to the east and southeast of Needle Point Spring (i.e. high reflectance saline soil areas shown in true color image). Conversely, anomalously high NDVI areas primarily occur in non-phreatophyte upland areas to the southwest of Needle Point Spring because of anomalously high rains that occurred during the summer of 2014. Future efforts for regional long-term and interannual GDE monitoring could combine spatial and temporal distributions of vegetation indices with vegetation type, and climatic, meteorologic, and hydrologic distributions to better isolate and analyze natural and anthropogenic impacts.

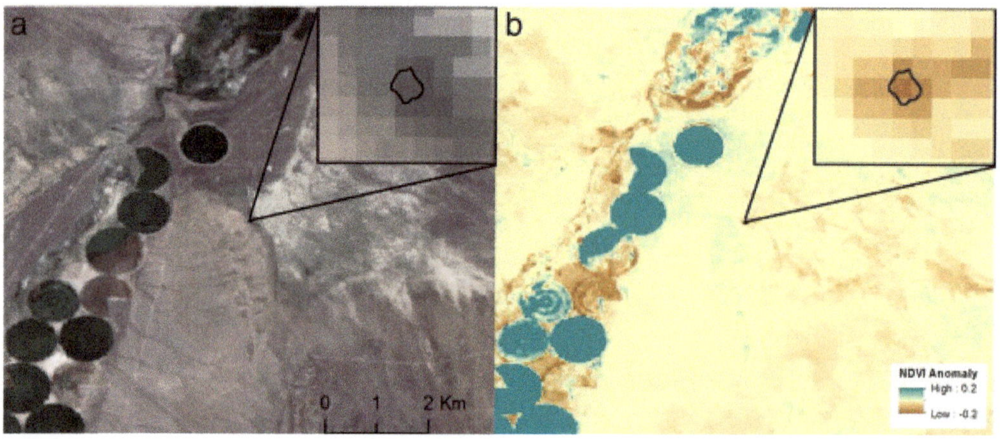

Figure 13: Needle Point Spring Landsat 8 median true color composite (a), and NDVI anomaly (b), for June–August 2014.

CONCLUSIONS

This paper highlights the use of the Landsat archive for monitoring groundwater dependent ecosystems (GDEs). These ecosystems provide critical habitat for many sensitive species in arid and semi-arid environments of phreatophyte shrub lands, meadows, spring areas, and riparian zones. The Landsat archive shows meaningful correlations between changes in annual vegetation vigor (NDVI), precipitation, evaporative demand, depth to groundwater, and land and water management within six GDE study areas. The study approach relied on sensor cross-calibration, cloud computing of Landsat and meteorological data, and statistical evaluation. This approach can provide rapid, useful interpretations of GDE conditions. Results indicate a benefit from the application of a consistent atmospheric correction method across different Landsat sensors. Special attention should be given to whether the observed level of inconsistency in at-surface reflectance for Landsat can be minimized in future operational products. Having demonstrated the ability of the current Landsat archive to detect trends in GDEs related to climate, groundwater, and resource management, that will ultimately fill critical information needs for science-informed decisions, it is clear that a compatible remote sensing data stream must be maintained into the future, and that the value of this consistent growing archive will compound over time. The longevity and continuity of measurements from sensors in the lineage of

Landsats provide important baseline and current conditions for ecosystem assessments that would not otherwise be attainable.

ACKNOWLEDGMENTS

Funding for this study was provided by the U.S. Bureau of Land Management grant #L13AC00169, U.S. Department of Agriculture Agricultural Research Service grant #59-5370-3-001, Desert Research Institute Maki Endowment, U.S. Geological Survey 2012–2017 Landsat Science Team, Department of the Interior Northwest Climate Science Center grant to evaluate new strategies for addressing climate change impacts on water and ecosystems in the Great Basin, and a Google Earth Engine faculty research grant. ESPA Landsat Surface Reflectance products used were courtesy of the U.S. Geological Survey. The authors would like to thank Stephen Maples of the U.S. Geological Survey, and four anonymous reviewers for their thoughtful comments and suggestions. They greatly improved the manuscript.

REFERENCES

1. Abatzoglou, J.T., 2013. Development of gridded surface meteorological data for ecological applications and modelling. International Journal of Climatology 33 (1), 121–131.
2. Aldridge, C.L., Boyce, M.S., 2007. Linking occurrence and fitness to persistence: Habitatbased approach for endangered greater sage-grouse. Ecological Applications 17 (2), 508–526.
3. ASCE-EWRI. (2005). The ASCE standardized reference evapotranspiration equation, report 0-7844-0805-X, ASCE task committee on standardization of reference evapotranspiration, Reston, Virginia, American Society of Civil Engineers. (Available at) http://www. kimberly. uidaho.edu/water/asceewri/
4. Beamer, J.P., Huntington, J.L., Morton, C.G., Pohll, G.M., 2013. Estimating annual groundwater evapotranspiration from phreatophytes in the great basin using landsat and flux tower measurements. JAWRA Journal of the American Water Resources Association 49 (3), 518–533.
5. Berger, D.L., 2000. Water budgets for pine valley, Carico lake valley, and upper Reese river valley hydrographic areas, middle Humboldt river basin, north-central Nevada methods for estimation and results: U.S. geological survey water-resources investigation report 99-4272 (45 p).
6. BLM (Bureau of Land Management). 2012. White paper on BLM and U.S. forest service greater Sage-Grouse preliminary habitat map. Available at http://www. blm.gov/nv/ st/en/prog/wildlife/greater_sage-grouse/ preliminary_habitat.html
7. Bredehoeft, J.D., 2002. The water budget myth revisited: Why hydrogeologists model. Groundwater 40 (4), 340–345.
8. Bredehoeft, J.D., Papadopulos, S.S., Cooper, H.J., 1982. Groundwater: The water budget myth. Scientific basis of water resource management, National Academy of Sciences Studies in Geophysics. pp. 51–57.
9. Brutsaert, W., Stricker, H., 1979. An advection-aridity approach to estimate actual regional evapotranspiration. Water Resources Research 15 (2), 443–450.
10. Burns, A.G., Drici, W., 2011. June. Hydrology and water resources of spring, cave, dry lake, and Delamar valleys, Nevada and vicinity. Presentation to the Office of the Nevada State Engineer: Southern Nevada Water Authority, Las Vegas, Nevada.
11. BWG (Biological Work Group), 2009. Biological monitoring plan for the spring valley stipulation. (February 2009). (Available at) http://www.fws.gov/ nevada/highlights/comment/ spring_valley/Biological_ Monitoring_Plan_Spring_Valley_Stipulation_Feb_09. pdf.
12. Childress, J.M., Smith, D.L., 2015. Exhibit 54, public administrative hearing before the state engineer in the matter of protested applications 78795, etc., February 2–6, 2015, of- ficial records in the Office of the State Engineer.
13. Chimner, R.A., Cooper, D.J., 2004. Using stable oxygen isotopes to quantify the water source used for transpiration by native shrubs in the San Luis Valley, Colorado USA. Plant and Soil 260 (1–2), 225–236.
14. Clark, R.N., Swayze, G.A., Wise, R., Livo, K.E., Hoefen, T.M., Kokaly, R.F., Sutley, S.J., 2007. USGS digital spectral library splib06a, U.S. geological survey, data series 231..
15. Cohen, W.B., Yang, Z., Kennedy, R., 2010. Detecting trends in forest disturbance and recovery using yearly landsat time series: 2. TimeSync—Tools for calibration and validation. Remote Sensing of Environment 114 (12), 2911–2924.
16. Cooper, D.J., Sanderson, J.S., Stannard, D.I., Groeneveld, D.P., 2006. Effects of long-term water table drawdown on evapotranspiration and vegetation in an arid region phreatophyte community. Journal of Hydrology 325 (1), 21–34.
17. Daly, C., Neilson, R.P., Phillips, D.L., 1994. A statistical-topographic model for mapping climatological precipitation over mountainous terrain. Journal of Applied Meteorology 33 (2), 140–158.
18. Dawson, T.E., Pate, J.S., 1996. Seasonal water uptake and movement in root systems of Australian phreatophytic plants of dimorphic root morphology—A stable isotope investigation. Oecologia 107, 13–20.
19. DOI (Department of the Interior), 2015. Endangered and threatened wildlife and plants; 12-month finding on a petition to list greater Sage-Grouse (Centrocercus urophasianus) as an endangered or threatened species; proposed rule. Fish and wildlife service, 50 CFR Part 17. Federal register Vol. 80. No. 191.
20. Donnelly, J.P., Naugle, D.E., Hagen, C.A., Maestas, J.D., 2016. Public lands and private waters: Scarce mesic resources structure land tenure and sage-grouse distributions. Ecosphere 7 (1).
21. Elliott, J., Haskins, R.L., Weller, G., 2004. Lahontan cutthroat trout species management plan for the upper Humboldt river drainage basin. State of Nevada, Department of Wildlife.

22. Elmore, A.J., Mustard, J.F., Manning, S.J., 2003. Regional patterns of plant community response to changes in water: Owens Valley, California. Ecological Applications 13 (2).

23. Elmore, A.J., Manning, S.J., Mustard, J.F., Craine, J.M., 2006. Decline in alkali meadow vegetation cover in California: The effects of groundwater extraction and drought. Journal of Applied Ecology 43 (4), 770–779.

24. Garcia, C.A., Huntington, J.M., Buto, S.G., Moreo, M.T., Smith, J.L., Andraski, B.J., 2014. Groundwater discharge by evapotranspiration, Dixie Valley, west-central Nevada, March 2009–September 2011 (No. 1805). US geological survey.

25. Glancy, P.A., Rush, F.E., 1968. Water-resources appraisal of smoke creek-San Emidio desert area, Nevada and California: Nevada Department of Conservation and Natural Resources, Water Resources–Reconnaissance series report. 44 (57 p).

26. Goward, S.N., Waring, R.H., Dye, D.G., Yang, J., 1994. Ecological remote sensing at OTTER: Satellite macroscale observations. Ecological Applications 322–343.

27. Groeneveld, D.P., 2008. Remotely-sensed groundwater evapotranspiration from alkali scrub affected by declining water table. Journal of Hydrology 358 (3), 294–303.

28. Groeneveld, D.P., Baugh, W.M., Sanderson, J.S., Cooper, D.J., 2007. Annual groundwater evapotranspiration mapped from single satellite scenes. Journal of Hydrology 344 (1), 146–156.

29. Halford, K.J., 2015. Exhibit 153, public administrative hearing before the state engineer in the matter of protested applications 78795, etc., February 2–6, 2015, official records in the Office of the State Engineer.

30. Helsel, D.R., Hirsch, R.M., 1992. Statistical methods in water resources. 49. Elsevier.

31. Hobbins, M.T., Ramírez, J.A., Brown, T.C., 2004. Trends in pan evaporation and actual evapotranspiration across the conterminous US: Paradoxical or complementary? Geophysical Research Letters 31 (13).

32. Homer, C.G., Xian, G., Aldridge, C.L., Meyer, D.K., Loveland, T.R., O'Donnell, M.S., 2015. Forecasting sagebrush ecosystem components and greater sage-grouse habitat for 2050: Learning from past climate patterns and landsat imagery to predict the future. Ecological Indicators 55, 131–145.

33. Huntington, J.L., Niswonger, R.G., 2012. Role of surface-water and groundwater interactions on projected summertime streamflow in snow dominated regions: An integrated modeling approach. Water Resources Research 48 (11).

34. Huntington, J.L., Szilagyi, J., Tyler, S.W., Pohll, G.M., 2011. Evaluating the complementary relationship for estimating evapotranspiration from arid shrublands. Water Resources Research 47 (5).

35. Irish, R.R., Barker, J.L., Goward, S.N., Arvidson, T., 2006. Characterization of the landsat-7 ETM+ automated cloud-cover assessment (ACCA) algorithm. Photogrammetric Engineering and Remote Sensing 72 (10), 1179–1188.

36. Jaksa, W.T., Sridhar, V., Huntington, J.L., Khanal, M., 2013. Evaluation of the complementary relationship using Noah land surface model and north American regional reanalysis (NARR) data to estimate evapotranspiration in semiarid ecosystems. Journal of Hydrometeorology 14 (1), 345–359.

37. Kennedy, R.E., Yang, Z., Cohen, W.B., 2010. Detecting trends in forest disturbance and recovery using yearly landsat time series: 1. LandTrendr—Temporal segmentation algorithms. Remote Sensing of Environment 114 (12), 2897–2910.

38. Kennedy, R.E., Andréfouët, S., Cohen, W.B., Gómez, C., Griffiths, P., Hais, M., ... Zhu, Z., 2014. Bringing an ecological view of change to landsat-based remote sensing. Frontiers in Ecology and the Environment 12 (6), 339–346.

39. Li, P., Jiang, L., Feng, Z., 2013. Cross-comparison of vegetation indices derived from landsat-7 enhanced thematic mapper plus (ETM+) and landsat-8 operational land imager (OLI) sensors. Remote Sensing 6 (1), 310–329.

40. Martínez-Beltrán, C., Jochum, M.O., Calera, A., Melia, J., 2009. Multisensor comparison of NDVI for a semi-arid environment in Spain. International Journal of Remote Sensing 30 (5), 1355–1384.

41. Masek, J.G., Vermote, E.F., Saleous, N.E., Wolfe, R., Hall, F.G., Huemmrich, K.F., ... Lim, T.K., 2006. A landsat surface reflectance dataset for North America, 1990–2000. Geoscience and Remote Sensing Letters IEEE 3 (1), 68–72.

42. McEvoy, D.J., Huntington, J.L., Abatzoglou, J.T., Edwards, L.M., 2012. An evaluation of multiscalar drought indices in Nevada and Eastern California. Earth Interactions 16 (18), 1–18.

43. McGwire, K., Minor, T., Fenstermaker, L., 1999. Hyperspectral mixture modeling for quantifying sparse vegetation cover in arid environments. Remote Sensing of Environment 72, 360–374.

44. Mitchell, K.E., Lohmann, D., Houser, P.R., Wood, E.F., Schaake, J.C., Robock, A., ... Bailey, A.A., 2004. The multi-institution North American Land Data Assimilation System (NLDAS): Utilizing multiple GCIP products and partners in a continental distributed hydrological modeling system. Journal of Geophysical Research - Atmospheres 109, D7 (1984–2012).

45. Mo, X., Liu, S., Lin, Z., Wang, S., Hu, S., 2014. Trends in land surface evapotranspiration across China with remotely sensed NDVI and climatological data for 1981–2010. Hydrological Sciences Journal.

46. Moreo, M.T., Laczniak, R.J., Stannard, D.I., 2007. Evapotranspiration rate measurements of vegetation typical of ground-water discharge areas in the basin and range carbonaterock aquifer system, white Pine County, Nevada, and adjacent areas in Nevada and Utah, September 2005–August 2006. US geological survey. pp. 443–460.

47. Naumburg, E., Mata-Gonzalez, R., Hunter, R.G., Mclendon, T., Martin, D.W., 2005. Phreatophytic vegetation and groundwater fluctuations: A review of current research and application of ecosystem response modeling with an emphasis on Great Basin vegetation. Environmental Management 35 (6), 726–740.
48. NDWR (Nevada Division of Water Resources), 2015. Water level database. accessed on April, 20, 2015 http:// water.nv.gov/data/waterlevel/.
49. Nguyen, U., Glenn, E.P., Nagler, P.L., Scott, R.L., 2014. Long-term decrease in satellite vegetation indices in response to environmental variables in an iconic desert riparian ecosystem: The Upper San Pedro, Arizona, United States. Ecohydrology.
50. Nichols, W.D., 2000. Regional ground-water evapotranspiration and ground-water budgets. Great Basin, Nevada (No. 1628).
51. NSEO (The Office of the State Engineer of the State of Nevada), 2012a. The ruling (#6164) in the matter of applications 54003 through 54021, inclusive, filed to appropriate the underground waters of the spring valley hydrographic basin (184). Nevada: Lincoln and White Pine Counties (218 pp.).
52. Patten, D.T., Rouse, L., Stromberg, J.C., 2008. Isolated spring wetlands in the Great Basin and Mojave deserts, USA: Potential response of vegetation to groundwater withdrawal. Environmental Management 41 (3), 398–413.
53. Pritchett, D., Manning, S.J., 2012. Response of an intermountain groundwater-dependent ecosystem to water table drawdown. Western North American Naturalist 72 (1), 48–59.
54. Prudic, D.E., Niswonger, R.G., Plume, R.W., 2006. Trends in streamflow on the Humboldt River between Elko and Imlay, Nevada, 1950–99. U.S. geological survey scientific investigation report 2005–5199 (58p).
55. Robinson, T.W. (1958). Phreatophytes: U.S. geological survey water supply paper 1423, 84 p., http://pubs.er.usgs.gov/publication/wsp1423
56. Roy, D.P., Wulder, M.A., Loveland, T.R., Woodcock, C.E., Allen, R.G., Anderson, M.C., ... Zhu, Z., 2014. Landsat-8: Science and product vision for terrestrial global change research. Remote Sensing of Environment 145, 154–172.
57. Roy, D.P., Kovalskyy, V., Zhang, H.K., Vermote, E.F., Yan, L., Kumar, S.S., Egorov, A., 2016. Characterization of landsat-7 to landsat-8 reflective wavelength and normalized difference vegetation index continuity. Remote Sensing of Environment.
58. Shafroth, P.B., Stromberg, J.C., Patten, D.T., 2002. Riparian vegetation response to altered disturbance and stress regimes. Ecological Applications 12 (1), 107–123.
59. Smith, J.L., Laczniak, R.J., Moreo, M.T., Welborn, T.L., 2007. Mapping evapotranspiration units in the basin and range carbonate-rock aquifer system, White Pine County, Nevada, and adjacent areas in Nevada and Utah: US Geological Survey Scientific Investigations Report 2007–5087 20 p., with downloadable video, accessed November 29, 2010. US Geological Survey Nevada Science Center, Carson City, Nevada.
60. Steven, M.D., Malthus, T.J., Baret, F., Xu, H., Chopping, M.J., 2003. Intercalibration of vegetation indices from different sensor systems. Remote Sensing of Environment 88 (4), 412–422.
61. Stromberg, J.C., 2001. Restoration of riparian vegetation in the south-western United States: Importance of flow regimes and fluvial dynamism. Journal of Arid Environments 49 (1), 17–34.
62. Summers, P., 2001. Hydrogeologic analysis of needle point spring Fillmore Field office, Utah. Bureau of Land Management report 19p.
63. Szilagyi, J., 2002. Vegetation indices to aid areal evapotranspiration estimations. Journal of Hydrologic Engineering 7 (5), 368–372.
64. Tasumi, M., Allen, R.G., Trezza, R., 2008. At-surface reflectance and albedo from satellite for operational calculation of land surface energy balance. Journal of Hydrologic Engineering 13 (2), 51–63.
65. Trezza, R., Allen, R.G., 2013. Report on developing albedo weighting coefficients and surface reflectance coefficients for the Landsat 8 operational land imager. Report by the University of Idaho Kimberly Research and Extension Center. Kimberly, Idaho, p. 15 (rev. 2016).
66. USGS (U.S. Geological Survey), 2010. LANDFIRE: LANDFIRE 1.1.0, existing vegetation type layer, U.S. Department of the Interior, U.S. geological survey. (Available at) http:// landfire.cr.usgs.gov/viewer/.
67. USGS (U.S. Geological Survey). (2015a). Product guide: Provisional landsat 8 surface re- flectance product, version 1.3. (Available at:) http://landsat.usgs.gov/documents/ provisional_l8sr_product_guide.pdf. Accessed July 15, 2015.
68. USGS (U.S. Geological Survey). (2015b). LANDSAT 8 (L8) data users handbook – Version 1, U.S. Department of the Interior, U.S. geological survey. (Available at) https:// landsat.usgs.gov/documents/ Landsat8DataUsersHandbook.pdf. (Accessed January 17, 2016).
69. Vogelmann, J.E., Helder, D., Morfitt, R., Choate, M.J., Merchant, J.W., Bulley, H., 2001. Effects of landsat 5 thematic mapper and landsat 7 enhanced thematic mapper plus radiometric and geometric calibrations and corrections on landscape characterization. Remote Sensing of Environment 78 (1), 55–70.
70. Williams, J.E., Neville, H.M., Haak, A.L., Colyer, W.T., Wenger, S.J., Bradshaw, S., 2015. Climate change adaptation and restoration of western trout streams: Opportunities and strategies. Fisheries 40 (7), 304–317.
71. Wu, W., 2014. The generalized difference vegetation index (GDVI) for dryland characterization. Remote Sensing 6, 1211–1233.
72. Yang, X., Smith, P.L., Yu, T., Gao, H., 2011. Estimating evapotranspiration from terrestrial groundwater-dependent ecosystems using landsat images. International Journal of Digital Earth 4 (2), 154–170.

CHAPTER 3

ASSESSING WATER ECOSYSTEM SERVICES FOR WATER RESOURCE MANAGEMENT

B. Grizzetti, D. Lanzanova, C. Liquete, A. Reynaud, A.C. Cardoso

Keywords: Water ecosystem services, Ecosystem services indicators, Ecosystem services valuation, Water Framework Directive

INTRODUCTION

Ecosystem services are defined as the benefits that people obtain from ecosystems (MEA, 2005a), and the direct and indirect contributions of ecosystems to human well-being (TEEB, 2010). The concept of ecosystem services is relevant for connecting people to nature. It makes visible the key role of ecosystem functioning and biodiversity to support multiple benefits to humans. Understanding the linkages between the natural and socio-economic systems can lead to improved and more sustainable management of ecosystems (Guerry et al., 2015).

In 2010 the parties of the Convention of Biological Diversity adopted a revised Strategic Plan for Biodiversity including the Aichi Biodiversity Targets (CBD, 2010), a reinforced action to halt the loss of biodiversity and ensure ecosystems are resilient and continue to provide essential services. In line with this international framework, in 2011 the European Union (EU) presented the European Biodiversity Strategy to 2020 (European Commission, 2011) that put emphasis on the protection and value of ecosystem services, setting a specific target on maintaining and restoring ecosystems and their services (Target 2).

Aquatic ecosystems (rivers, lakes, groundwater coastal waters, seas) support the delivery of crucial ecosystem services, such as fish production, water provisioning and recreation. Key ecosystem services are also connected to the hydrological cycle in the river basin, for example water purification, water retention and climate regulation. Most of these water related ecosystem services can be directly appreciated by people and quantified, but some, especially regulating and maintenance services, are less evident. Though, all ecosystem services have to be considered for the sustainable use and management of water resources.

In Europe, the development of River Basin Management Plans (RBMP) under the EU Water Framework Directive (WFD, Directive 2000/60/EC) is an actual situation where territorial planning for water management is needed, and where the concept of ecosystem services could be adopted to recognise the multi-functionality of the water systems and account for the benefits people receive from nature, justifying the costs of protection and restoration. The Blueprint to safeguard Europe's water resources (European Commission, 2012) indicated that natural water retention measures can greatly contribute to reduce the effects of floods and droughts ensuring the provisioning of ecosystem services, and these measures should be included in the RBMPs and in the Flood Risk Management Plans. In line with the Blueprint, some recent studies have been reflecting on the potential of the ecosystem service approach in the implementation of the

---

Citation: B. Grizzetti, D. Lanzanova, C. Liquete, A. Reynaud, A.C. Cardoso, Assessing water ecosystem services for water resource management, Environmental Science & Policy, Volume 61, July 2016, Pages 194-203, ISSN 1462-9011, http://dx.doi.org/10.1016/j.envsci.2016.04.008.

WFD, emphasising the opportunity of holistic system thinking to understand the co-benefits of measures and to integrate different sectoral policies (Vlachopoulou et al., 2014, COWI, 2014 and ESAWADI, 2010).

However, the lack of agreed definitions of ecosystem services and approaches for their quantification and valuation has limited the uptake by practitioners and policy makers (Polasky et al., 2015). The MAES Working Group (Mapping and Assessment of Ecosystems and their Services), established to support the implementation of the EU Biodiversity Strategy, has suggested an analytical framework for the implementation of the ecosystem service approach in the EU, and tested it in a pilot study on freshwater and marine ecosystems (Maes et al., 2016). Two FP7 projects, OpenNESS (2015) and OPERA (2015), are currently working on the general definition of concepts and methodologies for assessing and valuing ecosystem services, and on the operationalization of the concepts through real case studies. More specifically on water policy, the FP7 projects MARS (2015) and Globaqua (2015) aim to understand and quantify the impacts of pressures on the ecological status of EU waters and the consequent effects on the delivery of ecosystem services.

Understanding the relationship between multiple pressures, conditions and services of aquatic ecosystems would help design measures to achieve the target of good ecological status of water systems, foreseen by the WFD, by considering the benefits of investing in nature conservation and restoration. But the lack of clear definitions and practical methods to assess the water related ecosystem services could hamper the adoption of the approach (Kull et al., 2015 and Crossman et al., 2013). Also, while mapping of ecosystem services directly linked to land occupation is quite straightforward, for fresh water ecosystems the assessment is more complex, as the hydrological cycle and the land-water interactions have to be taken into consideration.

The objective of this study was to develop a practical methodology for assessing and valuing ecosystem services relevant for water resource management, considering the links between pressures, ecological status and ecosystem services. The work is based on literature review and scientific partners' consultation. It started from the experience of the MAES freshwater pilot and was developed within the EU FP7 project MARS.

The paper is structured as follows. The first part describes the methodological approach adopted in the study. The second part presents the results of our analysis in the form of a practical approach for assessing and valuing ecosystem services relevant for water resource management. The third part discusses the challenges in valuing ecosystem services and integrating biophysical and economic assessments.

METHOD

We analysed definitions and methods for assessing and valuing ecosystem services to synthesize the current knowledge and propose a practical and flexible approach relevant for water resource management. The use context of the approach is the study of the relationship between multiple pressures, ecological status and delivery of ecosystem services in water systems, with the overall goal to support the EU water policy (WFD).[1]The analysis was based on literature review and consultation of the scientific partners of the project MARS, from 24 research institutes across Europe.

The focus of the analysis is on inland waters and the spatial scale of interest ranges from the water body to the catchment/river basin and the European scale. While for water bodies the main focus is on specific ecosystem functions that support ecosystem services, and their alteration under different stressors, the catchment is the appropriate scale to observe and quantify processes related to the water cycle, and to implement monitoring and management plans to reduce multiple-pressures. The assessment and valuation of ecosystem services at the European scale allows us to address regional trends, identify hot spots in the delivery or degradation of services, test the effectiveness of regional policies (such as EU Directives) and conduct scenario analysis at the large scale. In the development of the methodology we considered these different spatial scales.

The approach that we developed is organised in four building steps: 1) definitions and scoping (Section 3.1); 2) framework (relations between pressures, ecological status and delivery of ecosystem services) (Section 3.2); 3) biophysical assessment of ecosystem services (Section 3.3); 4) economic valuation of ecosystem services (Section 3.4). In the following part of the paper we describe the results of our study proposing guidelines on how to develop these components.

RESULTS: APPROACH FOR ASSESSING AND VALUING WATER ECOSYSTEM SERVICES

(Step 1) Scoping – Water related ecosystem services

A large variety of ecosystem services have been addressed by assessments such as Millennium Ecosystem Assessment (MEA, 2005a), the Economics of Ecosystems and Biodiversity (TEEB, 2010), MAES (Maes et al., 2016), and national assessments (e.g. Pereira et al., 2006 and UK NEA, 2011). In this study we are interested in ecosystem services related to water and aquatic ecosystems. MAES analysed the ecosystem services per typology of ecosystem, considering the services delivered by rivers, lakes, groundwater and wetlands in the freshwater pilot study, and those provided by transitional waters, coastal waters, shelf waters and open oceanic water in the marine pilot study. With a slightly different approach, Brauman et al. (2007) discussed the 'hydrologic ecosystem services', defined as the ecosystem services that "encompass the benefits to people produced by terrestrial ecosystem effects on freshwater", each hydrological service being characterised by the hydrological attributes of quantity, quality, location and timing. Keeler et al. (2012) described in detail water-quality related ecosystem services. Recently, Guswa et al. (2014) have addressed more generally the 'water related ecosystem services', discussing the link between hydrological modelling and the ecosystem services relevant for river basin management. From these studies we can observe two approaches in the organisation of the analysis, one per ecosystem typology (Maes et al., 2016) and the other per hydrological relevant services (Brauman et al., 2007). Both approaches consider the integration of all the services, the first by accounting for all the ecosystems in the analysis, the second by integrating the processes in the river basin. The ecosystem services of relevance for the water management (and the WFD) are those related to the aquatic ecosystems and to the interaction of water and land in different ecosystems, such as forests, agricultural lands, riparian areas, wetlands, and water bodies. In this study we indicate all these services as 'water ecosystem services'.

For the assessment, the identification of the relevant ecosystem services is the first step. We propose a simplified classification of ecosystem services based on the Common International Classification of Ecosystem Services Version 4.3 (CICES, 2015), which is the framework adopted by the common implementation of the ecosystem assessment approach in the EU. The idea is to offer a coherent terminology relevant for scientists, sufficiently simple for stakeholders, and meaningful for river basin managers. The services (S1) can be related to the following ecosystems: lakes, rivers, transitional waters, coastal waters, groundwater, freshwater wetlands, coastal wetlands, riparian areas, floodplains. Providing a list of ecosystem services can support the practical implementation of the methodology, but of course the list is not to be considered exhaustive and more services can be included, especially hydrological services relevant for river basin planning and decision making.

(Step 2) Framework – Linking pressures, ecological status and ecosystem services for water management

Understanding the relationship between anthropogenic pressures and ecological status is the basis of the RBMP, in order to devise cost-effective measures to achieve a good ecological status for all water bodies. In particular, for planning sound restoration actions, it is necessary to consider the complex links between pressures combinations and the ecological response of aquatic systems (Teichert et al., 2016, Brown et al., 2013 and Halpern et al., 2008), as multiple pressures may have additive, synergetic or antagonistic effects (Nõges et al., 2015). The research project MARS studies the complex relationships between multiple stressors, status of aquatic ecosystems and the delivery of ecosystem services (see Hering et al., 2015 for the MARS conceptual model).

In the EU, water pollution, over-abstractions and hydromorphological alterations have been indicated as the major significant pressures for the European water bodies (European Commission, 2015). At the same time, concern over the increase of alien species exists (Butchart et al., 2010; Regulation EU No 1143/2014). Overall, the main pressures affecting the aquatic ecosystems can be summarised as alterations of water quantity and quality, and changes in the physical habitat and the biological components, as shown in Table 1.

Table 1: Stressors and pressures on water systems

Alteration of:
Water quantity
Flow modifications (hydrological alterations):
• Quantity and frequency (dams, water abstractions, irrigation, transfers)
• Groundwater abstractions
• Changes in precipitation and temperature
• Changes in runoff
Water quality
Diffuse and point pollution:
• Nutrients
• Chemicals (pesticides, endocrine disrupting compounds, nanoparticles, etc.)
• Metals
• Pathogens
• Litter
• Groundwater salinization
• Sediments, increased turbidity and brownification
Habitat
Hydromorphological alterations (physical alteration of channels, bed disruption, dams)
Biota and biological communities
Alien species, other changes in biological communities

Human activities are the major drivers for generating multiple pressures (Fig. 1). Pressures affect the biodiversity and the status of the aquatic ecosystem, which can result in a change in the ecosystem services and their economic value. The excessive exploitation of ecosystem services can turn into a pressure for an ecosystem. For this reason is important to consider the resilience of the system and to introduce the notion of sustainability when assessing the delivery of ecosystem services (Fig. 1). The interest of RBMP is to quantify the changes in the components of this system under remediation measures and scenarios of pressures.

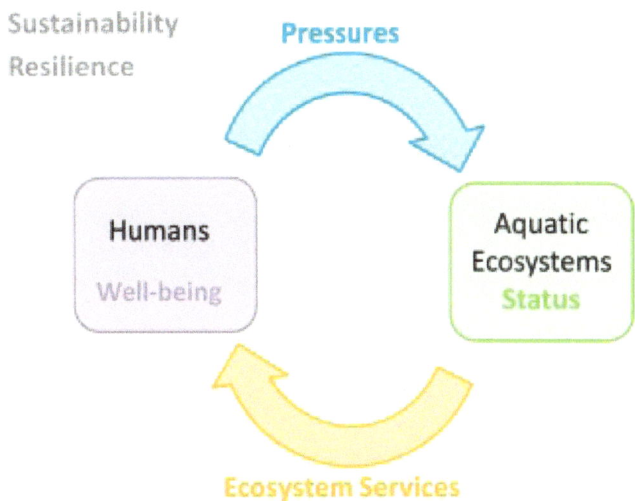

Figure 1: Schematic representation of the system under analysis. Relationships between humans and aquatic ecosystems under present and future scenarios (explanation in the text).

To support the analysis of these linkages, we developed a conceptual framework for the integrated assessment of water related services (presented in Fig. 2). In the framework, we identify the main pressures affecting aquatic ecosystems (according to Table 1) and the possible links to the alteration of four ecosystem/hydrologic attributes: 1) water quantity (including seasonality); 2) water quality; 3) biological quality elements; 4) hydromorphological & physical structure. The attributes are different from those proposed by Brauman et al. (2007), to include in the analysis the biological and hydromorphological aspects and to make the link to the WFD elements explicit (so that the relationship to ecological status should be in principle more easy). For each attribute we selected a number of representative indicators (as examples) and identified some possible relationships with the ecosystem services suggested in this study (S1).

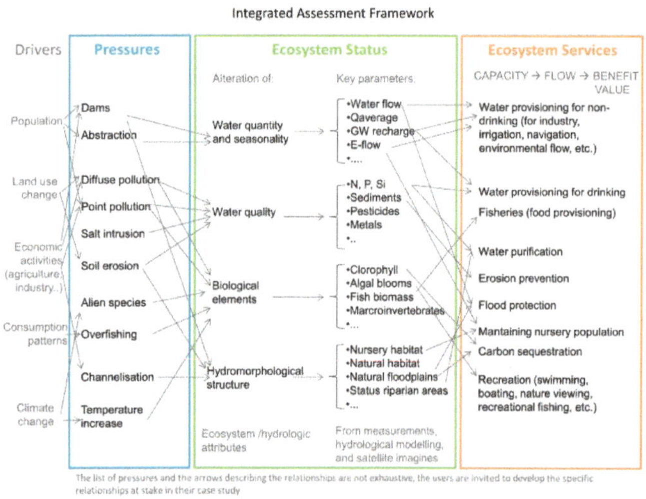

Figure 2: Integrated assessment framework for analysing the links between pressures, ecosystem status and ecosystem services. The list of pressures and the arrows describing the relationships are not exhaustive; the users are invited to develop the specific relationships at stake in their case study.

The purpose of this framework (Fig. 2) is to support the users in describing the relationships between pressures and ecosystem services and design a conceptual scheme of the assessment and scenario analysis. The arrows are examples. Each user can select the relationships under analysis and complete and adapt the framework to the specific case under study. The idea is to think about the links between the selected services and specific stressors. Fig. 3, which presents expected qualitative effects of stressors/pressures on different ecosystem services, could inspire this reflection.

	Ecosystem services	Flow modifications	Diffuse and point pollution	Groundwater salinization	Erosion/ Brownification	Hydromorphological alterations	Alien species	Overfishing
Provisioning	Fisheries and aquaculture	●	●	○	●	◍	●	●
	Water for drinking	●	●	●	●	◍	○	○
	Raw (biotic) materials	◍	●	○	◍	●	●	◍
	Water for non-drinking purposes	●	●	●	●	◍	○	○
	Raw materials for energy	●	●	○	○	●	◍	○
Regulation & Maintenance	Water purification	●	●	◍	●	●	◍	○
	Air quality regulation	●	●	○	○	◍	○	○
	Erosion prevention	●	○	○	○	●	◍	○
	Flood protection	●	○	○	○	●	◍	○
	Maintaining populations and habitats	●	●	○	◍	●	●	●
	Pest and disease control	●	●	○	◍	●	●	●
	Soil formation and composition	●	◍	○	●	●	○	○
	Carbon sequestration	◍	○	○	◍	●	◍	○
	Local climate regulation	●	○	○	○	●	○	○
Cultural	Recreation	●	●	○	●	●	◍	●
	Intellectual and aesthetic appreciation	●	●	●	●	●	●	●
	Spiritual and symbolic appreciation	●	●	●	●	●	●	●

Legend: Expected impact of each pressure over the ecosystem service: ● high, ◍ medium, ○ low.

Figure 3: Expected qualitative effect of stressors/pressures on different ecosystem services.

(Step 3) Biophysical assessment

Tools

Several approaches to assess and map ecosystem services are available in the literature, from land cover maps combined with scoring factors (Burkhard et al., 2009) to specific ecosystem service models based on ecological production functions (Sharp et al., 2015), and decision support tools (Bagstad et al., 2013 reviewed 17 tools for assessing and valuing ecosystem services). These tools usually combine ecology and economics, considering the spatial dimension. The EU FP7 project OpenNESS is studying methodologies for mapping and modelling the biophysical control of ecosystem services and approaches for the valuation of ecosystem services, for application in 27 real case studies. The biophysical methods include spreadsheet/GIS approaches (Burkhard et al., 2012 and Vihervaara et al., 2012); Quickscan (2015); Bayesian Belief Networks (Smith et al., 2014); State and Transition Models (Bestelmeyer et al., 2011); ESTIMAP (Zulian et al., 2013); InVEST (2015). These methods apply to all ecosystem services, not specifically to water ecosystem services. Most of them rely on the spatial mapping of the ecosystems and land use.

The water quantity and quality, and the water related ecosystem services, are affected by the complex interactions of climate, topography and geology, land cover and management, and other anthropogenic modification of the landscape. Incorporating water related ecosystem services in the decision making process requires the capacity to predict the effects of land use and climate changes on the water resources, which can be offered by the hydrological models (Guswa et al., 2014). Hydrological and biogeochemical catchment models are appropriate tools for dealing with water related ecosystem services (Guswa et al., 2014, Vigerstol and Aukema, 2011 and Brauman et al., 2007). They can represent the dynamic of the river basin (resilience) and the temporal (lag time) and spatial distance between beneficiaries and impacts, and they can be used in scenario analysis of multiple stressors. They also allow describing physical relationships between stressors, status and services as presented in the integrated assessment framework (Fig. 2).

Indicators

Following this line and considering the wealth of knowledge in hydrological modelling, we propose to base the biophysical assessment of ecosystem services on 'indicators' rather than 'tools'. Similar to Maes et al. (2014) and Layke et al. (2012), we suggest the selection of some suitable indicators or proxies of ecosystem services that are directly related to water bodies or to water-land interaction in the watershed, as a flexible and handy approach to measure ecosystem services.

To support the correct understanding and appropriate use of the indicators for ecosystem services, and more generally to structure the assessment, we have to analyse which dimension of the ecosystem service is captured by the indicators. To this purpose we propose a simplified conceptual framework for structuring the analysis and the classification of indicators of water ecosystem services. The framework, presented in Fig. 4, includes the *capacity* of the ecosystem to deliver the service, the actual *flow* of the service, and the *benefits*. Capacity refers to the potential of the ecosystem to provide ecosystem services, while flow is the actual use of the ecosystem services. The capacity relies on biophysical data, while flow requires the acquisition of socio-economic data. Benefits are associated with human well-being and the value system (for studies discussing the concepts of capacity and flow and cascade model of ecosystem services see Haines-Young and Potschin, 2010, Layke et al., 2012, Villamagna et al., 2013, Maes et al., 2014 and Schröter et al., 2014). Services are often associated with high exploitation of the ecosystem; the risk is an unsustainable use of nature. For this reason we are interested in looking at the sustainable flow of services. This is considered in the conceptual framework by including indicators informing about the *sustainability*, i.e. indicators combining capacity and flow (an example could be the Water Exploitation Index). In many cases, the information on capacity and flow is lacking, or the full capacity of the ecosystem is unknown or unaccountable. In these cases we can try to collect indicators about the efficiency of the processes (for example the removal rate of a pollutant per unit of input).

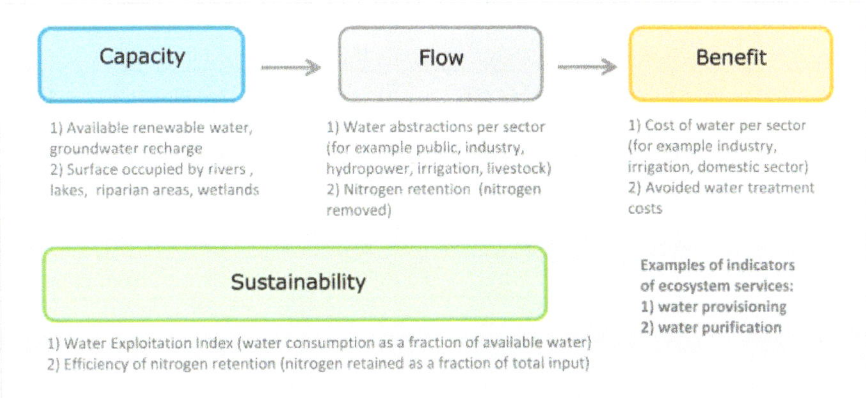

Figure 4: Conceptual framework to classify indicators of water ecosystem services. Some examples of indicators for the ecosystem services of 1) water provisioning and 2) water purification are reported. For studies using these indicators at the European scale see 1) De Roo et al., 2012 for water provisioning, and 2) Clerici et al., 2013 (*capacity*), Grizzetti et al., 2012 (*flow*), La Notte et al., 2015 (*benefit*), Liquete et al., 2015(*sustainability*) for water retention (nitrogen purification).

In this study we compiled a list of potential proxies/indicators for water ecosystem services based on the literature review. We classified them according to the categories of the conceptual framework: capacity, flow and benefit (the category of 'sustainability' and 'efficiency' were not explicitly used in the classification). Our compilation includes a total of 206 proxies and is based on Maes et al. (2014), Egoh et al. (2012), Layke et al. (2012), Russi et al. (2013) and Liquete et al. (2013) (minor modifications from the original authors like re-phrasing or re-allocation were required to avoid duplications and to respect our conceptual framework). Table 11 of Maes et al. (2014) comprises all the indicators proposed in the deliberative process of implementation of the EU Biodiversity Strategy around the freshwater pilot. The Ecosystem Service Indicators Database of the World Resources Institute (www.esindicators.org accessed in December 2015, Layke et al., 2012) compiles metrics and indicators from numerous sources that have been identified and applied by individuals from varied organizations. Russi et al. (2013) highlights the relevance of water and wetlands and links it to decision-making, providing a few examples of indicators for freshwater ecosystem services. We reviewed also Liquete et al. (2013), which includes a systematic compilation of 476 marine and coastal ecosystem services' indicators, in order to cover additional aspects specifically related to transitional and coastal waters. In the literature review, other studies were also considered (although some of them do not provide directly indicators for ecosystem services): the specific studies of the Millennium Ecosystem Assessment dealing with freshwater systems (MEA, 2005b and MEA, 2005c); UNEP (2009), UNEP-WCM (2011), Feld et al., 2009 and Feld et al., 2010, TEEB (2010), Vigerstol and Aukema (2011) and Clerici et al. (2014)

In Fig. 4 we provide some examples of indicators of ecosystem services for water provisioning and water purification, with references to studies that have used these indicators at the European scale. Other examples of application of the proposed conceptual framework for indicators can be found in Karabulut et al. (2016) for water provisioning, Rankinen et al. (submitted) for water purification, Vigiak et al. (submitted) for erosion prevention, and Liquete et al. (submitted) for several ecosystem services.

(Step 4) Economic valuation

Several methods are available in the literature to estimate economic values of freshwater ecosystem services (see for instance Koundouri et al., 2015). Overall, there are three categories of approaches: cost-based, revealed preferences and stated preferences approaches. Cost-based approaches consider the costs that arise in relation to the provision of services. Revealed preferences approaches refer to techniques that use actual data regarding individual's preferences for a marketable good which includes environmental attributes. Stated preferences approaches refer to methods based on structured surveys to elicit individuals' preferences for non-market environmental goods. Another practical way to value ecosystem services under non-availability of site-specific data or funding constraints is the benefit transfer approach. This approach consists of using economic estimates from previous studies to value services provided by the ecosystem of interest (Navrud and Ready, 2007).

For the economic assessment, the first step consists of identifying the benefits provided by the ecosystem service to be valued. Fisher et al. (2009), Fisher and Turner (2008) argued that it is the easiest way to perform a valuation exercise avoiding any double counting. Following this approach, only the services that have a direct impact on welfare are valued. The spatial scale of the assessment is also relevant for the selection of the method.

Economic valuation at the water body/catchment scale

The choice of the primary valuation method depends on the ecosystem service to be valued and on the beneficiary population. Table 2 reports valuation methods per ecosystem service based on the literature review.

Table 2: Freshwater ecosystem services, type of value and applied valuation methods. The classification of ecosystem services has been developed for fresh and transitional water (Reynaud and Lanzanova, 2015)

Ecosystem services	Category[a]	Value type	Valuation method[b]	Examples of economic good provided
1-Fisheries and aquaculture	Provisioning	Direct	MP, RC	fish catch
2-Water for drinking	Provisioning	Direct	MP, CV	water for domestic uses
3-Raw (biotic) materials	Provisioning	Direct	MP, RC	algae as fertilizers
4-Water for non-drinking purposes	Provisioning	Direct	MP,PF	water for industrial or agricultural uses
5-Raw materials for energy	Provisioning	Direct	RC	wood from riparian zones
6-Water purification	Regulation	Indirect	RC, CV	excess nitrogen removal by micro-organisms
7-Air quality regulation	Regulation	Indirect	RC	deposition of NOx on vegetal leaves
8-Erosion prevention	Regulation	Indirect	RC	vegetation controlling soil erosion
9-Flood protection	Regulation	Indirect	RC, CV	vegetation acting as barrier for the water flow
10-Maintaining populations and habitats	Regulation	Indirect	RC	habitats use as a nursery
11-Pest and disease control	Regulation	Indirect	RC, CV	natural predation of diseases and parasites
12-Soil formation and composition	Regulation	Indirect	RC	rich soil formation in flood plains
13-Carbon sequestration	Regulation	Indirect	RC, MP	carbon accumulation in sediments
14-Local climate regulation	Regulation	Indirect	RC, MP	maintenance of humidity patterns
15-Recreation	Cultural	Direct	CV, TC, DC, HP	swimming, recreational fishing, sightseeing
16-Intellectual and aesthetic appreciation	Cultural	Non-use	CV, DC	matter for research, artistic representations
17- Spiritual and symbolic appreciation	Cultural	Non-use	CV, TC, DC	existence of emblematic species
18-Raw abiotic materials	Extra abiotic	Direct	PF, MP	extraction of sand gravel
19-Abiotic energy sources	Extra abiotic	Direct	PF, MP	hydropower generation

[a]Provisioning, Regulation and maintenance, Cultural, Extra abiotic services.

[b]Contingent valuation (CV), Hedonic price (HP), Market price (MP), production function (PF), Replacement cost (RC), travel costs (TC).

One of the main difficulties in the economic valuation is to decide on the size of the benefiting population (beneficiaries). Aggregate benefits depend on estimates of both individual benefits and of the number of beneficiaries (Hanley et al., 2003). As a general rule, the beneficiaries should be the households/persons aggregated at the relevant geographic scale, and should include both users and non-users impacted by the ecosystem service considered (except for services of only local importance). In addition, for some services (for example recreational services), when spatially aggregating individual benefits, it is usually considered that the willingness to pay (WTP) decreases with the distance from water body providing ecosystem services, as the opportunities of the ecosystem service provision are expected to decrease with the distance, and concurrently the existence of possible substitutes is assumed to increase (Bateman and Langford, 1997, Georgiou et al., 2000 and Jørgensen et al., 2013). Generally a distance decay function is adopted to take into account the decrease of the willingness to pay with the distance from the water body providing the ecosystem services (Bateman et al., 2006). This distance determines the boundaries of the geographical area, or so-called economic jurisdiction, over which the individual WTP-values can be aggregated over the population of beneficiaries to calculate the total economic value of a proposed scenario of environmental change (Schaafsma et al., 2012). However, the specification of the distance decay relations has been highly debated among economists. A number of studies have examined in particular how the distance decay relation differs between users and non-users of the ecosystem service (Hanley et al., 2003 and Bateman et al., 2006).

Economic valuation at the large/continental scale

At the large/continental scale, such as the European scale, methodologies upscaling values of primary studies (value transfer) and accounting for the spatial heterogeneity of biophysical and socio-economic characteristics are more appropriate. This approach consists of a *meta*-analysis using the results of available past studies on ecosystem services valuation in water bodies to estimate a function that represents the relationship between the features of water ecosystems and the value of the services they provided (see Brander et al., 2006 and Brander et al., 2007 for examples of *meta*-analysis for the valuation of ecosystem services).

The first step of a *meta*-analysis consists of searching and selecting studies valuing services provided by ecosystems similar to the one of interest (the policy site), most often through systematic searches. All relevant data from primary studies are collected and organised in a *meta*-database, including information on methods applied, ecosystem services valued, biophysical characteristics of the ecosystem, and the characteristics of the beneficiaries of the ecosystem services. To enable a comparison across studies, the economic values reported using different metrics (i.e. WTP, marginal values, capitalized value) are standardized. However, this is a difficult and controversial task (Ghermandi et al., 2010). Purchasing power parity indexes are applied to the original values to account for differences in purchasing power among countries, and appropriate price deflators are used to deal with the difference in the years of observation (Ghermandi and Nunes, 2013). In addition, values issued by different methods are normalized. For example values are expressed in monetary units per area and time (Ghermandi et al., 2010 and Brander et al., 2012); or per visit and time (Brander et al., 2007) or per household/respondent and time (Brouwer et al., 1999 and Johnston et al., 2005). The following step of the *meta*-analysis is the estimation of the *meta*-values transfer function. A regression technique allows accounting for the biophysical or socio-economical differences between the primary study sites and the policy site. There are two popular panel-data models which can be used for estimating the *meta*-regression model, e.g. the fixed-effect model and the random-effect model. The random-effects model allows the true effect size to differ from study to study and this is the approach usually recommended. The values of ecosystem services that are estimated by the regression analysis are then transferred and aggregated at the larger geographic areas through a scaling-up procedure. The most appropriate transfer function among the different *meta*-regression specifications has to be selected based on the explanatory power of the model, sign and significance of the coefficients estimated. In addition, the appropriate geographic scale for transferring values has to be defined (Ghermandi and Nunes, 2013).

DISCUSSION

Considering the current and impellent challenge of the implementation, i.e. being able to translate the concepts of ecosystem services into practice (European Commission, 2011, European Commission, 2012 and Guerry et al., 2015), the need to be operational constituted one of the leading criteria in the development of the present review and proposed approach. We considered this research as a learning process. The application of the approach is on-going in the project MARS at the catchment and the European scale. This will provide the necessary feedback to improve and refine the approach.

Here we discuss some aspects that we consider important when assessing and valuing ecosystem service to support the WFD and RBMP, and how the proposed approach could address them.

The valuation of ecosystem services

The WFD refers to economic valuation in decision-making to support the RBMP in the identification and selection of cost-effective Programmes of Measures (PoM, WFD Article 11). Quantifying the benefits (ecosystem services) that nature provides to people would help justify the investments in conservation and restoration of aquatic ecosystems. In addition, the development of the PoM can be improved integrating all relevant ecosystem services, by considering the co-benefits of different measures and nature-based solutions on different ecosystem services (Liquete et al., submitted). The benefits of ecosystem services could also be included in the cost-benefit analysis to implement the cost-recovery principle in the water supply system (WFD Article 9) (Vlachopoulou et al., 2014 and COWI, 2014).

Yet the valuation of ecosystem services also involves some important risks, i.e. creating economic markets for provisioning, regulating and cultural services. We have to reflect on the notion of 'valuation'. Any decision involving trade-offs of ecosystem service implies valuation (Costanza et al., 2014). There are different values in the relationship of human and non-human nature, including inherent, fundamental, eudaimonistic and instrumental values (Jax et al., 2013). The values that are captured by the ecosystem service concept depend on how the concept is operationalised and implemented (approaches and methodologies used). Different stakeholders have different value systems and perspectives. Therefore involving all the stakeholders in the valuation process is necessary to consider the plurality of values, while neglecting some values would exclude the people who embrace these values (Jax et al., 2013).

The notion of value should not be restricted to the mere monetary value but embrace a larger range of values. If restricting the value of ecosystem services to economic value, we risk false accounting of all value dimensions and environmental components (trade-offs) of policy decision (Keeler et al., 2012). 'Value pluralism' refers to the idea that there are multiple values, including economic (monetary), sociocultural and ecological values. An integrated valuation should endorse the value pluralism (Gómez-Baggethun et al., 2014). The valuation techniques vary with the typology of values to be elicited and the scope of the valuation exercise, the geographical scale, spatial resolution, and reliability and accuracy required. The purpose of the valuation can range from awareness raising, to accounting, priority setting, instrument design and litigation (Gómez-Baggethun and Barton, 2013).

In the proposed approach, we think it is important to interpret the economic valuation in monetary terms sensu Costanza et al. (2014), i.e. for awareness raising about relative changes over a period in time. This excludes the intent of treating all ecosystem services as substitutable. We are interested mainly in the change of value as the result of the effects of multiple stressors changes or the implementation of measures.

Strengthen the connection human-nature

RBMPs are based on the principles of Integrated Water Resource Management (IWRM), "a process which promotes the coordinated development and management of water, land and related resources, in order to maximize the resultant economic and social welfare in an equitable manner without compromising the sustainability of vital ecosystems" (GWP, 2000). Before the ecosystem service approach, IWRM already stressed the need for connecting environment and human well-being and proposed the integration of multi-disciplinary knowledge from different sectors and stakeholders in the water management. The ecosystem services approach has significant similarities with IWRM. Cook and Spray (2012) argue that the two concepts are 'nearly identical'. Both aim at a management of natural resources that optimises the economic and social welfare and contemporary insures the ecological sustainability, integrating the knowledge of stakeholders and multiple disciplinary perspectives. Ecosystem services and IWRM both share the goal of negotiating the trade-offs between different human and ecosystem needs, while supporting sustainability, and require the involvement of stakeholders for making explicit the whole range of values (not only economic values). The ecosystem service approach offers a framework for analysing the trade-offs among different services and the links to beneficiaries (Brauman et al., 2007). Learning from the criticisms to IWRM would help improve the adoption of the ecosystem service approach. These criticisms are related to the lack of consistent definitions, the difficulty of developing a holistic approach, the risk of opposite interpretations of the concepts, and the failure to incorporate the principles in the governance (Cook and Spray, 2012). The 'implementation gap' is an important challenge for the ecosystem service approach and this study aims to contribute to this endeavour.

To address current sustainability challenges the recognition of the dependency of human well-being on natural capital is necessary (Guerry et al., 2015). Integrative frameworks such as the ecosystem service approach allow incorporation of natural components in the system analysis (Liu et al., 2015). The concept of a human-ecological system advocated by the ecosystem service approach is powerful in linking biophysical processes and human benefits, and allows ecosystem services to be valued and integrated in the river basin decision making process. However, economic models to value ecosystem services related to water quality are often poorly integrated with the biophysical models describing the underpinning natural processes (Keeler et al., 2012).

In this study we considered both dimensions of biophysical assessment and economic valuation and we suggest to perform them in collaboration. This is a crucial but evasive step of ecosystem service analyses, and remains one of the main challenges in this field research (Polasky et al., 2015). The integration also depends on the method used for the assessment. In the case of freshwater systems, many biophysical results (coming from models or measures) can be used as an input for economic valuations. Some examples are the improvements of juvenile fish to estimate the economic enhancement of commercial fish by seagrass (Blandon and zu Ermgassen, 2014), the nitrogen retention efficiency used to estimate the replacement cost of water purification (La Notte et al., 2015), the forecasted trend of fish biomass used to estimate future employment in the fishing sector, the erosion rate or level of degradation linked to loss in property values, or the climate records and forecasts linked to economic damages caused by floods.

CONCLUSIONS

The method proposed in this paper to assess and value water ecosystem services provides some knowledge basis for the enrichment of water management; in particular it proposes a more holistic view to the implementation of the EU WFD linking multiple pressures, ecological status and delivery of ecosystem services. Under this perspective, the analysis of cost-effective and remediation measures can be improved including all hidden benefits and beneficiaries from water ecosystem services.

The first part of the analysis should at least identify the ecosystem services of interest and frame the major effects of multiple drivers and pressures on the ecological status of water bodies. Then, the approach suggests a biophysical quantification of the natural capacity, actual flow and social benefit of water ecosystem services. We propose to use selected proxies/indicators based mainly on hydrological models or data for this part of the assessment. One important and novel point in this approach is to assess also some sustainability (or efficiency) index that estimates the flow of service that can be sustained with a certain capacity. This could avoid the overexploitation of certain services. The proposed method includes also the economic valuation of aquatic ecosystem, providing a list of techniques for each service and spatial scale of application. There is a large variety of valuation methods that have to be carefully selected. Valuing water ecosystem services could highlight hidden benefits for society and could raise awareness among users and stakeholders. Even if monetary values are probably the most appealing arguments for water management, we also advocate and describe the advantages of using a plurality of values. Overall, the proposed approach can be used for assessing the benefits of conservation and restoration of aquatic ecosystems in the implementation of the EU water policy.

There are opportunities by adopting the ecosystem services approach to capture and integrate all the effects (economic, environmental and social) associated with new water plans and investments. Performing biophysical assessment and economic valuation collaboratively could boost awareness and inclusion of the interdependence of nature and people for a sustainable management of water resources. The integration of biophysical and economic approaches and data remains one of the main challenges and key aspects of this approach.

ACKNOWLEDGEMENTS

This study has been funded by the EU FP7 project MARS (Managing Aquatic ecosystems and water Resources under multiples Stress; project number 603378). The authors would like to thank all the MARS partners for their feedback to the methodology.

REFERENCES

1. Bagstad, K.J., Semmens, D.J., Waage, S., Winthrop, R., 2013. A comparative assessment of decision-support tools for ecosystem services quantification and valuation. Ecosyst. Serv. 5, e27–e39.
2. Bateman, I.J., Langford, I.H., 1997. Non-users' willingness to pay for a National Park: an application and critique of the contingent valuation method. Reg. Stud. 31, 571–582.
3. Bateman, I.J., Day, B.H., Georgiou, S., Lake, I., 2006. The aggregation of environmental benefit values: welfare measures, distance decay and total WTP. Ecol. Econ. 60, 450–460.
4. Bestelmeyer, B.T., Goolsby, D.P., Archer, S.R., 2011. Spatial perspectives in state-andtransition models: a missing link to land management? J. Appl. Ecol. 48, 746–757.
5. Blandon, A., zu Ermgassen, P.S., 2014. Quantitative estimate of commercial fish enhancement by seagrass habitat in southern Australia. Estuarine Coastal Shelf Sci. 141, 1–8.
6. Brander, L.M., Florax, R.J.G.M., Vermaat, J.E., 2006. The empirics of wetland valuation: a comprehensive summary and a meta-analysis of the literature. Environ. Resour. Econ. 33, 223–250.
7. Brander, L.M., Van Beukering, P., Cesar, H.S.J., 2007. The recreational value of coral reefs: a meta-analysis. Ecol. Econ. 63, 209–218.
8. Brander, L.M., Wagtendonk, A.J., Hussain, S.S., McVittie, A., Verburg, P.H., de Groot, R. S., van der Ploeg, S., 2012. Ecosystem service values for mangroves in Southeast Asia: a meta-analysis and value transfer application. Ecosyst. Serv. 1, 62–69.
9. Brauman, K.A., Daily, G.C., Duarte, T.K., Mooney, H.A., 2007. The nature and value of ecosystem services: an overview highlighting hydrologic services. Annu. Rev. Environ. Resour. 67–98.
10. Brouwer, R., Langford, I.H., Bateman, I.J., Turner, R.K., 1999. A meta-Analysis of wetland contingent valuation studies. Reg. Environ. Change 1, 47–57.
11. Brown, C.J., Saunders, M.I., Possingham, H.P., Richardson, A.J., 2013. Managing for interactions between local and global stressors of ecosystems. PLoS One 8, e65765.
12. Burkhard, B., Kroll, F., Müller, F., Windhorst, W., 2009. Landscapes' capacities to provide ecosystem services a concept for land-cover based assessments. Landsc. Online 15, 1–22.
13. Burkhard, B., Kroll, F., Nedkov, S., Müller, F., 2012. Mapping ecosystem service supply, demand and budgets. Ecol. Indic. 21, 17–29.
14. Butchart, S.H.M., Walpole, M., Collen, B., Van Strien, A., Scharlemann, J.P.W., et al., 2010. Global biodiversity: indicators of recent declines. Science 328,1164–1168.
15. CBD, 2010. Convention on Biological Diversity. COP Decision X/2.
16. CICES, 2015. http://cices.eu/ (accessed December, 2015).
17. COWI, 2014. Support Policy Development for Integration of an Ecosystem Services Approach with WFD and FD Implementation. Resource document.
18. Clerici, N., Weissteiner, C.J., Paracchini, M.L., Boschetti, L., Baraldi, A., Strobl, P., 2013. Pan-European distribution modelling of stream riparian zones based on multisource Earth Observation data. Ecol. Indic. 24, 211–223.
19. Clerici, N., Paracchini, M.L., Maes, J., 2014. Land-cover change dynamics and insights into ecosystem services in European stream riparian zones. Ecohydrol. Hydrobiol. 14, 107–120.
20. Cook, B.R., Spray, C.J., 2012. Ecosystem services and integrated water resource management: different paths to the same end? J. Environ. Manage. 109, 93–100.
21. Costanza, R., de Groot, R., Sutton, P., van der Ploeg, S., Anderson, S.J., Kubiszewski, I., Farber, S., Turner, R.K., 2014. Changes in the global value of ecosystem services. Glob. Environ. Change 26, 152–158.
22. Crossman, N.D., Burkhard, B., Nedkov, S., Willemen, L., Petz, K., Palomo, I., Drakou, E. G., Martín-Lopez, B., McPhearson, T., Boyanova, K., Alkemade, R., Egoh, B., Dunbar, M.B., Maes, J., 2013. A blueprint for mapping and modelling ecosystem services. Ecosyst. Serv. 4, 4–14.
23. De Roo, A., Burek, P., Gentile, A., Udias, A., Bouraoui, F., Aloe, A., Bianchi, A., La Notte, A., Kuik, O., Tenreiro, J.E., Vandecasteele, I., Mubareka, S., Baranzelli, C., Der Perk, M.V., Lavelle, C., Bidoglio, G., 2012. A multi-criteria optimisation of scenarios for the protection of water resources in Europe. JRC Scientific and Policy Report. JRC75919, ISSN 1831-9424, EC-JRC-IES, Italy.
24. Directive 2000/60/EC, 2000. Establishing a Framework for Community Action in the Field of Water Policy. (OJ (2000) L327/1). (Water Framework Directive).
25. ESAWADI, 2010. http://www.iwrm-net.eu/node/7 (accessed December, 2015).
26. Egoh, B., Drakou, E.G., Dunbar, M.B., Maes, J., Willemen, L., 2012. Indicators for Mapping Ecosystem

Services: a Review. Publications Office of the European Union, Luxembourg.

27. European Commission, 2011. COM(2011) 244 final. Our life insurance, our natural capital: an EU biodiversity strategy to 2020.

28. European Commission, 2012. COM(2012) 673 final. A Blueprint to Safeguard Europe's Water Resources.

29. European Commission, 2015. COM(2015) 120 final. The Water Framework Directive and the Floods Directive: Actions towards the 'good status' of EU water and to reduce flood risks.

30. Feld, C.K., Da Silva, P.M., Sousa, J.P., De Bello, F., Bugter, R., Grandin, U., Hering, D., Lavorel, S., Mountford, O., Pardo, I., Pärtel, M., Römbke, J., Sandin, L., Bruce Jones, K., Harrison, P., 2009. Indicators of biodiversity and ecosystem services: a synthesis across ecosystems and spatial scales. Oikos 118, 1862–1871.

31. Feld, C.K., Sousa, J.P., da Silva, P.M., Dawson, T.P., 2010. Indicators for biodiversity and ecosystem services: towards an improved framework for ecosystems assessment. Biodiver. Conserv. 19, 2895–2919.

32. Fisher, B., Turner, K.R., 2008. Ecosystem services: classification for valuation. Biol. Conserv. 141, 1167–1169.

33. Fisher, B., Turner, R.K., Morling, P., 2009. Defining and classifying ecosystem services for decision making. Ecol. Econ. 68, 643–653.

34. Gómez-Baggethun, E., Barton, D.N., 2013. Classifying and valuing ecosystem services for urban planning. Ecol. Econ. 86, 235–245.

35. Gómez-Baggethun, E., Martin-Lopez, B., Barton, D.N., Braat, L., Kelemen, E., Llorente, M.G., Saarikoski, H., van den Bergh, J., 2014. State of the art report on integrated valuation of ecosystem services. Deliverable D4.1/WP4 OpenNESS project.

36. GWP, 2000. Integrated water resource management. Technical Advisory Committee Background Paper Number 4. Global Water Partnership, Stockholm.

37. Georgiou, S., Bateman, I., Cole, M., Hadley, D., 2000. Contingent ranking and valuation of water quality improvements. CSERGE Discussion Paper 2000–18. University of East Anglia.

38. Ghermandi, A., Nunes, P.A.L.D., 2013. A global map of coastal recreation values: results from a spatially explicit meta-analysis. Ecol. Econ. 86, 1–15.

39. Ghermandi, A., Van Den Bergh, J.C.J.M., Brander, L.M., De Groot, H.L.F., Nunes, P.A.L. D., 2010. Values of natural and human-made wetlands: a meta-analysis. Water Resour. Res. 46.

40. Globaqua, 2015. http://www.globaqua-project.eu/en/home/ (accessed December, 2015).

41. Grizzetti, B., Bouraoui, F., Aloe, A., 2012. Changes of nitrogen and phosphorus loads to European seas. Glob. Change Biol. 18, 769–782.

42. Guerry, A.D., Polasky, S., Lubchenco, J., Chaplin-Kramer, R., Daily, G.C., Griffin, R., Ruckelshaus, M., et al., 2015. Natural capital and ecosystem services informing decisions: from promise to practice. Proc. Natl. Acad. Sci. 112, 7348–7355.

43. Guswa, A.J., Brauman, K.A., Brown, C., Hamel, P., Keeler, B.L., Sayre, S.S., 2014. Ecosystem services: challenges and opportunities for hydrologic modeling to support decision making. Water Resour. Res. 50, 4535–4544.

44. Haines-Young, R., Potschin, M., 2010. The links between biodiversity, ecosystem services and human well-being. In: Raffaelli, D.G., Frid, C.L.J. (Eds.), Ecosystem Ecology: A New Synthesis. Cambridge University Press, Cambridge, pp.110–139.

45. Halpern, B.S., McLeod, K.L., Rosenberg, A.A., Crowder, L.B., 2008. Managing for cumulative impacts in ecosystem-based management through ocean zoning. Ocean Coastal Manage. 51, 203–211.

46. Hanley, N.D., Schläpfer, F., Spurgeon, J., 2003. Aggregating the benefits of environmental improvements: distance-decay functions for use and non-use values. J. Environ. Manage. 68, 297–304.

47. Hering, D., Carvalho, L., Argillier, C., Beklioglu, M., Borja, A., Cardoso, A.C., Duel, H., Ferreira, T., Globevnik, L., Hanganu, J., Hellsten, S., Jeppesen, E., Kodeš, V., Solheim, A.L., Nõges, T., Ormerod, S., Panagopoulos, Y., Schmutz, S., Venohr, M., Birk, S., 2015. Managing aquatic ecosystems and water resources under multiple stress an introduction to the MARS project. Sci. Total Environ. 503–504,10–21.

48. InVEST, 2015. http://www.naturalcapitalproject.org/invest/ (accessed December, 2015).

49. Jørgensen, S.L., Olsen, S.B., Ladenburg, J., Martinsen, L., Svenningsen, S.R., Hasler, B., 2013. Spatially induced disparities in users' and non-users' WTP for water quality improvements-Testing the effect of multiple substitutes and distance decay. Ecol. Econ. 92, 58–66.

50. Jax, K., Barton, D.N., Chan, K.M.A., de Groot, R., Doyle, U., Eser, U., et al., 2013. Ecosystem services and ethics. Ecol. Econ. 93, 260–268.

51. Johnston, R.J., Besedin, E.Y., Iovanna, R., Miller, C.J., Wardwell, R.F., Ranson, M.H., 2005. Systematic variation in willingness to pay for aquatic resource improvements and implications for benefit transfer: a meta-analysis. Can. J. Agric. Econ. 53, 221–248.

52. Karabulut, A., Egoh, B.N., Lanzanova, D., Grizzetti, B., Bidoglio, G., Pagliero, L., Bouraoui, F., Aloe, A., Reynaud, A., Maes, J., Vandecasteele, I., Mubareka, S., 2016. Mapping water provisioning services to support the ecosystem-water-foodenergy nexus in the Danube river basin. Ecosyst. Serv. 17, 278–292.

53. Keeler, B.L., Polasky, S., Brauman, K.A., Johnson, K.A., Finlay, J.C., O'Neille, A., Kovacs, K., Dalzell, B., 2012. Linking water quality and well-being for improved assessment and valuation of ecosystem services. Proc. Natl. Acad. Sci. U. S. A. 109, 18619–18624.

54. Koundouri, P., Ker Rault, P., Pergamalis, V., Skianis, V., Souliotis, I., 2015. Development of an integrated methodology for the sustainable environmental and socio-economic management of river ecosystems. Sci. Total Environ. 540, 90–100.

55. Kull, C.A., Arnauld de Sartre, X., Castro-Larrañaga, M., 2015. The political ecology of ecosystem services. Geoforum 61, 122–134.

56. La Notte, A., Liquete, C., Grizzetti, B., Maes, J., Egoh, B.N., Paracchini, M.L., 2015. An ecological-economic approach to the valuation of ecosystemservices to support biodiversity policy. A case study for nitrogenretention by Mediterranean rivers and lakes. Ecol. Indic. 48, 292–302.

57. Layke, C., Mapendembe, A., Brown, C., Walpole, M., Winn, J., 2012. Indicators from the global and sub-global millennium ecosystem assessments: an analysis and next steps. Ecol. Indic. 17, 77–87.

58. Liquctc, C., Piroddi, C., Drakou, E.G., Gurney, L., Katsanevakis, S., Charef, A., Egoh, B., 2013. Current status and future prospects for the assessment of marine and coastal ecosystem services: a systematic review. PLoS One 8.

59. Liquete, C., Kleeschulte, S., Dige, G., Maes, J., Grizzetti, B., Olah, B., Zulian, G., 2015. Mapping green infrastructure based on ecosystem services and ecological networks: a Pan-European case study. Environ. Sci. Policy 54, 268–280.

60. Liquete, C., Udias, A., Conte, G., Grizzetti, B., Masi, F., 2016. Integrated valuation of a nature-based solution for water pollution control. Highlighting Hidden Benefits. (submitted).

61. Liu, J., Mooney, H.A., Hull, V., Davis, S.J., Gaskell, J., Hertel, T., Lubchenco, J., Seto, K.C., Gleick, P., Kremen, C., Li, S., 2015. Systems integration for global sustainability. Science 347, 1258832.

62. MARS, 2015. http://www.mars-project.eu/ (accessed December, 2015).

63. MEA, 2005a. Millennium ecosystem assessment. Ecosystems and Human WellBeing: Synthesis. Island Press, Washington, DC.

64. MEA, 2005b. Millennium ecosystem assessment. Current State & Trends Assessment. Freshwater Ecosystem Services (Chapter 7).

65. MEA, 2005c. Millennium ecosystem assessment. Ecosystems and Human Wellbeing: Wetlands and Water. Synthesis. World Resources Institute, Washington, DC.

66. Maes, J., Teller, A., Erhard, M., Murphy, P., Paracchini, M.L., Barredo, J.I., Grizzetti, B., Cardoso, A.C., et al., 2014. Mapping and Assessment of Ecosystems and their Services. Indicators for ecosystem assessments under Action 5 of the EU Biodiversity Strategy to 2020. 2nd Report Final, February 2014. Technical Report 2014-080.

67. Maes, J., Liquete, C., Teller, A., Erhard, M., Paracchini, M.L., Barredo, J.I., Grizzetti, B., Cardoso,A., et al.,2016. An indicator frameworkfor assessing ecosystem services in support of the EU Biodiversity Strategy to 2020. Ecosyst. Serv. 17, 14–23.

68. Nõges, P., Argillier, C., Borja, A., Garmendia, J.M., Hanganu, J., Kodeš, V., Pletterbauer, F., Sagouis, A., Birk, S., 2015. Quantified biotic and abiotic responses to multiple stress in freshwater, marine and ground waters. Sci. Total Environ. 540, 43–52.

69. Navrud, S., Ready, R.C., 2007. Environmental Value Transfer: Issues and Methods. Springer, Dordrecht.

70. OPERAs, 2015. http://operas-project.eu/ (accessed December, 2015).

71. OpenNESS, 2015. http://www.openness-project.eu/ (accessed December, 2015).

72. Pereira, H.M., Domingos, T., Vicente, L., 2006. Assessing ecosystem services at different scales in the Portugal Millennium Ecosystem Assessment. In: Reid, W., Berks, F., Wilbanks, T., Capistrano, D. (Eds.), Bridging Scales and Knowledge Systems. Island Press, Washington DC, pp. 59–80.

73. Polasky, S., Tallis, H., Reyers, B., 2015. Setting the bar: standards for ecosystem services. Proc. Natl. Acad. Sci. 112, 7356–7361.

74. Quickscan, 2015. http://quickscan.pro/ (accessed December, 2015).

75. Rankinen, K., Butterfield, D., Faneca Sànchez, M., Grizzetti, B., Whitehead, P., Pitkänen, T., Uusi-Kämppä, J., Leckie, H., 2016. The INCA-pathogens model: an application to the Loimijoki River basin in Finland. Sci. Total Environ. submitted.

76. Regulation (EU), 2014. No 1143/2014 on the prevention and management of the introduction and spread of invasive alien species.

77. Reynaud, A., Lanzanova, D., 2015. A global meta-Analysis of ecosystem services values provided by lakes. Presented at the 2nd Annual Conference of the French Association of Environmental and Resource Economists, 10–11 September in Toulouse.

78. Russi, D., ten Brink, P., Farmer, A., Badura, T., Coates, D., Förster, J., Kumar, R., Davidson, N., 2013. The Economics of Ecosystems and Biodiversity for Water and Wetlands. IEEP, London and Brussels; Ramsar Secretariat Gland.

79. Schaafsma, M., Brouwer, R., Rose, J., 2012. Directional heterogeneity in WTP models for environmental valuation. Ecol. Econ. 79, 21–31.

80. Schröter, M., Barton, D.N., Remme, R.P., Hein, L., 2014. Accounting for capacity and flow of ecosystem services: a conceptual model and a case study for Telemark, Norway. Ecol. Indic. 36, 539–551.

81. Sharp, R., Tallis, H.T., Ricketts, T., Guerry, A.D., Wood, S.A., Chaplin-Kramer, R., Nelson, E., Ennaanay, D., et al., 2015. InVEST +VERSION+ User's Guide. The Natural Capital Project. Stanford University, University of Minnesota, The Nature Conservancy, and World Wildlife Fund.

82. Smith, R., Madsen, A.L., Haines-Young, R., Barton, D., 2014. WP3 Methodological Guidelines for Bayesian Belief Networks. . (accessed December, 2015) http://openness.hugin.com/huginprog/documentation/ WP3_Method_Guidelines_BBNs_050314.pdf.

83. TEEB, 2010. The Economics of Ecosystems and Biodiversity: Ecological and Economic Foundation. Earthscan, London and Washington.

84. Teichert, N., Borja, A., Chust, G., Uriarte, A., Lepage, M., 2016. Restoring fish ecological quality in estuaries:

implication of interactive and cumulative effects among anthropogenic stressors. Sci. Total Environ. 542, 383–393.

85. UK NEA, 2011. The UK National Ecosystem Assessment Technical Report. UNEP/ WCMC, Cambridge.

86. UNEP, 2009. Water security and ecosystem services: the critical connection. Nairobi, Kenya.

87. UNEP-WCM, 2011. Developing Ecosystem Service Indicators: Experiences and Lessons Learned from Sub-global Assessments and Other Initiatives. Secretariat of the Convention on Biological Diversity, Montréal, Canada.

88. Vigerstol, K.L., Aukema, J.E., 2011. A comparison of tools for modeling freshwater ecosystem services. J. Environ. Manage. 92, 2403–2409.

89. Vigiak, O., Malagó, A., Bouraoui, F., Grizzetti, B., Weissteiner, C.J., Pastori, M., 2016. Impact of current riparian land on sediment retention in the Danube River Basin. J. Sustainability Water Qual. Ecol. submitted.

90. Vihervaara, P., Kumpula, T., Ruokolainen, A., Tanskanen, A., Burkhard, B., 2012. The use of detailed biotope data for linking biodiversity with ecosystem services in Finland. Int. J. Biodivers. Sci. Ecosyst. Serv. Manage. 8, 169–185.

91. Villamagna, A.M., Angermeier, P.L., Bennett, E.M., 2013. Capacity, pressure, demand, and flow: a conceptual framework for analyzing ecosystem service provision and delivery. Ecol. Complexity 15, 114–121.

92. Vlachopoulou, M., Coughlin, D., Forrow, D., Kirk, S., Logan, P., Voulvoulis, N., 2014. The potential of using the ecosystem approach in the implementation of the EU water framework directive. Sci. Total Environ. 470–471, 684–694.

93. Zulian, G., Paracchini, M.L., Maes, J., Liquete, C., 2013. ESTIMAP: Ecosystem Services Mapping at European Scale. JRC Report EUR 26474 EN.

CHAPTER 4 ASSESSMENT OF HYDROLOGICAL IMPACTS

Ouko M. Chrisphine, Odhiambo A. Maryanne and Boitt K. Mark

Keywords: Mau complex, Hydrological issues, Lesser flamingos, Lake nakuru

INTRODUCTION

The Mau Forest Complex is the largest closed-canopy montane ecosystem in Eastern Africa. It encompasses seven forest blocks within the Mau Narok, Maasai Mau, Eastern Mau, Western Mau, Southern Mau, South West Mau and Transmara regions. The area is thus the largest water tower in the region, being the main catchment area for 12 rivers draining into Lake Baringo, Lake Nakuru, Lake Turkana, Lake Natron and the Trans-boundary Lake Victoria [1].

However, in the past three decades or so, the Mau Forest Complex has undergone significant land use changes due to increased human population demanding land for settlement and subsistence agriculture. The encroachment has led to drastic and considerable land fragmentation, deforestation of the headwater catchments and destruction of wetlands previously existing within the fertile upstream parts. Today, the effects of the anthropogenic activities are slowly taking toll as is evident from the diminishing river discharges during periods of low flows, and deterioration of river water qualities through pollution from point and non-point sources [2].

The Mau Complex is drained mainly by 12 rivers including Rivers Njoro, Molo, Nderit, Makalia, Naishi, Kerio, Mara, Ewaso Nyiro, Sondu, Nyando, Yala and Nzoia. In the last three decades, physical evidence has revealed that the rivers in the complex have had significant decline in discharges, coupled by dwindling water quality. Other studies have also highlighted the changing hydrological response of the area are as a result of the land use/land cover changes [3].

Citation: Ouko M. Chrisphine, Odhiambo A. Maryanne and Boitt K. Mark, Assessment of Hydrological Impacts, doi: 10.4172/2157-7587.1000223.

The previous studies carried out in Mau have always demonstrated the relationship between deforestation and rate of forest degradation, but the effects on water quality and the impact on tourism resulting from the flamingoes migration has not been addressed adequately. For instance Landsat images from 1986 to 2000 have been used to address land degradation of Mau forest [1].

Apart from being the largest closed-canopy montane in Eastern Africa, the catchment basin has two important biodiversity zones namely the forests shrouding the catchment's upper reaches and Lake Nakuru national park. Interposed between these two zones are human habitations with less biodiversity value, but which are dependent either directly or indirectly on ecological services provided by high biodiversity zones [4]. Lake Nakuru is the home to various flamingoes. Currently the number of flamingoes in the lake has been declining with some migrating to other saline lakes due to urbanization and industrialization within the nearby Nakuru town.

Lake Nakuru national park

Lake Nakuru National park occupies 188km^2 [4]. It has very high concentration of wildlife consisting of 70 mammal species, 400 bird species and over 200 plant species. Lake Nakuru, which forms the centerpiece of this showcase of wildlife is internationally renowned for the large concentrations of lesser flamingos that use it for feeding, displaying and occasionally, for breeding. The water-bird assemblage, scenic beauty and central location of the town make this lake a tourist destination. The park has also contributed to the socio-economic development of Nakuru town and its environs through tourism development, hotel accommodation, food, curio sales and other entrepreneur activities.

Lesser flamingos have unpredictable spontaneous movements between the alkaline lakes, whose ecological triggers and forcing functions are not well known. This can be attributed to the following factors ranging from changes in food quality and quantity, breeding migrations, alkalinity changes, fresh water requirements and predation pressure. Lake Nakuru has witnessed dramatic changes in land use over the recent decades resulting in the destruction of critical watersheds. The most important threats relate to the sustained supply of water to the lake which is threatened by deforestation and degradation of the Mau catchment areas and by upstream water abstraction. The massive population of lesser flamingoes depend on a network of at least ten 'flamingo lakes', so any significant ecological change in any of the other lakes (in Kenya, Ethiopia and Tanzania) could have a devastating impact on the entire system.

Major indications of a stressed Lake Nakuru ecosystem dates back to 1990 when there was mass fish deaths followed by the 1993 flamingo mortalities and the 1994 lake drying, a response to cumulative effect of disastrous negative impact on the catchment that resulted into lake environmental degradation. Since 1995, the, Arthrospira fusiformis, (the preferred food for the lesser flamingoes) population in the lake has only made occasional and very transient appearances. Flamingo occupation of the lake over the same period has also shown a declining.

Dumping site

Nakuru Municipality, the region's largest town, has expanded its geographic reach more than four and half times since 1970. The expansion is due to high urban migration in the area resulting to high volume of waste collected by the municipality. The town is also located on a higher ground, which means all the waste from the town and dumping site easily finds their way to the lake.

Two miles to the North West of Nakuru municipality along Nakuru- Kabarak road is the Giotto dumpsite on a large site of about 50 hectares in a place known as Kiamunyi and it is an open dumping site. It was started in 1972 to help with the dumping of collected waste from the town. It is the main dumpsite for the domestic waste, agriculture waste, hospital waste and electronic waste. Each day the town generates an estimated 240 tons of domestic solid waste. About 60% of this is removed to approved dumping sites, the rest accumulates in the environment and is eventually deposited into the lake by storm water and wind. With the population increase, it has resulted to the growing burden of waste generated.

The previous studies carried on Lake Nakuru have not paid sufficient attention to the dumping site and industrial waste from Nakuru town. While exaggerating the effects of human activities on Mau forest which is the main source of Lake Nakuru, they have neglected the effects of the dumping site. Moreover, they have exaggerated the positive effects of human activities (ecological engineering) while neglecting the ecological risks of industrialization and urbanization. These two areas of exaggeration can potentially lead to misinterpretations of the outcomes of appropriate policies that support restoration projects.

RESEARCH QUESTIONS

The following questions are posed to help guide the study

1. What are some of the changes in forest cover within Mau complex for the past 20yrs?
2. What are some of the impacts on Lake Nakuru due to deforestation and various agricultural activities within the catchment area?
3. What impacts does the dumping site at Nakuru town has on Lake Nakuru?
4. What will be the impacts on lesser flamingoes and tourism sector if Lake Nakuru diminishes due to siltation?

STUDY OBJECTIVES

The main objective of this study is to determine forest cover within changes Mau complex for the past 20 years.

Specific objectives

To determine some of the impacts on Lake Nakuru given the current rate of deforestation and various agricultural activities at the source of various its rivers.

To determine the impacts of the dumping site at Nakuru town on lake Nakuru.

To determine the impacts on tourism sector and the lesser flamingoes if Lake Nakuru diminishes due to siltation and other environmental effects.

SCOPE OF THE STUDY

The study area has been limited to a section of Mau catchment area, this is because the whole catchment covers a large area. Therefore the study only focusses on south western and western parts of the catchment area as shown in Figure 1 below.

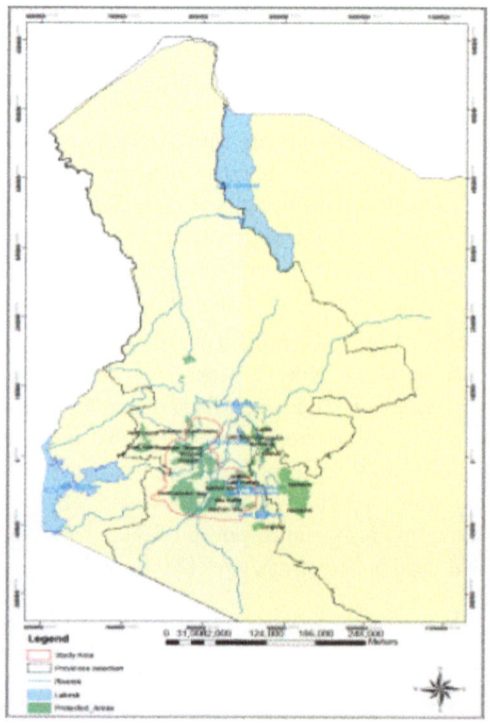

Figure 1: Map showing Mau catchment area.

STUDY METHODOLOGY

The study was divided into three parts, image analysis, GIS integration and theoretical analysis and concepts. Landsat images acquired from 1984 to 2015 were used for the analysis. The processing and analysis were done using Erdas Imagine, ENVI and ArcGIS. Figure 2 below shows the outline of the study.

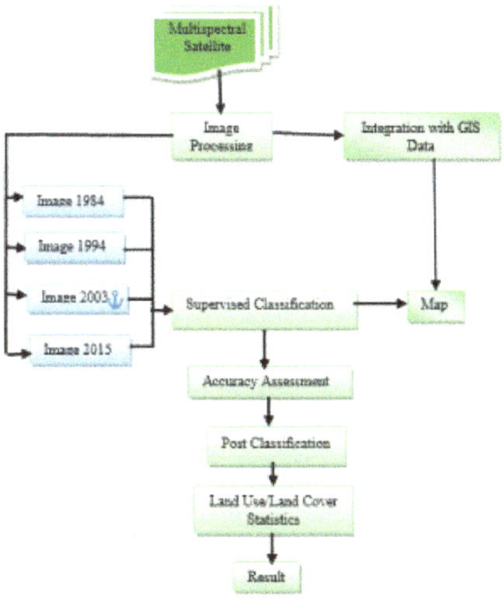

Figure 2: Study outline.

STUDY AREA LOCATION

The catchment is located in Rift valley province and is situated at 0° 34' 23".90 S and 35° 39' 26".61 E. Figure 3 below shows the map of the study area.

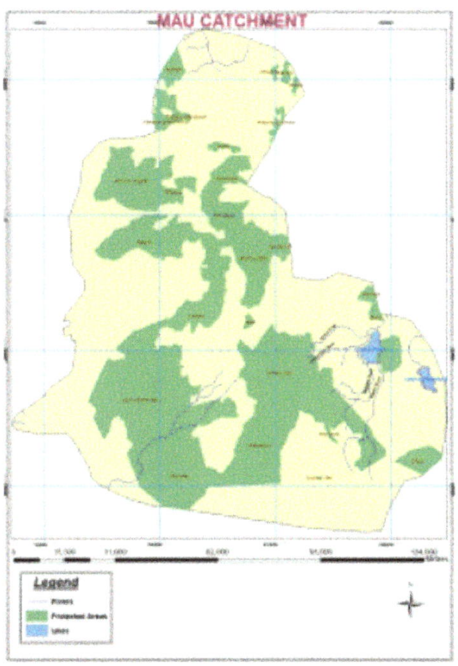

Figure 3: Mau catchment map.

LITERATURE REVIEW

Early studies in the Amazon basin investigated the degradation mapping potential of Landsat imagery by applying visual interpretation [5,6]. Extensive studies in Central Africa used Landsat imagery from three decades to derive area estimates of land cover change by combining a systematic regional sampling scheme, based on high-spatial-resolution imagery with object-based unsupervised classification techniques [7] to track the progression of logging roads. Higher resolution data has been used to track skid trails and tree felling [8]. Basic investigation work in the Congo Basin on the temporal and spatial resolution needed to detect selective logging from remote sensing data has been performed in the Central African Republic [9]. In Cameroon and the Republic of Congo, there has been some degradation monitoring work, based on SAR imagery, e.g., from CosmoSkyMed data using a 3D approach [9], with promising results.

Also, several studies focusing mainly on time analysis series for forest monitoring has been carried out in various areas around the world. For instance, drought forecasting in Somalia [10] or mapping selective logging activities in Brazil [11]. Same has also been applied in non-tropical areas, like forest disturbance detection in Northern Maine [12]. With regard to tropical forest monitoring, a supervised classification approach was followed in Indonesia [13]. individual epochs were classified and the changes in terms of tree cover were classified [14]. Considerable thematic developments and mapping work was done using a denser time series (bi-annual to annual) [15-17].

DATA ANALYSIS AND PRESENTATION

The analysis of the data was carried out in Erdas Imagine, ENVI and ArcGIS softwares. Several steps in image processing and analysis were carried out in order to obtain the final results.

Image processing

Remote-sensing-based detection of forest degradation is difficult, as these subtle degradation signals are not easy to detect in the first place and quickly lost over time due to fast re-vegetation. To overcome these shortcomings, a time series analysis has been developed to map and monitor forest degradation over a longer period of time, with frequent updates based on Landsat data [18].

Four Landsat TM satellite images of Mau were acquired for 1984, 1994, 2003 and 2015 with a spatial resolution of 30m. The images for the same months were not easy to obtain due to cloud coverage by some. Processing of these images was done using ERDAS Imagine and ENVI. ERDAS Imagine was used to layer stack the images and to subset. Classification, post classification and accuracy determination was carried out in ENVI.

Layer stacking: Landsat images are often distributed as Geotiff files (with separate Geotiff files for each band). A Geotiff is like a standard .tiff image file but has geographical information (coordinates) embedded. Each spectral band may be given as a separate Geotiff file and must be overlaid into one master file before analysis [19].

The Geotiff files were imported into ERDAS imagine and converted to .img files. Each band was then layer stack to form a new multiband image in ERDAS as shown in Figure 4.

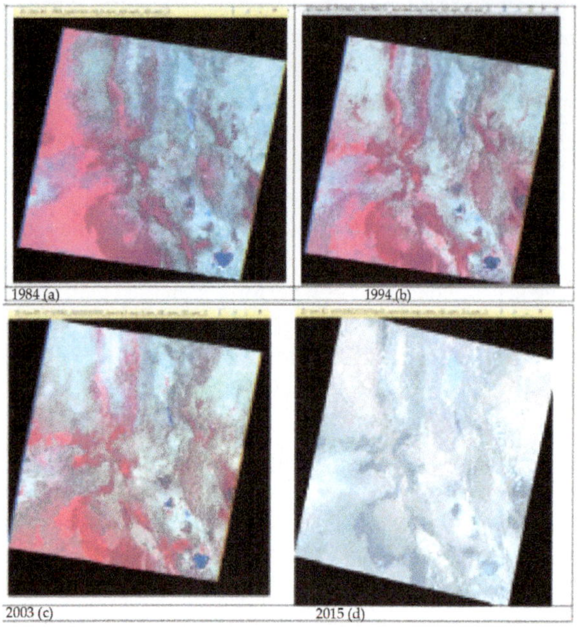

Figure 4: Layer Stacks for 1984, 1994, 2003 and 2015.

Subsetting: Subsetting involves clipping of the datasets for instance the layer stack images using by either defining the boundary from the image or by use of a polygon. It removes data outside the area of interest reducing the file size and improving the processing time for many operations [20-22].

Mau forest covers a large area; therefore a subset of the area was used based on a shape file generated in ArcMap. The shape file was used to create a subset of the study area. All the four images were clipped using the same shape file as shown in Figure 5 below.

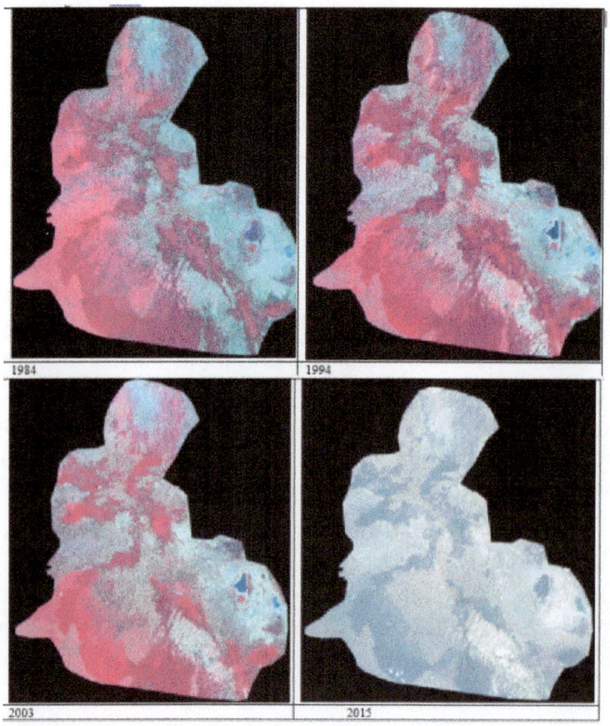

Figure 5: Subsets for 1984, 1994, 2003 and 2015 images.

Classification: Classification involves dividing the study area into distinct classes as determined from spectral reflectance of the classes. This is done in the software. There are two types of classification, unsupervised and supervised classification that can be performed with any Remote Sensing software, this study was based on supervised classification.

Supervised classification involves selecting representative samples for each land cover class (training sites) in the digital image. The training sites are used by the software to identify land cover classes in the entire image. The classification of land cover is based on the spectral signature defined in the training set. The digital image classification software determines each class on what it resembles most in the training set. Most common type of supervise classification is maximum likelihood and minimum distance. This study utilized maximum likelihood classification technique to come up with seven classes as shown in Table 1a below. The process was carried out using ENVI 4.7. The signature file was generated for all the regions of interest taking into account homogeneous representation of all areas. The signatures were then used to obtain the various classes for the image.

Table 1a: Land Use/Land cover Distribution for Year 1984, 1994, 2003 and 2015 in km²

	1984(km²)	1994(km²)	2003(km²)	2015(km²)
Water	72.03	67.32	57.5	44.12
Dense Forest	4138.408	2971.94	2695.328	2100.384
Plantation	191.6964	324.8019	619.8174	498.2762
Bare Ground	636.2676	1736.625	1493.375	1136.573
Cultivated Land	1347.354	1449.505	1984.38	1093.728
Vegetation	1108.332	1012.982	1758.144	3046.075
Trees and Shrubs	3674.442	3565.999	2500.987	3165.849
Total	11168.53	11129.172	11109.531	10586.729

Table 1b: Land Use/Land cover Percentage Distribution for Year 1984, 1994, 2003 and 2015

	1984 (%)	1994(%)	2003 (%)	2015 (%)
Water	0.64	0.6	0.52	0.42
Dense Forest	37.05	26.7	24.26	19.84
Plantation	1.716	2.92	5.58	4.71
Bare Ground	5.7	15.6	13.44	10.74
Cultivated Land	12.06	13.02	17.86	10.33
Vegetation	9.92	9.1	15.82	28.77
Trees and Shrubs	32.9	32.04	22.51	29.9

Post classification: In order to obtain the class statistics for all the classes generated, post classification was carried out in ENVI.

The values obtained were used to compute changes for the classes.

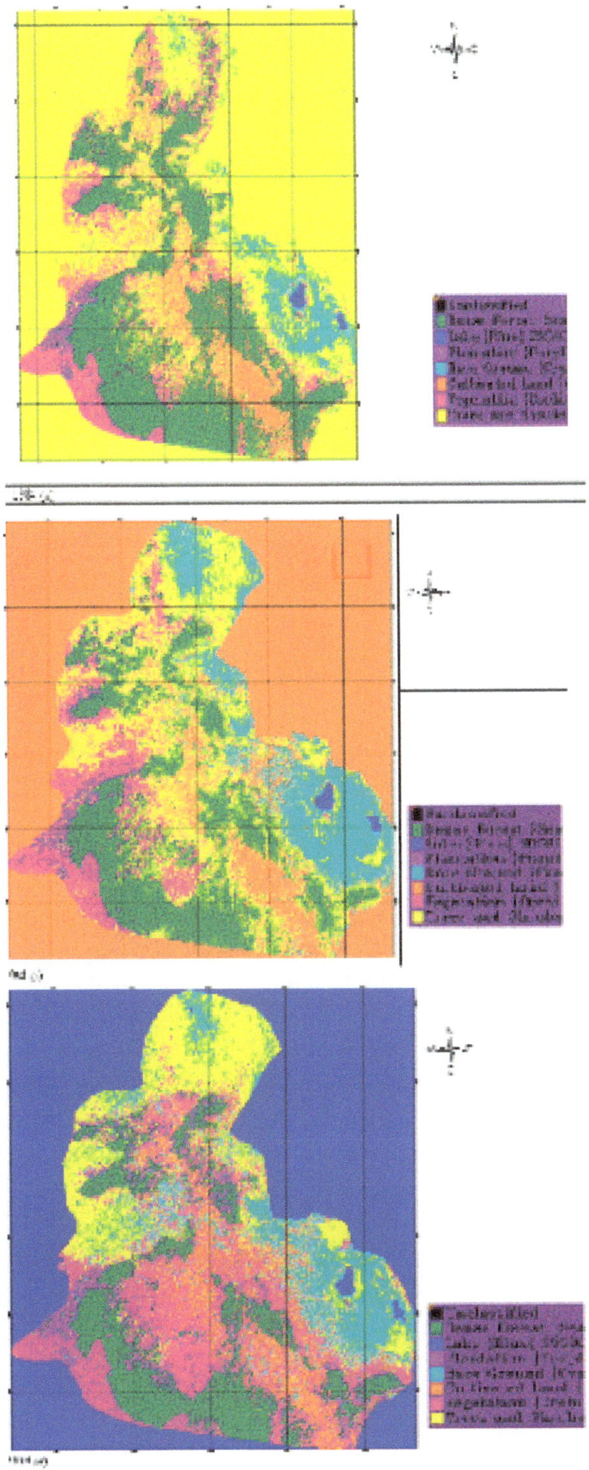

Figure 6: Classified Images for 1984, 1994, 2003 and 2015.

Integration of satellite images with GIS data

Integration was done in ArcGIS. The processed images were opened in ArcMap and various shape files such as rivers and districts overlaid on the image. Lakes and forest were digitized from the processed image to show the changes that have taken place.

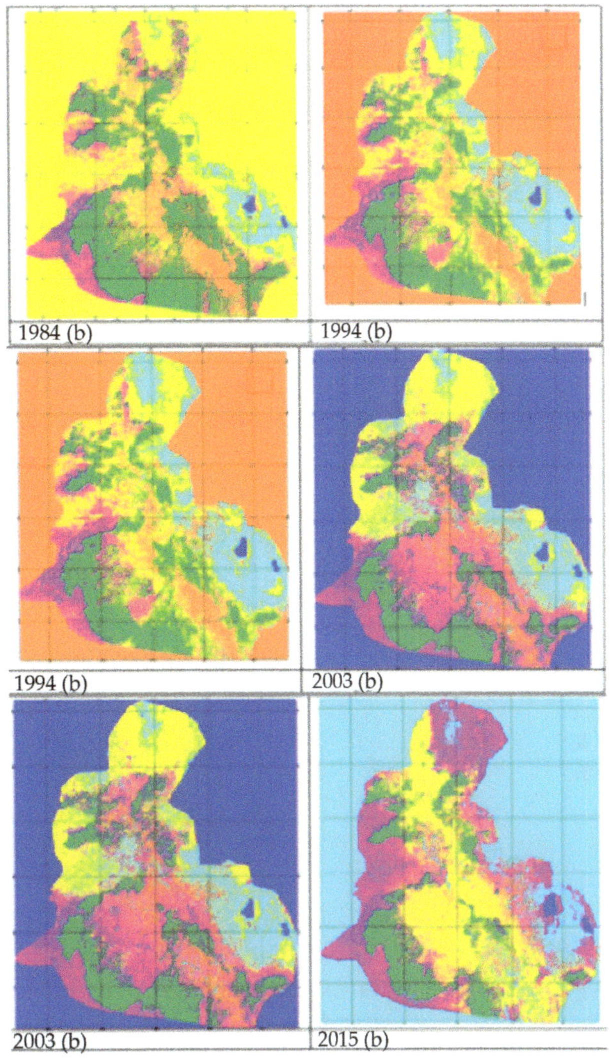

Figure 7: Comparison of Each Two Successive Image.

DATA ANALYSIS AND DISCUSSION OF RESULTS

Image analysis

Supervised classification: A supervised (full Gaussian) maximum likelihood classification was implemented for the four images and the final classification products provided an overview of the major land use/land cover features of Mau forest reserve for the years 1984, 1994, 2003 and 2015. Classification was carried out in ENVI by defining seven homogeneous training classes (Region of Interest) for the four images. The training classes were then used to obtain the classes from classification [23].

Seven categories of land use/land cover were identified; these are: Dense forest, cultivated land, forest, plantation, vegetation, bare land and water body. Figures 6 and 7 below illustrate the land use/land cover map of Mau forest reserve for the year 1984, 1994, 2003 and 2015 and comparison between each two successive epoch. In each image, atleast one feature class shared a color with the background; however this was accounted for in post classification by subtracting the background the area of the background from the total area covered by the color.

Post classification: In order to obtain the area extent of the resulting land use/land cover type for each study year and for subsequent comparison, post classification was carried out in ENVI to obtain class statistics in area in meters square. A constant figure of 8387.796 km^2 was subtracted from all the classes that had the same color as the background color in order to obtain the exact area of the feature class. The percentage change for every year was also computed and results shown in Tables 1a and 1b below.

Confusion matrix: Confusion Matrix shows the accuracy of a classification result by comparing a classification result with ground truth information. Confusion matrix was calculated in ENVI using ground truth Region of Interest (ROIs). This was in order to produce the overall accuracy, producer and user accuracies as shown below for each year (Figure 8).

Class	Prod. Acc. (Percent)	User Acc. (Percent)	Prod. Acc. (Pixels)	User Acc. (Pixels)
Unclassified	0.00	0.00	0/0	0/0
Dense Forest	100.00	100.00	4598231/4598231	4598231/4598231
Lake [Blue] 2	100.00	100.00	49018/49018	49018/49018
Plantation [P	100.00	100.00	212996/212996	212996/212996
Bare Ground [	100.00	100.00	706964/706964	706964/706964
Cultivated La	100.00	100.00	1497060/1497060	1497060/1497060
Vegetation [O	100.00	100.00	1231480/1231480	1231480/1231480
Trees and Shr	100.00	100.00	13402487/13402487	13402487/13402487

Class	Prod. Acc. (Percent)	User Acc. (Percent)	Prod. Acc. (Pixels)	User Acc. (Pixels)
Unclassified	0.00	0.00	0/0	0/0
Water [Blue]	100.00	100.00	63888/63888	63888/63888
Dense Forest	100.00	100.00	3302156/3302156	3302156/3302156
Plantation [P	100.00	100.00	360891/360891	360891/360891
Bare Ground [	100.00	100.00	1929583/1929583	1929583/1929583
Cultivated La	100.00	100.00	10930335/10930335	10930335/10930335
Vegetation [O	100.00	100.00	1125535/1125535	1125535/1125535
Trees and Shr	100.00	100.00	3962221/3962221	3962221/3962221

Class	Prod. Acc. (Percent)	User Acc. (Percent)	Prod. Acc. (Pixels)	User Acc. (Pixels)
Unclassified	0.00	0.00	0/0	0/0
Water [Blue]	100.00	100.00	9394575/9394575	9394575/9394575
Dense Forest	100.00	100.00	2994809/2994809	2994809/2994809
Plantation [P	100.00	100.00	688686/688686	688686/688686
Bare Ground [	100.00	100.00	1659305/1659305	1659305/1659305
Cultivated La	100.00	100.00	2204867/2204867	2204867/2204867
Vegetation [O	100.00	100.00	1953493/1953493	1953493/1953493
Trees and Shr	100.00	100.00	2778874/2778874	2778874/2778874

Class	Prod. Acc. (Percent)	User Acc. (Percent)	Prod. Acc. (Pixels)	User Acc. (Pixels)
Unclassified	0.00	0.00	0/0	0/0
Lake [Blue] 1	100.00	100.00	320159/320159	320159/320159
Dense Forest	100.00	100.00	9335040/9335040	9335040/9335040
Plantation [P	100.00	100.00	2214561/2214561	2214561/2214561
Bare Ground [	100.00	100.00	42330529/42330529	42330529/42330529
Cultivated La	100.00	100.00	4861014/4861014	4861014/4861014
Vegetation [O	100.00	100.00	13538113/13538113	13538113/13538113
Trees and Shr	100.00	100.00	14070440/14070440	14070440/14070440

Figure 8: Over producer and user accuracies from 1984-2015.

DISCUSSION OF THE RESULTS

The 1984, 1994, 2003 and 2000 land use/land cover practice in this depleting Mau Forest Reserve were determined in order to ascertain the causes of deforestation and various impacts on Lake Nakuru. Seven major classes were identified and classified as land use/cover of all the four images as follows: Lakes, bare ground, cultivated land, dense forest, plantation, vegetation and trees and shrubs. However, amongst these seven major classes, cultivated land and plantation were found to be the practices leading to forest depletion. Vegetation and trees and shrubs are the results of the depleted dense forest while bare ground describes either areas that had been cleared or could not sustain growth of any kind of vegetation.

The sizes of dense forest and lakes have been declining significantly from 1984 to 2015 as shown in the Figure 9 below. This is from the fact that there has been a lot of illegal cutting down of trees and clearing of land for cultivation within. As a result, the impact of depleting forest has negatively impacted on various lakes that draw their water from Mau forest and the surrounding as a whole.

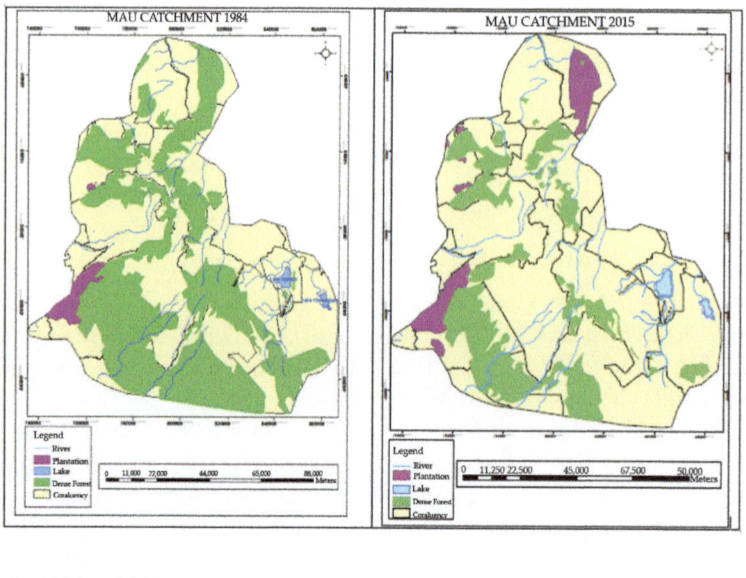

Figure 9: Mau Catchment Map in 1984 and 2015.

Impacts of deforestation and agricultural activities on Lake Nakuru

When forest cover bonding the soil in a lake basin is destroyed, the result is usually an increase in land erosion and sediment transport, which in turn leads to reduced lake water quality which limits the penetration of light into the water column. The soil being of volcanic origin and due to its high porosity, permeability and its loose structure is highly susceptible to erosion, land sinking and fractures during or after heavy rain. Deforestation would therefore decreases natural seepage into underground aquifers, leading to accelerated surface runoff.

Soils have only a limited capacity to absorb nutrients used to increase crop production. The nutrients which are not taken up by the soil/plant system leach into the surface and are carried into the lake by runoff resulting in water quality problems related to Eutrophication. In addition the amount of organic nitrogen from animal waste also increases due to high livestock densities. It has been estimated that the average sediment yield in Lake Nakuru watershed is of the order 100 to 300 tons/km^2/year. Nutrient enrichment is a large-scale problem in Lake Nakuru. It results in reduced productivity of the lake's natural primary producer Arthospira fusiformis and frequent occurrence of toxic blue-green algae blooms. This phenomenon easily leads to desertion of the lake by lesser flamingos.

The destruction of forests in the Lake Nakuru catchment basin not only has consequences for the biodiversity they support but also cumulative impacts on the catchment hydrology. The hydrology of the lake is dependent on catchment supply through rivers, demanding for catchment integrity to sustain the lake levels. The shallow depth of 4.5m, the high evaporation rates and seasonal rivers make the lake a hydrologically impacted ecosystem, since it does not have any buffering capacity to withstand hydrological impacts driven by catchment processes [4].

Impacts of environmental effects and siltation to the tourism sector

Water pollution is a significant threat. The presence of pollutants in the lake has resulted in an imbalance in the ecosystem that has caused a food shortage for the flamingos. The diminishing flamingo food could also be the result of the introduction of tilapia that led to competition for the planktonic blue-green alga, (Spirulina platensis) fed on by the flamingos. This could be a possible cause for migration of flamingos to other lakes including Lake Simbi Nyaima in Western Kenya among others.

In the place of Arthrospira fusiformisthe algal composition of the lake has been largely dominated by undesirable species of blue-green algae such as Microcystis sp. and Anabaena sp. Both these species are known to flourish in nutrient rich waters with high organic content and both are capable of elaborating potent hepatic and neuro toxins to which birds and fish are susceptible.

Encroachment on wetlands and flamingo breeding areas has resulted to flamingo and fish mortality due to diseases, parasites, predation and toxic pollution. The ecological effect of this has been the loss of biodiversity through migration of the birds to other water bodies within the rift valley where complimentary food is available. All these cumulatively result in the reduced revenues associated with ecotourism [4].

Impacts of the dumping site to Lake Nakuru

Like most cities, Nakuru is a huge consumer of resources, and a prodigious producer of waste. Of the municipal waste generated in the urban centers, 21% emanates from industrial areas and 61% from residential area. Generally, about 40% of the total waste generated in urban centers is collected and disposed of at the designated disposal sites. About 60% of this is removed to approved dumping sites, the rest accumulates in the environment and is eventually deposited into the lake by storm water and wind. With the population increase, it has resulted to the growing burden of waste generated. As a result of these developments, water balance and water quality of Lake Nakuru and its feeder streams have been adversely affected. Studies done on the site reveal high concentration of heavy metals in soil sample. This indicates how dangerous the site might be to human health and the surrounding environment.

CONCLUSION

The study have demonstrated that the size of forest cover in Mau have been changing from 1984 to present. This is due to deforestation and agricultural activities taking place within the forested area.

When forest cover bonding the soil in a lake basin is destroyed, the result is usually an increase in land erosion and sediment transport, which in turn leads to reduced lake water quality which limits the penetration of light into the water column. Deforestation would therefore decreases natural seepage into underground aquifers, leading to accelerated surface runoff causing siltation in the lake. Therefore the size and depth of Lake Nakuru is expected to reduce if the current situation proceeds.

Nakuru dumping site is located on higher ground meaning water from the dumping site do find its way easily to the lake hence affecting water quality of Lake Nakuru and its feeder streams.

The presence of pollutants in the lake has resulted in an imbalance in the ecosystem that has caused a food shortage for the flamingos. The diminishing flamingo food could also be the result of the introduction of tilapia that led to competition for the planktonic blue-green alga, (Spirulina platensis) fed on by the flamingos. This could be a possible cause for migration of flamingos to other lakes including Lake Simbi Nyaima in Western Kenya among others. This will greatly affect tourism in the area since flamingoes are the main tourism attraction in the area.

RECOMMENDATIONS

Deforestation is not an unstoppable or irreversible process. Increased and concerted efforts in forest plantation 'rebirth' and rejuvenation will bring to use the type of forest reserve we envisaged. In order to reduce the effects of deforestation in Mau forest, the study has the followings as its recommendations:

Government by way of policy should be strict in preserving forest reserves from illegal occupation.

The dumping site in Nakuru town should be relocated to a lower ground where water from the site cannot find its way to Lake Nakuru.

Promotion of alternative energy source for fire wood in order to reduce the pressure on the forest.

Development and promotion of trade in non-timber forest product to reduce the pressure on timber resources and to enhance rural livelihood.

It is strongly recommended that any form of forest plantation degradation should be stopped forthwith, having realized the purpose for which the reserve was meant for.

The available vegetation area and the farmland must be converted into forest plantation of exotic fast growing species.

Lastly, the technology of remote sensing and GIS should be employed in major studies, concerning national issue such as deforestation, desertification etc.

REFERENCES

1. Olang LO, Peter Musula, Kundu (2010) Land Degradation of the Mau Forest Complex in Eastern Africa (A Review for Management and Restoration Planning).
2. Baldyga TJ, Miller NS, Driesse LK, Gichaba NC (2007) Assessing land cover change in Kenya's Mau Forest region using remotely sensed data. African Journal of Ecology 46: 46-54.
3. Owido SFO, Chemelil CM, Nyawade FO, Obadha WO (2003) Effects of Induced Soil compaction on Bean (Phaseolus Vagaries) Seedling Emergence from a Haplic phaeozen soil. Agricultura Tropica ET subtropica 36: 65-69.
4. Eric O Odada, Jackson Raini, Robert Ndetei (2006) Lake Nakuru. Experience and lesions learned Brief for Lake Nakuru.
5. Stone TA, Lefebvre P (1998) Using multi-temporal satellite data to evaluate selective logging in Para, Brazil. Int J Remote Sens 19: 2517-2526.
6. Nepstad DC, Verissimo A, Alencar A, Nobre C, Lima E (1999) Large-scale impoverishment of Amazonian forests by logging and fire. Nature 398: 505-508.
7. Duveiller G, Defourny P, Desclée B, Mayaux P (2008) Deforestation in central africa: Estimates at regional, national and landscape levels by advanced processing of systematically-distributed landsat extracts. Remote Sens Environ 112: 1969-1981.
8. Laporte NT, Stabach JA, Grosch R, Lin TS, Goetz SJ (2007) Expansion of industrial logging in Central Africa. Science 316: pp1451.
9. De Wasseige C, Defourny P (2004) Remote sensing of selective logging impact for tropical forest management. For Ecol Manag 188: 161-173.
10. Verbesselt J, Zeileis A, Herold M (2012) Near real-time disturbance detection using satellite image time series: Drought detection in Somalia. Remote Sens Environ 123: 98-108.
11. Koltunov A, Ustin SL, Asner GP, Fung I (2009) Selective logging changes forest phenology in the Brazilian Amazon: Evidence from MODIS image time series analysis. Remote Sens Environ 113: 2431-2440.
12. Jin S, Sader SA (2005) MODIS time-series imagery for forest disturbance detection and quantification of patch size effects.Remote Sens Environ 99: 462-470.
13. Margono BA, Turubanova S, Zhuravleva I, Potapov P, Tyukavina A, et al. (2012) Mapping and monitoring deforestation and forest degradation in sumatra (Indonesia) using Landsat time series data sets from 1990 to 2010. Environ Res Lett 7: 1-16.
14. Manuela H, Martin S, Gallaun H, Schardt M (2014) Mapping Forest Degradation due to Selective Logging by Means of Time Series Analysis" Case Studies in Central Africa.Remote Sens 6: 756-775.
15. Healey S, Moisen R, Masek J, Cohen W, Goward Sister, et al. (2005) Measurement of Forest Disturbance and Regrowth with Landsat and Forest Inventory and Analysis Data: Anticipated Benefits from Forest and Inventory Analysis? In Proceedings of the Seventh Annual Forest Inventory and Analysis Symposium 6: 171-178.
16. Kennedy RE, Cohen WB, Schroeder TA (2007) Trajectory-based change detection for automated characterization of forest disturbance dynamics.Remote Sens Environ 110: 370-386.
17. Kennedy RE, Yang Z, Cohen WB (2010) Detecting trends in forest disturbance and recovery using yearly Landsat time series: 1. LandTrendr-Temporal segmentation algorithms.Remote Sens Environ 114: 2897-2910.
18. Nabutola W (2010) The Mau Forest in the Rift Valley: Kenya's Largest Water Tower: a Perfect.
19. Renee AR (2014) Model for the Challenges and Opportunities of a Sustainable Development Project. QScience Connect 22: 1-14.
20. Laura N (2010) Buffer Zone Plans for Lake Nakuru National Park and for Njoro River in Kenya 19: 1-66.
21. Asiyanbola RA (2014) Remote Sensing in Developing Country-Nigeria 6: 1-19.
22. Ayoola AA, Oloyede SOA, Kosoko, Aborisade DK (2012) Remote Sensing and GIS Application for Forest Reserve Degradation Prediction and Monitoring.
23. Deutscher J, Perko R, Gutjahr K, Hirschmugl M, Schardt M (2013) Mapping tropical rainforest canopy disturbances in 3D by COSMO-SkyMed spotlight InSAR-Stereo data to detect areas of forest degradation. Remote Sens 5: 648-663.

CHAPTER 5

ASSESSING THE POTENTIAL OF AUTONOMOUS SUBMARINE GLIDERS FOR ECOSYSTEM MONITORING ACROSS MULTIPLE TROPHIC LEVELS (PLANKTON TO CETACEANS) AND POLLUTANTS IN SHALLOW SHELF SEAS

Lavinia Suberg , Russell B. Wynn, Jeroen van der Kooij, Liam Fernand, Sophie Fielding, Damien Guihen, Douglas Gillespie, Mark Johnson, Kalliopi C. Gkikopoulou, Ian J. Allan, Branislav Vrana, Peter I. Miller, David Smeed, Alice R. Jones

Keywords: Autonomous underwater vehicles, Submarine glider, Slocum, Ecosystem monitoring, Multiple trophic levels

INTRODUCTION

Shelf and adjacent coastal seas host highly productive ecosystems and are shared by an increasing variety of stakeholders utilizing limited space, e.g. shipping, fishing, aquaculture, recreation, hydrocarbon and aggregate extraction, and renewable energy (Collie and Adamowicz et al., 2013 and Sharples and Ellis et al., 2013). These potentially conflicting demands require appropriate management, e.g. through Marine Spatial Planning, a complex task that is dependent upon high quality data and evidence (Gilman, 2002 and Douvere and Ehler, 2011).

In addition to the management of multiple stakeholders to ensure that ecosystem health and services are maintained, additional data from shelf seas are required to meet international statutory obligations such as establishment of Marine Protected Areas (MPAs) and implementation of the EU Marine Strategy Framework Directive (MSFD) (European Union, 2008 and Brennan and Fitzsimmons et al., 2013). However, marine mapping and monitoring using dedicated research and survey vessels is expensive, and offshore operations can be hindered due to weather constraints (Schofield and Glenn et al., 2013). In addition, the spatial and temporal resolution of vessel-based data are often insufficient to fully capture ecosystem dynamics, including the linkage of physical and biological processes, predator–prey interactions, community structure, and the spatio-temporal variability of different ecosystem components (Day, 2008). Satellite remote sensing of the oceans can provide useful supporting data at large spatial scales, but is restricted to the uppermost layers of the sea surface. Fixed moorings and profiling floats may provide long time series, but the former only collects data at a single point and the latter are difficult to spatially control (L'Heveder and Mortier et al., 2013).

Submarine (buoyancy-driven) gliders are a type of Autonomous Underwater Vehicle (AUV) that oscillates through the water column and can remain unattended at sea for several weeks to months (Rudnick and Crowley et al., 2012). Gliders carrying appropriate sensors can simultaneously

Citation: Lavinia Suberg, Russell B. Wynn, Jeroen van der Kooij, Liam Fernand, Sophie Fielding, Damien Guihen, Douglas Gillespie, Mark Johnson, Kalliopi C. Gkikopoulou, Ian J. Allan, Branislav Vrana, Peter I. Miller, David Smeed, Alice R. Jones, Assessing the potential of autonomous submarine gliders for ecosystem monitoring across multiple trophic levels (plankton to cetaceans) and pollutants in shallow shelf seas, Methods in Oceanography, Volume 10, September 2014, Pages 70-89, ISSN 2211-1220, http://dx.doi.org/10.1016/j.mio.2014.06.002.

monitor a range of physical and biological parameters, and regular surface communications with satellite allow their movement to be controlled and data to be uploaded in near real-time. However, gliders are relatively slow moving (20–40 cm/s horizontally), making them prone to drift in areas of strong currents (Leonard et al., 2007 and Davis and Leonard et al., 2009). Their sensor load is limited and each mission necessitates a balance between battery life, mission duration, sampling frequency, and data quality (Willcox and Bellingham et al., 2001). Despite these limitations, the scientific research community is increasingly focusing on gliders as a tool for monitoring of features at the meso- and sub-mesoscale, including highly variable and dynamic phenomena such as oceanic fronts, eddies and upwelling regions (Davis and Ohman et al., 2008).

Traditionally, gliders have been deployed with a basic set of sensors that enable measurement of physical oceanographic parameters such as temperature, salinity or currents (e.g. Perry and Sackmann et al., 2008, Ruiz and Pascual et al., 2009, Bouffard and Pascual et al., 2010, Merckelbach and Smeed et al., 2010 and Todd and Gawarkiewicz et al., 2013) and lower trophic levels of the ecosystem (i.e. phytoplankton and zooplankton, e.g. Baumgartner and Fratantoni, 2008, Niewiadomska and Claustre et al., 2008, Fox and Gower et al., 2009, Frajka-Williams and Rhines et al., 2009 and Guihen et al., 2014). More recently, new sensors have been integrated onto gliders that can measure abundance of higher trophic level organisms, e.g. fish and cetaceans (e.g. Baumgartner and Fratantoni, 2008, Ferguson and Lo et al., 2010, Klinck and Mellinger et al., 2012, Meyer-Gutbrod and Greene et al., 2012, Baumgartner and Fratantoni et al., 2013 and Send and Regier et al., 2013). In addition, glider 'fleets' are increasingly used to establish ocean monitoring networks rather than single platform deployments (e.g. English and Chuanmin et al., 2009, Alvarez and Mourre, 2012, Bouffard and Renault et al., 2012 and Alvarez and Chiggiato et al., 2013). However, until now, there has been limited effort devoted to the simultaneous measurement of physical parameters and multiple biological components of the ecosystem using gliders.

This contribution describes a deployment of gliders carrying sensor loads capable of simultaneously monitoring multiple marine ecosystem components, from physical parameters and chlorophyll a fluorescence (CTD and fluorometer) to zooplankton and fish (echosounder), and cetaceans (hydrophone). This 'proof-of-concept' study involved the deployment of two shallow-water gliders off southwest UK in autumn 2013, targeting a known tidal-mixing front in a productive inner-shelf environment. By targeting a frontal area, the gliders were expected to encounter steep vertical and horizontal gradients in physical parameters and potentially elevated levels of biomass. In addition, fronts in UK shelf waters are considered as potential targets for spatial protection measures (e.g. MPAs; (Miller and Christodoulou, 2014)) and are therefore a particular target for multi-trophic-level monitoring.

The aims of the paper are therefore to: (1) describe the various sensors that were deployed on the gliders, (2) provide an overview of glider and sensor operations during the autumn 2013 deployment, (3) present some initial scientific results and examples of collected data, and (4) discuss some of the benefits and issues that arose from the glider missions. The intention is that this study will aid future assessment of submarine gliders as a suitable platform for cost-effective, long-term monitoring of shelf sea ecosystems.

SURVEY AREA

The survey was conducted west of the Isles of Scilly, off southwest UK (Fig. 1), where shelf waters are <100 m deep. Here, tidal flows to the northeast and southwest peak at ~75 cm/s during spring tides, and are weakest towards the northwest and southeast. Residual currents flow northwest and northwards with speeds of up to 5 cm/s (Pingree and Lecann, 1989). Shelf waters to the west of the Isles of Scilly are seasonally stratified (typically from late spring to late autumn), whereas inshore waters around the islands themselves remain mixed as a result of tidal–topographic interactions; a series of seasonal bottom and surface tidal-mixing fronts occur at the boundary between these stratified and mixed water masses (Simpson, 1981). Frontal dynamics lead to cooler, nutrient-rich waters being transferred into the photic zone and thus enhancing primary production (Pingree and Holligan et al., 1976 and Simpson and Tett et al., 1982), which can in turn lead to fronts (including the Isles of Scilly fronts) becoming hotspots for higher trophic levels such as marine mammals (Beaumont and Austen et al., 2007 and Leeney and Witt et al., 2012). The recognition of a wide variety of important marine habitats around the Isles of Scilly has led to the designation of the Isles of Scilly Marine Conservation Zone (MCZ) in 2013 (Defra, 2013). In addition, the waters around the Isles of Scilly are valuable for marine recreation and tourism (Beaumont and Austen et al., 2007). There is also significant commercial fishing activity in the area, including a nearshore industry for shellfish and crustaceans (Beaumont and Austen et al., 2007). There are three traffic separation zones (TSZ) surrounding the islands due to high levels of shipping traffic (Fig. 1).

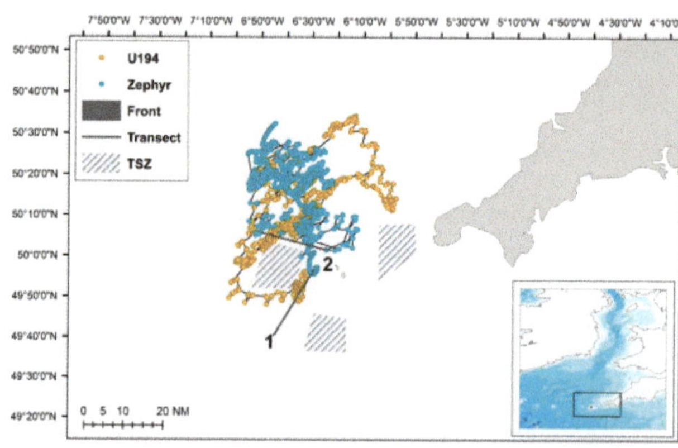

Figure 1: Map of survey area and glider positions. Inset shows geographic overview of the survey area and depth contours (dark blue = deeper water; white = land). Enlarged map shows land in grey (Isles of Scilly and west Cornwall). Dots represent glider GPS positions (blue: *Zephyr*; orange: *U194*); thin black lines indicate glider tracks between positions, based on linear interpolation. Large grey arrows indicate the approximate location of the Isles of Scilly tidal mixing front determined by remote sensing Fig. A.1. Numbered dark grey lines indicate the initial (1) and revised (2) transects. Shaded areas represent traffic separation zones (TSZ). Note that *U194* locations are only based upon GPS fixes received following re-deployment on 04 October 2013 (*cf.* Table 1 and Fig. 2). (For interpretation of the references to colour in this figure legend, the reader is referred to the web version of this article.)

Table 1: Key events in chronological order

Date	Event
12/09/2013	Deployment
14/09/2013	*Zephyr* d-tag memory card full (no data stored hereafter) until re-deployment
17/09/2013	*U194* "beached"
21/09–03/10/2013	Recovery and repairs *U194*
04/10/2013	Re-deployment *U194*; exchange *Zephyr* d-tag
10/10/2013	*Zephyr* d-tag stops recording: no power supply due to damaged cable
21/10/2013	Retrieval of *Zephyr*
13/11/2013	Retrieval of *U194*

METHODS

Gliders

We used two shallow-water *Slocum* gliders, *Zephyr* and *U194* (Fig. 2), developed by Teledyne Webb Research Corporation, which are specifically designed for shallow-water operations (<200 m). A detailed description of the glider can be found in Webb and Simonetti et al. (2001), but here we focus on aspects relevant to this paper. *Slocums*move through the water column in a saw-tooth motion induced by changes in their buoyancy. A pump transfers seawater in and out of a holding chamber, which results in a change to the vehicle's density; this leads to a sequence of sinking and rising, which is translated into a forward motion by the attached wings. The average horizontal speed is 20–40 cm/s and vertical motion is 10–20 cm/s. Dive depth can be regulated by either a pressure sensor or an altimeter. Due to the shallow depth and varying bathymetry, we used the altimeter with an inflexion height of 10 m above the bottom to maximize dive depth. Forward and backward shifting of the battery packs controls dive angle; a rudder in the tail fin controls the yaw. When at the surface for communication, an air bladder at the tail is filled for additional buoyancy. Two-way communication is via Iridium satellite connection, allowing for near real-time transfer of subsets of data and manipulation of mission settings. Navigation is by Global Positioning System (GPS) when the vehicle is at the surface, and via dead reckoning during dives.

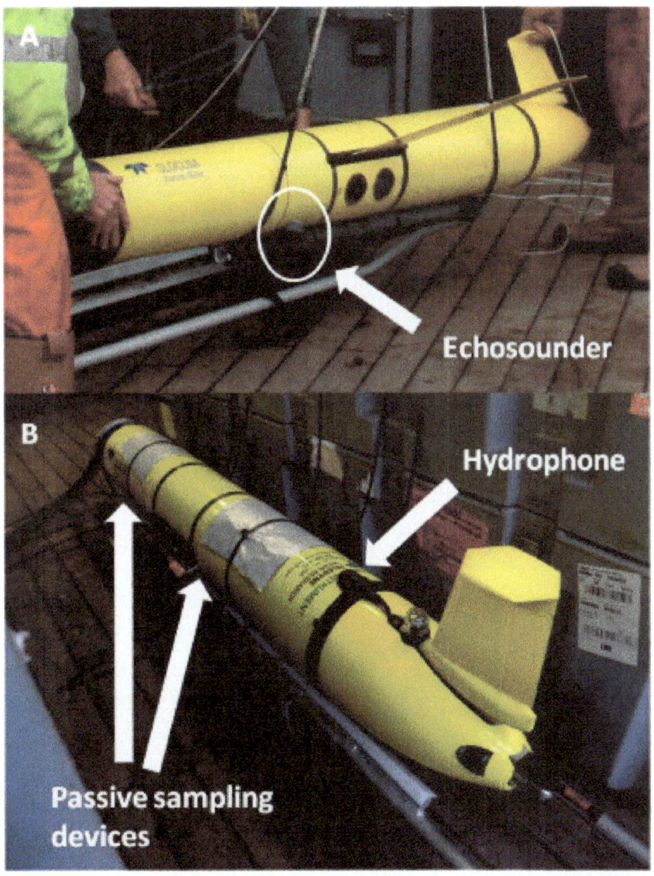

Figure 2: (A) Picture of *U194* showing ES853 echosounder on the underside of the glider and (B) picture of *Zephyr*(wings not yet attached), showing the d-tag hydrophone and silicone rubber passive sampling sheets attached to the glider body.

Glider sensors

Oceanographic sensors

Both gliders were equipped with the standard *Slocum* science package, which includes a non-pumped, low-drag, continuous profile Sea-bird SBE-41 CTD (temperature, conductivity and pressure) and WET Labs ECO pucks (BB2FLS, *Zephyr*; FlbbCD-SLK, *U194*), providing combined measurements of chlorophyll a fluorescence, coloured dissolved organic matter (CDOM), and backscatter. The excitation wavelengths were 470, 370 and 700 nm, the emission wavelengths 695, 460 nm. The ECO pucks were mounted in the science bay, facing downwards. Also included are an Aanderaa Oxygen Optode 4835 and an altimeter. CDOM and backscatter provide information on organic and inorganic material in the water-column, which would be useful for an in-depth examination on ecosystem dynamics along a tidal mixing front. However, in this study, only temperature and chlorophyll a fluorescence measurements are considered for demonstration purposes, which were obtained on a 1 Hz sampling frequency. All acquired data were stored internally and a subset transmitted via Iridium each time the glider surfaced.

Ensuring data quality from autonomous vehicles is an important, but potentially difficult task. The issues can be reduced by using stable instruments (such as ship-borne CTD rosettes) with calibrations pre- and post-deployment. Therefore, we used two ship-based CTD casts, one taken directly after *Zephyr* retrieval (21st October 2013) and one taken at day 38 (19th October 2013) of the *U194* mission to correct the glider data (temperature and chlorophyll a) where necessary. Due to the problems of "in-mission" calibration (e.g. aligning the ship and glider position), we were unable to take CTD casts at the same time and location as *U194*. Therefore, we present the ship-based profile with the nearest spatio-temporal *U194* match (Δ time: 27 min; distance: 7.65 km). Additional water samples at surface and bottom were taken for chlorophyll a measurement verification. A subset of oceanographic data was stored on the National Oceanography Centre (NOC) sftp server and made available to the Met Office via the British Oceanographic Data Centre (BODC), for inclusion in their Forecasting Ocean Assimilation Models (FOAM).

Echosounder

In addition to the standard science payload, glider *U194* was equipped with an Imagenex ES853: a self-contained, single-beam 120 kHz echosounder (Fig. 2A). The ES853 was mounted in a science bay, aligned centrally along the short axis of the glider and at an angle of 64° from the long axis (towards the nose). This ensured that when the glider was in a downward glide (typical dive angle of −26°), the transducer pointed directly downwards, analogous with the downward-looking orientation of a ship's echosounder. The echosounder operates with a pulse length of 100 µs, beam angle of 10°, range of 100 m, configurable gain of either 20 or 40 dB, and measures mean volume backscattering (*Sv*, dB re 1 m^{-1}) per range bin interval of 0.5 m. It requires power of 0.25 W, drawn from a 24 V DC supply, and can output values to a PC or record data to internal storage. The ES853 has a dynamic range of 120 dB and records signals as integer values, thus the resolution in signal strength is reduced compared with typical floating-point recording of larger, ship-based echosounders.

The glider-integrated ES853 was controlled via serial port and polled at a frequency of 0.25 Hz, with data stored on the glider's internal memory. Upon retrieval of the glider the data were downloaded and processed following Guihen et al. (2014), where raw echo-intensity data were converted to mean volume backscattering strength (*Sv*), using the active version of the SONAR equation (Urick, 1983) and calibration and manufacturers' constants for echosounder receiving response and source level. Glider position data are used to locate individual pings in time, depth and aspect. These data were then processed using Myriax Echoview software (version 4.80), including subtraction of time-varied gain amplified background noise (after Watkins and Brierley, 1996), and accounting for variability in aspect of the transducer (after Dunford, 2005) and depth of the echosounder. Due to operational difficulties, the ES853 was not calibrated using standard sphere methods (Foote et al., 1987), so data shown here are relative. However, Guihen et al. (2014) showed that a calibrated ES853, mounted on a glider, provided quantitative estimates of zooplankton distribution comparable to a ship-borne echosounder, providing a suitable sampling strategy is employed. As a result we are confident that the ES853 data provide a relative index of zooplankton distribution.

Passive acoustic monitoring

Glider *Zephyr* was equipped with a Passive Acoustic Monitoring (PAM) system based on a modified d-tag sensor (Johnson and Tyack, 2003). The d-tag was mounted in the glider's aft wet space close to the buoyancy bladder, with the hydrophone on top of the glider just in front of the rudder (Fig. 2B). Power was taken from the glider's main batteries. The sensor was configured to sample at 480 kHz, and an automatic detector was implemented to detect the narrow-band high-frequency clicks of harbour porpoise *Phocoena phocoena* (Villadsgaard et al., 2007). A 2 ms waveform clip of detected clicks was stored to flash memory. Recordings were also made of acoustic data decimated to a sample rate of 96 kHz for offline detection of other cetacean species and the measurement of noise. Although recordings were made using a lossless compression format which typically gives a compression ratio of 4:1 (Johnson et al., 2013), continuous recording would have filled available storage in approximately two weeks. The recorder was therefore programmed to operate for 10 out of every 40 s and only at water depths >40 m.

Recovered data were processed offline with PAMGuard software (Gillespie et al., 2008) to detect cetacean clicks and whistles and to measure noise in third octave bands between 22 Hz and 45 kHz. An operator (KG) viewed all detections manually, viewing click waveforms and whistle time frequency contours, and listening to sections of data to confirm detections and classify groups of transient sounds as either 'cetacean clicks' or as 'pump noise'.

Passive sampling devices for trace organic contaminant monitoring

Passive sampling devices made of a thin layer of polymer were deployed on *Zephyr* (Fig. 2B). These polymer sheets are made of AlteSil®silicone rubber (24 cm × 28 cm × 0.5 mm thick), and were Soxhlet extracted with ethyl acetate before further soaking in methanol. Performance reference compounds (PRC; perdeuterated polycyclic aromatic hydrocarbons and fluorinated polychlorinated biphenyls) were then uniformly spiked into the batch of samplers according to a method similar to that described by Booij et al. (2002). Samplers were kept in closed containers at −20 °C until deployment. Control samplers were used to assess possible contamination during transport, deployment and retrieval operations, and to measure initial PRC concentrations. Two sampling sheets were placed on top of the glider and fastened using cable ties (Fig. 2B). A thin sheet of aluminium foil (muffle furnaced prior to exposure) was placed between the samplers and the glider body to minimize possible contamination of the samplers by direct diffusion from the glider hull itself.

After retrieval, the surface of the samplers was thoroughly cleaned in the laboratory to remove any bio-fouling before static batch extraction with pentane (three times 300 mL over 72 h). Extracts were combined and reduced in volume. The solvent was changed to dichloromethane before clean-up by gel permeation chromatography. The extracts were then reduced in volume and analysed by gas chromatography–mass spectrometry for polycyclic aromatic hydrocarbons (PAHs), polychlorinated biphenyls (PCBs) and other chlorinated organic compounds. Field and laboratory procedures have been described elsewhere in more detail (e.g. Allan et al., 2010 and Allan et al., 2013).

Glider mission

Both gliders were deployed southwest of the Isles of Scilly from the RV *CEFAS Endeavour* on 12 September 2013, in order to conduct repeated transects over the targeted tidal-mixing front (Fig. 1, 'transect 1'). The approximate location of the front was monitored prior to the survey via satellite-derived front maps (Miller, 2009). A ship-based pre-deployment transect was performed in order to ensure the area was free of static fishing gear (which is common in this area and represents a potential hazard to any glider) and to take repeated CTD and water samples to affirm the transect was located over the target front. A ship-based transect was performed on 21 October 2013 along the glider survey line (Fig. 1, 'transect 2'), in order to (1) compare the ES853 echosounder data with those collected using the calibrated multi-frequency (38, 120 and 200 kHz) split-beam Simrad EK60 on the RV *CEFAS Endeavour*, and (2) to collect CTD and water samples for *U194* calibration purposes. Due to glider control problems encountered during the mission, the initial glider transect (Fig. 1, 'transect 1') was shifted to the north of the Isles of Scilly on 24 September 2013 (Fig. 1, 'transect 2'). *Zephyr* was recovered from the RV *CEFAS Endeavour* on 21 October 2013 and on this occasion, a CTD cast was conducted for glider-sensor calibration. Due to quickly deteriorating weather, retrieval of *U194* had to be abandoned and the glider was picked up via fast-boat three weeks later on 13 November 2013.

RESULTS

Mission summary

Strong tidal flows and northwards density currents deflected both gliders to the northwest of the Isles of Scilly soon after deployment on 12 September 2013 (Hill et al., 2008). Unable to counteract these currents, *U194* was pushed into very shallow waters West of the islands on 17 September 2013, where manoeuvring was impossible and the glider needed to be recovered. Emergency recovery and subsequent repairs were carried out between 20 September and 04 October 2013 (Table 1).

U194 was redeployed on 04 October 2013 via fast-boat northwest of the Isles of Scilly. This also provided an opportunity to update faulty software on the *Zephyr* d-tag; both gliders then resumed their mission. The planned glider transect was repositioned to the north of the western TSZ (Fig. 1), where tidal flows and currents were weaker and where *Zephyr* had been located for most of the previous three weeks. Using a more dynamic piloting approach (e.g. regularly adjusting waypoints according to tidal flows, changing current correction setting), control of the flight path of *U194* was much improved. In contrast, piloting of *Zephyr* became increasingly challenging following redeployment, possibly due to the hydrocarbon sensor sheets loosening and adding extra drag to the vehicle (making it more susceptible to current drift). Nevertheless, both gliders executed the mission without any further problems until retrieval. A summary of key events and a timeline of sensor activity is given in Table 1 and Fig. 3, respectively.

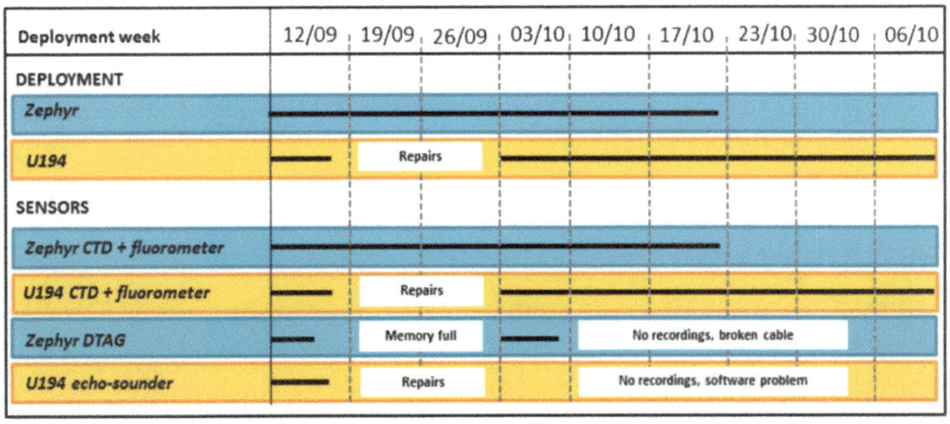

Figure 3: Timeline glider deployments (blue shading: *Zephyr*; yellow: *U194*) and sensor activity, represented by black lines. Dates indicate the beginning of each week of the deployment, white boxes provide information on sensor malfunctions. (For interpretation of the references to colour in this figure legend, the reader is referred to the web version of this article.)

Over the six-week deployment, the two gliders performed a total of 5474 dives (each dive comprising one up and downcast, based on dives of ≥10 m) to a maximum depth of 101 m (*Zephyr*) and 104 m (*U194*) (Table 2). The total horizontal distance covered was 2389 km, with an average distance of ~0.9 km between GPS fixes.

Table 2: Mission statistics. Statistics only considers dives of ≥10 m (to exclude non-mission dives) and GPS fixes of ≥100 m apart. *U194* statistics based on post-redeployment dives only

Statistic	*Zephyr*	*U194*
Total deployment days	38	39
Total number of dives	2654	2821
Total horizontal distance (km)	1080	1309
Mean horizontal distance b/w GPS fixes (km)	0.9	0.85
Max distance b/w GPS fixes (km)	13.7	10.3
Mean dive depth (m)	44.55	40
Max dive depth (m)	101.49	103.89

Physical environment

Fig. 4 shows the glider data is in good agreement with measurements obtained from ship-based CTD-casts. However, *U194* chlorophyll a displayed a significant offset (Fig. A.2) and data shown here are corrected based on the vessel-profile by applying an ordinary linear model and adjusting the *U194* chlorophyll a data by the slope and offset ($U194_{corrected}=U194_{raw}\times1.33+0.084$). Although there is a slight offset between Zephyr and ship chlorophyll a, no correction was applied, because the water samples suggest that the glider fluorometer gives a better reflection of the true chlorophyll than the vessel data. No correction of temperature measurements was necessary.

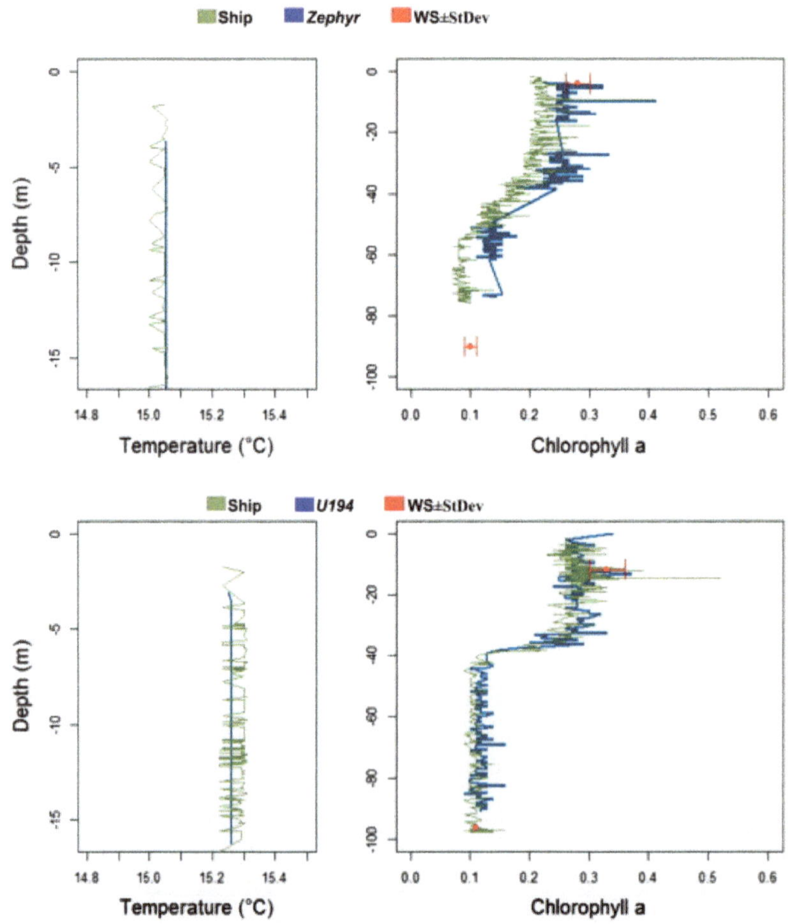

Figure 4: Comparison between ship and glider derived temperature and chlorophyll for *Zephyr* (upper panel) and *U194* (lower panel). Green: Ship; blue: glider; red points show chlorophyll a from water samples ± standard deviation (N=3). (For interpretation of the references to colour in this figure legend, the reader is referred to the web version of this article.)

Glider *Zephyr* predominantly sampled thermally stratified waters during its mission, although it crossed the targeted tidal-mixing front several times in the first 500 h (Fig. 5A, Fig. A.3). A strong thermocline at 20–40 m depth is particularly evident after ~500 h(Fig. 5A). Elevated levels of chlorophyll a fluorescence are visible in the water column towards the base of the thermocline, particularly around 300 h (Fig. 5B). Fluorescence increases notably where the thermocline crops towards the surface in the frontal region, e.g. during a frontal crossing at ~200 h (Fig. 5B). High levels of chlorophyll afluorescence remain visible even after the equinox (21 September 2013; ~200 h), which is usually taken as the end of the primary production period elsewhere on the UK shelf (Weston et al., 2005).

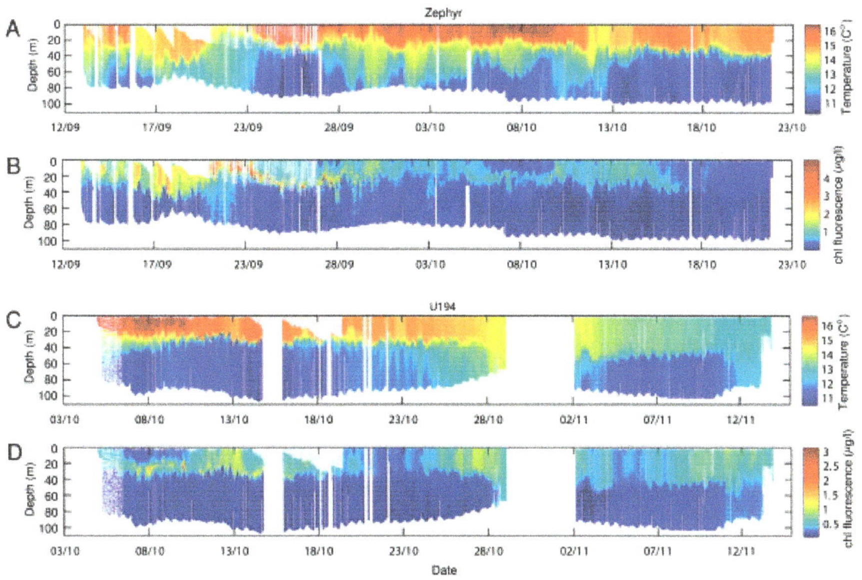

Figure 5: Time series plots of temperature and chlorophyll concentration of both gliders against depth (panels A–B, *Zephyr*; panels C–D, *U194*). Time refers to hours since the mission start on 12 September 2013 at 1200 h. Note shifted time axis for *U194*, which only considers data since re-deployment from 04 October 2013.

Glider *U194* also predominantly sampled thermally stratified waters throughout its mission, but spent less time in frontal regions (Figs. 1 and 5C). Subsurface chlorophyll maxima at ~30 m depth, associated with the thermocline, are clearly visible in *U194* profiles until ~800 h into the mission (Fig. 5D). The deployment of *U194* was substantially longer than that of *Zephyr*(Fig. 3), and included a major storm event on 27 October 2013 (1080 h into the mission, Fig. 5C). Interestingly, the water mass remained stratified (although much weaker) after this storm event, although surface waters had cooled significantly and the thermocline depth increased to ~50 m (Fig. 5C).

An interesting feature visible on the *Zephyr* temperature profile is the lack of an obvious relationship between bottom temperature gradients and surface gradients, particularly around 400-600 h (Fig. 5A). Consistently high surface temperatures were experienced throughout this period (>15 °C), whereas the bottom temperature alternated between ~11–13 °C. In general, the transition zone along the bottom appears wider than at the surface (Fig. 5A, Fig. A.3). In addition, clear bottom front crossings occurred on nine occasions, whereas the surface front was intersected six times only (Fig. 5A, Fig. A.3).

Echosounder

The ES853 120 kHz echosounder, integrated onto *U194* (Fig. 2A), successfully recorded acoustic data from the water column during the first five-day deployment (Fig. 3 and Fig. 6). Unfortunately, after redeployment a technical error prevented communication between the echosounder and the glider and, as a consequence, acoustic data for the rest of the mission were not stored on the internal memory of glider and echosounder. Focusing on the downcast, when the echosounder is vertically orientated and most comparable to a ship-based echosounder, typical features observed in echograms include the seabed and fish shoals (Fig. 6A). Similar acoustic features were observed in an echogram derived from the ship-based echosounder in the same area at the same time of year (Fig. 6B).

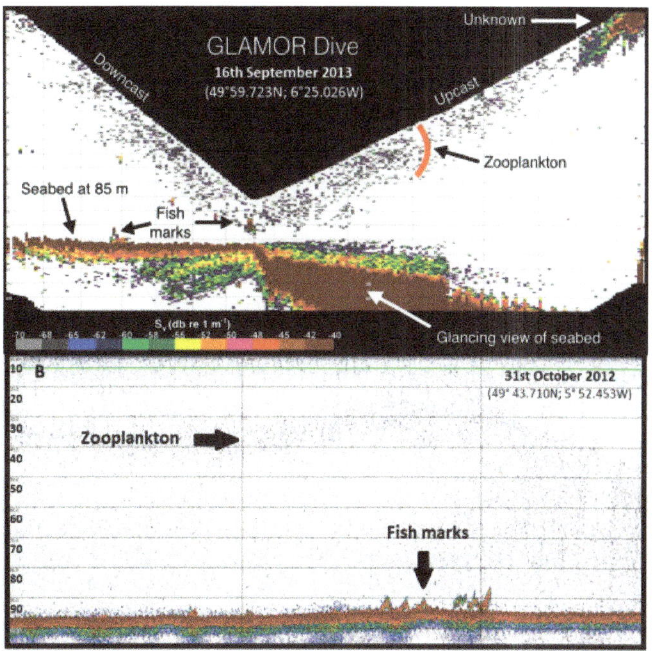

Figure 6: Echograms of glider-integrated ES853 at 120 kHz (A) and equivalent vessel based EK60 120 kHz (B). Distance between the two sites 43.3 km. Features such as the seabed, fish shoals and zooplankton scattering have been indicated. Gain thresholds for both echograms were set to −70 dB.

A band of scattered targets, consistently present within 20–25 m of the glider echosounder during both down- and upcasts (Fig. 6A), was interpreted to represent small organisms such as zooplankton. The vessel-based echogram at 120 kHz revealed similar scattered targets throughout the water column, with some evidence for elevated concentrations at ~30–40 m depth associated with the thermocline (Fig. 6B). Due to the reduced signal-to-noise ratio of the glider echosounder, these small targets were only observed within a limited range of the transducer. In contrast, the stronger water column targets related to fish shoals were recorded throughout the water column on glider data (Fig. 6A). These targets appeared similar to those recorded from RV *CEFAS Endeavour* during its annual pelagic fish survey in the area (Fig. 6B), and were thought to consist of boarfish (*Capros aper*). Due to differing ping rates (4 vs 0.25 s^{-1}) and vehicle speeds (5 ms^{-1} vs. 0.3 ms^{-1}), the horizontal resolution of acoustic data obtained from the glider was nearly twice as high compared to the vessel (1.2 m vs. 2.5 m per ping), which suggests that the fish shoals imaged on the glider echogram are possibly small examples of these boarfish shoals.

A noticeable feature on glider acoustic profiles is the presence of high backscatter patches in the uppermost water column, observed down to ~20 m water depth during upcasts (Fig. 6A). Adverse weather conditions, with large waves or swell, are known to introduce bubbles into surface waters; however, weather conditions were favourable during the period that these features were observed so this cause is doubtful. Technical issues, such as side-lobe detection of the sea surface, are also thought to be unlikely as these patches were not always present, absent during previous trial deployments, and no changes had been applied to the echosounder settings. Vessel-derived echograms did not show the same features, although the transducer depth of the vessel-based echosounders was 8.2 m below the surface and so did not cover the surface layer. It is therefore possible that these high backscatter patches have a biological origin, representing organisms restricted to surface layers. However, their unusually high backscatter, widespread occurrence, and consistent appearance throughout the deployment, also make this an unlikely source. Therefore, further investigation is required to determine the true origin.

Passive acoustic monitoring

The power cable to the d-tag acoustic sensor was damaged during or soon after installation, so PAM data were only collected between 04 and 10 October (Fig. 3). These data included 2413 10s recordings over 291 separate dive cycles, with an average of eight recordings per dive. Harbour porpoise clicks were detected during two separate dives, although one of these consisted of only a single click. Dolphin clicks and whistles were detected during 194 dives: 145 of these were clicks only, 49 whistles only, and 42 both (for map of locations of recordings see Fig. A.4). Fig. 7 shows the waveforms and power spectra of a typical dolphin and porpoise click. No other obvious noise sources, such as the pitch battery motor were apparent in the data.

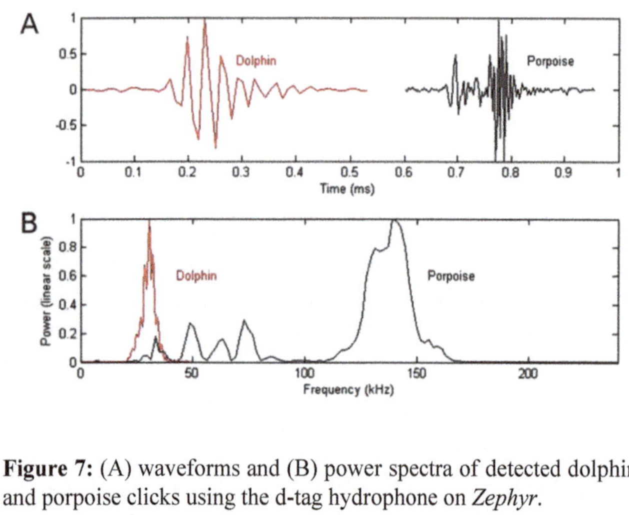

Figure 7: (A) waveforms and (B) power spectra of detected dolphin and porpoise clicks using the d-tag hydrophone on *Zephyr*.

Noise attributable to the glider pump was recorded on 169 occasions. The times of these noises corresponded to the times when the pump was 'on' in the glider log files, typically occurring for only a few seconds at the bottom of each dive. Fig. 8 shows the distribution of noise in third octave bands with the pump on and the pump off. Median noise levels with the pump off are generally low indicating that the system can be used to make accurate measurements of ambient noise. Noise levels at all frequencies are considerably higher when the pump is in operation. However since the duration of these noisy periods is relatively short, it will have little effect on overall survey effort.

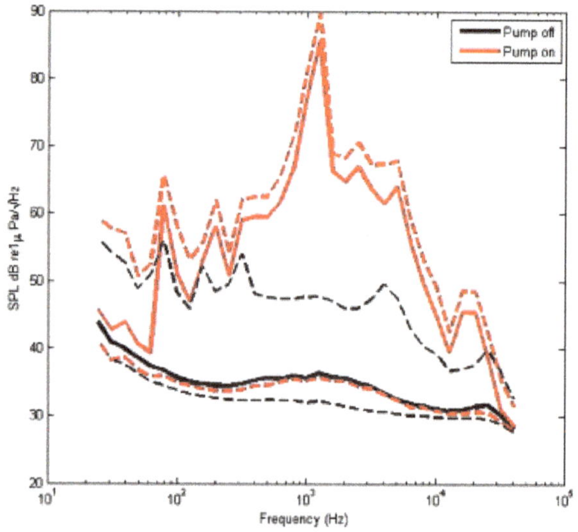

Figure 8: Spectrum-level noise measurements at third octave intervals for times when the Zephyr pump was on and when the pump was off. The three lines for each situation represent the median values and the lower and upper 90% intervals.

Passive sampling devices

Recovered silicone rubber sample sheets were visually inspected for bio-fouling before analysis (as this can affect the uptake of contaminants into passive samplers); bio-fouling levels were found to be very low compared to static exposures. Significant dissipation of performance reference compounds (PRC) was observed from sample sheets. PRC dissipation was used to estimate sampling rates using a published procedure (Booij and Smedes, 2010). These rates were in the range of 3–7 L d^{-1} depending on the chemical. Sampling rates were normalized to a standard surface area (Huckins et al., 1993) and compared with those observed during static or other mobile deployments of passive samplers e.g. Booij et al., 2007, Allan and Harman, 2011 and Lohmann et al., 2012); Table 3). Sampling rates for the glider-mounted passive samplers used here are in a similar range to those commonly observed for static deployments, whereas mobile exposures with high velocities tend to achieve high sampling rates (Table 3). The PRC data from this passive sampler exposure (i.e. 50% dissipation for d_{10}-phenanthrene) indicate that sampling was time-integrative for substances with octanol–water partition coefficient (log K_{ow}) above 4.5. Since gliders are launched for periods of weeks to months, these exposures can still detect polycyclic aromatic hydrocarbons (PAH) and polychlorinated biphenyls (PCB) in the low pg L^{-1} range, i.e. levels at which these compounds typically occur in oceanic waters.

Table 3: Passive sampling rates, typical exposure times, and limits of detection for various passive sampler deployment modes

Deployment mode	Sampling *Rs* (L d^{-1})[a]	Exposure time	Limits of detection (pg L^{-1})[b]	Refs.
Static	1–20	0.5–3 months	0.3–30	(Vrana and Schuurmann, 2002, Allan et al., 2010, Prokes et al., 2012 and Allan et al., 2013)
Mobile[c]	18–27	4–6 days	3.0–7.0	(Booij et al., 2007)
Mobile[d]	70–200	5 h and 48 h	1.0–30	(Allan and Harman, 2011, Allan et al., 2011 and Lohmann et al., 2012)
Calibration[e]	60–200	15 days	0.2–0.6	(Booij et al., 2003)
Glider[f]	3–7[g]	39 days	2.0–4.0	Present study

[a]For a standard semipermeable membrane device sampling surface area of 460 cm^2.

[b]Limits of detection in water for PAHs/PCBs in the linear phase of uptake ($C_{w,lim} = m_{lim}/[R_s t]$) with a arbitrarily set m_{lim} of 0.5 ng per sample.

[c]Ship-based measurement using the ship's continuous water supply (water velocity in the pipe of 15 cm s^{-1}).

[d]Samplers towed behind a benthic trawl net (1.2–1.4 knot); towed behind a research vessel.

[e]During sampler calibration with water velocity of 90 cm s^{-1} and water temperature of 30 °C.

[f]Average glider velocity through water of 20–40 cm s^{-1} (horizontal velocity).

[g]Sampling rates corrected to a surface area of 460 cm^2.

Freely dissolved concentrations of PAHs ranged from just over 1 ng L^{-1} for phenanthrene down to below 10 pg L^{-1} for higher molecular weight PAHs. Polychlorinated biphenyls (PCBs) were found at levels below 10 pg L^{-1} (except for PCB congener 28 with a concentration of 32 pg L^{-1}). The freely dissolved concentration of hexachlorobenzene was 45 pg L^{-1}. These concentrations were not corrected for temperature or salinity, but are generally in the range of those measured in open waters of the North Atlantic.

DISCUSSION

In this 'proof-of-concept' study, two submarine gliders were equipped with sensors capable of simultaneously measuring physical properties of the water column and multiple trophic levels, in order to test their potential for ecosystem monitoring. The results highlight the advantages and current limitations of utilizing gliders as autonomous platforms for the outlined purpose. The key outcomes relate to operational aspects, current sensor technology, instrument calibration and data validation.

Glider operations

The two gliders successfully completed their missions, which lasted for just under 40 days and which each comprised >2500 dives to depths of up to 100 m over total horizontal travel distances of 1000–1300 km. The gliders continued to collect data in weather conditions that stopped research vessel activity in the region, including a particularly severe storm on 27 October 2013 (Fig. 5C).

The two major operational challenges were flight control and alignment of the two glider trajectories in space and time. Although previous deployments have demonstrated that gliders are capable of following a proposed transect in areas of strong tidal flow (e.g. Leonard et al., 2007), this proved extremely challenging in the present study (Fig. 1). This is likely due to the combination of strong tidal and non-tidal residuals acting on the vehicles together. In order to fully understand the effect of hydrodynamics on our gliders, a detailed analysis is required including currents, tides and meteorological forcing in relation to the glider tracks and flight settings, which is beyond the scope of this paper.

The flight of *U194* was significantly improved by applying a more dynamic piloting approach, including e.g. increased monitoring, modifying dive angle and current correction settings, and adjusting waypoints depending on currents and tidal state. Hybrid *Slocums* equipped with a thruster are also now commercially available (Jones, 2012), which should further aid future deployments in tidally dominated environments. Furthermore, ongoing modelling and simulation research, dealing with optimal path planning and influences on glider trajectories, can be used to aid the survey design (e.g. Fernandez-Perdomo and Hernandez-Sosa et al., 2011 and Ting and Mujeebu et al., 2012).

Aligning the flight of multiple gliders is a key requirement in whole-ecosystem monitoring, as the sensors required to simultaneously measure multiple ecosystem components cannot currently be integrated onto a single vehicle due to limited payload capacity and energy constraints. In addition, certain instruments need to go onto different platforms due to acoustic interference problems (e.g. active echosounder and passive hydrophone). In order to obtain meaningful data on multiple parameters simultaneously, the vehicles should be aligned in time and space; ideally the maximum distance between the vehicles should not exceed the scale on which the variables change. This is a challenging task, because gliders equipped with different sensors will display specific flight behaviours, e.g. due to different ballasting and external sensor configurations. For example, the hydrocarbon sheets used on *Zephyr* in this study are likely to have added extra drag to the vehicle resulting in different flight behaviour compared to *U194* (*cf.*Fig. 1). If time and budget allow, extended trials focusing on simultaneous flight prior to a mission could improve piloting for individual gliders and, therefore, trajectory alignment.

Sensors

Throughout the mission, CTD and fluorometer sensors on both gliders provided water-column data at a spatio-temporal resolution and frequency not attainable through ship-based surveys or satellite imagery. For example, *Zephyr* crossed the Isles of Scilly tidal-mixing front several times (Fig. 5A), with collected data highlighting the spatial offset of the surface and bottom fronts (Fig. 5) as well as chlorophyll maxima associated with the thermocline (Fig. 5B); these features would not be detected using remote sensing data. Glider *U194* dominantly sampled stratified waters, and monitored change in the water column from seasonally stratified to mixed over a period of six weeks (Fig. 5C); within this period the glider continued collecting data during the severe storm event on 27 October 2013. Despite the high resolution and amount of information collected, file sizes were small enough for a subset of data to be transmitted via Iridium during the mission.

The newly integrated ES853 echosounder and d-tag passive acoustic monitoring system provided promising data on the spatial distribution of higher trophic-level organisms. The hydrophone recorded numerous cetacean clicks and whistles from different species during the six days of operation (Fig. 7 and Fig. A.4). The echosounder was capable of detecting targets, including fish and zooplankton, similar to targets detected by vessel-based data echosounders (Fig. 6).

The novel hydrocarbon sheets trialed here can easily be attached to gliders and AUVs, and collect supplementary information on water quality. This has particular application for regulatory monitoring, e.g. in response to European legislation such as the Water Framework Directive and the Marine Strategy Framework Directive. The sampling scheme can be extended from hydrophobic compounds (as used in this study) to samplers specifically designed to sample hydrophilic substances and metals.

Limitations of glider-integrated instrumentation include the need for data calibration and validation, particularly over long-term surveys. CTD and fluorometer calibration procedures are generally undertaken during glider deployment and retrieval using calibrated ship-based CTDs and by taking water samples. However, depending on survey area and season, a considerable amount of bio-fouling can affect the glider-integrated instruments; this causes sensor drift in a non-linear fashion, which is difficult to account for with calibration procedures at the beginning and end of a mission. In addition, collection of water samples remains essential for phytoplankton biomass estimations or species identification.

Although the echosounder and PAM system are capable of collecting data on the distribution of higher trophic level organisms over large spatio-temporal scales, they are significantly harder to calibrate compared with the physical sensors. Single-beam single-frequency echosounders provide a limited capability for target identification, which are normally undertaken using multi-frequency discrimination methods (e.g. Korneliussen and Ona, 2003). However, these instruments are currently too large and powerful to be integrated into underwater gliders. To establish fish species with absolute certainty, verification via net sampling also remains essential. Size–length relationships needed for target strength model parameterization for conversion to biomass and information on age-class distribution are additional parameters that can only be acquired with dedicated net sampling from vessel-based surveys. There has been some success at classifying cetacean whistles collected by PAM systems to species level (Gillespie et al., 2013), although considerable problems remain in estimating absolute animal abundance from glider-based data. Critical data for the estimation of abundance are the range at which animals are detected and either the rate of vocalization or the ability to isolate and count individuals (Marques et al., 2013). Gliders are too slow moving to use target motion analysis to localize animals in the water columns, as is commonly done for sperm whales (e.g. Lewis et al., 2007) and vocalization rate is poorly known for most species. Finally, echosounder and PAM system development should focus on methods for summarizing the complex data to enable transmission of a subset of collected data, rather than after retrieval. This would enable the glider to target, on the fly, hotspots of animals.

CONCLUSIONS AND RECOMMENDATIONS

This proof-of-concept study has highlighted some of the advantages and current limitations of utilizing submarine gliders for the purpose of ecosystem monitoring. Advantages include cost efficiency, capability of working in adverse weather, and collection of 3D water-column data at high spatio-temporal resolution over periods of weeks to months. This is not achievable with other single platforms, but is essential in order to detect change and its effect on an ecosystem, e.g. breakdown of seasonal stratification, impact of short-duration storm events. All of the novel glider-integrated sensors used in this study delivered useable data, although they could only provide information on distribution of biological indicators rather than accurate estimates of abundance or biomass. More broadly, glider sensor calibration and data validation remain challenging and dependent upon supporting infrastructure (ships, moorings), so acquired data are often only suitable for qualitative analysis. In addition, for the purpose of multi-vehicle surveys, aligning glider trajectories and flight control are significant issues that require further improvement (especially in dynamic tidally-dominated environments). Nevertheless, the promising results achieved in this study have led to a further deployment targeting oceanic fronts off southwest UK (planned for autumn 2014) using unmanned surface vehicles (USVs) as an in addition to submarine gliders.

For effective whole-ecosystem monitoring in the future, a range of autonomous platforms will need to be deployed within a marine monitoring network, including USVs, AUVs and gliders. Successful implementation of an autonomous monitoring network has already been demonstrated in the California Current Ecosystem, where satellite imagery, vessel surveys, gliders, floats and moorings are used in combination to provide compatible data on ecosystem dynamics at various spatio-temporal scales (Ohman et al., 2013). Satellite remote sensing provides information on large-scale surface processes, which is supplemented by submarine gliders measuring water column properties. Moorings and targeted vessel surveys utilize more powerful sensors working at high frequency, and are used to calibrate satellite and glider data; vessels are also used to run structured experimental surveys, which require ship-based equipment. Together, the sampling network efficiently provides a synoptic view of an ecosystem at multiple scales, which can significantly advance our understanding of marine ecosystem functioning and drivers of change.

ACKNOWLEDGEMENTS

This project would not have been possible without the hard work and professionalism of the Marine Autonomous and Robotic Systems (MARS) staff at NOC, particularly David White, James Burris and Sam Ward. We also thank to the ship staff of the RV *CEFAS Endeavour*and the Scilly Inshore Fisheries and Conservation Authority for their support in glider deployment, recovery and calibration activities.

The Gliders and AUVs for Marine Observation and Research (GLAMOR) project was primarily funded by UK Department for the Environment, Food and Rural Affairs (Defra), through their Strategic Evidence and Partnerships Fund. Damien Guihen was funded by Natural Environment Research Council (NERC) grant NE/H014756/1. Additional financial support came through Defra project MF1112 (POSEIDON). Ian Allan (NIVA) and Branislav Vrana (RECETOX) acknowledge funding from the European Union Seventh Framework Programme (FP7/2007–2013) under grant agreement no. 262584, JERICO (TNA project GLISS). Passive sampler analyses were also carried out through the JERICO project (Work Package 10). We thank James Bowcott (PML) for providing the satellite front bulletins throughout the cruise.

The authors thank the two anonymous reviewers and guest editor Gwyn Griffiths for their detailed and helpful comments, which substantially improved the manuscript.

APPENDIX.

See Fig. A.1, Fig. A.2, Fig. A.3 and Fig. A.4.

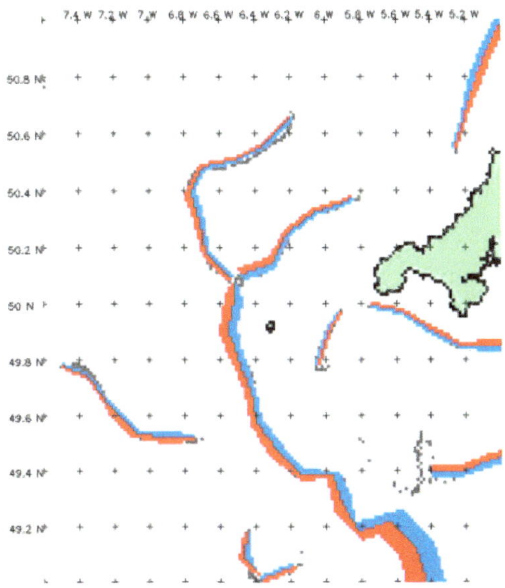

Figure A.1: Composite satellite-derived front map of the survey area off southwest UK, based on data from 12 September–17 October 2013. Red indicates warm-water (stratified) side of tidal-mixing fronts, blue the cold-water (mixed) side. Line thickness equates to the strength of the front (the thicker the line, the stronger the front). Note the strong tidal-mixing front west of the Isles of Scilly, separating stratified waters to the west from tidally mixed waters to the east (*cf.* Fig. 1). (For interpretation of the references to colour in this figure legend, the reader is referred to the web version of this article.)

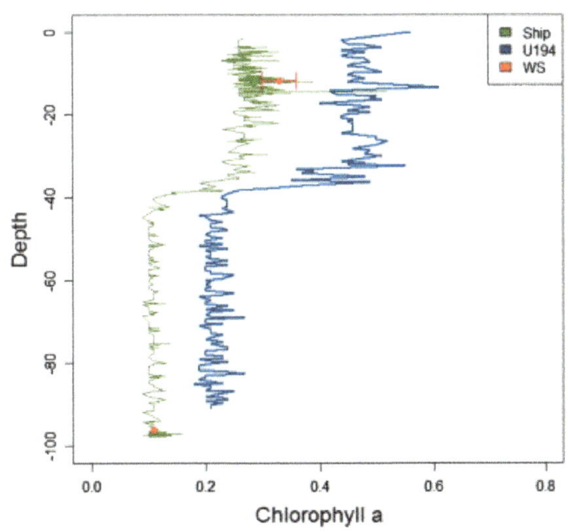

Figure A.2: Calibration profiles between ship (green) and U194 (blue), showing uncorrected chlorophyll *a* data. Red points show values from water samples ± standard deviation (N=3). (For interpretation of the references to colour in this figure legend, the reader is referred to the web version of this article.)

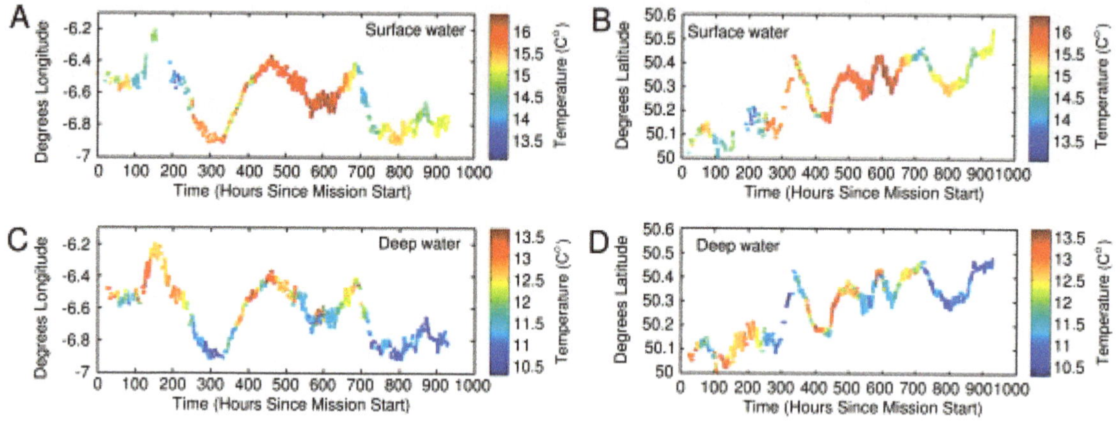

Figure A.3: Temperature of surface waters (A and B, between 5–10 m depth) and deeper waters (C and D, between 60–70 m depth) as a function of time and longitude (A and C) and latitude (B and D).

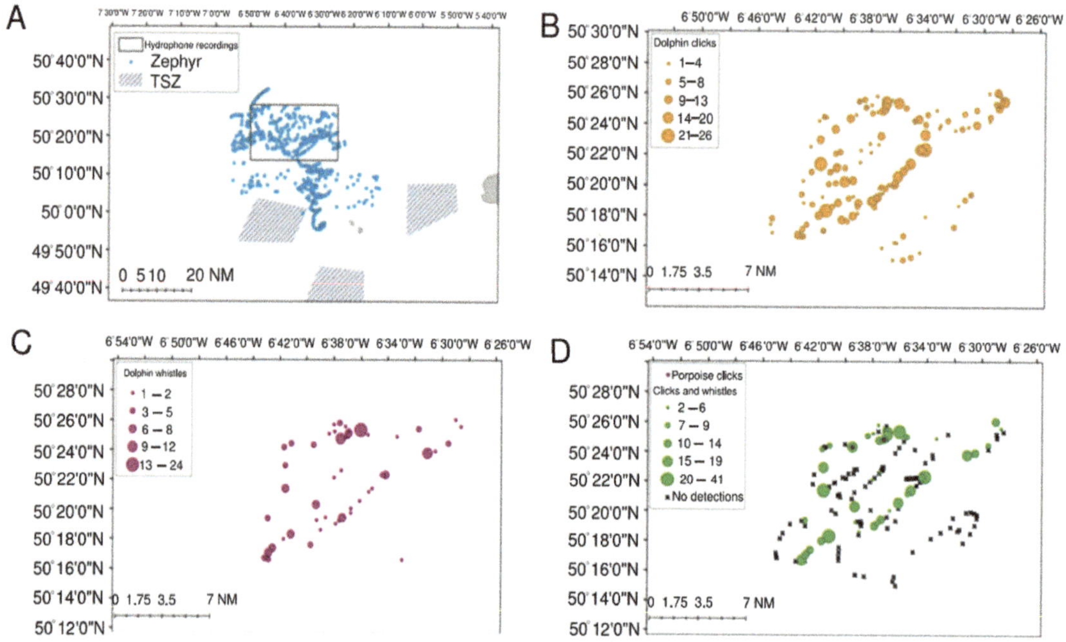

Figure A.4: Maps showing the area where the d-tag hydrophone on *Zephyr* was active (A) and recorded dolphin clicks (B), whistles (C), and dives with both and porpoise clicks (D). Black crosses refer to locations of dives without recordings.

REFERENCES

1. Allan, I.J., Harman, C., 2011. Global aquatic passive sampling: maximizing available resources using a novel exposure procedure. Environ. Sci. Technol. 45 (15), 6233–6234.
2. Allan, I.J., Harman, C., et al., 2010. Effect of sampler material on the uptake of PAHs into passive sampling devices. Chemosphere 79 (4), 470–475.
3. Allan, I.J., Harman, C., et al., 2013. Passive sampling for target and nontarget analyses of moderately polar and nonpolar substances in water. Environ. Toxicol. Chem. 32 (8), 1718–1726.
4. Allan, I.J., Nilsson, H.C., et al., 2011. Mobile passive samplers: Concept for a novel mode of exposure. Environ. Pollut. 159 (10), 2393–2397.
5. Alvarez, A., Mourre, B., 2012. Oceanographic Field Estimates from Remote Sensing and Glider Fleets. J. Atmos. Ocean. Technol. 29 (11), 1657–1662.
6. Alvarez, A., Chiggiato, J., et al., 2013. Mapping subsurface geostrophic currents from altimetry and a fleet of gliders. Deep-Sea Res. Oceanogr. Res. Papers 74, 115–129.
7. Baumgartner, M.F., Fratantoni, D.M., 2008. Diel periodicity in both sei whale vocalization rates and the vertical migration of their copepod prey observed from ocean gliders. Limnol. Oceanogr. 53 (5), 2197–2209.
8. Baumgartner, M.F., Fratantoni, D.M., et al., 2013. Real-time reporting of baleen whale passive acoustic detections from ocean gliders. J. Acoust. Soc. Am. 134 (3), 1814–1823.
9. Beaumont, N.J., Austen, M.C., et al., 2007. Identification, definition and quantification of goods and services provided by marine biodiversity: Implications for the ecosystem approach. Mar. Pollut. Bull. 54 (3), 253–265.
10. Booij, K., Hofmans, H.E., et al., 2003. Temperature-dependent uptake rates of nonpolar organic compounds by semipermeable membrane devices and low-density

polyethylene membranes. Environ. Sci. Technol. 37 (2), 361–366.

11. Booij, K., Smedes, F., 2010. An Improved Method for Estimating in Situ Sampling Rates of Nonpolar Passive Samplers. Environ. Sci. Technol. 44 (17), 6789–6794.
12. Booij, K., Smedes, F., et al., 2002. Spiking of performance reference compounds in low density polyethylene and silicone passive water samplers. Chemosphere 46 (8), 1157–1161.
13. Booij, K., van Bommel, R., et al., 2007. Air-water distribution of hexachlorobenzene and 4,4'-DDE along a North-South Atlantic transect. Mar. Pollut. Bull. 54 (6), 814–819.
14. Bouffard, J., Pascual, A., et al., 2010. Coastal and mesoscale dynamics characterization using altimetry and gliders: A case study in the Balearic Sea. J. Geophys. Res.-Oceans 115.
15. Bouffard, J., Renault, L., et al., 2012. Sub-surface small-scale eddy dynamics from multi-sensor observations and modeling. Prog. Oceanogr. 106, 62–79.
16. Brennan, J., Fitzsimmons, C., et al., 2013. EU marine strategy framework directive (MSFD) and marine spatial planning (MSP): Which is the more dominant and practicable contributor to maritime policy in the UK? Marine Policy 43, 359–366.
17. Collie, J.S., Adamowicz, W.L., et al., 2013. Marine spatial planning in practice. Estuar. Coastal Shelf Sci. 117, 1–11.
18. Davis, R.E., Leonard, N.E., et al., 2009. Routing strategies for underwater gliders. Deep-Sea Res. Topical Stud. Oceanogr. 56 (3–5), 173–187.
19. Davis, R.E., Ohman, M.D., et al., 2008. Glider surveillance of physics and biology in the southern California Current System. Limnol. Oceanogr. 53 (5), 2151–2168.
20. Day, J., 2008. The need and practice of monitoring, evaluating and adapting marine planning and management - lessons from the Great Barrier Reef. Marine Policy 32 (5), 823–831.
21. Defra,, (2013). The Isles of Scilly Marine Conservation Zones Designation Order 2013. Ministerial Order. No 10, London: Department of Environment, Food and Rural Affairs.
22. Douvere, F., Ehler, C.N., 2011. The importance of monitoring and evaluation in adaptive maritime spatial planning. J. Coastal Conserv. 15 (2), 305–311.
23. Dunford, A.J., 2005. Correcting echo-integration data for transducer motion (L). J. Acoust. Soc. Am. 118 (4), 2121–2123. English, D., Chuanmin, H., et al., 2009. Observing the 3-dimensional distribution of bio-optical properties of West Florida Shelf waters using gliders and autonomous platforms. Oceans 7.
24. European Union, 2008. Directive 2008/56/EC of the European Parliament and the Council of 17 June 2008. Establishing a framework for community action in the field of marine environmental policy (Marine Strategy Framework Directive). Official J. Eur. Union L164, 19–40.
25. Ferguson, B.G., Lo, K.W., et al., 2010. Sensing the underwater acoustic environment with a single hydrophone onboard an undersea glider. Oceans.
26. Fernandez-Perdomo, E., Hernandez-Sosa, D., et al., 2011. Single and multiple glider path planning using an optimization-based approach. Oceans.
27. Foote, K., Knudsen, H., et al., 1987. Calibration of acoustic instruments for fish density estimation: a practical guide. ICES Cooperative Research Report 144, 69.
28. Fox, R.D., Gower, J.F.R., et al., 2009. Slocum glider observations during the spring bloom in the Strait of Georgia. Oceans 9.
29. Frajka-Williams, E., Rhines, P.B., et al., 2009. Physical controls and mesoscale variability in the Labrador Sea spring phytoplankton bloom observed by Seaglider. Deep-Sea Research Part I-Oceanographic Research Papers 56 (12), 2144–2161.
30. Gillespie, D., Caillat, M., et al., 2013. Automatic detection and classification of odontocete whistles. J. Acoust. Soc. Am. 134 (3), 2427–2437.
31. Gillespie, D., Mellinger, D., et al., 2008. PAMGUARD: semiautomated, open source software for real-time acoustic detection and localisation of cetaceans. J. Acoust. Soc. Am. 30 (5), 54–62.
32. Gilman, E., 2002. Guidelines for coastal and marine site-planning and examples of planning and management intervention tools. Ocean & Coastal Manag. 45 (6–7), 377–404.
33. Guihen, D., Fielding, S., et al., 2014. An assessment of the use of ocean gliders to undertake acoustic measurements of zooplankton: the distribution and density of Antarctic krill (Euphausia superba) in the Weddell Sea. Limnol. Oceanogr.- Methods 12.
34. Hill, A.E., Brown, J., et al., 2008. Thermohaline circulation of shallow tidal seas. Geophys. Res. Lett. 35 (11).
35. Huckins, J.N., Manuweera, G.K., et al., 1993. Lipid-containing semipermeable-membrane devices for monitoring organic contaminants in water. Environ. Sci. Technol. 27 (12), 2489–2496.
36. Johnson, M., Partan, J., et al., 2013. Low complexity lossless compression of underwater sound recordings. J. Acoust. Soc. Am. 133 (3), 1387–1398.
37. Johnson, M.P., Tyack, P.L., 2003. A digital acoustic recording tag for measuring the response of wild marine mammals to sound. IEEE J. Ocean. Eng. 28 (1), 3–12.
38. Jones, C.P., 2012. Slocum glider: persistent oceanography. 2012 IEEE/OES Autonomous Underwater Vehicles (AUV 2012) 6.
39. Klinck, H., Mellinger, D.K., et al., 2012. Near-Real-Time Acoustic Monitoring of Beaked Whales and Other Cetaceans Using a Seaglider (TM). Plos One 7 (5).
40. Korneliussen, R.J., Ona, E., 2003. Synthetic echograms generated from the relative frequency response. ICES J. Marine Sci. J. Conseil 60 (3), 636–640.
41. Leeney, R.H., Witt, M.J., et al., 2012. Marine megavertebrates of Cornwall and the Isles of Scilly:

relative abundance and distribution. J. Marine Biol. Assoc. United Kingdom 92 (08), 1823–1833.

42. Leonard, N.E., Paley, D.A., et al., 2007. Collective motion, sensor networks, and ocean sampling. Proc. IEEE 95 (1), 48–74.
43. Lewis, T., Gillespie, D., et al., 2007. Sperm whale abundance estimates from acoustic surveys of the Ionian Sea and Straits of Sicily in 2003. Journal of the Marine Biological Association of the United Kingdom 87 (1), 353–357.
44. L'Heveder, B., Mortier, L., et al., 2013. A Glider network design study for a synoptic view of the oceanic mesoscale variability. J. Atmos. Ocean. Technol. 30 (7), 1472–1493.
45. Lohmann, R., Klanova, J., et al., 2012. PCBs and OCPs on a east-to-west transect: the importance of major currents and net volatilization for PCBs in the Atlantic Ocean. Environ. Sci. Technol. 46 (19), 10471–10479.
46. Marques, T.A., Thomas, L., et al., 2013. Estimating animal population density using passive acoustics. Biol. Rev. 88 (2), 287–309.
47. Merckelbach, L., Smeed, D., et al., 2010. Vertical water velocities from underwater gliders. J. Atmos. Ocean. Technol. 27 (3), 547–563.
48. Meyer-Gutbrod, E., Greene, C., et al., 2012. Long term autonomous fisheries survey utilizing active acoustics. Oceans. Miller, P., 2009. Composite front maps for improved visibility of dynamic sea-surface features on cloudy SeaWiFS and AVHRR data. J. Mar. Syst. 78 (3), 327–336.
49. Miller, P.I., Christodoulou, S., 2014. Frequent locations of ocean fronts as an indicator of pelagic diversity: application to marine protected areas and renewables. Marine Policy 45, 318–329.
50. Niewiadomska, K., Claustre, H., et al., 2008. Submesoscale physical-biogeochemical coupling across the Ligurian Current (northwestern Mediterranean) using a bio-optical glider. Limnol. Oceanogr. 53 (5), 2210–2225.
51. Ohman, M.D., Rudnick, D.L., et al., 2013. Autonomous ocean measurements in the California Current Ecosystem. Oceanography 26 (3), 18–25.
52. Perry, M.J., Sackmann, B.S., et al., 2008. Seaglider observations of blooms and subsurface chlorophyll maxima off theWashington coast. Limnol. Oceanogr. 53 (5), 2169–2179.
53. Pingree, R.D., Holligan, P.M., et al., 1976. Influence of physical stability on spring, summer and autumn phytoplankton blooms in the Celtic Sea. Journal of the Marine Biological Association of the United Kingdom 56 (4), 845–873.
54. Pingree, R.D., Lecann, B., 1989. Celtic and Armorican slope and shelf residual currents. Prog. Oceanogr. 23 (4), 303–338.
55. Prokes, R., Vrana, B., et al., 2012. Levels and distribution of dissolved hydrophobic organic contaminants in the Morava river in Zlin district, Czech Republic as derived from their accumulation in silicone rubber passive samplers. Environ. Pollut. 166, 157–166.
56. Rudnick, D.L., Crowley, M., et al., 2012. A National Glider Network for Sustained Observation of the Coastal Ocean. IEEE, New York.
57. Ruiz, S., Pascual, A., et al., 2009. Mesoscale dynamics of the Balearic Front, integrating glider, ship and satellite data. Journal of Marine Systems 78, S3–S16.
58. Schofield, O., Glenn, S., et al., 2013. The Robot Ocean Network. Am. Sci. 101 (6), 434–441.
59. Send, U., Regier, L., et al., 2013. Use of underwater gliders for acoustic data retrieval from subsurface oceanographic instrumentation and bidirectional communication in the deep ocean. J. Atmos. Ocean. Technol. 30 (5), 984–998.
60. Sharples, J., Ellis, J.R., et al., 2013. Fishing and the oceanography of a stratified shelf sea. Prog. Oceanogr. 117, 130–139.
61. Simpson, J.H., 1981. The shelf-sea fronts- Implications of their existence and behavior. Phil. Trans. R. Soc. Lond. Ser. A Math. Phys. Eng. Sci. 302 (1472), 531.
62. Simpson, J.H., Tett, P.B., et al., 1982. Mixing and phytoplankton growth around an island in a stratified sea. Cont. Shelf Res. 1 (1), 15–31.
63. Ting, M.C., Mujeebu, M.A., et al., 2012. Numerical Study on Hydrodynamic Performance of Shallow Underwater Glider Platform. Indian J. Geo-Mar. Sci. 41 (2), 124–133.
64. Todd, R.E., Gawarkiewicz, G.G., et al., 2013. Horizontal Scales of variability over the middle Atlantic bight shelf break and continental rise from finescale observations. J. Phys. Oceanogr. 43 (1), 222–230.
65. Urick, R., 1983. Principles of Underwater Sound. McGraw-Hill. Villadsgaard, A., Wahlberg, M., et al., 2007. Echolocation signals of wild harbour porpoises, Phocoena phocoena. J. Exp. Biol. 210 (1), 56–64.
66. Vrana, B., Schuurmann, G., 2002. Calibrating the uptake kinetics of semipermeable membrane devices in water: Impact of hydrodynamics. Environ. Sci. Technol. 36 (2), 290–296.
67. Watkins, J.L., Brierley, A.S., 1996. A post-processing technique to remove background noise from echo integration data. ICES J. Mar. Sci. 53 (2), 339–344.
68. Webb, D.C., Simonetti, P.J., et al., 2001. SLOCUM: an underwater glider propelled by environmental energy. IEEE J. Ocean. Eng. 26 (4), 447–452.
69. Weston, K., Fernand, L., et al., 2005. Primary production in the deep chlorophyll maximum of the central North Sea. J. Plankton Res. 27 (9), 909–922.
70. Willcox, J.S., Bellingham, J.G., et al., 2001. Performance metrics for oceanographic surveys with autonomous underwater vehicles. IEEE J. Ocean. Eng. 26 (4), 711–725.

CHAPTER 6

ANALYZING PEATLAND DISCHARGE TO STREAMS IN WATERSHED

Michael B. Gracz, Mary F. Moffett, Donald I. Siegel, Paul H. Glaser

Keywords: Peatlands, Stream flow, EMMA, Evapotranspiration

INTRODUCTION

The IPCC (2013) predicts a global warming trend that began in the late 19th century will continue to warm northerly regions by as much as 2–6 °C by 2100. Rising air temperatures will probably perturb stream ecosystems particularly during droughts when low flow rates are less capable of buffering stream temperatures (Cowx et al., 1984 and Jones and Petreman, 2013). In Southcentral Alaska, stream temperatures have already exceeded the threshold for spawning king salmon (Oncorhynchus tshawytscha) during the yearly dry season (Mauger, 2005), and this type of environmental stress may be pervasive elsewhere. Since dry-season flow is dependent on groundwater inputs from different landscape elements, identifying the relative contributions from different elements is critical to understanding stream ecology.

Peatlands cover approximately 25% of the land surface in northern regions above 45°N latitude but are especially prominent in coastal areas and continental lowlands (Kivinen and Pakarinen, 1981, Wieder et al., 2006 and Rydin and Jeglum, 2006). Despite their abundance and the high water-holding capacity of peat (e.g. Clymo, 1983), the evidence for peatland contributions to stream flow remains equivocal. Two thirds of the studies reviewed by Bullock and Acreman (2003) concluded that wetlands are associated with reduced stream flow during dry seasons within a wide range of physiographic settings. Although these studies were largely based in Europe and North America they are supported by overwhelming evidence that evapotranspiration rates are higher in wetlands than in non-wetlands in the same watershed (Bullock and Acreman, 2003).

Other explanations for the relationship between peatlands and lower stream flows during dry seasons are (a) insufficient water storage in the relatively porous upper layers of peat deposits (Bay, 1969, Ingram, 1983 and Evans et al., 1999) and (b) poor drainage related to the low hydraulic gradient and permeability of peat deposits (Boelter and Verry, 1977, Siegel, 1988a and Burt, 1995). In contrast, Panu (1988) reported higher dry-season flows in streams from Newfoundland in which the watersheds contained a high cover of relatively pristine peatlands. Other studies of peatlands in paired watersheds from Minnesota, Great Britain, and Sweden associate a high cover of peatlands with relatively high stream flows during droughts (Ackroyd et al., 1967, Newson, 1980 and Brandesten, 1988).

One reason for the absence of a consensus among these studies may be the varied hydrogeologic settings of the peatland watersheds (Siegel, 1988a, Johansson and Seuna, 1994, Burt, 1995 and Spence and Woo, 2006). Boelter and

Citation: Michael B. Gracz, Mary F. Moffett, Donald I. Siegel, Paul H. Glaser, Analyzing peatland discharge to streams in watershed, Journal of Hydrology, Volume 530, November 2015, Pages 667-676, ISSN 0022-1694, http://dx.doi.org/10.1016/j.jhydrol.2015.09.072.

Verry (1977), for example, suggest that while flow may be straightforward to quantify from peatlands in small depressions that have a single outlet, these small peatlands may be poor contributors to streamflow because they lack sufficient storage or comprise only a small portion of a watershed. In contrast, a more common setting for peatlands in many boreal watersheds are broad lowlands overlying gently sloping deposits of glacio-lacustrine sediment or glacial till (Gore, 1983 and Rydin and Jeglum, 2006). The extensive peatlands that spread over these deposits commonly lack well-defined outlet streams and may only produce diffuse discharge from pore waters.

Quantifying peatland contributions to streamflow presents an array of challenges. Studies based solely on water budgets are prone to compounding measurement and estimation errors particularly when terms are calculated as residuals (Winter, 1981). Although advances in instrumentation have permitted more precise estimates of ET using tower based instruments (e.g. energy balance and eddy covariance) sources of error still remain (e.g. Twine et al., 2000, Wilson et al., 2002 and Drexler et al., 2004). Holden et al. (2004) therefore identified a need for process-based investigations to understand the dynamics of peatland contributions to stream flow. An alternative approach is provided by end member mixing analysis (EMMA), which has been used to assess end-member contributions to event flows in a range of watersheds (Christophersen et al., 1990, Christophersen and Hooper, 1992, Hooper et al., 1990 and Liu et al., 2008). EMMA uses the chemical signature of water originating from potential end-members within a watershed to determine the percent that each contributes to a final mixture. We therefore compared an end member mixing analysis with a water budget approach to quantify peatland-stream interactions in a small watershed from Southcentral Alaska. This watershed is typical of many in Southcentral Alaska and serves as a useful template to characterize the climatic sensitivity of these ecologically important streams, which provide spawning habitat for salmon.

STUDY REGION

The 1516 ha Limpopo Creek watershed lies in the Cook Inlet Basin of Southcentral Alaska. The two tributaries of this 17.3 km-long creek flow down a gradient of 5–7% from their headwaters near tree line at an elevation of 250 m through alder (Alnus viridis(Chaix) DC.) and open meadows overlying weakly-lithified sedimentary deposits. The tributaries then flow at a gradient of 1–2% through a landscape of lutz spruce forest (Picea X lutzii Little) and peatlands that developed on glacial deposits. The tributaries eventually join about 2.4 km above the creek's confluence with the Anchor River. Both tributaries are confined to a single channel along most of their length except for a reach of a few hundred meters where the northern tributary anastomoses as un-channelized flow over peat (Fig. 1). Peatlands cover about 22% of the watershed and consist of fens or poor fens supporting either a lutz spruce woodland, or non-forested assemblages dominated by ericaceous shrubs and sedges interspersed with pools.

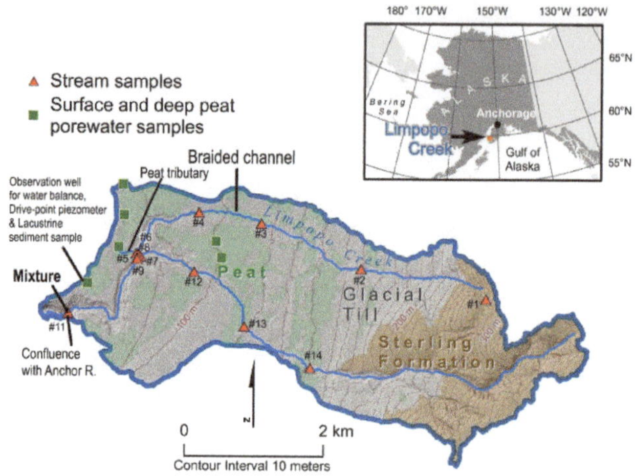

Figure 1: The Limpopo Creek Watershed in Southcentral Alaska. The steeper eastern third of the watershed is underlain by the Sterling Formation (tan), whereas the western two-thirds is composed of glacial till (gray) and lacustrine deposits overlain by peatlands (green). Sample locations are numbered and marked by triangles for stream samples and squares for EMMA end-members. The northern tributary of Limpopo Creek flows as a braided channel across peat where indicated.

The upper third of the watershed generally lacks glacial deposits, and is primarily underlain by alluvial sedimentary deposits, carbonaceous shale, and lignite beds of the thick Sterling Formation (Flores et al., 1997). These deposits were eroded from the surrounding mountains that support diverse rock lithologies, including: sandstone, arkose, argillite, greywacke, slate, granodiorite, breccia, and intermediate-to-felsic volcanic rocks (Beikman, 1994). The lower watershed is underlain by glacio-lacustrine and poorly-sorted till deposits of the last glacial advances (Reger et al., 2007 and Petrik, 1993). Peatlands are primarily restricted to these low-permeability, surficial materials. In addition, the entire watershed has frequently been blanketed by volcanic ash for at least 10.5 ma (Fournelle et al., 1994). Ash deposition has created tephra layers whose composition ranges from high-silica andesite through low-silica dacite to calc-alkaline glass (Riehle, 1985). Mineral soils are generally entisols where wet, and andisols and humicryods where mesic to well-drained (Van Patten, 2005). Two gravel roads cross the watershed, which is inhabited by fewer than a dozen families.

The cool temperate climate of the watershed is moderated by its proximity to the Gulf of Alaska. Annual precipitation averages 625 mm at the nearest station with a long record (Homer), although a station near the headwaters of the Limpopo watershed reports an average of 748 mm (Utah Climate Center, 2013). More than half of the precipitation falls late in the year (August–December), whereas less than 20% falls during the yearly dry-period (April–July). Average annual temperature is 3.1 °C, and the average July maximum is 16.0 °C. The ratio of precipitation to potential evapotranspiration is 1.27 by the Thornthwaite method.

METHODS

Two independent methods were used to estimate peatland contributions to streamflow in the Limpopo watershed during dry periods in order to assess the reliability of their results. Water budget surpluses were first calculated for a representative peatland within this watershed using a sensitivity analysis and these results were then compared to EMMA calculations. The water budget was based on the drawdown of an observation well within this peatland during a well-defined dry period, whereas the EMMA provides a snapshot of geochemical mixing of end-members in the stream during conditions of low flow.

Water budget

Stream flow was measured three times in Limpopo Creek in order to compare these values with the results of a water budget for a shrub-dominated peatland during a dry season. The first measurement was made on July 13, 2010 at the end of the normal-summer dry period and two days prior to the stream sampling for water chemistry. Flow was re-measured a week later on July 22 following a storm, and also on September 23, 2010, at the end of an unseasonable late-summer dry period. Measurements were made with a Pygmy current meter along a 14-point transect across the 3 meter-wide channel 400 m above the confluence with the Anchor River.

To evaluate changes in peatland water storage during a dry period, an observation well was installed to a depth of 98 cm just above the base of a representative peatland in the watershed and instrumented with a U20-series water-level logger in 2005 (Fig. 1). The observation well was calibrated upon installation and changes in barometric pressure were compensated for by an additional logger suspended in the wellhead. The drawdown of water levels in this well during the longest rainless period (August 5–12, 2005) of the study was used to estimate the quantity of water potentially available for streamflow using the following water budget:

$$Q = P + GW_i - GW_o + SW_i \pm \Delta S - ET \qquad (1)$$

where Q is discharge from peatland porewater, P is precipitation, GW_i is input from deeper groundwater, GW_o is output to deeper groundwater, SW_i is surface water input, ΔS is the change in storage, and ET is evapotranspiration. All the components of Eq.(1) are considered in units of $m^3\ s^{-1}$ to facilitate the comparison to actual flows measured in Limpopo Creek. P was set to 0 for calculating the peatland water budget during a dry period marked by no precipitation. We assumed that $GW_i = GW_o$ and $SW_i = 0$. These assumptions are reasonable given: the topographically high setting of the peatland, the low permeability of the underlying lacustrine sediments, the absence of streams flowing into or out of this peatland, and the form of the water level hydrograph during the dry period, which suggested only discharge from porewater (Laine, 1984). The hydrograph indicates a steeper decline during the day than at night (Fig. 2). Eq. (1) thus simplifies to:

$$Q = \Delta S - ET \qquad (2)$$

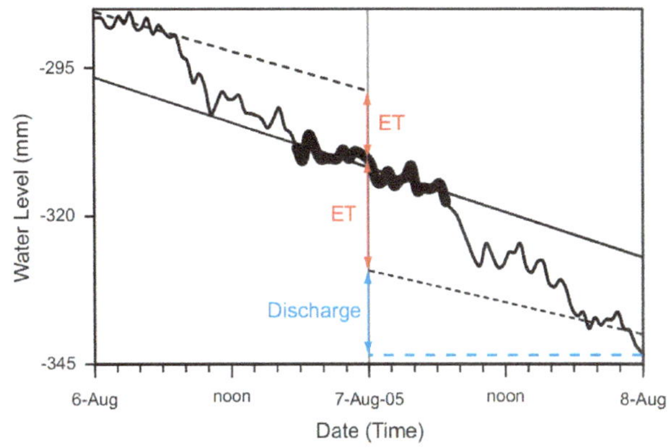

Figure 2: Estimation of ET using the diurnal variation of a declining water-level hydrograph in the absence of recharge. Linear regressions (straight sloping lines) are fitted to the slower rate of decline during the evening and morning when only discharge is active (thicker black line crossing at midnight on day 2) and then extended to predict the water level at midnight on both the previous and following day. The average difference between the predicted water level and the actual water level is attributed to ET (red arrows), which is multiplied by S_y to obtain the actual amount available for discharge (blue arrows). The partial regression lines from day 1 and day 3 are also shown (dotted sloping lines). Data shown are from the observation well between 6 and 8 August 2005. (For interpretation of the references to color in this figure legend, the reader is referred to the web version of this article.)

The effect of ET on storage was estimated by two methods. First, the Thornthwaite method (Thornthwaite, 1948) was calculated using an MS Excel spreadsheet (Lehre, 1994). The Thornthwaite method has been widely used as a complementary method to estimate Potential Evapotranspiration (PET) from peatlands in both Europe and North America (e.g. Bay, 1968, Ingram, 1983, Bridgham et al., 1999 and Brooks et al., 2011). Although this method tends to overestimate PET at upland sites (e.g. Shaw and Riha, 2011) by not considering the effects of changing vegetation, soil, or hydrogeologic setting, PET may still provide a reasonable estimate for peatlands, in which the peat mass moderates water table fluctuations. To calculate fluxes by this method (Eq. (2)), the change in water level is first multiplied by specific yield (S_y) after accounting for recharge (White, 1932, Laine, 1984 and Mitsch and Gosselink, 2007). S_y is the proportion of water in a saturated soil column that freely drains between 0 and ~−0.1 atm (Freeze and Cherry, 1979). This method of calculation produces a reliable estimate for the proportion of drawdown that is actually available in wetlands where water levels are close to the surface, or where recharge is sufficient to raise the water level each night. However, because the water level was falling steadily from deeper in the peat profile in the absence of recharge, a significant amount of vadose water (stored between field capacity and the wilting point) is available to meet ET demands. It is possible that all of the ET demand could be supplied from the vadose zone as it is for plants in non-wetland ecosystems (White, 1932). In recognition of this possibility, ET was subtracted from the total drawdown (accounting for porosity) before multiplying by S_y. In order to obtain the flow produced by similar peatlands in the Limpopo watershed the following equation was used:

$$Q = ((\Phi \Delta W) - ET) S_y A t^{-1} \qquad (3)$$

where Φ is porosity (dimensionless proportion), ΔW is the water level decline (m), ET is evapotranspiration (m), S_y is a dimensionless proportion, A is the area of similar peatlands in the watershed (m^2), t is the time period of water level decline (s), and Q is in $m^3\ s^{-1}$.

The second method uses the diurnal pattern in the water table hydrograph to estimate the portion of ΔS due to ET. During the dry period, the hydrograph from the well was characterized by a lower slope at night that steepened during the day. The slopes differ because only discharge is assumed to be active at night, while both ET and discharge occur during the day. To quantify the difference, the nighttime drawdown rate is projected by linear regression to midnight on the following and preceding days. The average difference between the actual water level at midnight and the level predicted by the regression (multiplied by S_y) is attributed to ET (Fig. 2). The two estimates for each day are averaged to account for an inconstant rate of drawdown (Laine, 1984).

We evaluated the sensitivity of Q in Eq. (3) by varying ET, S_y, and Φ. For the Thornthwaite method S_y was set to 0.05, 0.14, 0.25, and 0.45 with porosity equal to both 0.8 and 0.9 while Q was allowed to vary. We then set Q equal to 0 and allowed porosity to vary to estimate the porosity value necessary for the total measured decline in the water level to be due to Thornthwaite Potential Evapotranspiration (PET). We then set Q equal to 0.06 and porosity to 0.8 (following Boelter, 1972) to determine the value for S_y that would produce all of the lowest flow measured in Limpopo Creek. For the diurnal method we set S_y = 0.05, 0.1, 0.14, and 0.45 and allowed Q to vary. We then set Q = 0.06 and 0.03 to determine if a reasonable value for S_y could produce all or half the lowest streamflow measured in the creek.

Antecedent moisture

Antecedent moisture conditions were evaluated by installing a shallow drive-point piezometer near the observation well from 16 June until 9 October 2010. The shallow drive-point piezometer was installed at depth of 30 cm and instrumented with a water-level logger as above except that it was calibrated monthly against an arbitrary datum driven into the underlying lacustrine sediments. The elevation of the datum was estimated from 1.23-m resolution bare-earth LiDAR data collected in 2009.

EMMA sampling

Water samples were collected from Limpopo Creek for the EMMA snapshot on July 15, 2010 at 12 points distributed downgradient from the headwaters to just above the confluence with the Anchor River where the streamflow was measured. This date was selected as the most likely time to sample the lowest stream flow of the summer based on observations from past years. The extreme–downstream sampling point was located just above the confluence with the Anchor River and represented the final mixture in the EMMA. On the same day, surface water samples were collected from other potential end-members. These samples were collected from (a) near the headwaters of the stream where only groundwater from the Sterling Formation could contribute to stream flow, (b) a spring originating in the glacial till just above the creek, and (c) a short rivulet originating from within a peatland. During July and August in 2010, samples were also collected from two additional end-members (the deep peat and the glacio-lacustrine sediment) from seven piezometers installed in the watershed (Fig. 1). The peatland piezometers were installed near the base of the peat profile (at depths of 98–210 cm) or into the underlying lacustrine sediment (140 cm). Pointed inserts facilitated pounding the piezometers to depth (Chason and Siegel, 1986). All piezometers were bailed until they produced clear water and then were left to equilibrate overnight before being sampled the next day. The piezometer in the lacustrine sediment required five days of equilibration to produce an adequate sample volume. Water from an additional end-member, the peat surface, was also collected near the piezometers. All samples were filtered through a 0.45μ capsule filter using a peristaltic pump and kept cool until analysis.

Samples were analyzed for $\delta^{18}O$ vs. VSMOW at the Stable Isotope Laboratory at University of Alaska, Anchorage (UAA) using a Picarro L-1102i WS-CRDS analyzer. The samples were then acidified and analyzed on an Agilent inductively-coupled plasma mass-spectrometer at the Applied Science Engineering & Technology lab at UAA for cation concentrations. Samples collected in separate bottles were analyzed at the Mid-Continent Ecology Division lab of the United States Environmental Protection Agency (EPA) for chloride and sulfate using EPA method 300.0 with NaOH eluent on a Dionex-DX-600 ion chromatograph using Chromeleon v6.6 software.

EMMA calculations

EMMA was used to estimate the percent contribution from each end-member to the ultimate geochemical mixture at the mouth of the stream. An end-member is defined as the waters originating from a discrete watershed element, which ideally could be distinguished by a distinct geochemical signature (Fig. 1). Samples collected along the length of the creek would be expected to show the shifting chemistry of the stream in response to inputs from different end-members. In EMMA, the shifting chemistry of the stream samples is analyzed by Principal Components Analysis (PCA) to determine the appropriate tracers, the concentrations of which are used to project stream sample points onto PCA axes. Residual analysis guides the evaluation of conservative mixing and the number of principal components identifies the number of end-members responsible for the variance in the chemistry of the stream samples. The contributing end-members are identified by plotting end-members and stream samples on a mixing diagram using the prediction equation for the stream samples. Percent contribution is then solved for each contributing end-member by using matrix algebra to solve the simultaneous equations of complex mixing. The conceptual model was that water originating in the Sterling Formation at the headwaters mixed with other end-members to form the final mixture in the sample collected near the confluence with the Anchor River.

A series of centered PCAs were performed on the correlation matrix of stream samples (rows) by tracer concentrations (columns), in the software PC-ORD™ 6 (McCune and Mefford, 2011) to find the combination of tracers that explained the most variance in the stream samples with the fewest principal components. Residual plots (which are in units of concentration) were examined for non-linearity and deviation from errors greater than laboratory detection limits, both of which are diagnostic of non-conservative mixing (Hooper, 2003).

Once the criteria in the residual analysis were satisfied, the mixing diagram was constructed. On an adequate mixing diagram, the points representing the end-members should define a convex mixing space that completely encloses the stream sample points. Missing end-members are indicated if any stream values lay far outside the mixing space (Hooper, 2003). After an adequate mixing space was defined, the percentage contribution of each end-member to the mixture was calculated by post-multiplying the inverse matrix of PCA scores for the end-members by the column vector of the mixture scores:

$$\mathbf{S_e^{-1} S_m} = \mathbf{f_e} \quad (4)$$

where $\mathbf{S}_e$ is the 3 × 1 matrix of principal component scores (rows) for each end-member (columns), with the first row vector consisting of 1 s (ones); $\mathbf{S}_m$ is the 1 × 3 matrix (column vector) of mixture scores, with the first row element = 1; and $\mathbf{f}_e$ is the 1 × 3 vector of the fractional contribution of each end-member (multiplied by 100 to obtain percentages). Values of unity were used in the first row of each matrix to force 100% contribution from the end-members (Hooper et al., 1990 and Liu et al., 2008). For comparison with the values for peatland discharge produced in the water budget analysis, the percentage contribution from peatland end-member(s) was multiplied by the flow measured in the creek two days prior to EMMA sampling.

RESULTS

Water budget

The peatland observation well provided an estimate for water losses driven by ET in the dry year of 2005. From August 5–12, 2005 the water level in the well fell by 228 mm, from a depth of 256 to 484 mm. Total Thornthwaite PET was estimated at 29.29 mm for this eight-day period. Using this Thornthwaite estimate, contributions from peatlands to Limpopo Creek varied widely when different values for S_y were used, whereas different values for Φ had a smaller effect (Fig. 3). The lowest values for S_y (0.05) and Φ (0.8) presented by Eggelsmann (1971) (as cited by Ingram, 1983 and Siegel, 1988a) produced a surplus equal to 48% of the lowest stream flow measured in the creek. When the highest values for S_y (0.45) and Φ (0.9), reported by Boelter (1972) were used, the surplus equaled approximately two times the highest flow measured in the creek. An unrealistically low Φ (0.131) was required for the Thornthwaite PET method to explain all of the drawdown observed in the peatland well in August, 2005. However, sufficient surplus remained to support all of the lowest flow measured in Limpopo Creek when Φ = 0.8 and S_y = 0.10.

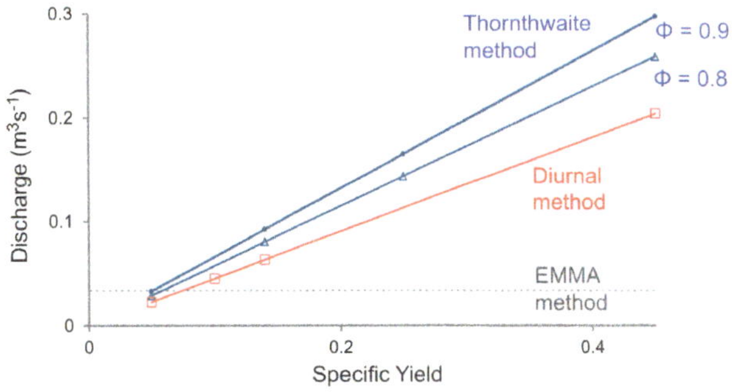

Figure 3: Sensitivity analysis of the water surplus (Q) in the Limpopo peatlands that is available for stream flow. The uppermost two lines (blue) show surplus calculated with the Thornthwaite method using different estimates of porosity (Φ). The lower line (red) shows the amount calculated using the diurnal method, which does not depend on porosity. For comparison, the lower horizontal dashed line shows the amount of streamflow predicted by the EMMA. (For interpretation of the references to color in this figure legend, the reader is referred to the web version of this article.)

Using the diurnal method for ET estimation, a higher value for S_y was needed to produce the same flow as the Thornthwaite method (Fig. 3). When S_y is equal to 0.132, all of the lowest stream flow (0.06 m^3 s^{-1}) could be produced, and with S_y equal to 0.066 about half of the low flow could be produced. The diurnal method could not attribute all of the observed drawdown to ET alone, however, because discharge was evident at night, in the presumed absence of ET (Fig. 2).

Stream flow and antecedent moisture

The lowest flow in the creek (0.06 m^3 s^{-1}) was measured at the beginning of a minor rain event on July 13, 2010 that appeared to have little effect on stream flow by the time of the EMMA sampling, on July 15, 2010 (Fig. 4). The highest flow (0.144 m^3 s^{-1}) was measured a week later on July 22 following a rain event that was observed to produce even greater rainfall at the creek than at the nearby Homer climate station. Another low flow (0.07 m^3 s^{-1}) was measured near the end of an unseasonably long dry-period (23 September 2010).

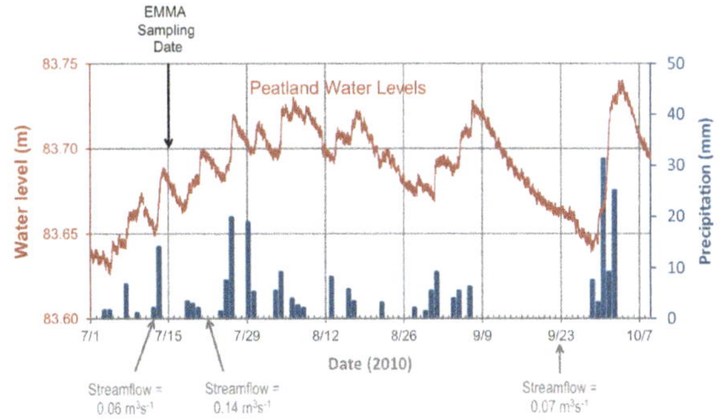

Figure 4: Peatland hydrograph from the Limpopo watershed with respect to other measurements from 2010. Peatland water levels recorded in the shallow drive-point piezometer (red line) are compared to precipitation (blue bars), stream flow and sample times (gray arrows); and the EMMA sampling period (black arrow). The precipitation data are from the Homer Airport, 20 km to the southeast. The peat surface elevation was 83.73 m according to a LiDAR survey. (For interpretation of the references to color in this figure legend, the reader is referred to the web version of this article.)

EMMA

The final PCA in the EMMA used five tracers (^{18}O, SO_4^{2-}, K, Ni, and Ba) to explain 95.8% of the variance with two principal components. Plots of the residuals revealed random errors that were smaller in magnitude than laboratory detection limits, supporting the assumption that the tracers mixed conservatively (Fig. 5). Because two principal components were retained, a model with three end-members is required to solve the mixing equation (Eq. (4)). The three end-members that enclose the mixing space were waters from the Sterling Formation, the surface peat, and the glacial till. These end members enclose all the water samples with the exception of two stream samples that lie somewhat outside the mixing space (Fig. 6). However, the shape of the convex mixing space (Renner, 1993) is affected by sampling and lab errors (Christophersen and Hooper, 1992). As a result it would be reasonable to assume that all of the points lie within a suitably convex mixing space, although we consider the effects on the analysis if two of the points (#4 and #6) in Fig. 6 were to lie outside of it. In addition, the coordinates of samples representing potential end-members from the deep peat (−10.9, 5.8) and the lacustrine sediments (113.6, 76.5) plotted far from the stream values.

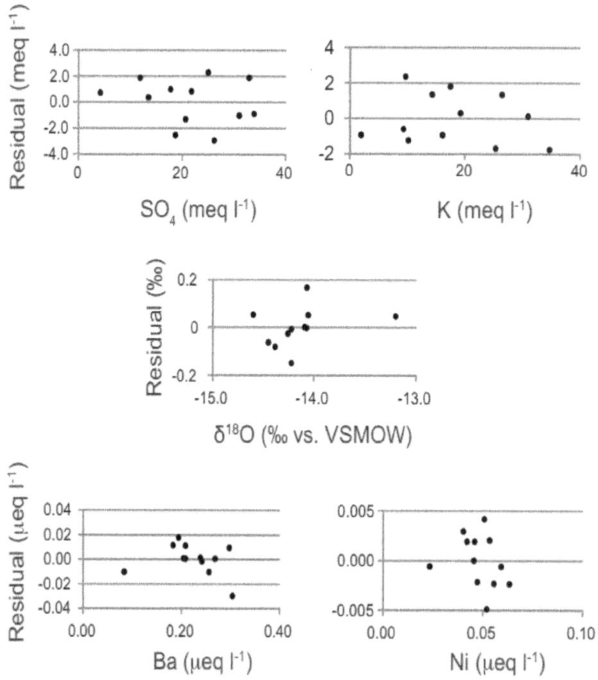

Figure 5: Residual plots of observed versus predicted concentration values for the tracers used in the EMMA. The absence of pattern and the low values of the residuals (within the detection limits of the laboratory analyses) is an indication of conservative mixing.

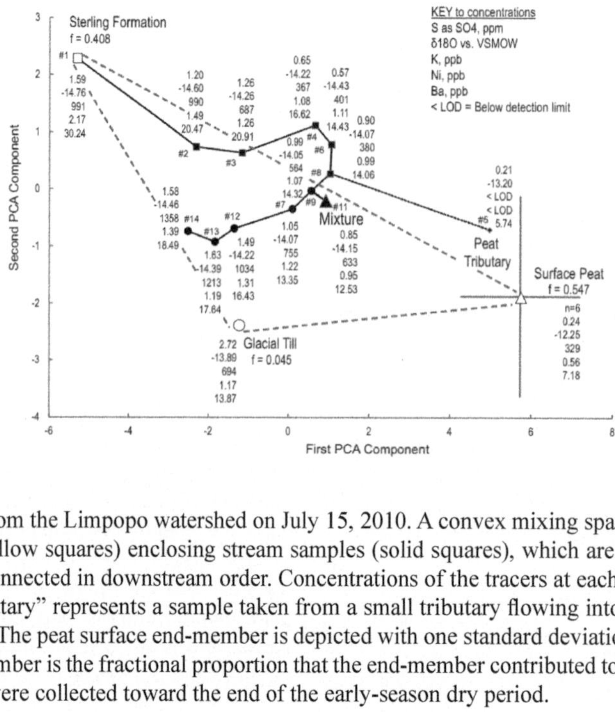

Figure 6: EMMA mixing diagram from the Limpopo watershed on July 15, 2010. A convex mixing space (approximated by the dashed lines) is defined by three end-members (hollow squares) enclosing stream samples (solid squares), which are numbered according to their location along Limpopo Creek (Fig. 1) and connected in downstream order. Concentrations of the tracers at each point are given in the order shown in the key. The point labeled "peat tributary" represents a sample taken from a small tributary flowing into the creek from a peatland just above the confluence of the two tributaries. The peat surface end-member is depicted with one standard deviation in its projected values (large cross). The f-value listed under each end-member is the fractional proportion that the end-member contributed to the final mixture located at the mouth of the stream. The EMMA samples were collected toward the end of the early-season dry period.

The EMMA calculations show that the end member represented by the peat surface contributed 54.7% to the composition of the final mixture, the Sterling Formation contributed 40.8% and the glacial till contributed 4.5%. In contrast, the potential contributing volume of peatlands comprises 0.5% of the watershed, whereas the potential contributing volume of the Sterling Formation is 35%, and the till 65%. Contributing volumes are defined as the volume of each potential end-member within a 3D conceptual model of the watershed. The volumes were determined by multiplying the surface area of each end-member by its depth relative to the level of the Limpopo Creek outlet, which is the lowest point in the watershed. The percentage contribution from the peatland surface multiplied by the flow in the stream two days before the EMMA sampling gives a value of 0.03 m^3s^{-1}, which equals the estimate produced by the diurnal water budget when S_y equals 0.072 (Fig. 3).

DISCUSSION

Peatlands are typically confined to topographic depressions or poorly-drained lowlands where saturated soils favor the accumulation of organic matter (Rydin and Jeglum, 2006). However, in northern maritime regions peatlands have also spread over sloping terrain creating many challenges for quantifying peatland contributions to streams that may be important for spawning fish populations. These sloping peatlands usually lack a well-defined discharge point that can be monitored for hydrological responses to climatic fluctuations. The peatlands in Limpopo Creek watershed, for example, lack visible outlet pipes that characterize the blanket bogs of the British Isles (Gilman and Newson, 1980, Holden and Burt, 2002 and Evans and Warburton, 2007) and similar peatlands in the Maritime provinces of Canada (Glaser and Janssens, 1986). In addition, an extensive peat cover obscures the underlying topography of the mineral substratum that can be responsible for routing subsurface runoff into discrete discharge zones (Allan and Roulet, 1994). In these complex settings end-member mixing analysis provides an alternate means to estimate peatland contributions to stream flow. This approach appears to be well suited to the Limpopo Creek watershed, which is composed of five principal watershed elements, which are distinguishable on the basis of the chemical composition of their waters. The results of the mixing analysis can then be compared to a sensitivity analysis of a water budget in order to independently assess the potential of peatland storage to contribute to stream flow during dry periods.

Estimates from the sensitivity analysis of the water budget indicate that evapotranspiration is sufficiently low and storage capacity and yield is sufficiently high for peatlands to support approximately half the flow measured in Limpopo Creek during dry periods. For example, if streamflow during the EMMA sampling was set to 0.06 $m^3 s^{-1}$, then S_y would need to equal 0.072 (diurnal method) or 0.057 or 0.049 (Thornthwatie method) for the water budget to produce an equivalent flow (Fig. 3). These estimates of S_y generally agree with Ingram's (1983) suggestion for using Eggelsmann's (1971)values of between 0.03 and 0.10. However, they are lower than the range of values (0.26–0.125) that Letts et al. (2000) applied for more humidified layers of peat. Although these findings are supported by studies of paired-watersheds from other regions (Ackroyd et al., 1967, Newson, 1980, Brandesten, 1988 and Panu, 1988), they differ from the conclusions of Bullock and Acreman (2003) that peatlands have minimal effect on flow during dry periods. The contrast between these concepts of peatland-stream interaction may be due to differences in methodology, hydrogeologic setting, or antecedent conditions.

Differences in methodology

EMMA provides an alternative approach for estimating contributions to streamflow from different landscape components that avoids the sources of error related to a water budget approach such as the estimation of recharge, ET, and groundwater flow. Mixing analysis should yield a reliable estimate of peatland contributions to stream flow so long as four fundamental assumptions are met: (1) distinct composition; (2) hydrologic feasibility; (3) conservative mixing; (4) fixed composition.

The first two assumptions are satisfied because the composition of the end member is sufficiently distinct in the Limpopo watershed to bound the stream samples on the mixing diagram and the mixing diagram reproduces the down-gradient locations of the sample points along the stream tributaries showing that mixing is hydrologically feasible (Fig. 6). The third assumption is at least partially satisfied because three of the five tracers (barium, nickel, and $\delta^{18}O$) should mix conservatively at the concentrations measured. Barium and nickel will not form precipitates at the observed concentrations (Stumm and Morgan, 1996 and Snodgrass, 1980). Complexation of these two elements with larger organic molecules is possible but was not likely to be a major control because their respective stability constants differ by orders of magnitude (Stumm and Morgan, 1996) yet their concentrations behaved similarly in the stream (Fig. 5). Complexes with smaller organic molecules and with inorganic ligands would have remained soluble. Furthermore, since kinetic considerations favor the formation of Ba and Ni complexes at much higher rates than their dissociation, any complexes should have formed rapidly near the headwaters, where the concentrations of these two elements was highest, with little effect on mixing further downstream. The stable oxygen isotope (^{18}O) is not fractionated during transpiration (Clark and Fritz, 1997). However, the other two tracers (potassium and sulfate) may mix non-conservatively. Potassium is an essential nutrient for plants and other organisms, whereas sulfur can degas as hydrogen sulfide under anaerobic conditions. The fourth assumption of fixed composition may have been violated by the end-member represented by the pore waters at the peat surface. Therefore, only the non-fixed composition of the peatland end-member and the potential non-conservative mixing of potassium and sulfate may depart from the fundamental assumptions of mixing analysis. Below, we consider the effects that these departures may have had on the EMMA calculations.

Potassium was apparently present at concentrations that exceeded the limiting values in local ecosystems, which are limited by nitrogen and/or phosphorus (Shaftel et al., 2012). Furthermore, when potassium is excluded from the EMMA analysis, there was little change in the final proportions of contributing end-members (Sterling Formation 44%, surface peat 57%, till 0%). The small amount of uptake of potassium compared to its concentration likely did not measurably affect the EMMA.

Eliminating sulfate as a tracer in the EMMA also produces little change in the contributing proportions from the original end-members (Sterling Formation 42%, surface peat 58%, till 0%). Non-conservative mixing could only be a problem with sulfate if it excluded the two end-members that were sampled under anaerobic conditions (glacial-lacustrine sediment and deep peat). However, the concentrations of other tracers in the lacustrine sediment and deep peat were incompatible with the composition of the final mixture. With respect to the lacustrine sediment end-member, $\delta^{18}O$ was substantially more depleted (−15.98‰) and the concentration of barium was higher (373 ppb) than that of samples from both the headwaters (−14.76‰ and 30.2 ppb respectively) and the final mixture (−14.15‰ and 7.2 ppb respectively). In the deep peat porewater, the barium concentration was also too high (47.8 ppb) for this end-member to be a significant contributor to the stream due to the excessive dilution that would be required from other end-members. Non-conservative mixing from the anaerobic samples likely did not affect the EMMA.

Two sample points lying outside of the mixing space suggest that the peatland end-member may not be of fixed composition (Fig. 6). Points lying outside of the convex mixing space (e.g. points #4 and #6 in Fig. 6) indicate that additional end-members may be unaccounted for in an EMMA (Hooper, 2003). On the Limpopo mixing diagram, however, a small shift in the coordinates of the peatland end-member within one standard deviation of its range of variability (e.g. to the coordinates of the peat tributary) could create a triangular space encompassing all of the stream samples (Fig. 6). Alternatively, the range of variation of the peat surface water might indicate that it is actually composed of two different end-members, e.g. fens and poor fens. Splitting the peatland end-member might produce a three-dimensional mixing space with four end-members completely bounding the stream samples, yet would not alter the conclusion that peatlands, whether from two types or one, contribute substantially to baseflow.

Hydrogeologic setting

Three major characteristics of the hydrogeologic setting of the Limpopo Creek watershed are most likely responsible for enhancing peatland contributions to streamflow during dry periods. (1) moderate annual precipitation and a relatively high P/ET ratio, (2) extensive peat cover over sloping terrain, and (3) restriction of peatlands to relatively impermeable glacial deposits. In contrast, studies that determined peatlands were not substantial contributors to downgradient streams were largely based in watersheds with (1) relatively high summer rainfall (Ingram, 1983 and Evans et al., 1999), (2) a small area of peatlands (Siegel, 1988a and Siegel, 1988b), or (3) peatlands confined to small topographic depressions with a single outlet stream (Boelter and Verry, 1977).

The large volume of relatively impermeable glacial till in the Limpopo Creek watershed may be especially important for determining the relative contributions of each landscape component to stream flow (Fig. 1). The low conductivity of this till deposit precludes substantial contributions from this end-member to streamflow. For example, if the area of till traversed by the stream (14,438 m long × 2 m wide) is multiplied by a mean value of hydraulic conductivity for till (10^{-8} m/s) (Freeze and Cherry, 1979) a flow rate of 0.0029 $m^3\ s^{-1}$ would result. A comparison of this value to the flow measured in the creek (0.06 $m^3\ s^{-1}$) suggests that the saturated till in the Limpopo watershed could directly produce only 4.8% of the lowest stream flow measured. This low contribution to stream flow is consistent with the value from the mixing analysis (4.5%) and other reports. Ackroyd et al. (1967), for example compared watersheds across Minnesota, USA and found that watersheds with a greater proportion of glacial till produced less runoff during dry periods. In addition, Brandesten (1988) compared eight catchments with different proportions of either bogs or glacial till in Sweden. Bog catchments provided stream flow during a drought, whereas in the till catchments stream flow sometimes ceased altogether.

Antecedent moisture conditions

Different antecedent conditions could alter the contributing proportions of the end-members to stream flow in the Limpopo watershed. Rain fell two days previous to the EMMA sampling period in 2010 although the previous 2.5 months experienced lower-than-average precipitation (56 vs. 69 mm) (Utah Climate Center, 2013). With even drier antecedent conditions the proportion of contributions from the peat end-member to stream flow may decrease as pore water storage within the peatlands becomes depleted. However, the lowest water-levels measured in the drive-point piezometer at the peatland study site suggest that ample storage remains in the peat during dry periods. Furthermore, the long, steady drawdown in September 2010 along with the measurements of stream flow at this time shows that discharge from peat at this site continues during even longer dry periods (Fig. 4). In other regions, however, discharge from peat may dry up during a severe drought, as Newson (1980) reported in Wales.

Remaining challenges

The variability of natural landscapes in space and time poses a persistent challenge for estimating diffuse discharge from peatlands to nearby streams. This fundamental property of all landscapes introduces an element of uncertainty for determining water budgets for watersheds, particularly with respect to evapotranspiration and the validity of using solutes to determine sources of stream flow.

This study relied on two different methods to estimate ET, which represents one of the more problematic values in most hydrological water budgets. The relative simplicity of the Thornthwaite method for estimating potential ET has facilitated its application to a wide range of peatlands (Ingram, 1983) including those in regions subject to dry periods (e.g. Bay, 1968, Bridgham et al., 1999 and Brooks et al., 2011). The diurnal method, in contrast, has also been widely used but provides an estimate of actual ET on the basis of water level fluctuations in wells (Todd, 1980, Mitsch and Gosselink, 2007 and Lautz, 2008). This method was originally developed by White (1932) and Troxell (1936) but later applied to peatlands by Heikurainen (1971) and Laine (1984). Both the Thornthwaite and diurnal methods lack the precision of more rigorous methods (e.g. energy balance and eddy covariance) that require intensive instrumentation. However, even these instrumental approaches are not free from errors that can propagate when scaled beyond the footprint of the instruments (e.g. Twine et al., 2000, Wilson et al., 2002, Field et al., 1992 and Drexler et al., 2004). As a result we relied on a sensitivity analysis of both the Thornwaite and diurnal methods to provide a probable upper and lower range of values for ET from the Limpopo peatland during a dry period.

We then used EMMA to estimate the proportional contribution of peatlands to streamflow in the Limpopo watershed during a dry period. However, a series of studies in Canada and Wales have raised a serious challenge to the theoretical foundations of end-member mixing analysis. James and Roulet (2006) for example noted variable results from their EMMA analysis of a watershed in southern Quebec that were probably caused by temporal changes in the partitioning of water and solute fluxes from different portions of the watershed to a nearby stream. Kirchner and Neal (2013) and Kirchner et al., 2000 and Kirchner et al., 2001 have reported spurious trends in stream chemistry over time from the Plynlimon watershed of upland Wales that are related to the complexity of solute transport across a sloping landscape. Non-linear trends in the time series of stream chemistry were probably related to the time-dependent variation in flow paths, localized residence times of individual solutes, and relative importance of advection and dispersion as solute transport mechanisms. The complexity of these interactions acted as "fractal filters" producing constantly increasing variance (1/f noise) in the time series of stream chemistry. Our sampling plan was designed to avoid this problem by substituting spatial sampling at one point in time for the temporal sampling at one point in the stream that is traditionally used in EMMA. Therefore, rather than showing shifting stream chemistry in response to mixing from different storm events over the course of several years, our mixing diagram provides a snapshot of the shifting chemistry of the stream along its length, as waters from different end-members mix on a single day during a dry period.

CONCLUSIONS

Peatlands represent an important source for stream flow in the Limpopo Creek watershed of Southcentral Alaska probably accounting for more than half of the flow during dry seasons. Most of the remaining dry-season flow probably originates in the bedrock of the Sterling Formation, whereas little additional discharge was contributed from the poorly-sorted glacial till. Peatlands may also be important contributors to streamflow in other northern watersheds that have similar hydrogeologic settings. The key features are (1) an extensive peat cover over sloping terrain, (2) an underlying mineral substratum composed of impermeable till or other unconsolidated deposits, and (3) a moist regional climate with a pronounced dry period. The disruption of this sensitive linkage between peatlands and streams by drainage or reclaimation operations could therefore magnify the effects of global warming on the ecohydrology of streams, which provide important spawning habitats for salmon in Southcentral Alaska.

ACKNOWLEDGEMENTS

Funding was provided by the U.S. Environmental Protection Agency (EPA grant 8348260) through a cooperative agreement with the Kenai Watershed Forum, and also by the Conservation Biology Program of the University of Minnesota (summer fellowship to M. Gracz), and the U.S. National Science Foundation (Award-0628647 to P. Glaser). The views, trade names, and commercial products mentioned above do not constitute endorsement or recommendation for use by the EPA.

We would like to thank Matt LaCroix, Phil North, Sue Mauger, Marcella Dent, and Iris Fletcher for field assistance and B. Hagedorn (ASET), J. Welker, M. Rogers (ENRI), L. Anderson, and C. Elonen (EPA-MED) for lab analyses. We also thank the editors, Nigel Roulet, and three anonymous reviewers for their many helpful insights for improving the text.

REFERENCES

1. Ackroyd, E.A., Watson, W.C., Hills, D.L., 1967. Groundwater contributions to streamflow an its relation to basin characteristics in Minnesota. Report of Investigations No. 6, University of Minnesota, USA.
2. Allan, C., Roulet, N., 1994. Runoff generation in zero order Precambrian shield catchments: the stormflow response of a heterogenous landscape. Hydrol. Process. 8, 369–388.
3. Bay, R.R., 1968. Evapotranspiration from two peatland watersheds. In: IAHS, (Ed.), Geochemistry, Precipitation, Evapotranspiration, Soil-moisture, Hydrometry. General Assembly of Bern, UVESCO/IAHS 78, pp. 300–307.
4. Bay, R.R., 1969. Runoff from small peatland watersheds. J. Hydrol. 9, 90–102. http:// dx.doi.org/10.1016/0022-1694(69)90016-X.
5. Beikman, H.M., 1994. Geologic map of Alaska. In: Plaker, G., Berg, H.C. (Eds.), The Geology of Alaska. The Geological Society of America, Boulder.
6. Boelter, D.H., 1972. Water table drawdown around an open ditch in organic soils. J. Hydrol. 15, 329–340. http://dx.doi.org/10.1016/0022-1694(72)90046-7.
7. Boelter, D.H., Verry, E.S., 1977. Peatland and Water in the Northern Lake States. USDA Forest Service, North Central Expiremental Station, St. Paul, Minnesota, USA. 22pp.
8. Brandesten, C.O., 1988. Seasonal variation in streamflow recessions in the mire complex Komosse, Southern Central Sweden. In: Symposium on the Hydrology of Wetlands in Temperate and Cold Regions. Academy of Finland, Helsinki, pp. 84–91.
9. Bridgham, S.D., Pastor, J., Updegraff, K., Malterer, T.J., Johnson, K., Harth, C., Chen, J., 1999. Ecosystem control over temperature and energy flux in northern peatlands. Ecol. Appl. 9, 1345–1358.
10. Brooks, K.N., Verma, S.B., Kim, J., Verry, E.S., 2011. Scaling up evapotranspiration estimates from process studies to watersheds. In: Kolka, R., Sebestyen, S., Verry, E.S., Brooks, K.N. (Eds.), Peatland Biogeochemistry and Watershed Hydrology at the Marcell Experimental Forest. CRC Press, New York, pp. 177–192.
11. Bullock, A., Acreman, M., 2003. The role of wetlands in the hydrological cycle. Hydrol. Earth Syst. Sci. 7, 358–389. http://dx.doi.org/10.5194/hess-7-358-2003.
12. Burt, T.P., 1995. The role of wetlands in runoff generation from headwater catchments. In: Hughes, J., Heathwaite, L. (Eds.), Hydrology and Hydrochemistry of British Wetlands. Wiley, Chichester, UK, pp. 21–38.

13. Chason, D.B., Siegel, D.I., 1986. Hydraulic conductivity and related physical properties of peat, Lost River peatland, northern Minnesota. Soil Sci. 142, 91–99.

14. Christophersen, N., Neal, C., Hooper, R.P., Vogt, R.D., Andersen, S., 1990. Modelling streamwater chemistry as a mixture of soilwater end-members – a step towards second-generation acidification models. J. Hydrol. 116, 307–320.

15. Christophersen, N., Hooper, R.P., 1992. Multivariate analysis of stream water chemical data: the use of principal components analysis for the end-member mixing problem. Water Resour. Res. 28, 99–107. http://dx.doi.org/10.1029/ 91wr02518.

16. Clark, I., Fritz, P., 1999. Environmental Isotopes in Hydrogeology. CRC Press, Boca Raton.

17. Clymo, R.S., 1983. Peat. In: Gore, A.J.P. (Ed.), Ecosystems of the World 4A: Mires: Swamp, Bog, Fen and Moor, General Studies. Elsevier, Amsterdam.

18. Cowx, I.G., Young, W.O., Hellawell, J.M., 1984. The influence of drought on the fish and invertebrate populations of an upland stream in Wales. Freshw. Biol. 14, 165–177. http://dx.doi.org/10.1111/j.1365-2427.1984.tb00030.x.

19. Drexler, J.Z., Snyder, R.L., Spano, D., Paw U, K.T., 2004. A review of models and micrometeorological methods used to estimate wetland evapotranspiration. Hydrol. Process. 18, 2017–2101. http://dx.doi.org/10.1002/hyp.1462.

20. Eggelsmann, R., 1971. Über den hydrologidchen Einfluss der Moore. Telma 1, 37–48. Evans, M.G., Burt, T.P., Holden, J., Adamson, J.K., 1999. Runoff generation and water table fluctuations in blanket peat: evidence from UK data spanning the dry summer of 1995. J. Hydrol. 221, 141–160. http://dx.doi.org/10.1016/S0022-1694(99)00085-2.

21. Evans, M., Warburton, J., 2007. Geomorphology of Upland Peat: Erosion, Form, and Landscape Change. Blackwells, Malden, Massachusetts.

22. Field, R.T., Fritschen, L.J., Kanemasu, E.T., Smith, E.A., Stewart, J.B., Verma, S.B., Kustas, W.P., 1992. Calibration, comparison, and correction of net radiation instruments used during FIFE. J. Geophys. Res. Atmos. 97 (D17), 18681–18695.

23. Flores, R.M., Stricker, G.D., Bader, L.R., 1997. Stratigraphic architecture of the Tertiary alluvial Beluga and Sterling Formations, Kenai Peninsula, Alaska. In: Karl, S.M. (Ed.), 1997 Guide to the Geology of the Kenai Peninsula, Alaska, pp 128. Alaska Geological Society, Anchorage.

24. Fournelle, J.H., Marsh, B.D., Meyers, J.D., 1994. Age, character, and significance of Aleutian arc volcanism. In: Plaker, G., Berg, H.C. (Eds.), The Geology of Alaska. The Geological Society of America, Boulder.

25. Freeze, R.A., Cherry, J.A., 1979. Groundwater. Prentice Hall, Englewood Cliffs, NJ.

26. Gilman, K., Newson, M.D., 1980. Soil pipes and pipeflow – a hydrological study in upland Wales. Geobooks, Norwich, 114 pp.

27. Glaser, P.H., Janssens, J.A., 1986. Raised bogs in eastern North America: transitions in landforms and gross stratigraphy. Can. J. Bot. 64, 395–415.

28. Gore, A.J.P., 1983. Introduction. In: Gore, A.J.P. (Ed.), Ecosystems of the World 4A: Mires: Swamp, Bog, Fen and Moor, General Studies. Elsevier, Amsterdam.

29. Heikurainen, L., 1971. Pohjavesipinta ja sen mittaaminen ojitetuilla soilla. Summary: Ground water table in drained peat soils and its measurement. Acta Forest. Fenn. 113, 1–23.

30. Holden, J., Burt, T.P., 2002. Piping and pipeflow in a deep peat catchment. Catena 48, 163–199.

31. Holden, J., Chapman, P.J., Labadz, J.C., 2004. Artificial drainage of peatlands: hydrological and hydrochemical process and wetland restoration. Prog. Phys. Geogr. 28, 95–123.

32. Hooper, R.P., Christophersen, N., Peters, N.E., 1990. Modelling streamwater chemistry as a mixture of soilwater end-members – an application to the Panola Mountain catchment, Georgia, U.S.A. J. Hydrol. 116, 321–343.

33. Hooper, R.P., 2003. Diagnostic tools for mixing models of stream water chemistry. Water Resour. Res. 39 (3), 1055. http://dx.doi.org/10.1029/2002wr001528.

34. Ingram, H.A.P., 1983. Hydrology. In: Gore, A.J.P. (Ed.), Ecosystems of the World 4A Mires: Swamp, Bog, Fen and Moor. Elsevier, Amsterdam, pp. 67–158.

35. IPCC, 2013. Climate Change 2013: The Physical Science Basis. Contribution of Working Group I to the Fifth Assessment Report of the Intergovernmental Panel on Climate Change. In: Stocker, T.F., Qin, D., Plattner, G.-K., Tingnor, M., Allen, S. K., Boschung, J., Nauels, A., Xia, Y., Bex, V., Midgley, P.M. (Eds.). Cambridge University Press, New York. http://dx.doi.org/10.1017/CBO9781107415324, 1535 pp.

36. James, A.L., Roulet, N.T., 2006. Investigating the applicability of end-member mixing analysis (EMMA) across scale: A study of eight small, nested catchments in a temperate forested watershed. Water Resour. Res. 42, WO8434. http://dx.doi. org/10.1029/2005WRR004419.

37. Johansson, B., Seuna, P., 1994. Modelling the effects of wetland drainage on high flows. Aqua Fennica 24, 59–68.

38. Jones, N.E., Petreman, I.C., 2013. Relating extremes of flow and air temperature to stream fish communities. Ecohydrology 6, 826–835. http://dx.doi.org/10.1002/eco.1305.

39. Kirchner, J.W., Neal, C., 2013. Universal fractal scaling in stream chemistry and its implications for solute transport and water quality trend detection. Proc. Nat. Acad. Sci. 110, 122213–122218.

40. Kirchner, J.W., Feng, X., Neal, C., 2000. Fractal stream chemistry and its implications for contaminant transport in catchments. Nature 403, 524–527. http://dx.doi.org/10.1038/35000537.

41. Kirchner, J.W., Feng, X., Neal, C., 2001. Catchment-scale advection and dispersion as a mechanism for fractal scaling in stream tracer concentrations. J. Hydrol. 254,

82–101. Kivinen, E., Pakarinen, P., 1981. Geographical distribution of peat resources and major peatland complex types in the world. Ann. Acad. Sci. Fen. A. III, 1–29. Laine, J., 1984. Estimation of Evapotranspiration from Peatlands by Means of Daily Water Table Hydrographs. University of Helsinki, Helsinki, Agriculture & Forestry.

42. Kirchner, J.W., Feng, X., Neal, C., 2001. Catchment-scale advection and dispersion as a mechanism for fractal scaling in stream tracer concentrations. J. Hydrol. 254, 82–101.
43. Kivinen, E., Pakarinen, P., 1981. Geographical distribution of peat resources and major peatland complex types in the world. Ann. Acad. Sci. Fen. A. III, 1–29.
44. Laine, J., 1984. Estimation of Evapotranspiration from Peatlands by Means of Daily Water Table Hydrographs. University of Helsinki, Helsinki, Agriculture & Forestry.
45. Lautz, L.K., 2008. Estimating groundwater evapotranspiration rates using diurnal water-table fluctuations in a semi-arid riparian zone. Hydrogeol. J. 16, 483–497.
46. Lehre, A., 1994. Thornthwaite estimate of potential evapotranspiration (template). Humboldt State University. (accessed 7 June 2011).
47. Letts, M.G., Roulet, N.T., Comer, N.T., Skarupa, M.R., Verseghy, D.L., 2000. Parameterization of peatland hydraulic properties for the Canadian land surface scheme. Atmos. Ocean 38, 141–160. http://dx.doi.org/10.1080/ 07055900.2000.9649643.
48. Liu, F., Bales, R.C., Conklin, M.H., Conrad, M.E., 2008. Streamflow generation from snowmelt in semi-arid, seasonally snow-covered, forested catchments, Valles Caldera, New Mexico. Water Resour. Res. 44, W12443. http://dx.doi.org/ 10.1029/2007wr006728.
49. Mauger, S., 2005. Lower Kenai Peninsula's salmon streams: annual water quality assessment. Homer Soil and Water Conservation District and Cook Inlet Keeper, Homer, Alaska. p. 62.
50. McCune, B., Mefford, M.J., 2011. PC-ORD multivariate analysis of ecological data. Version 6. MjM Software, Gleneden Beach.
51. Mitsch, W.J., Gosselink, J.G., 2007. Wetlands. John Wiley & Sons Inc., New York. Newson, M.D., 1980. Water balance of selected sites. In: Doornkamp, J.C., Gregory, K.J., Burn, A.S. (Eds.), Atlas of Drought in Britain, 1975–76. Institute of British Geographers, London, pp. 37–38.
52. Panu, U.S., 1988. Environmental assessment of peatland development operations in Newfoundland. In: International Symposium on the Hydrology of Wetlands in Temperate and Cold Regions, Academy of Finland, Joensuu, Finland, pp. 291– 300.
53. Petrik, W.A., 1993. Aquifers and Hydrology at Anchor Point, Alaska. Public-Data File 93–50a. Division of Geological & Geophysical Surveys, Fairbanks.
54. Reger, R.D., Sturmann, A.G., Berg, E.E., Burns, P.A.C., 2007. A guide to the late Quaternary history of northern and western Kenai Peninsula, Alaska. Division of Geological and Geophysical Surveys, Fairbanks.
55. Renner, R.M., 1993. The resolution of a compositional data set into mixtures of fixed source compositions. Appl. Stat. 42, 615–631.
56. Riehle, J.R., 1985. A reconnaissance of the major Holocene tephra deposits in the upper Cook Inlet region, Alaska. J. Volcanol. Geoth. Res. 26, 37–74.
57. Rydin, H., Jeglum, J., 2006. The Biology of Peatlands. Oxford University Press, Oxford.
58. Shaftel, R., King, R., Back, J., 2012. Alder cover drives nitrogen availability in Kenai lowland headwater streams, Alaska. Biogeochemistry 107, 135–148. http://dx. doi.org/10.1007/s10533-010-9541-3.
59. Shaw, S.B., Riha, S.J., 2011. Assessing temperature-based PET equations under a changing climate in temperate, deciduous forests. Hydrol. Process. 25, 1466–1478. http://dx.doi.org/10.1002/hyp.7913.
60. Siegel, D.I., 1988a. The recharge–discharge function of wetlands near Juneau, Alaska: Part I. Hydrogeological investigations. Ground Water 26, 427–434. http://dx.doi.org/10.1111/j.1745-6584.1988.tb00408.x.
61. Siegel, D.I., 1988b. Evaluating cumulative effects of disturbance on the hydrologic function of bogs, fens, and mires. Environ. Manage. 12, 621–626. http://dx.doi.org/10.1007/bf01867540.
62. Snodgrass, W.J., 1980. Distribution and behavior of nickel in the aquatic environment. In: Nriagu, J.O. (Ed.), Nickel in the Environment. John Wiley & Sons, New York.
63. Spence, C., Woo, M-k., 2006. Hydrology of subarctic Canadian shield: heterogeneous headwater basins. J. Hydrol. 317, 138–154. http://dx.doi.org/10.1016/j.jhydrol.2005.05.014.
64. Stumm, W., Morgan, J.J., 1996. Aquatic Chemistry: Chemical Equilibria and Rates in Natural Waters. John Wiley & Sons Inc., New York.
65. Thornthwaite, C.W., 1948. An approach toward a rational classification of climate. Geogr. Rev. 38, 55–94. http://dx.doi.org/10.2307/210739.
66. Todd, D.K., 1980. Groundwater Hydrology, second ed. John Wiley, New York.
67. Troxell, H.C., 1936. The diurnal fluctuation in the ground water and flow of the Santa Ana River and its meaning. Transactions-Am. Geophys. Union 17, 496– 504.
68. Twine, T.E., Kustas, W.P., Norman, J.M., Cook, D.R., Hourser, P.R., 2000. Correcting eddy-covariance flux underestimates over a grassland. Agric. For. Meteorol. 103, 279–300.
69. Utah Climate Center. Utah State University. Utah MAPS-Utah Climate Center: (accessed 9 April 2013).
70. Van Patten, D.J., 2005. Soil Survey of the Western Kenai Peninsula Area, Alaska. Natural Resources Conservation Service, Homer.
71. White, W.N., 1932. A method of estimating ground-water supplies based on discharge by plants and evaporation from soil – results of investigations in Escalante Valley,

Utah. Water Supply Paper 659-A. U.S. Department of the Interior, Geological Survey, Boulder.

72. Wieder, R.K., Vitt, D.H., Benscoter, B.W., 2006. Peatlands and the boreal forest. In: Wieder, R.K., Vitt, D.H. (Eds.), Boreal Peatland Ecosystems. Springer-Verlag, Berlin, pp. 1–8.

73. Wilson, K., Goldstein, A., Falge, E., Aubinet, M., Baldocchi, D., Berbigier, P., Bernhofer, C., Ceulemans, R., Dolman, H., Field, C., Grelle, A., Ibrom, A., Law, B.E., Kowalski, A., Meyers, T., Moncrieff, J., Monson, R., Oechel, W., Tenhunen, J., Valentini, R., Verma, S., 2002. Energy balance closure at FLUXNET sites. Agric. For. Meteorol. 113, 223–243. http://dx.doi.org/10.1016/S0168-1923(02)00109-0.

74. Winter, T.C., 1981. Uncertainties in estimating the water balance of lakes. JAWRA J. Am. Water Resour. Assoc. 17, 82–115. http://dx.doi.org/10.1111/j.1752-1688.1981.tb02593.x.

CHAPTER 7

A NOVEL METHOD TO EVALUATE THE EFFECT OF A STREAM RESTORATION ON THE SPATIAL PATTERN OF HYDRAULIC CONNECTION OF STREAM AND GROUNDWATER

C. Lehr, F. Pöschke, J. Lewandowski, G. Lischeid

Keywords: Groundwater-stream water interactions, Principal component analysis, Signal propagation, Hydraulic connectivity, Clogging, Riparian zone

INTRODUCTION

In the past decades there has been an increasing effort on research and practice according to the restoration of rivers and their floodplains. The main reasons for that are the valuation of river ecosystems as place for species conservation and habitat diversity, recreational and aesthetic purposes, flood protection, enhancing the potential of contaminant deposition and nutrient degradation (Bernhardt et al., 2007, Kondolf et al., 2007, Hester and Gooseff, 2010, Pander and Geist, 2013 and Schirmer et al., 2013). This is also reflected in a growing body of legislative directives (Pander and Geist, 2013 and Schirmer et al., 2013), e.g. the EU Water Framework Directive demands a good chemical and ecological status of groundwater and surface water (European Commission, 2000). The chemical and ecological status of surface waters is impacted by the adjacent connected aquifer and vice versa. Hence both waters have to be considered when assessing water qualities of either of them. Nevertheless, in river restoration practice the measures most often focus solely on surface waters, whereas the connection of the river and the groundwater below the river bed and the adjacent floodplain is often neglected (Boulton, 2007, Boulton et al., 2010 and Hester and Gooseff, 2010).

Previous studies identified the transition zone between stream water and groundwater, the hyporheic zone, as highly relevant for mass exchange, residence time of water and substances in the stream or in the sediment, the chemical and metabolic turnover and in general as crucial for water quality (Brunke and Gonser, 1997; Sophocleous, 2002 and Boulton, 2007). The spatial extent of the hyporheic zone is mainly determined by two drivers, the hydraulic gradient between the river and the groundwater and the sediment structure (Kasahara et al., 2009), especially the permeability of the stream bed and aquifer sediments (Woessner, 2000 and Kalbus et al., 2009). Therefore, clogging of the stream bed, i.e. the sealing of the stream bed with sediments of very low hydraulic conductivity, has been identified as major problem for exchange of surface water and groundwater and the related ecological functions of the hyporheic zone (Sophocleous et al., 1995, Brunke and Gonser, 1997 and Sophocleous, 2002).

Fluxes are spatially and temporally heterogeneous due to the spatial heterogeneity of hydraulic conductivity of the sediments and spatial and temporal variability of hydraulic gradients (Woessner, 2000, Malard et al., 2002, Krause et al., 2011 and Binley et al., 2013). Different methods are available to estimate fluxes across the interface in a river-groundwater system (Kalbus et al., 2006). Selective approaches, such as vertical temperature profiles (e.g. Schmidt et al., 2006 and Anibas et

Citation: C. Lehr, F. Pöschke, J. Lewandowski, G. Lischeid, A novel method to evaluate the effect of a stream restoration on the spatial pattern of hydraulic connection of stream and groundwater, Journal of Hydrology, Volume 527, August 2015, Pages 394-401, ISSN 0022-1694, http://dx.doi.org/10.1016/j.jhydrol.2015.04.075.

al., 2009), heat pulse sensors (e.g. Lewandowski et al., 2011), hydraulic gradients (e.g. Krause et al., 2012) or seepage meters (e.g. Rosenberry and LaBaugh, 2008) are able to monitor the flux over time for a specific point, but it is not possible to draw conclusions for a whole river section. A method to capture larger areas is distributed temperature sensing (DTS) (e.g. Selker et al., 2006a, Selker et al., 2006b and Krause and Blume, 2013), which is able to detect spots with intense groundwater ex- and infiltration. Another option is to use natural or artificial tracers to determine the degree of interactions (e.g. Négrel et al., 2003 and Cox et al., 2007). Beside the different measuring techniques, numerical modelling was often used to examine groundwater-surface water interactions (e.g. Nützmann et al., 2013). One advantage of the latter method is that it does not have to be restricted on the hyporheic zone itself and can include the adjacent floodplain. All methods have in common the large temporal and monetary effort and in most case the restriction to certain areas or certain seasons. Furthermore, most approaches rely on information about hydraulic conductivity or related parameters, which are hard to estimate and result in uncertainties.

Another method which directly estimates the spatial distribution of the hydraulic properties of the sediments is hydraulic tomography (e.g. Yeh and Liu, 2000 and Zhu and Yeh, 2005). In a network of spatially distributed wells the response to an artificial pressure signal induced by a pump at one well is recorded at all other wells. The procedure is repeated by sequentially circulating the pump through the other wells. With packers each well can be segregated in different depth intervals and by circulating the pump through the depth intervals at all wells the depth integrated estimation of hydraulic properties can be enhanced to a 3D-tomography (Yeh and Liu, 2000 and Cardiff and Barrash, 2011). With an inverse model the spatial distribution of the hydraulic properties of the aquifer is estimated from the interplay of all the observed hydraulic head series. Up to now most of the non-numerical hydraulic tomography studies aimed to map the small scale variability of the hydraulic properties of the sediments on the lab to plot scale and used artificial pressure pulses (Yeh et al., 2009 and Cardiff and Barrash, 2011). Recently there were attempts to extend hydraulic tomography to the groundwater basin scale and to use natural pressure signals, e.g. river stage fluctuations as signal (Yeh et al., 2009).

Similarly we use in the present study river stage fluctuations as natural pressure signals and study their propagation in the aquifer before and after a stream restoration measure to study groundwater-stream water interactions. Time series of hydraulic head reflect effects of different causes, like river stage fluctuations, groundwater recharge, precipitation, evapotranspiration, measurement errors, etc. (Yeh et al., 2009). In contrast to the aforementioned approaches we decomposed the hydraulic head series into independent components using a principal component analysis in order to disentangling the different effects.

The study was conducted at a section of the river Spree and its floodplain in the east of Berlin. Here an island is formed by an artificial stream channel (shortcut) and an oxbow. As restoration measure the shortcut was detached from the river at its upstream end, the former oxbow was reconnected to the stream and its clogging layer was excavated. The site was equipped with 15 groundwater observation wells, 2 river stages and 2 hyporheic wells, where data loggers measured every hour two years before and after the restoration. Please note that in our study we do not focus on the small scale heterogeneity of the hydraulic properties of the sediments as it would be important e.g. for estimations of the flowpaths of contaminants or the study of biogeochemical processes in the hyporheic zone. Instead we investigated the effect of the removal of the clogging layer and the change of the river course on the hydraulic connection of stream and groundwater.

Our analysis is based only on hydraulic head data and does not require any additional information. Please note that therefore our analysis is restricted to the transmission of pressure waves. We use the term "hydraulic connectivity" in contrast to the broader concept of "hydrologic connectivity" which is defined by Pringle (2001) as "water-mediated transfer of matter, energy, and/or organisms within or between elements of the hydrologic cycle" to account for that. The presented approach does not allow direct conclusions on related mass fluxes, flowpaths and water exchange rates (Lewandowski et al., 2009 and Page et al., 2012). Instead the permeability for pressure signals is a necessary prerequisite for the exchange of mass fluxes. Thus, the hydraulic connectivity between the observation wells can be used as proxy for the relative differences in effective hydraulic conductivity of the floodplain sediments between the observation wells. With this integrative measure the problem of measuring the small scale variability of hydraulic conductivity in the floodplain is avoided.

To that end, we followed the approach presented by Lewandowski et al. (2009) for a time period where the river section was not restored and applied a principal component analysis on time series of groundwater heads and stream water levels. Based on the findings of Lewandowski et al. (2009) we hypothesized that (1) due to the restoration the hydraulic connectivity between the oxbow and the nearby groundwater will increase and that (2) in the shortcut the river bed will be clogged due to the reduced stream velocity, resulting in decreasing hydraulic connectivity between the shortcut and the adjacent groundwater.

METHODS

Study Site

The Freienbrink site is situated in the floodplain of the lowland river Spree about 30 km east of the centre of Berlin (52°22′06″N, 13°48′25″E). The discharge of the river Spree is regulated by the Weir Grosse Tränke located 10 km upstream and varies usually between 5 and 20 $m^3 s^{-1}$ (Nützmann et al., 2013). At the site, a straight, artificial channel (shortcut) and an old meander (oxbow) form an artificial island. The shortcut was constructed in the 1960s to increase the flow velocity within the river and lower the surrounding groundwater table for agricultural purposes. In the first two years of the monitoring period, the shortcut served as the main stream channel and the oxbow was nearly completely blocked at its upstream end with a dam (Fig. 1). Some pipes inside the dam that connected the oxbow to the main stream were blocked with fine sediments. Therefore, the flow velocity in the oxbow was almost zero and as a consequence an organic silt layer had developed in the oxbow with thicknesses varying spatially between zero and more than 1 m (Nützmann and Lewandowski, 2009).

The organic silt layer had a hydraulic conductivity k_f between 10^{-6} and 10^{-5} ms^{-1} and an effective porosity n_e of 0.5. Compared to the surrounding aquifer with k_f of 10^{-4} to $5 * 10^{-5}$ and n_e of 0.15–0.2 (Nützmann et al., 2013), the organic layer can be attributed as a clogging layer.

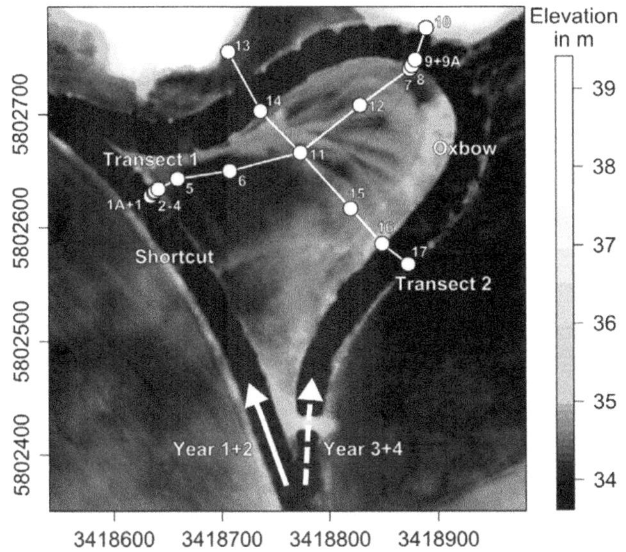

Figure 1: Elevation map based on a LIDAR-Scan from 3rd December 2009 with 1 m gridsize and 0.3 m resolution for altitude in projection ETRS89 UTM Zone 33. First transect of groundwater observation wells in west-northeast direction and second transect in northwest-southeast direction. At both ends of the first transect there was a water level gauge situated in the stream (1A + 9A). The filled arrow marks the main stream flow in the first two years of the study, the dashed arrow the main stream flow in the third and fourth year. The flow through the other reach was blocked in both situations.

At the end of the second year of measurements, in October/November 2008, the clogging mud in the oxbow was excavated with a suction dredger. The oxbow was reconnected as main stream channel on the 27th November 2008. After the reconnection of the oxbow, a dam was built at the upstream end of the shortcut and closed on the 9th December 2008. Hence, there was a switch in the river course from the shortcut to the oxbow (Fig. 1). In the subsequent two years of the monitoring period the former shortcut was blocked with the dam at the upstream end and the former oxbow was reactivated. Only at high water levels the dam was overflown.

The river intersects an unconfined aquifer of about 20 m thickness. The floodplain consists mainly of medium to fine grained sandy sediments of glacial and fluvio-glacial origin (Lewandowski et al., 2009). In the northern part the floodplain adjoins a steep hillslope to a 5 m higher plateau (Lewandowski et al., 2009). The topography of the floodplain is the result of morphological work of the river (Fig. 1). The western part of the floodplain is characterized by a depression with fine grained sand and silt, where the groundwater level repeatedly exceeded the surface (Fig. 1). Starting from the middle of the floodplain, a ridges-swales structure with about 0.4 m difference in elevation established due to the meandering oxbow. Here, coarser sediments with intermediate organic layers are present (Pöschke et al., 2014).

Most of the time groundwater is exfiltrating into the stream. A mean groundwater exfiltration rate of 233 L m^{-2} d^{-1} with a groundwater flow velocity between 10^{-7} and 10^{-6} ms^{-1} was estimated by Nützmann et al. (2013). Maximum lateral infiltration of river water into the aquifer is less than 4 m (Lewandowski et al., 2009). Velocity of pressure wave propagation (celerity) from the stream into the aquifer was found to be about 1550 md^{-1}, thereby three to four orders of magnitude higher than the velocity of groundwater mass flux (Lewandowski et al., 2009). For a general elaboration on velocity of pressure waves (celerity) vs. velocity of water particles (mass fluxes) in hydrology, see McDonnell and Beven (2014).

Water level measurements

Groundwater heads at 15 groundwater observation wells, hydraulic head at 2 hyporheic wells and stream water levels at 2 river stages were measured at hourly intervals with data loggers (Aquatronic, Kirchheim/Teck, Germany, ±1 mm) along two transects. Most of the groundwater wells had filter screens of 2 m length and the upper end of the filter screen at approximately 25–70 cm below the ground, with the exception of observation well No. 11 with 9 m filter length. The first transect consisted of 10 shallow groundwater wells crossing the island in southwest-northeast direction from the shortcut to the oxbow (Fig. 1). Measurements were conducted from 1st October 2006 to 1st October of 2010. Additionally, at both ends of the first transect surface water level was measured in the oxbow and in the shortcut (gauges no. 1A and 9A). Adjacent to the stream water gauges two hyporheic wells no. 1 and 9 were installed that screened at 0.5 to 1.5 m below the riverbed at gauge no. 1A and between 1.5 and 2.5 m at gauge no. 9A (Lewandowski et al., 2009). The second transect consisted of five additional groundwater observation wells that crossed the island in northwest-southeast direction and the first transect at observation well no. 11 (Fig. 1) and measured groundwater level from 12th November 2007 to 1st October of 2010.

Pre-processing of the data

Short data gaps of up to seven hours, which resulted from the biweekly water quality sampling and maintenance of the pressure transducers and loggers, were interpolated using natural splines (less than 0.5% of the readings per year at all wells). Apparent offsets between subsequent periods that could have been due to inexact reinstallation of the devices were adjusted with the offset of the linear extrapolation before and after the corruption. Lengthening of the cable of the devices in the wells were identified by comparison with biweekly manual readings of water levels and groundwater heads and corrected with simple linear regression.

Longer gaps in the data or periods with distorted data were interpolated with the best multiple linear regression model (package 'leaps' in R (R Core Team, 2013)). Two abrupt shifts in the water level of the oxbow compared to the water level in the shortcut and the nearby groundwater wells at 27th November 2008 12:00 and 9th December 2008 15:00 were attributed to the reconnection of the oxbow and the later decoupling of the shortcut with the new dam and hence were not corrected. At maximum, correction comprised 3026 consecutive readings, that is, 34.5% of the readings of the fourth year at groundwater well no. 5. Second most affected was well no. 11 with 538 consecutive readings (6.1%) in the third and 421 consecutive readings (4.8%) in the second year.

At another ten events correction was up to 250 consecutive readings or 2. 9% of the readings per year.

Principal component analysis

Time series of groundwater and river water level show a very close linear correlation (r = 0.98), which indicates a good hydraulic connectivity between both water bodies. Nevertheless, there is no perfect correlation. Aside from measurement noise and possible artefacts, these differences presumably have to be ascribed to different factors that affect different sites to different degrees. Among these, pressure signals induced by river water level fluctuations as well as groundwater recharge, depending on vegetation and the soil properties of the overlying vadose zone, are considered to play a major role (Lewandowski et al., 2009). Thus every water level series can be regarded to be the result of different superimposing effects. Our analysis aimed at extracting the effect of river water level fluctuations from these mixed signals by following the approach presented by Lewandowski et al. (2009).

Firstly, the deviation from the mean of groundwater heads and river water level was calculated for each time step. Afterwards, each of these residual time series was normalized to zero mean and unit variance to ensure equal weighting. Then, a principal component analysis (PCA) was applied to the prepared data. PCA performs an eigenvalue decomposition of a data matrix, yielding a series of independent components. We considered these components as depicting different drivers of hydraulic head fluctuations.

In order to analyse for long-term shifts PCA was performed for single hydrologic years separately (October throughout September yielding 8760 readings at each observation well and 8784 readings in the second year, respectively). For the first transect four years of water level measurements were available since 2006, whereas measurements at the second transect started one year later. To enable comparison between the two transects, a joint analysis with data from both transects was performed for the second, the third and the fourth year. Since the data acquisition at the second transect started at the 24. November 2007 (7476 readings at each observation well), the second year does not cover the full hydrological year 2007/2008. The documentation of the changes in the system in annual resolution is a compromise between comparability with the preceding study of Lewandowski et al. (2009), precision of the effect of the restoration we wanted to measure and comparability of the different PCAs among each other. Furthermore, the PCA was performed for each quartile of the hydrologic year to account for inter-annual variability.

Loadings on a component are the expression of a component at the different sites. They were calculated as Pearson correlation coefficients of the z-normalized residuals of the water level series and the values (scores) of the component (compare Lewandowski et al., 2009). The stability of the loadings in each observation year was estimated with the mean of the loadings of the 4 quartiles of the hydrologic year and their corresponding confidence intervals. The areal expression of the first component was estimated as thin plate splines – a kriging variant implemented in the package 'fields' in R (R Core Team, 2013) – based on the mean of loadings of the 4 quartiles of the third observation year, as it was the year with the lowest inter-annual variability.

All the statistics and calculations were done with the free software package R, version 3.0.1 (R Core Team, 2013).

RESULTS

Characteristics of water level dynamics and floods

The mean amplitude of groundwater level fluctuations was approximately 0.9 m for the first three years of the monitoring and approximately 1.3 m in the last year of the study. The groundwater observation wells, except the two hyporheic wells, have a mean distance of groundwater level to the surface of approximately 60 cm with a standard deviation of approximately 20 to 25 cm. The mean difference in water level from one hour to the next hour was between two to three mm. The mean difference in water level among the groundwater observation wells on the island was in the magnitude of a few cm. The mean groundwater level increased from the first year to the second year by approximately 15 cm and from the third year to the fourth year by approximately 13 cm. The increase in mean groundwater level and amplitude in the last year is due to several flood events: one week of flooding in December 2009, three weeks in January 2010 and 1 month of flooding from the mid of August to the mid of September 2010. For more details including graphs of the water level series characteristics please see Lewandowski et al.'s (2009) examination of the first year of the observation period.

Principal component analysis

For the data set of the first transect the first principal component depicted 70% of the total variance in the first year and 70%, 65% and 63% in the subsequent years (Fig. 2). Please note, that the range of loadings on the first component describes the *relative* differences in the correlations of the first component with the original water level series at the observation wells because the PCA was applied on the z-normalized residuals of the water level series. Hence, we refer to loadings close to one as high loadings and loadings close to minus one as low loadings.

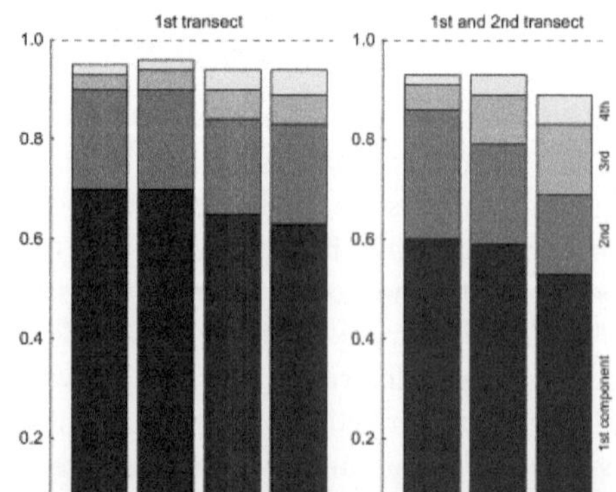

Figure 2: Ratio of overall variance explained by the first four principal components of the PCA of the data set of the first transect (left) and of the PCA of the joint data set from both transects (right).

At the stream water gauges no. 1A in the shortcut and no. 9A in the oxbow loadings were very high on the first component throughout the observation period with the exception of lower loadings for no. 9A in the third year (Fig. 3). At the groundwater observation wells next to the shortcut (no. 2, 3, 4) loadings were very high on the first component and exhibited approximately the same loadings compared to the gauges in the stream (Fig. 3). The hyporheic well no. 1 below the stream bed depicted substantial lower loadings than observation wells no. 2 to 4 in the second year, and slightly lower loadings in the first and fourth year (Fig. 3). Only in the third year loadings at observation well no. 2 were slightly lower than loadings at hyporheic well no. 1.

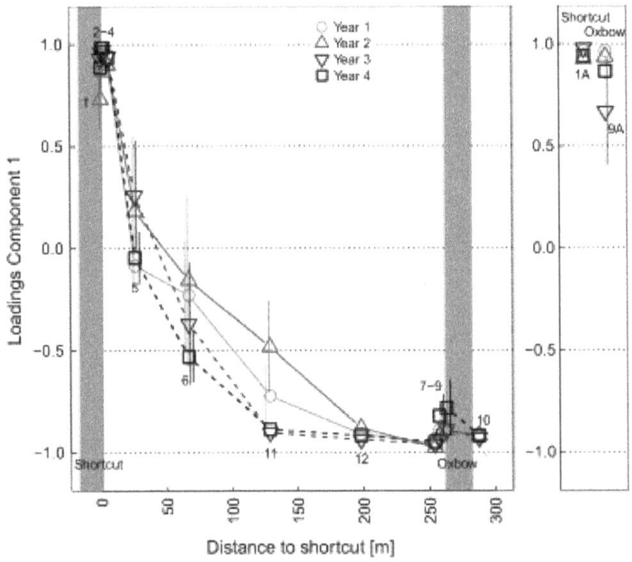

Figure 3: Loadings of the groundwater observation wells of the first transect on the first principal component vs. distance to shortcut (left) and of the stream water gauges (right). Mean of loadings of the 4 quartiles of the hydrologic year (Oct.–Sept.) and their corresponding confidence intervals. Only confidence intervals >0.2 are shown. On the left panel the grey bars indicate the position of the stream.

Groundwater wells next to the oxbow showed very low loadings with slightly increasing loadings on the last meter to the oxbow from observation well no. 7 to 9 (Fig. 3). Only in the fourth year loadings were substantially higher at observation wells no. 8 and 9 than in the years before. Observation well no. 12 was on the same level as the gauges next to the oxbow. Observation wells no. 5, 6 and 11 in the middle part of the island showed decreasing loadings along the first transect from the shortcut to the oxbow. The loadings in this middle part showed the highest variability between the years and also within their quartiles. From the first to the second year all loadings in the middle part increased and decreased thereafter until the fourth year.

For the joint data set (water level data from both transects) the first principal component depicted 60% in the second year and 59% and 53% in the third and fourth year (Fig. 2). The main pattern in the first transect with high loadings next to the shortcut, and decreasing loadings to the east was similar to that of the separate analysis based on the first transect. The loadings of the two variants were correlating in the three common years with an r^2 of at least 0.97.

In the northern part of the second transect groundwater well no. 13 was loading constantly low and groundwater well no. 14 constantly loading high on the first principal component over the entire observation period (Fig. 4). In the southern part of the second transect groundwater wells no. 11, 15, 16 and 17 were jointly shifting from slightly positive loadings in the second year towards negative loadings in the third year and stayed on this level in the fourth year (Fig. 4). Inter-annual variability was prominent in the second year for wells no. 11, 15, 16 and 17 and in the last year for wells no. 13, 16 and 17 (Fig. 4). There was no regular seasonal pattern along both transects. The overall spatial pattern remained relative stable throughout the observation period (Fig. 3, Fig. 4 and Fig. 5).

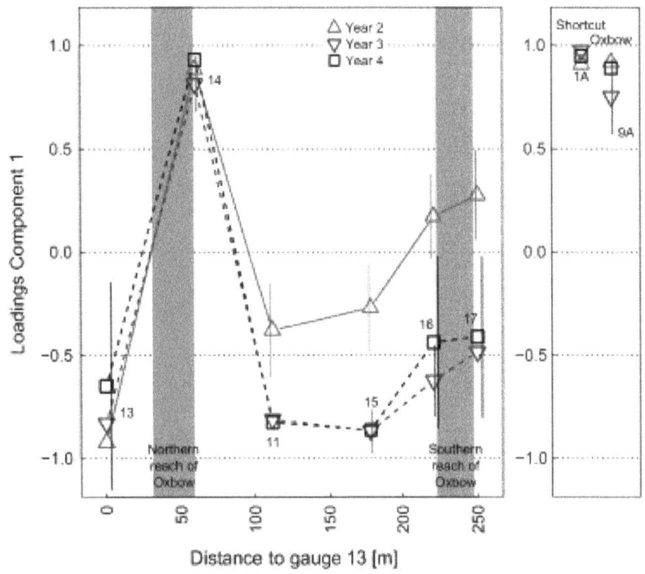

Figure 4: Loadings of the groundwater observation wells of the first principal component of the second transect vs. distance to the northern end of transect two at observation well 13 (left) and of the stream water gauges (right). Mean of loadings of the 4 quartiles of the hydrologic year (Oct.–Sept.) and their corresponding confidence intervals. Only confidence intervals >0.2 are shown. On the left panel the grey bars indicate the position of the stream.

DISCUSSION

Interpretation of the first component

In a more detailed analysis of the first year of the dataset, Lewandowski et al. (2009) identified the first component as dampening and delay of the fluctuations of the water level in the river. We follow their interpretation as loadings of the stream water level gauges exhibit loadings close to one in all of our analyses (Fig. 3 and Fig. 4). Thus the first component explains almost 100% of the deviation of the stream gauges from the mean behaviour of all observation wells. This implies that processes that impact the groundwater head in the wells could hardly have any additional effect on the stream water level gauges. Thus, the loadings on the first component can be used as a quantitative measure for the hydraulic connectivity between river and groundwater.

High positive loadings on the first component imply that the respective time series of the groundwater head deviates from the spatial mean in the same way as the stream water level gauges. That allows the conclusion that groundwater head at the respective site is strongly affected by the stream water level fluctuations. In contrast, high negative loadings point to a weak impact *relative* to the other observation wells, and zero loadings to intermediate effects.

Our prior assumption was that the observed water level dynamics at the single sites are a mixture of different superimposing effects. In fact the first component explains only 53–70% of the spatial variance, indicating that other factors have substantial effects on the observed groundwater heads as well. To investigate in particular the hydraulic connection of the river and the groundwater by analysing the original time series, e.g. using cross-correlation, would therefore not have been satisfactory.

Spatial patterns of hydraulic connectivity

The oxbow and the shortcut have been connected at their downstream end throughout the entire study period and exhibited the same water level. The slightly lower loadings at hyporheic well no. 1 compared to observation wells no. 2 to 4 on the western end of the first transect are interpreted as slightly lower hydraulic conductivity of the river bed compared to the river bank. Along the first transect (Fig. 1) the impact of river level fluctuations decreased with distance from the stream up to well no. 11 and remained approximately stable for the eastern half of the transect (Fig. 3). As the loadings of the groundwater wells in the eastern part of the island are substantial lower than the loadings of the stream wells and the loadings of the groundwater wells in the western part of the island, we conclude that groundwater wells in the eastern part of the first transect are hydraulically disconnected from the oxbow relative to their hydraulic connection to the shortcut (Fig. 5). Only groundwater wells no. 8 and 9 adjacent or even underneath the oxbow are slightly more influenced by the water level fluctuations in the oxbow (Fig. 3). The same pattern was already found by Lewandowski et al. (2009) who suggested that the slight increase is either due to the heterogeneity of the aquifer or due to a very low - but not zero - hydraulic connectivity of the clogging layer.

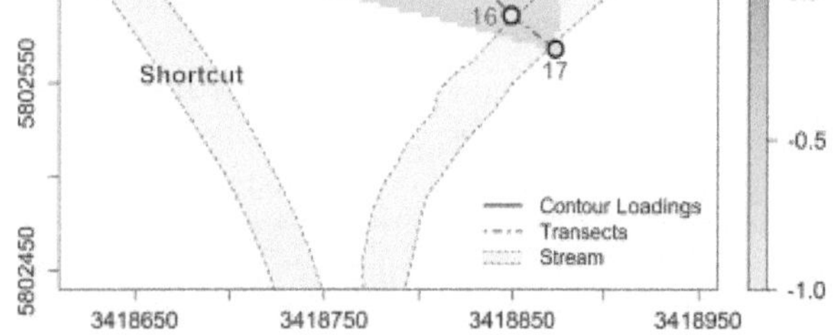

Figure 5: Spatial interpolation of loadings of the first component based on the mean of the loadings of the 4 quartiles of the third observation year (Oct.–Sept.). Projection is ETRS89 UTM Zone 33. Values < -1 are artefacts and excluded.

Results from the second transect show high connectivity with the oxbow only for well no. 14 in its northern reach, which is also not affected by the stream channel restoration, since the pattern stays stable over the four years (Fig. 4). This specifies the hypothesis of areas of relatively better hydraulic connection in the north-western part of the island (Fig. 5). Observation wells no. 16 and 17 close to the southern reach of the oxbow show higher hydraulic connectivity with the stream than the observation wells no. 11 and 15 in the middle of the island and also than the observation wells in the north-eastern part of the first transect. This specifies the area of lowest hydraulic connectivity in the north-eastern part of the island, which is the area where the inner bank of the meander is and was located. Since the flow velocity is lowest at this position during the formation of the meander, material of finer grain sizes has been accumulated in that area. This explanation is also supported by investigations of Pöschke et al. (2014), who conducted a ground penetrating radar survey to characterize the sediment composition. They could show that the north-eastern part of the island is characterized by finer grain sizes close to the meander (well no. 7) in comparison to the western part, where the sediment adjacent to the channel bed is much coarser.

The low hydraulic connectivity of the observation wells no. 10 and 13 in the hillslope of the adjoining plateau is not due to the hydraulic gradient induced by the topographic gradient, because the first component captures the pressure-wave signals of the river.

Temporal variation of hydraulic connectivity

The analysis was performed for single years separately in order to capture changes of connectivity. Loadings of the stream gauges were constantly close to one as it is mandatory for the interpretation of the first component as influence of river level fluctuations on groundwater levels. The lower loading at stream gauge no. 9A in the third year (Fig. 3) could be ascribed to the reconnection of the oxbow and the construction of the new dam in the shortcut.

Loadings of observation wells close to the Spree river at the western end of the first transect, and of groundwater well no. 14 close to the oxbow in its northern reach were close to one in all years of our study, pointing to a constant high connectivity between river and aquifer. Clogging of the stream bed in the shortcut due to the reduced stream velocity was not detected during the two years after the restoration. We assume that the first clogging sediments that settled throughout the first year have been removed due to the intense floods in the fourth year. Nonetheless we expect clogging of the shortcut in successive years.

In the central and south-easterly part of the island wells no. 5, 6, 11, 15, 16 and 17 exhibit substantial shifts between single years, although the general spatial pattern remains approximately the same (Fig. 3 and Fig. 4). There the loadings are highest for the second year and smallest in the fourth year, whereas those of the third year are similar to those of the fourth year. We did not find an unequivocal explanation for that temporal shift. Such systematic shift can hardly be explained by artefacts of the measurements. In addition, a corresponding change of the properties of the aquifer can be excluded. All of these wells are located close to the same former river bed, although this had not been intended. The associated interbedding of different substrates such as gravel, silt, former clogging layers and peat might still affect groundwater flowpaths and lead to a closely coupled behaviour (Nützmann and Lewandowski, 2009).

Wells no. 8 and 9 close to the north-easterly reach of the oxbow exhibit almost no shift in time. Loadings are close to minus one for the first three years, indicating low hydraulic contact with the oxbow. This suggests that the clogging layer is not the only reason for the lower hydraulic connectivity of the hyporheic well no. 9 below the oxbow compared to the hyporheic well no. 1 below the shortcut. Instead we suggest that the effective hydraulic conductivity of the sediments around the shortcut in the north western part of the island is higher than of the sediments around the oxbow (Fig. 5). This can be attributed to the construction of the shortcut. The shortcut itself and the surroundings are situated in sandy sediments, which were dislocated artificially. In contrast, the sediments around the oxbow are characterized by inclined layers of sand and organic material of the recent and former point bars. On the other hand the loadings at well no. 14 are close to one for the whole observation period of the second transect, suggesting a strong impact of natural spatial heterogeneity of effective hydraulic conductivity of the floodplain sediments (Fig 4).

Only in the fourth year loadings at wells no. 8 and 9 are slightly higher (but not significantly higher), pointing to increased, although still fairly low connectivity (Fig. 3). Likewise a minor increase of loadings between the third and fourth year is observed at wells no. 13, 16 and 17 close to the oxbow (Fig. 4). This is consistent with our prior assumptions of increasing connectivity due to stream channel restoration. However, this is more a little piece of evidence rather than a proof and the development has to be checked in the following years. Anyhow, the effect is much weaker than assumed.

CONCLUSIONS

The data driven PCA approach used in this study is easy to apply and requires only time series of hydraulic heads in the aquifer and in the stream. It splits up the water table dynamics into independent components which represent different drivers of the hydraulic head dynamics in the system and the spatial expression of those drivers can be analysed. In the present study, the first component captures the propagation of river stage fluctuations into the aquifer and can therefore be used to describe the hydraulic connectivity of groundwater and stream water. Areas of relative low and high hydraulic connectivity could be identified, although it is not possible to quantify mass fluxes and water exchange rates. Despite the small differences in water table of a few cm along the groundwater transects and intense periods of flooding in the fourth year, it is possible to derive throughout the observation period spatial and temporal consistent patterns of one distinct hydraulic process, namely the propagation of the river water table fluctuations, on the hectare-scale (Fig. 3, Fig. 4 and Fig. 5). One benefit of the presented approach is that hydraulic head devices are long established and often already permanent part of hydrologic monitoring networks. Once installed the operating effort is very low. The hydraulic head data are usually measured with relative high temporal resolution of minutes to hours and stored with data loggers.

The presented approach can be used to optimize the monitoring network and monitor further developments of the groundwater-stream water system (Page et al., 2012), as in our example after a stream channel restoration. It can also provide a framework for further modelling of groundwater-stream water interactions (Lewandowski et al., 2009), help to identify the hydrological active functional properties on the landscape scale (Lischeid et al., 2010), regionalize the different hydrological contributions to the aquifer dynamics (Longuevergne et al., 2007), or be used as explorative data analysis tool to develop hypothesis such as the spatial patterns of areas of relative low and high effective hydraulic conductivity of floodplain sediment as done in the present study.

Our results demonstrate that even two years after the restoration hydraulic connectivity of stream water and the groundwater next to the new main stream channel (oxbow) had not reached the level of connectivity next to the old artificial stream channel (shortcut). We identified three factors to explain this finding. That is (1) that the change in stream velocity and sedimentation rate in the oxbow and in the shortcut due to the switch of the main stream channel might need more time to effectively change the stream bed permeability in such a scale that it alters the spatial pattern of hydraulic connectivity in the floodplain. The two other options are that hydraulic connectivity around the shortcut is in the long run higher than around the oxbow due to the spatial heterogeneity of the sediments in the floodplain, whether the heterogeneity is (2) natural or (3) induced by the digging of the artificial shortcut directly into the sandy floodplain sediments. To distinguish between this three factors and their relative contribution to the observed spatial pattern of hydraulic connectivity further analysis of the spatial distribution of the sediments around the stream channels and their hydraulic conductivities, as well as the further monitoring of the development of hydraulic connectivity after the restoration is necessary.

With this study the importance of comprehensive monitoring of restoration measures even several years after the restoration and the explicit identification of specific properties of the system to be restored is underpinned (Kondolf, 1995). Option one of our suggested set of influencing factors exemplifies that restoration of connectivity of groundwater and surface water is a process that might last for several years after the initial restoration of the surface water, while on the surface the restoration might appear already successfully completed. The proposed differences in effective hydraulic conductivity of the alluvial sediments in different parts of the floodplain (option two and three) highlight the importance of pre-studies that identify the specific natural conditions before the restorations (Kondolf and Micheli, 1995, Woolsey et al., 2007 and Pander and Geist, 2013). Focusing only on the hydrogeomorphology of the surface waters or individual target species in restoration practice might be insufficient to restore the functional properties of the river ecosystem (Kondolf et al., 2006 and Woolsey et al., 2007). If groundwater-stream water interactions are understood as substantial features of the system to be restored, this has to be taken into account in the restoration and monitoring practice, as well as in the evaluation of restorations.

ACKNOWLEDGEMENTS

We thank Christine Sturm and Grit Siegert from IGB for the collection of the data. We thank Tobias Hohenbrink and Steven Böttcher for fruitful discussions on data analysis, especially PCA, and Marcus Fahle for very useful tricks in R. In addition, we thank T. Mehner and the participants of the workshop

'Scientific Writing' at the Leibniz-Institute of Freshwater Ecology and Inland Fisheries for helpful discussion on an early stage of the manuscript. In addition, we would like to acknowledge the constructive and helpful comments of three anonymous reviewers and the associated editor which helped to improve this manuscript. The work was founded by the Leibniz association (SAW-2012-IGB-4167) within the international Leibniz graduate school: Aquatic boundaries and linkages in a changing environment – Aqualink (http://www.igb-berlin.de/aqualink.html).

REFERENCES

1. Anibas, C., Fleckenstein, J.H., Volze, N., Buis, K., Verhoeven, R., Meire, P., Batelaan, O., 2009. Transient or steady-state? Using vertical temperature profiles to quantify groundwater–surface water exchange. Hydrol. Process. 23, 2165–2177. http:// dx.doi.org/10.1002/hyp.7289.
2. Bernhardt, E.S., Sudduth, E.B., Palmer, M.A., Allan, J.D., Meyer, J.L., Alexander, G., Follastad-Shah, J., Hassett, B., Jenkinson, R., Lave, R., Rumps, J., Pagano, L., 2007. Restoring rivers one reach at a time: results from a survey of U.S. river restoration practitioners. Restor. Ecol. 15, 482–493. http://dx.doi.org/10.1111/j.1526 100X.2007.00244.x.
3. Binley, A., Ullah, S., Heathwaite, A.L., Heppell, C., Byrne, P., Lansdown, K., Trimmer, M., Zhang, H., 2013. Revealing the spatial variability of water fluxes at the groundwater-surface water interface. Water Resour. Res. 49, 3978–3992. http:// dx.doi.org/10.1002/wrcr.20214.
4. Boulton, A.J., 2007. Hyporheic rehabilitation in rivers: restoring vertical connectivity. Freshwater Biol. 52, 632–650. http://dx.doi.org/10.1111/j.1365-2427.2006.01710.x.
5. Boulton, A.J., Datry, T., Kasahara, T., Mutz, M., Stanford, J.A., 2010. Ecology and management of the hyporheic zone: stream-groundwater interactions of running waters and their floodplains. J. N. AM. Benthol. Soc. 29, 26–40. http:// dx.doi.org/10.1899/08-017.1.
6. Brunke, M., Gonser, T., 1997. The ecological significance of exchange processes between rivers and groundwater. Freshwater Biol. 37, 1–33. http://dx.doi.org/ 10.1046/j.1365-2427.1997.00143.x.
7. Cardiff, M., Barrash, W., 2011. 3-D transient hydraulic tomography in unconfined aquifers with fast drainage response. Water Resour. Res. 47, W12518. http:// dx.doi.org/10.1029/2010WR010367.
8. Cox, M.H., Su, G.W., Constantz, J., 2007. Heat, chloride, and specific conductance as groundwater tracers near streams. Ground Water 45, 187–195. http:// dx.doi.org/10.1111/j.1745-6584.2006.00276.x.
9. European Commission, 2000: Directive 2000/60/EC of the European Parliament and of the Council establishing a framework for Community action in the field of water policy. Off. J. Eur. Commun., vol. L327, pp. 1–72.
10. Hester, E.T., Gooseff, M.N., 2010. Moving beyond the banks: hyporheic restoration is fundamental to restoring ecological services and functions of streams. Environ. Sci. Technol. 44, 1521–1525. http://dx.doi.org/10.1021/es902988n.
11. Kalbus, E., Reinstorf, F., Schirmer, M., 2006. Measuring methods for groundwatersurface water interactions: a review. Hydrol. Earth Syst. Sci. 10, 873–887. http://dx.doi.org/10.5194/hess-10-873-2006.
12. Kalbus, E., Schmidt, C., Molson, J.W., Reinstorf, F., Schirmer, M., 2009. Influence of aquifer and streambed heterogeneity on the distribution of groundwater discharge. Hydrol. Earth Syst. Sci. 13, 69–77. http://dx.doi.org/10.5194/hess- 13-69-2009.
13. Kasahara, T., Datry, T., Mutz, M., Boulton, A.J., 2009. Treating causes not symptoms: restoration of surface-groundwater interactions in rivers. Mar. Freshwater Res. 60, 976–981. http://dx.doi.org/10.1071/MF09047.
14. Kondolf, G.M., 1995. Five elements for effective evaluation of stream restoration. Restor. Ecol. 3, 133–136. http://dx.doi.org/10.1111/j.1526- 100X.1995.tb00086.x. Kondolf, G., Micheli, E., 1995. Evaluating stream restoration projects. Environ. Manage. 19, 1–15. http://dx.doi.org/10.1007/BF02471999.
15. Kondolf, G.M., Boulton, A.J., O'Daniel, S., Poole, G.C., Rahel, F. J., Stanley, E.H., Wohl, E., Bång, A., Carlstrom, J., Cristoni, C., Huber, H., Koljonen, S., Louhi, P., Nakamura, K., 2006: Process-Based Ecological River Restoration: Visualizing Three-Dimensional Connectivity and Dynamic Vectors to Recover Lost Linkages, Ecol. Soc., vol. 11. .
16. Kondolf, G.M., Anderson, S., Lave, R., Pagano, L., Merenlender, A., Bernhardt, E.S., 2007. Two decades of river restoration in california: what can we learn? Restor. Ecol. 15, 516–523. http://dx.doi.org/10.1111/j.1526-100X.2007.00247.x.
17. Krause, S., Blume, T., 2013. Impact of seasonal variability and monitoring mode on the adequacy of fiber-optic distributed temperature sensing at aquifer-river interfaces. Water Resour. Res. 49, 2408–2423. http://dx.doi.org/10.1002/ wrcr.20232.
18. Krause, S., Hannah, D.M., Fleckenstein, J.H., Heppell, C.M., Kaeser, D., Pickup, R., Pinay, G., Robertson, A.L., Wood, P.J., 2011. Inter-disciplinary perspectives on processes in the hyporheic zone. Ecohydrology 4, 481–499. http://dx.doi.org/ 10.1002/eco.176.
19. Krause, S., Blume, T., Cassidy, N.J., 2012. Investigating patterns and controls of groundwater up-welling in a lowland river by combining Fibre-optic Distributed Temperature Sensing with observations of vertical hydraulic gradients. Hydrol. Earth Syst. Sci. 16, 1775–1792. http://dx.doi.org/10.5194/ hess-16-1775-2012.
20. Lewandowski, J., Lischeid, G., Nützmann, G., 2009. Drivers of water level fluctuations and hydrological exchange between groundwater and surface water at the lowland River Spree (Germany): field study and statistical analyses. Hydrol. Process. 23, 2117–2128. http://dx.doi.org/10.1002/hyp.7277.
21. Lewandowski, J., Angermann, L., Nützmann, G., Fleckenstein, J.H., 2011. A heat pulse technique for the determination of small-scale flow directions and flow velocities in the streambed of sand-bed streams. Hydrol.

Process. 25, 3244– 3255. http://dx.doi.org/10.1002/hyp.8062.

22. Lischeid, G., Natkhin, M., Steidl, J., Dietrich, O., Dannowski, R., Merz, C., 2010. Assessing coupling between lakes and layered aquifers in a complex Pleistocene landscape based on water level dynamics. Adv. Water Resour. 33, 1331–1339. http://dx.doi.org/10.1016/j.advwatres.2010.08.002.

23. Longuevergne, L., Florsch, N., and Elsass, P., 2007: Extracting coherent regional information from local measurements with Karhunen-Loève transform: Case study of an alluvial aquifer (Rhine valley, France and Germany), Water Resour. Res., vol. 43, WO4430. http://dx.doi.org/10.1029/2006WR005000.

24. Malard, F., Tockner, K., Dole-Olivier, M.-J., Ward, J.V., 2002. A landscape perspective of surface–subsurface hydrological exchanges in river corridors. Freshwater Biol. 47, 621–640. http://dx.doi.org/10.1046/j.1365-2427.2002.00906.x.

25. McDonnell, J.J., Beven, K., 2014. Debates onWater Resources: The future of hydrological sciences: A (common) path forward? A call to action aimed at understanding velocities, celerities and residence time distributions of the headwater hydrograph. Water Resour. Res. http://dx.doi.org/10.1002/2013WR015141.

26. Négrel, P., Petelet-Giraud, E., Barbier, J., Gautier, E., 2003. Surface water– groundwater interactions in an alluvial plain: chemical and isotopic systematics. J. Hydrol. 277, 248–267. http://dx.doi.org/10.1016/S0022-1694(03)00125-2.

27. Nützmann,G.,Lewandowski,J.,2009.Wechselwirkungen zwischen Grundwasser und Oberflächenwasser an einem Tieflandfluss (Spree). Grundwasser 14, 195– 205. http://dx.doi.org/10.1007/s00767-009-0110-4.

28. Nützmann, G., Levers, C., Lewandowski, J., 2013. Coupled groundwater flow and heat transport simulation for estimating transient aquifer–stream exchange at the lowland River Spree (Germany). Hydrol. Process. http://dx.doi.org/10.1002/ hyp.9932.

29. Page, R.M., Lischeid, G., Epting, J., Huggenberger, P., 2012. Principal component analysis of time series for identifying indicator variables for riverine groundwater extraction management. J. Hydrol. 432–433, 137–144. http:// dx.doi.org/10.1016/j.jhydrol.2012.02.025.

30. Pander, J., Geist, J., 2013. Ecological indicators for stream restoration success. Ecol. Indic. 30, 106–118. http://dx.doi.org/10.1016/j.ecolind.2013.01.039.

31. Pöschke, F., Lewandowski, J., Nützmann, G., 2014. Impacts of alluvial structures on small-scale nutrient heterogeneities in near-surface groundwater. Ecohydrology. http://dx.doi.org/10.1002/eco.1535.

32. Pringle, C.M., 2001. Hydrologic connectivity and the management of biological reserves: a global perspective. Ecol.Appl. 11, 981–998. http://dx.doi.org/ 10.1890/1051-0761(2001) 011[0981:HCATMO]2.0.CO;2.

33. R Core Team (2013). R: A language and environment for statistical computing. R Foundation for Statistical Computing, Vienna, Austria. .

34. Rosenberry, D.O. and LaBaugh, J.W.: Field techniques for estimating water fluxes between surface water and ground water: U.S. Geological Survey Techniques and Methods 4–D2, 128 p., 2008.

35. Schirmer, M., Luster, J., Linde, N., Perona, P., Mitchell, E.A.D., Barry, D.A., Cirpka, O.A., Schneider, P., Vogt, T., Durisch-Kaiser, E., 2013.

36. River restoration: morphological, hydrological, biogeochemical and ecological changes and challenges. Hydrol. Earth Syst. Sci. Discuss. 10, 10913–10941. http:// dx.doi.org/10.5194/hessd-10-10913-2013.

37. Schmidt, C., Bayer-Raich, M., Schirmer, M., 2006. Characterization of spatial heterogeneity of groundwater-stream water interactions using multiple depth streambed temperature measurements at the reach scale. Hydrol. Earth Syst. Sci. 10, 849–859. http://dx.doi.org/10.5194/hess-10-849-2006.

38. Selker, J., van de Giesen, N., Westhoff, M., Luxemburg, W., Parlange, M.B., 2006a. Fiber optics opens window on stream dynamics. Geophys. Res. Lett. 33, L24401. http://dx.doi.org/10.1029/2006GL027979.

39. Selker, J.S., Thévenaz, L., Huwald, H., Mallet, A., Luxemburg, W., van de Giesen, N., Stejskal, M., Zeman, J., Westhoff, M., Parlange, M.B., 2006b. Distributed fiberoptic temperature sensing for hydrologic systems. Water Resour. Res. 42, W12202. http://dx.doi.org/10.1029/2006WR005326.

40. Sophocleous, M., 2002. Interactions between groundwater and surface water: the state of the science. Hydrogeol. J. 10, 52–67. http://dx.doi.org/10.1007/s10040- 001-0170-8.

41. Sophocleous, M., Koussis, A., Martin, J.L., Perkins, S.P., 1995. Evaluation of simplified stream-aquifer depletion models for water rights administration. Ground Water 33, 579–588. http://dx.doi.org/10.1111/j.1745-6584.1995.tb00313.x.

42. Woessner, W.W., 2000. Stream and Fluvial plain ground water interactions: rescaling hydrogeologic thought. Ground Water 38, 423–429. http:// dx.doi.org/10.1111/j.1745-6584.2000.tb00228.x.

43. Woolsey, S., Capelli, F., Gonser, T., Hoehn, E., Hostmann, M., Junker, B., Paetzold, A., Roulier, C., Schweizer, S., Tiegs, S.D., Tockner, K., Weber, C., Peter, A., 2007. A strategy to assess river restoration success. Freshwater Biol. 52, 752–769. http://dx.doi.org/10.1111/j.1365-2427.2007.01740.x.

44. Yeh, T.-C.J., Liu, S., 2000. Hydraulic tomography: development of a new aquifer test method. Water Resour. Res. 36, 2095–2105. http://dx.doi.org/10.1029/2000WR900114.

45. Yeh, T.-C.J., Xiang, J., Suribhatla, R.M., Hsu, K.-C., Lee, C.-H., Wen, J.-C., 2009. River stage tomography: a new approach for characterizing groundwater basins. Water Resour. Res. 45, W05409. http://dx.doi.org/10.1029/2008WR007233.

46. Zhu, J., Yeh, T.-C.J., 2005. Characterization of aquifer heterogeneity using transient hydraulic tomography. Water Resour. Res. 41, W07028. http://dx.doi.org/10.1029/2004WR003790.

CHAPTER 8

PERSISTENCE AND THE PREDICTABILITY OF HYDRO-ECO-GEOMORPHOLOGICAL SYSTEMS

Keith Beven

Keywords: Event persistence, Self-organisation, Event sequences, Shallow landslide model, Epistemic uncertainty

INTRODUCTION

Landscapes are structured. They show spatial organisation that is the result of their development over time. In that landscapes are open systems, with a throughput of mass and energy, that development is necessarily evolutionary, even if certain aspects of pattern and form might appear (at least for certain periods of time) to be in some steady or quasi-equilibrium state consistent with the distribution of external forcing. That external forcing has, for much of geological time, been due to natural agents but in the anthropocene, man has had an increasing influence on both process and pattern in the landscape. Until the anthropocene, the landscape was, necessarily, self-organising in ways that have led to some general emerging patterns (climatic zones, river basins, natural vegetation patterns) but now there is a co-evolution of man and the landscape that leads to new emergent structures (and subjects of research as reflected in the concept of ecohydrology, sociohydrology and the new IAHS Panta Rhei decadal programme; Sivapalan et al., 2012, Montanari et al., 2013, Ehret et al., 2013 and Lane, 2014).

The landscape we see now, both in our qualitative interpretation of form and process, and in the patterns of measurements we might make in space and time, is an integration of past temporal and spatial processes and events, with different time and space scales of effectiveness and impact. The system is open and the dynamics are nonlinear, even if we have to close the system and specify both initial and boundary conditions for a particular period of study. It has therefore been rather attractive to borrow from the concepts of nonlinear systems and apply concepts such as self-organised criticality to environmental systems. In brief, systems that tend to evolve to critical states are unpredictable in that small forcing events might, in the right circumstances, lead to significant and rather unpredictable consequences or emergent properties, in particular resulting in power law magnitude-frequency relationships. Such concepts have proven useful in explanations of environmental systems, including the fractal nature of river networks and catchments, the magnitudes of earthquakes, the form of debris cones, and the areas of forest fires. This form of explanation has been advocated as providing behavioural principles that govern environmental systems and should be included in environmental models (Schaefli et al., 2011) but also criticised as a general form of explanation (e.g. Frigg, 2003) and in specific applications.

One of the features of this approach is that history is important, which makes closure somewhat problematic if what we see now depends on unspecified initial and boundary conditions from the past. As systems evolve towards critical states, the ordering of events becomes important in that the sensitivity of the system to significant change will depend on the rather arbitrary occurrence of external forcing events in space or time (e.g. Beven, 1981). Significant change might be triggered at a particular point in time (or not) dependent on the current state of the system and the magnitude of the event. Thus, history is important, but its effect might be indeterminate. We can see now only the net effect of sequences of events, some of which might be seen as having persistence in what we see now.

Citation: Keith Beven, Persistence and the predictability of hydro-eco-geomorphological systems, Ecological Modelling, Volume 298, 24 February 2015, Pages 4-15, ISSN 0304-3800, http://dx.doi.org/10.1016/j.ecolmodel.2014.07.019.

Herein lies one of the problems with explanations based on self-organisation. Simulation models show that in order for the characteristics of the self-organisation to become fully apparent, a very large number of critical events are necessary in part because the different realisations of events will develop in different ways, even if they have a tendency to similar self-organisational forms. This requires time and a very large number of potential sites in space (as, for example, in earthquakes and river basins). Thus, even if there is a tendency towards self-organised criticality in such systems, will it be apparent in what we see now given this requirement, or will we see only transient characteristics when changes are occurring more rapidly than sufficient critical events can occur? Such changes might themselves be the result of self-organised criticality operating on some other processes at some other scale (see examples below). How does what we see now reflect this balance of changes within the overall evolution of the system and what are the implications for the predictability of the system of interest?

Here we would like to suggest that another viewpoint is richer in explanatory power. This is not to deny that concepts of self-organisation do not provide valuable insights but what we see now is, in very many landscapes, the result of past events that have had the effect of resetting the initial conditions for the processes currently operating. This can happen frequently in some systems; and at Holocene or orogenic scales for others. What we see now is a (non-linear) superposition of the effects of distributions of events that have occurred at different time scales into the past (including anthropogenic events) and of the dispersion of the effects of those events into different parts of the system that defines their persistence under the forcing of particular sequences of events in time at any particular site in space.

There is then a question of what self-organised criticality means in this situation (except in some rather trivial sense that an event necessarily has consequences). That depends very much on how sensitive a system is to small forcing events having large consequences, i.e. how close that system is to criticality or how quickly it moves towards criticality after an event. But that is not what we see in many systems. We certainly see history resulting from evolution in the nature of soil profiles, in the succession of vegetation communities, the form of river basins etc. but that history seems to be rather easily reset by external forcing events rather than internal organisation. What we see is in evolution but with evidence of the persistence of some past extreme forcing events that did have a dramatic effect on the nature of the system, changing the initial conditions for the time evolution at that site. We shall avoid the use of the word relaxation following such a critical event (e.g. Anderson and Calver, 1977, Culling, 1986, Ahnert, 1994, Calver and Anderson, 2004 and Phillips, 2009) because, even if there might be a return to some particular quasi-stationary form, the system is not actually stationary in any way. What we see now (over the limited time scale of a typical research project or research career) will then be a particular state in that evolution, as dependent on a particular sequence of events when the ordering of events might be important. Lacking data from the past, origins beyond that time scale are necessarily the stuff of speculation. Hence, the attraction of finding general concepts and theoretical principles for extending our knowledge and understanding over those longer time scales.

It has always been thus, of course. We naturally tend towards generalised theories, but the drive today is to find theoretical principles that have quantitative utility, rather than only qualitative explanatory power. The questions that consequently arise include how far it is possible to distinguish between competing explanations in the light of limited data and the particular contingencies of individual events (and anthropogenically induced change) in shaping what we see now. For the hydrologist this is an extension of the continuing discussion of equifinality of models and testing models as hypotheses (e.g. Beven, 1996, Beven, 2002, Beven, 2006, Clark et al., 2011 and Beven et al., 2012). For the geomorphologist it is an extension of the discussion of the concepts of equifinality, equilibrium, and nonlinear dynamics to landform systems (Culling, 1957, Culling, 1987, Culling, 1988, Culling and Datko, 1987, Phillips, 1997, Renwick, 1992, Ahnert, 1994, Beven, 1996, Phillips, 1997 and Phillips, 2011). For the ecohydrologist it is an extension of the discussion of behavioural principles to the landscape (Schaefli et al., 2011).

There is an interesting aspect to the original concept of equifinality in geomorphology (see Culling, 1957; and his later discussion in Culling, 1987 and Culling, 1988) that has relevance here. That is that looking backwards into the past, it is impossible to know all the details of the events and processes that formed a particular landform feature. Thus there might be an equifinality of explanation. This is a form of epistemic uncertainty that will hold for any open system under study within which history and sequences of events are important. But, it is particularly severe for all those events and processes that do not have persistence to the state we see now. Where self-organised criticality offers something new in this respect is to extend the concept of persistence to the net effect of long sequences of events producing recognisable organisation in the landscape that may not have its origin in some extreme event. Where it does not necessarily help is in shedding light on the impacts of events in producing that organisation. We are limited to seeing the effects of events that have persistence extending to the period of study or what we see now.

Thus, the palimpsest of landscape will be the result of an evolution that includes a variety of different forcing events. In general, the effects of large events will have longer persistence, and the effects of small events will dissipate more rapidly, but there is the possibility that for systems close to some critical threshold, a small event will induce some impact with persistence such that the ordering of events will be important; a form of conditioning and triggering (Phillips, 2009). All events contribute to the throughput of energy through the system, and consequently the evolution of the system, that may be gradual, punctuated by sudden changes rather than demonstrating some dynamic equilibrium. These features of nonlinear open systems now seem conceptually uncontroversial. It should also then not be controversial that what we see in the landscape is the persistence of events that have changed the boundary conditions for smaller scale processes.

TURBULENCE: ORGANISATION AND BOUNDARY CONDITIONS

Many fluxes of water and air in and above the landscape are turbulent. Turbulent flows are often cited as an example of nonlinear dynamics and self-organisation in practice, albeit one in which it has proven impossible to resolve completely (hence the resort to various closure schemes in predicting turbulent flows). Turbulence shows complex structures in both air and water, such as the horseshoe vortices that have been studied in rivers, and the boundary layers that build up around obstructions and changes in local internal boundary conditions for the flow domain of interest.

For the case of homogeneous turbulence, far from any obstruction or boundary, the complete description of turbulence remains unresolved. Even direct numerical simulations of the Navier–Stokes equations need some closure scheme for how momentum dissipates at the level of the Kolmogorov length scale (e.g. Meneveau and Sreenivasan, 1987). The issue is worse for inhomogeneous turbulence induced by boundary condition effects (Hunt et al., 2006). The length scales over which changes in boundary conditions can have an impact can be very long. But the effects of changes in boundary conditions in time and space will be partly organised by the flow itself (as in dune and other features in mobile bed rivers) and partly imposed on the flow by totally independent factors (such as geological structures in bedrock rivers and large boulders from rock falls or glacial deposition). There is a range of intermediate scales from gravel bars to flood plain meanders within which the organisation will depend on a sequence of external forcing events. The effects of such events may be more or less persistent dependent on the magnitude of that event and the integral effect of the subsequent effects that create what we see now.

FORCING EVENTS AND SELF-ORGANISATION – WHEN DOES EVOLUTION BECOME A CHANGE IN BOUNDARY CONDITIONS?

This is just one example of how landscape systems are only partly self-organising because of their dependence on external forcing events that drive the evolution of the landscape by changing the constraints on smaller scale processes. In fact, it could be suggested that self-organisation is really only a part of the mechanism for dissipating the effects of such singular events. This is quite clear when we look at the landscape in many parts of the world. What we see now is clearly a reflection of the past and, in particular, the result of glacial and periglacial events and processes. These have had long persistence in the forms of hillslopes and catchments and, less obviously, in soil structures and the occurrence of landslides and debris flows in moraine and head deposits that have been "waiting to fail" for thousands of years.

This is a modern form of a long time discussion in geomorphology between catastrophist and dynamic equilibrium views of landscape evolution. The balance between the two views depends on the magnitude of a forcing event relative to the rate of dissipation of the impact of that event. Extreme events, such as the last glaciation, can change the constraints for the dissipative processes to such an extent that the effects of such an event are persistent. The effects of small events will dissipate much more quickly, but might still be persistent to the observer depending on when the observations take place. The occurrence of events, including anthropogenic events, might be quite arbitrary even if contingent on the overall evolution. The balance is then complicated further by the dependence of the response to an event on the antecedent state of the system and the possibility that dissipative processes might act to make the system more sensitive to smaller events (including the instigation of catastrophic events at larger time scales). The occurrence of shallow landslides and debris flows are an interesting example in that respect.

EVENT-PERSISTENCE CONCEPTS AND CONTINGENCY

Thus, in an event-persistence view of hydro-eco-geomorphological evolution, how the system responds depends on the antecedent states and the nature of the event in space and time. The persistence of that event will have an effect on the antecedent conditions for the next event. After an event the system continues to evolve in ways that might make it more or less sensitive to future events.

Consider the case of the occurrence of a shallow landslip or debris flow. Debris flows can have locally dramatic effects on hillslides and valley bottoms, and any communities and human activities that they impact. In July 2007 a small debris flow occurred in the Mallerstang Valley, Cumbria, UK as a result of intense rainfall over Mallerstang Edge resulting in the transport of large amounts of sediment into the Outhgill Beck and the River Eden, blocking of the channel and road bridge in the hamlet of Outhgill. Fortunately the debris flow only caused damage to some field boundary walls; the channel and bridge were soon cleared; no properties were damaged or lives lost (although a lot of coarse sediment deposited upstream of Outhgill is still being transported in later events). Elsewhere in the world the impacts of debris flows have been much greater (e.g. Carrara et al., 2008).

In this case, the debris flow originated from glacial lateral moraines left high on the sides of the valley. There is evidence of similar failures elsewhere in the valley and similar locations in the UK. They are, however, rather arbitrary in occurrence, contingent on a combination of local occurrence of intense rainfalls and antecedent conditions (and possibly long term evolution of the cohesion of the materials at a site). In other small tributaries of Outhgill Beck, less than 100 m from the impacted channel, there was no sign of erosion or major change. There was, however, another similar debris flow that blocked a bridge on another tributary to the River Eden on the same day some 3 km downstream. In this location, the combination of circumstances producing such events seems to be rather rare in both time and space.

Once such a flow has occurred of course, it might have persistence in a number of different ways: visually as a qualitatively distinguishable feature of the landscape; in making sediments available from stream transport; in delaying the potential future occurrence of a debris flow at that same site for long into the future (relative to what we see now). Newson (1980) reports on a similar complexity of response from 2 events at the Plynlimon experimental catchments in mid-Wales. Both events were estimated to have return periods of the order of 100 years (although they were rather different in their rainfall patterns) but occurred within 5 years in 1973 and 1977. The first event was characterised by a landslide and debris flows; the second by transport of sediments in the river network. Effectively the first event changed the antecedent conditions for the second. Similar coupling sequences have been described for the Howgill Hills in Cumbria by Harvey (2001) with some evidence that there have been a number of distinct periods of hillslope activity in the last 2500 years (Chiverrell et al., 2008).

Clearly, once a slope has failed it will generally require a period of time to evolve towards being a potential failure site in the future (e.g. Fig. 1). This will also depend on the sequence of events in relation to the accumulation of soil at the site. Where a debris flow has exposed a bedrock surface then subsequent events might actually keep this clear of soil for significant periods of time. In Nelson County, Virginia, the scouring of valley bottoms to granite bedrock by debris flows during Hurricane Camille in 1969 has had decadal persistence (Bechtel, 2006). How quickly soil and vegetation starts to build up again will depend on geology and the surrounding topography and soils. Vegetation developing at a site will help retain soil and later, root strength will help maintain stability relative to the magnitude of forcing events. The non-occurrence of a failure, and subsequent increase in soil depth might also reduce the potential for failure. A change in rainfall regime with an increase in rainfall intensities might increase the potential for failure, depending on the development of the system states between events. Anthropogenic events, such as forest clearing, might change the potential for failure significantly (e.g. Montgomery et al., 2000). This is one example where the ordering of events at a site might be important on whether a failure occurs or not (Beven, 1981). Prediction of such events remains difficult, if only because of lack of adequate rainfall data in time and space, lack of adequate soil property data and the persistence of effects from past failures (e.g. Casadei et al., 2003).

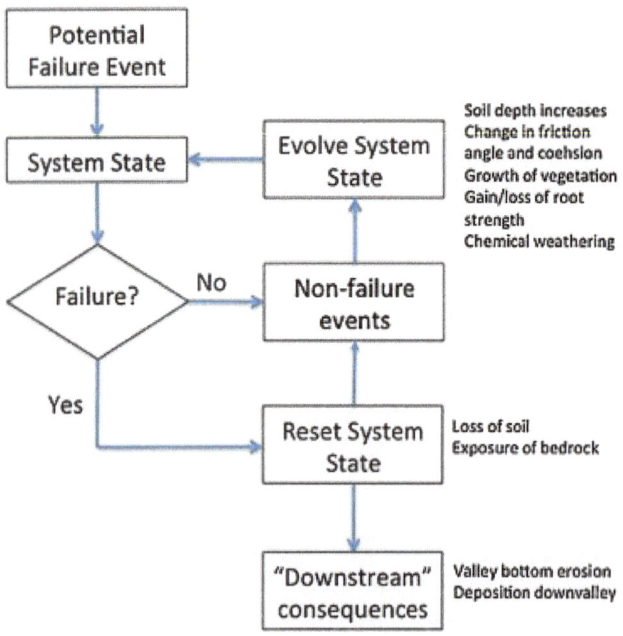

Figure 1: Simple event-relaxation concepts for a potential debris flow site.

Some attempts have been made to look at the long-term implications of debris flow events in time and space on catchment morphology (e.g. Benda and Dunne, 1997 and Stock and Dietrich, 2006). This change of scale reduces the focus on individual sites, to the integral effect over long periods of time. We do not then know, of course, what the particular distribution of events in the past might have been, nor generally what the past impacts of man might have been. What we see now is the persistence of (some) individual local events in the past superimposed on long term evolution which provides an antecedent condition for the local; to "see" the integral effect requires some theoretical interpretation and some auxiliary assumptions about distributions of events and their impacts.

ORGANISING PRINCIPLES AND QUANTITATIVE PREDICTION

Perhaps more controversial is what new concepts can be introduced within this framework that will result in additional insights, over and above information gained from purely empirical study of local persistence. Here again we should distinguish between qualitative concepts, such as self-organised criticality, that can be demonstrated on simple theoretical systems but which might be much more difficult to verify as hypotheses in a complex landscape because of the sheer number of degrees of freedom and uncertainties inherent at landscape scales (Frigg, 2003); and those that might be useful as deductive quantitative constraints of system behaviour. Within the first category we can include concepts such as the complex attractor, bifurcation and chaotic behaviour, dynamic equilibrium and nonequilibrium, and self-organised criticality. Within the latter category are principles of fractal scaling, maximum entropy production, minimum energy dissipation, maximising boundary energy gradients, and dissipation of Helmholz free energy (see Rodriguez-Iturbe and Rinaldo, 2001, Kleidon and Schymanski, 2008, Schymanski et al., 2010, Zehe et al., 2010, Schaefli et al., 2011, del Jesus et al., 2012 and Kleidon et al., 2012).

As Reggiani et al., 1998 and Reggiani et al., 2000 and Kleidon et al. (2012) amongst others have pointed out there has often been a lack of rigour in applying physical principles in studies of environmental processes and landscape. Such systems are subject to the requirements of thermodynamics and the conservation of mass, energy, and momentum. In particular, Kleidon et al. (2012) emphasise the open nature of such systems and the way in which the free energy supplied as net boundary fluxes must be dissipated according to the laws of thermodynamics (see also Porada et al., 2011). This has suggested to some that these types of organisational principles can provide constraints on models that might be useful in making predictions (Schaefli et al., 2011). Similar arguments have been made by Bejan (2007, Bejan and Lorente, 2010) in his theory of constructal networks. These are, however, purely theoretical conjectures that we have little hope of proving unequivocally because of the limitations on how we can account for mass, energy and momentum in any complex system (Beven, 2002 and Beven, 2012). Similar points have been made more generally by the philosophers Morton (1993) and Cartwright (1999). Cartwright suggests that such principles might be better perceived as potential *capacities* rather than quantitative constraints. Indeed, she suggests that even the thermodynamic laws must be treated in this way when applied to any particular system.

These principles are generally quantified and deemed applicable by empirical back-calculation of coefficients. This type of determination of empirical calibration is, of course, traditional in hydrology and other environmental sciences.

Many other process representations were developed and continue to be justified in this way. Equivalent roughness lengths at a flow boundary can always be calculated given a velocity profile (even if in some circumstances that calculation yields values orders of magnitude greater than the sand grains that Nikuradse›s original experiments involved, e.g. Babaeyan-Koopaei et al., 2002). Darcy›s law will hold when a hydraulic conductivity can be back-calculated from observations of profiles of soil water tension and moisture content, even where preferential flows would mean that it is not the appropriate theoretical framework (Beven and Germann, 2013).

The application of theory where it is not appropriate is clearly not new (even if it should be considered as bad practice) and in many areas of environmental modelling the aphorism of George Box that all models are wrong but some might be useful is generally accepted. It might then be helpful, however, to have some indication of the types of conditions under which certain theoretical concepts might be useful in this sense. In the case of the organisational principles discussed above, this would appear to depend on some balance between the rate at which the potential towards observable organisation evolution takes place and the persistence of boundary controls that apply at different time and length scales.

Consider the development of river networks. Rodriguez-Iturbe and Rinaldo (2001) argue that water draining from the land will take a path of least resistance and will consequently minimise energy dissipation in developing a drainage network and patterns of slope and catchment areas. The result is a set of fractal relationships characteristic of self-organised processes with exponents in the range found by the empirical analysis of actual networks. Bejan (2007) extends this optimisation argument to explain channel-cross sectional areas and flux in other tree forms, Kleidon et al. (2012) to critical zone fluxes, and Sidle et al., 2000 and Sidle et al., 2001 to drainage networks within hillslope soils.

Drainage networks develop over time under the influence of a (changing) distribution of events large and small. There are critical event thresholds at a variety of scales, from the initiation of motion of individual sediment particles, to the collapse of undercut banks, to the cut-off of river meanders and to major avulsions of channels from one position to another. The way in which momentum is dissipated in the network will depend on the capacity of the stream to erode and transport the material of the bed and banks, which itself will depend on the magnitude of a particular spate event and the disposition of the sediment prior to that event (bed armouring and the way in which phytoplankton, macrophytes and bankside vegetation locally affect the supply, cohesion, mobilisation, and deposition of sediments). As noted earlier, what is available to be transported by an event, might depend on past events. Transport within an event might be transport limited in some circumstances and supply limited in others, even at the same site. There may be a scale effect as well, in that persistent events at small scales might lead to the breakdown of scaling laws apparent at larger scales (e.g. Moody and Kinner, 2006).

This will be cases, however, where what we see now at the local scale will often show little in the way of clear persistence from past events. Where a major event has stripped vegetation and reworked gravel deposits then the persistence will be greater (and can sometimes be dated from vegetation ages, etc.). Even then, however, the channel modifications are largely constrained by what was there before, representing the integral effect of all past events.

It is clear that even in this case the network cannot be entirely self-organising. Persistence is still often apparent in prior landforms, the influence of underlying geology and the effects of changing base levels, etc. This will be most evident at the external links to the network when the uniqueness of particular places comes most into play (Beven, 2002). It will be less evident at the scales where a river is reworking its own past alluvial deposits. At smaller catchment scales, the scope for arbitrary effects and contingencies is greater.

One particular example of this is the effect of the outbreak of chestnut blight in the southern Applachians in the early part of the 20th Century. Although it is thought that this was introduced to the US by the importation of Asian chestnut trees around 1900, by 1940 some 4 billion trees had been lost. In parts of the Appalachians 25% of the trees were chestnuts. Many of the affected trees were cut for timber, some after damage or toppling by wind. The blight has been suggested as a major reason for the expansion of gully systems in the southern Appalachians in the 1930s that persist to this day. Fire (Moody and Kinner, 2006) and blowdown due to wind have also been invoked as forms of contingent events (Phillips et al., 2008 and Phillips and Park, 2009). In Virginia, in October 1979, an early snowstorm led to about 30% of trees in some places in Shenandoah National Park falling over. The arroyo expansion in the American south-west in the late 19th century might also have an anthropogenic origin with the introduction of cattle grazing. In this case, there may have been earlier, more "natural" periods of gullying induced by occasional droughts (Antevs, 1952, Tuan, 1966 and Pelletier et al., 2011). There is clear persistence, but less information about causes, initial conditions and changes.

PREFERENTIAL FLOW IN SOILS

Another consequence of chestnut blight in the Appalachians was the death and decay of root channels resulting in significant networks of preferential flow pathways through the soil. In this case a critical event has had a consequence, but in this case it is far more of an arbitrary event external to the preferential flow process. There are other factors that affect preferential flows, such as worms and burrowing animals, and cracking during drought periods, that are also external to the preferential flow process. Such processes might actually create opportunities for preferential flow where there was little or none before, even though there may be little change in the hydrological regime. Other external processes, such as the use of ploughing rather than no-till agricultural management, might reduce the opportunities for preferential flow. These external effects are changing the boundary conditions for the preferential flow process, rather than being the product of self-organisation of the flow process itself. What we see now at any site, clearly depends on the rather arbitrary occurrence of such events and their time scale of persistence. But they may have persistence over long periods of time. Beven and Germann (2013) note that, even in the case of cracking in a clay soil, the cracks may persist as preferential flow pathways even after a wet winter (in part due to roots that lie on cracked ped surfaces rather than penetrating the clay). It is therefore difficult to believe in the general validity of the suggestion that where there is "excess water" on hillslope a fractal preferential flow network will develop so as to evacuate that water (e.g. Sidle et al., 2000, Sidle et al., 2001 and Weiler and McDonnell, 2007).

It is interesting to think through the concept of excess water on hillslopes. There is a long history of such a concept. Robert Horton (1933) suggested that the soil surface was a separating surface, causing water in excess of the infiltration capacity of the soil to become surface runoff (though his concepts were actually somewhat more sophisticated, see Beven, 2004). He recognised that cracks in the soil could be important, but concentrated on the downslope movement of excess water through crack systems or "sun checks" which he called "concealed surface runoff" (Horton, 1942 and Beven, 2004). Later, Dunne and others introduced the concept of saturation excess surface runoff (Dunne, 1978) and Sidle et al., 2000 and Sidle et al., 2001 have suggested that preferential flow lines could develop in the soil to deal with water fluxes in excess of storage in the matrix. The generality here is that water in excess of what can be stored locally finds some way of finding its way downslope towards a recognisable channel. In some cases, of soils that are easily dispersed or eroded, that may cause a feedback into the development of subsurface pipe systems, rills or gullies with a consequent increase in the effective drainage density. Where the flow can have a direct effect on developing the flow pathways this can be a form of self-organisation, similar to that which leads to dendritic drainage patterns in general. It is not so clear, however, that this would be evident in cases where the soil surface is not easily eroded and the hillslope simply sheds any excess water as surface runoff, except perhaps in the most extreme events or after the surface has been disturbed by anthropogenic action.

It is also not so clear in the case of preferential flow within the soil, when the flow pathways due to roots, fauna, drying and other causal mechanisms that are largely independent of preferential flow of the water itself. The water might still locally take a path of least resistance, and some of the networks in the soil that are self-organisational (such as root networks, mycorrhizal mycelial networks and cracks and fissures) might have fractal scaling characteristics (e.g. Van Noordwijk et al., 1994, Deurer et al., 2003 and Boddy and Donnelly, 2008). It might also be the case that some flow pathways might be enhanced over others, for example by the translocation of clay particles and build-up of cutans on the surfaces of some preferential flow pathways, or long-term chemical weathering. This is also thought to happen in the shallow pipe networks in peat soils found in the UK and elsewhere (Chappell, 2010 and Jones, 2010). These networks appear to have their origins in cracking in rare drought summers, but then parts of the network are enhanced over others by the flow process itself. The network can also be changed by roof collapse and blocking (Gilman and Newson, 1980). On balance, however, this would appear to be one case where an event-persistence concept might dominate self-organisation in controlling the local boundary conditions for the flow.

VEGETATION PRODUCTIVITY AND DROUGHT

Another example of this balance between event persistence and organisational principles is provided by the organisation of vegetation patterns. One issue in looking at vegetation patterns is that there are very few vegetation communities that have not been affected by man, either directly or (as in the case of chestnut blight) indirectly or (as in the case of beetle infestations following the impacts of acid rain on trees in the north of the Czech Republic) both. Self-organisational principles have been invoked to explain some vegetation patterns, such as the banded patterns of the "tiger bush" in the semi-arid Sahel region of Africa as a response to limited water supply (Galle et al., 1999 and Valentin et al., 1999). Banded vegetation patterns are also found elsewhere (e.g. Dunkerley and Brown, 2002 and Pelletier et al., 2012).

An interesting example is the suggestion that more generally natural vegetation is adapted to maximise net carbon profit within the constraint of maximising entropy in the local energy and moisture regime. Schymanski et al., 2007, Schymanski et al., 2009 and Schymanski et al., 2010 have shown how maximisation of entropy might be used as a quantitative constraint in models of evapotranspiration processes. The interaction of vegetation, atmosphere, and the hydrological regime has been explored in some detail as a problem of nonlinear dynamics, including the possibility that changes in one area might have consequences in another area at distance. Such consequences might be part of the overall non-equilibrium response but would be seen in the affected area as a change in boundary condition with some degree of persistence.

However, consider the piñon pine–juniper community found in areas of the American south-west. This community has been the subject of significant research (e.g. Huffman et al., 2012 and Limousin et al., 2013). In 2002–2003 an external forcing event, in this case and extended drought, caused the death of many of the piñon trees, at sites that had already suffered mortality in the 1996 drought (Mueller et al., 2005). Such events are persistent. Other contingent events such as fire can also have a significant impact on water use and productivity, even where the area is recolonised by the same species, such as in the Mountain Ash (*Eucalyptus regnans*) that is found in the water supply catchments for Melbourne, Australia. In these catchments it has been suggested that the effects of fire on the water yield of these catchments can have a persistence of over a century (see Jayasuriya et al., 1993 and Watson et al., 1999). What we see now is the persistence of individual contingent events.

In these particular cases the impact and persistence of an event can often be distinguished by aging the vegetation using tree rings or other techniques. Where trees survive, a history of fires and droughts can sometimes be distinguished at a site (McBride, 1983, Swetnam, 1993, Buechling and Baker, 2004, Li et al., 2006 and Liang et al., 2006). Such studies have also shown that death can be a function of the stand characteristics and local soil properties (e.g. Ogle et al., 2000, Floyd et al., 2009 and Peterman et al., 2012) and might also involve other mechanisms such as increased sensitivity to insect damage (Clifford et al., 2008). The occurrence of contingent persistent events is not necessarily inconsistent with some organising efficiency or optimality principle applying at different stages during regrowth, but it must do so under continuously, and sometimes dramatically, changing constraints.

A SIMPLE CASE STUDY: SURFACE EROSION AND SHALLOW SLOPE FAILURE IN A HILLSLOPE HOLLOW

We can illustrate some of these issues by a simple model of soil accumulation and erosion in a single hillslope hollow where soil cohesion is a function of vegetation succession with increasing soil depth, and surface erosion is also affected by the vegetation cover (as in Fig. 1). To keep the example very simple we assume that the infinite slab model of shallow landslip failure holds (e.g. Montgomery and Dietrich, 1994 and Borga et al., 2002). In this case, the slope will fail if the ratio of water table depth to soil depth satisfies the inequality

$$\rho_s dg \sin\alpha \cos\alpha > C + (dg\rho_s - hg\rho_w - h_e g\rho_w \cos^2\alpha \tan\varphi) \tag{1}$$

where d is soil depth (m), h is the water table depth above the base of the soil (m), *he*excess pressure head (m), C is the total cohesion of the soil, (kPa) and the other variables are defined in Table 1.

Table 1: Parameters of the simple soil depth model

Parameter	Symbol	Base value	Units
Initial soil depth	*do*	0.1	m
Slope angle		0.36	rad
Effective upslope contributing area	*Ac*	1000	$m^2\ m^{-1}$
Density of water	*w*	1000	$kg\ m^{-3}$
Density of soil	*s*	1500	$kg\ m^{-3}$
Density of bedrock	*s*	2500	$kg\ m^{-3}$
Maximum cohesion of soil	*Co*	1	kPa
Maximum cohesion of roots	*Cr*	3.5	kPa
Time constant to reach maximum cohesion of roots and equilibrium soil depth	*Te*	50	y
Depth at which root cohesion reaches a maximum	*a*	0.8	m
Rate parameter in logistic root cohesion function	*b*	50	y
Equilibrium soil production rate over upslope contributing area	*Cp*	0.001	$m\ y^{-1}$
Friction angle		0.576	rad
Minimum effective event specific discharge	*qmin*	25	$m\ d^{-1}$
Maximum effective event specific discharge	*qmax*	300	$m\ d^{-1}$
Beta shape parameter 1 for effective specific discharge	*B1*	1	–
Beta shape parameter 2 for effective specific discharge	*B2*	12	–
Minimum inter-event time	*Imin*	0.01	y
Mean inter-event time	*Imean*	0.5	y
Transmissivity at soil depth of 1 m	*To*	150	$m^2\ d^{-1}$
Power of soil depth/transmissivity function	*f*	5	m
Surface erosion coefficient	*Ce*	0.025	$d\ m^{-3}$
Excess pressure coefficient	*Cx*	0.1	–

The water table depth will depend on the effective discharge from upslope and a transmissivity function that we will assume to be a simple power law that is constant in time. Implicitly therefore we assume that as the soil depth increases the transmissivity increases as a result of root and other channels in the soil as the vegetation develops on the site. Thus:

$$T = T_o d(i)^f \tag{2}$$

where T_o ($m^2\ s^{-1}$) is the transmissivity at a scaling depth of 1 m and *f* controls the nonlinearity. In a similar way, cohesion due to roots be a logistic function of soil depth, reaching a maximum strength at a soil depth *a*, reached at a time scale of *b* years.

$$C = C_o + \frac{C_r}{1 + \exp(-b\{d - a\})} \tag{3}$$

Once the slope has failed, we assume that the soil depth starts to increase as a function of both weathering and the dispersive transport of soil from the same effective contributing area upslope. It will, however, take time to reach an equilibrium soil depth. The accumulated soil depth is specified as a function of the rate of soil production over the upslope catchment area, the current soil depth and time to reach an equilibrium soil depth as:

$$d(i) = d(i-1) + \left(\frac{\rho_r}{\rho_s}\right) d_e \left(1 - \frac{d(i-1)^2}{d_e^2}\right)\left(\frac{I(i)}{T_e}\right) \tag{4}$$

where $d(i)$ is the soil depth (m) prior to event i after an event inter-arrival time $I(i)$ (years) and *de* is the equilibrium soil depth due to supply from the upslope contributing area of *AcCp* (m^2). Potential failure events are generated randomly with effective event magnitudes of specific discharge over the upslope area generated as a beta distribution with specified minimum and maximum values, and event inter-arrival times as an exponential distribution, shifted by a minimum event inter-arrival time.

Where the discharge supplied from upslope exceeds the downslope flux capacity of the current soil depth then it is assumed that there is some potential for surface erosion. This is represented as a simplified form of the Morgan–Morgan–Finney model (Morgan, 2001, Vigiak et al., 2006 and Morgan and Duzant, 2008) as:

$$G = C_e(q(i) - q_o(d))^2 \sin\theta; \quad q(i) > q_o(d) \tag{5}$$

where $q(i)$ is the specific discharge for event I, and $qo(d)$ is the maximum specific discharge at the current soil depth, *d*. The effects of bedrock fractures on the potential for excess pore water pressures in inducing failures is often revealed only after an event, (e.g. as at the Coos Bay site, see Montgomery et al., 2002) but might be important in controlling the occurrences of failures. This has been included here also in a very simple way by making excess pressure proportional to the excess water depth (as predicted by an extrapolation of the transmissivity function) over the soil depth as:

$$h_e = C_x(q(i) - q_o(d)); \quad q(i) > q_o(d) \tag{6}$$

The model considered here allows for the evolution of soil depth over time but considers only potential failures at that site and not potential failures at other sites in the same drainage, i.e. we assume we have chosen the most critical site in that drainage. However, it already requires a large number of parameters to be specified (Table 1), it assumes that those parameters are constant over time, and it shows rather complex behaviours (Fig. 2) that are sensitive to both realisation effects (Fig. 3) and changes in parameter values (Fig. 4). In Fig. 3, the only difference between runs is the random seed that starts the sequence of events. In Fig. 4 only the value of the parameter *b* has been changed, all other parameters and the sequence of events is the same. While this simple landslip model could be further complicated (e.g. by introducing state dependent parameters) this would not fundamentally change the inferences that can be drawn from the complex simulated responses.

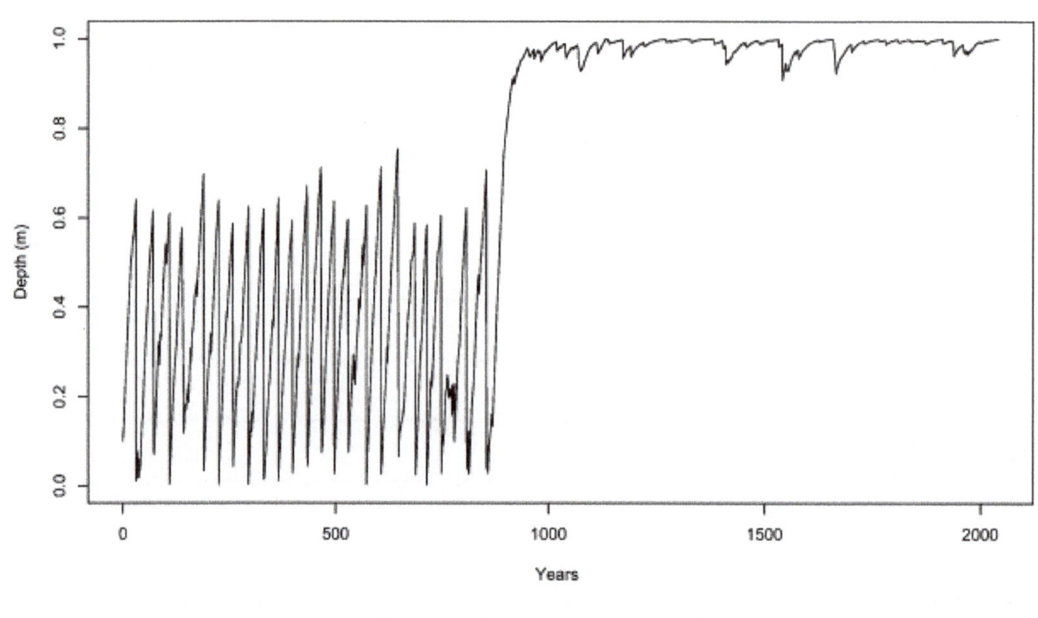

Figure 2: The evolution of soil depths over time at a hillslope hollow failure site for one realisation of the base parameter values of Table 1.

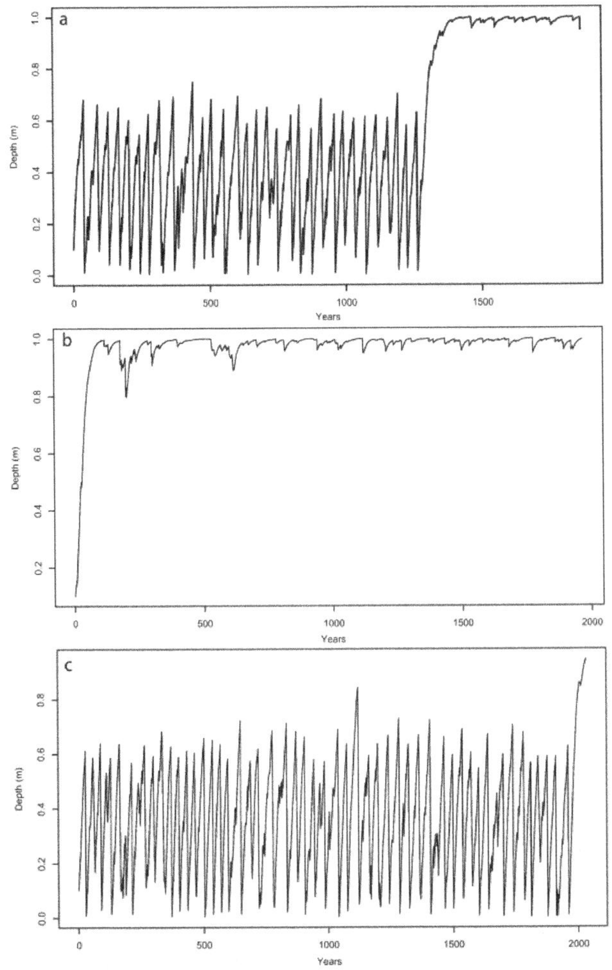

Figure 3: Different realisations of the evolution of soil depths over time in the slope failure model for the same base parameter values shown in Table 1. The runs vary only in the random seed used to initialise the forcing events.

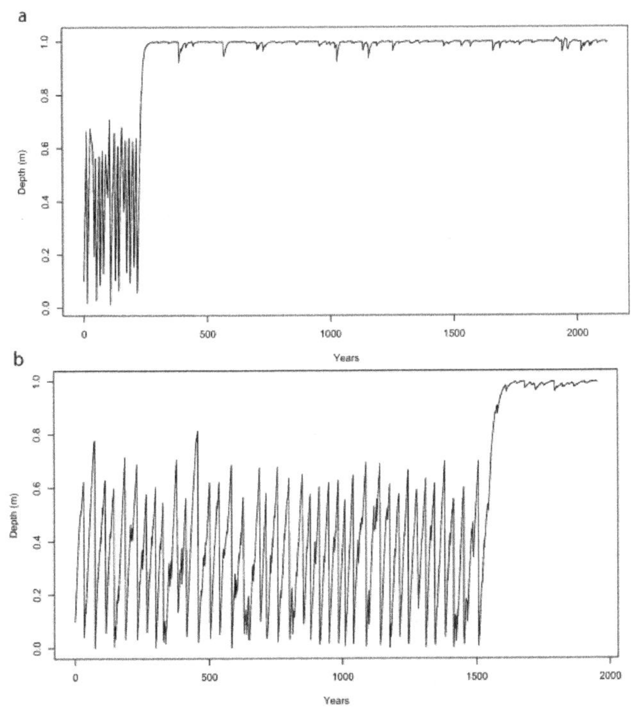

Figure 4: Different runs of the slope failure model for different values of the soil depth increase time scale parameter. Upper panel: $b = 25$ years; lower panel: $b = 50$ years. All other parameters as in Table 1.

This is undoubtedly a very simple model of soil depth evolution at a site. Similar models have been employed to assess the potential for failures in space (Montgomery et al., 2000) and, in some cases have been linked to equilibrium soil depth models (Dietrich et al., 1995), but many factors have been left out. It does not take any account of the effects of man or disease on the soil and vegetation characteristics and consequent cohesion at the site (e.g. Montgomery et al., 2000); it does not allow for the effects of geology and climate on weathering rates; it does not take any explicit account of the effects of vegetation changes on flows and antecedent conditions in the upslope contributing area; it does not allow for the effects of wetting and drying sequences on weathering and soil creep from upslope; and it does not allow for any non-stationarity of the parameters of the functional relationships incorporated in the model.

The time scales of these plots are long. What we see now at a site will depend on where we are on the trajectory at that site. We might see a site that is sensitive to failure with relatively shallow soil depths; we might equally see a site that is resistant to failure. Fig. 5shows the evolution of the factor of safety for one run of the model, at the point where there is a transition from a reduction of the factor of safety as soil depth increase, to where increasing soil depth results in an increasing factor of safety if a failure event does not occur (for those particular parameter values). The system exhibits a change in the mode of behaviour at this point. What we see now will depend very much on whether observations are made on one side of that transition or on the other.

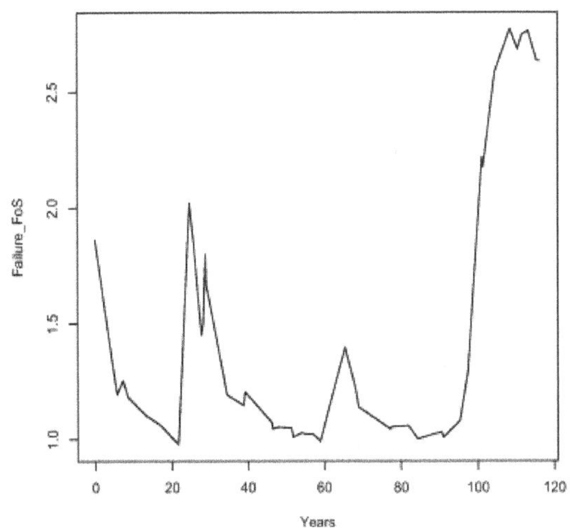

Figure 5: Factors of safety for part of a model run showing a transition from sequences of failures to relative stability once depth exceeds about 0.7 m.

An ergodic argument could be made at this point in that consideration of multiple sites in space might give a better reflection of the underlying nonstationarity and variability in the processes by reflecting the potential distribution of expressions of soil depth in any time window. This is clearly true to some extent, but there is also the potential for variability of basic parameters between sites, as well as other contingent auxiliary conditions that would increase the potential dimensions of the joint distribution to be considered. The number of sites required for robust inference might take on the magnitudes of samples needed to estimate a fractal dimension with precision (see Theiler, 1990 and Gallant et al., 1994).

MEDIATING MODELS IN UNDERSTANDING AND PREDICTING THE FUTURE

The point of all these examples is to show that it is not always possible to apply general constraining principles to explain the local detail of what we see now, nor to validate those general principles on the basis of short periods of observation. An event-persistence approach to explanation is not, of course, inconsistent with organisational principles that integrate over time or space, but both might contribute to a complete explanation of what we see now. This is analogous to the discussion of dynamic equilibrium and nonequilibrium in geomorphology (see for example, Renwick, 1992, and the simplexity arguments of Phillips, 1997) and more recent state and transition modelling methods that recognise the historical contingencies of system evolution (e.g. Phillips, 2007, Phillips, 2009, Phillips, 2012 and Phillips, 2013).

This is clearly the case with landscape and the environment. We have little reason to question the principles or capacities of mass balance and thermodynamics, but given the limitations of measurement techniques it is really not possible to disaggregate those balances or apply thermodynamical principles in any other than a very approximate way subject to significant uncertainties. The models that we use then will be mediating models, mediating between theory and observations, and subject to many auxiliary assumptions that may not be verifiable (Morton, 1993 and Gooding and Addis, 2008). I have argued elsewhere that in many systems, including catchments, it may then not be possible to distinguish between different models as hypotheses about system function (e.g. Beven, 2002, Beven, 2006, Beven, 2009 and Beven, 2012).

So there is scope for different types of mediating models, including false models. Some of the current (sometimes conflicting) organisational constraints that have been suggested might fall into this category. This is not necessarily a bad thing. As Frigg (2003) points out in his discussion of self-organised criticality concepts, false models can act as useful ways of changing entrenched thinking: "*to think differently about certain problems, motivate new questions, shed a different light on some issues, and finally make it easier to adopt an altogether different point of view. In doing so, the model acts as an antipode to stagnant assumptions, undercuts too readily accepted hypotheses, and helps to defamiliarise deeply entrenched styles of reasoning. In short, a false model can indicate alternative ways to deal with a phenomenon*" (Frigg, 2003).

The question that then arises, of course, is how best to test models given limited boundary condition and evaluation data that are subject to epistemic as well as statistical error. This is particularly important if we wish to use those models to make future predictions on which decisions will be based. Once it is accepted that epistemic errors are a significant constraint on model evaluation, it means that it is very difficult to make predictions associated with a realistic probability or likelihood that can be incorporated into risk-based decision making (Beven, 2012 and Beven and Young, 2013). We might then need to develop some non-statistical methods of model evaluation, such as in the latest applications of the GLUE methodology (Beven and Binley, 2013 and Beven and Smith, 2014). Global climate models are of this type; it is of some concern that they have not generally been rigorously evaluated even for short term decadal projections (Suckling and Smith, 2013). In this case, recognition of the epistemic uncertainties associated with the scenario projections might lead to decisions being made in different ways (Beven, 2011) but, this may not necessarily be a sufficient protection against an arbitrary future surprise (Beven, 2013).

Summary: the contingent, the arbitrary, and what we see now

What we see now in the landscape is the result of a long history of events with varying degrees of persistence. We have only limited access to much of that history and we know that many current events have only a minimal impact on what we see. Even rather extreme events may have impacts that are not very long-lasting but can have the effect of changing the constraints for future events. That means that sampling of sequences of events might be important in understanding the evolution of catchment hydro-eco-geomorphology. In some cases, however, extreme events can have persistence over hundreds or thousands of years. Any evolution of the landscape is then constrained by the effects of those events, however much it might be also constrained by self-organisational principles at shorter time scales. It might be difficult to verify those principles given the epistemic uncertainties that are generic to the studies that are possible within a research project or career. Such uncertainties also imply that it might therefore be difficult to test different models of development as hypotheses about how catchments and landscape systems function and even more difficult to assess future evolution when that might be dependent on arbitrary events in the future that could then have persistence. We search for general principles and theories that will support predictability but should recognise that what we see now may be subject to the effects of such contingent persistent events that may instigate new organisational structures.

Consequently predictability will be an issue for time scales between the very long, where organisation is integrated over many such contingent events, and the very short, in between contingent events that modify the short term local organisation that integrates over many small non-persistent events. The evidence presented by persistence in the landscape suggests that this situation is not uncommon such that an event-persistence conceptual framework might often be useful.

As the results presented in this paper show, however, we can also conceptualise that we might see rather different modes of behaviour at different points in the time trajectory of a sequence of events. Multiple sites might also be in quite different modes of behaviour at any given time, making it difficult to generalise and demonstrating a certain of uniqueness of place that is a result of both spatial and temporal variations (see discussion of Beven, 2000). Drawing general inferences in such situations might be difficult but we can at least propose some interesting topics for further study. This should include:

- What are the scales of persistence in any hydro-eco-geomorphological system?
- How far is any short term process organisation dependent on the effects of past persistent events?
- What are the potential future events, including anthropological interventions that might have the effect of changing the organisation of those processes?

These questions suggest a promising line of research that should be considered as complementary to the type of self-organisation and short-term behavioural principles of such systems being advocated elsewhere.

ACKNOWLEDGEMENT

This work is a contribution to the CREDIBLE consortium funded by the UK Natural Environment Research Council (Grant NE/J017299/1).

REFERENCES

1. Ahnert, F., 1994. Equilibrium, scale and inheritance in geomorphology. Geomorphology 11, 125–140.
2. Anderson, M.G., Calver, A., 1977. On the persistence of landscape features formed by a large flood. Trans. Inst. Br. Geogr. NS 2, 243–254.
3. Antevs, E., 1952. Arroyo-cutting and filling. J. Geol. 60 (4), 375–385.
4. Babaeyan-Koopaei, K., Ervine, D.A., Carling, P.A., Cao, Z., 2002. Velocity and turbulence measurements for two overbank flow events in River Severn. J. Hydraul. Eng. 128 (10), 891–900.
5. Bechtel, S., 2006. Roar of the Heavens. Kensington Publishing, New York.
6. Bejan, A., 2007. Constructal theory of pattern formation. Hydrol. Earth Syst. Sci. 11 (2), 753–768.
7. Bejan, A., Lorente, S., 2010. The constructal law of design and evolution in nature. Phil. Trans. R. Soc. B365, 1335–1347, http://dx.doi.org/10.1098/rstb.2009.0302.
8. Benda, L., Dunne, T., 1997. Stochastic forcing of sediment routing and storage in channel networks. Water Resour. Res. 33 (12), 2865–2880.
9. Beven, K.J., 1981. 'The effect of ordering on the geomorphic effectiveness of hydrologic events'. Proceedings of the International Conference on Erosion and Sediment Transport in Pacific Rim Steeplands, Christchurch, New Zealand. Int. Assoc. Sci. Hydrol. Pub. 132, 510–526.
10. Beven, K.J., 1996. Equifinality and uncertainty in geomorphological modelling. In: Rhoads, B.L., Thorn, C.E. (Eds.), The Scientific Nature of Geomorphology. Wiley, Chichester, pp. 289–313.
11. Beven, K.J., 2000. Uniqueness of place and process representations in hydrological modelling. Hydrol. Earth Syst. Sci. 4 (2), 203–213.
12. Beven, K.J., 2002. Towards a coherent philosophy for environmental modelling. Proc. R. Soc. Lond. A 458, 2465–2484.
13. Beven, K.J., 2004. Robert Horton's perceptual model of infiltration. Hydrol. Process. 18, 3447–3460.
14. Beven, K.J., 2006. A manifesto for the equifinality thesis. J. Hydrol. 320, 18–36.
15. Beven, K.J., 2009. Environmental Modelling: An Uncertain Future? Routledge, London.
16. Beven, K.J., 2011. I believe in climate change but how precautionary do we need to be in planning for the future? Hydrol. Process. 25, 1517–1520, http://dx.doi.org/10.1002/hyp.7939.
17. Beven, K.J., 2012. Causal models as multiple working hypotheses about environmental processes. C. R. Geosci. Acad. Sci. Paris 344, 77–88, http://dx.doi.org/10.1016/j.crte.2012.01.005.
18. Beven, K.J., 2013. So how much of your error is epistemic? Lessons from Japan and Italy. Hydrol. Process. 27 (11), 1677–1680, http://dx.doi.org/10.1002/hyp.9648.
19. Beven, K.J., Binley, A.M., 2013. GLUE, 20 years on. Hydrol. Process., http://dx.doi.org/10.1002/hyp.10082.
20. Beven, K.J., Germann, P.F., 2013. Macropores and water flow in soils revisited. Water Resour. Res. 49, http://dx.doi.org/10.1002/wrcr.20156.
21. Beven, K.J., Smith, P.J., 2014. Concepts of information content and likelihood in parameter calibration for hydrological simulation models. ASCE J. Hydrol. Eng. (in press).
22. Beven, K.J., Young, P., 2013. A guide to good practice in modeling semantics for authors and referees. Water Resour. Res. 49, http://dx.doi.org/10.1002/wrcr.20393.
23. Beven, K., Smith, P.,Westerberg, I., Freer, J., 2012. Comment on "Pursuing the method of multiple working hypotheses for hydrological modeling" by P. Clark et al. Water Resour. Res. 48, W11801, http://dx.doi.org/10.1029/2012WR012282.
24. Boddy, L., Donnelly, D.P., 2008. Fractal geometry and microorganisms in the environment. In: Senesi, N., Wilkinson, K.J. (Eds.), Biophysical Chemistry of Fractal Structures and Processes in Environmental Systems. JohnWiley and Sons, Chichester, pp. 239–272.
25. Borga, M., Dalla Fontana, G., Gregoretti, C., Marchi, L., 2002. Assessment of shallow landsliding by using a physically based model of hillslope stability. Hydrol. Process. 16 (14), 2833–2851.
26. Buechling, A., Baker, W.L., 2004. A fire history from tree rings in a highelevation forest of Rocky Mountain National Park. Can. J. For. Res. 34 (6), 1259–1273.
27. Calver, A., Anderson, M.G., 2004. Conceptual framework for the persistence of floodinitiated geomorphological features. Trans. Inst. Br. Geogr. NS 29, 129–137.
28. Carrara, A., Crosta, G., Frattini, P., 2008. Comparing models of debris-flow susceptibility in the alpine environment. Geomorphology 94 (3), 353–378.
29. Cartwright, N., 1999. The Dappled World: a Study of the Boundaries of Science. Cambridge University Press, Cambridge, UK.
30. Casadei, M., Dietrich, W.E., Miller, N.L., 2003. Testing a model for predicting the timing and location of shallow landslide initiation in soil-mantled landscapes. Earth Surf. Process. Landf. 28 (9), 925–950.
31. Chappell, N.A., 2010. Soil pipe distribution and hydrological functioning within the humid tropics: a synthesis. Hydrol. Process. 24 (12), 1567–1581.
32. Chiverrell, R.C., Harvey, A.M., Millington, J., Richardson, N.J., 2008. Late Holocene environmental

change in the Howgill Fells, northwest England. Geomorphology 100 (1), 41–69.

33. Clark, M.P., Kavetski, D., Fenicia, F., 2011. Pursuing the method of multiple working hypotheses for hydrological modeling. Water Resour. Res. 47 (9), W09301.

34. Clifford, M.J., Rocca, M.E., Delph, R., Ford, P.L., Cobb, N.S., 2008. Drought Induced Tree Mortality and Ensuing Bark Beetle Outbreaks in Southwestern Pinyon–Juniper Woodlands. In: Ecology, management, and restoration of pinon-juniper, and ponderosa pine ecosystems: Combined proc. of the 2005 St. George, Utah and 2006 Albuquerque, New Mexico workshops, USDA For. Serv., Proc. RMRS-P-51, Rocky Mountain Research Station, Fort Collins, CO, pp. 39–51.

35. Culling, W.E.H., 1957. Multicyclic streams and the equilibrium theory of grade. J. Geol. 65, 259–274.

36. Culling, W.E.H., 1986. Hurst phenomena in the landscape. Trans. Jpn. Geomorphol. Union 7, 1–23.

37. Culling, W.E.H., 1987. Equifinality: modern approaches to dynamical systems and their potential for geographical thought. Trans. Inst. Br. Geogr. NS 12 (1), 57–72.

38. Culling, W.E.H., 1988. A new view of the landscape. Transactions of the Institute of British Geographers NS 13 (3), 345–360.

39. Culling, W.E.H., Datko, M., 1987. The fractal geometry of the soil covered landscape. Earth Surf. Proc. Landf. 12, 369–385.

40. del Jesus, M., Fotia, R., Rinaldo, A., Rodriguez-Iturbe, I., 2012. Maximum entropy production, carbon assimilation, and the spatial organization of vegetation in river basins. Proc. Nat. Acad. Sci. 109, 20837–20841.

41. Deurer,M., Green, S.R., Clothier, B.E., Böttcher, J., Duijnisveld,W.H.M., 2003. Drainage networks in soils. A concept to describe bypass-flow pathways. J. Hydrol. 272 (1), 148–162.

42. Dietrich, W.E., Reiss, R., Hsu, M.L., Montgomery, D.R., 1995. A process-based model for colluvial soil depth and shallow landsliding using digital elevation data. Hydrol. Process. 9 (3–4), 383–400.

43. Dunkerley, D.L., Brown, K.J., 2002. Oblique vegetation banding in the Australian arid zone: implications for theories of pattern evolution and maintenance. J. Arid Environ. 51 (2), 163–181.

44. Dunne, T., 1978. Field studies of hillslope flow processes. In: Kirkby, M.J. (Ed.), Hillslope Hydrology. John Wiley & Sons, Chichester, pp. 227–293.

45. Ehret, U., Gupta, H.V., Sivapalan, M., Weijs, S.V., Schymanski, S.J., Blöschl, G., Gelfan, A.N., Winsemius, H.C., 2013. Advancing catchment hydrology to deal with predictions under change. Hydrol. Earth Syst. Sci. Discuss. 10 (7), 8581–8634.

46. Floyd, M.L., Clifford, M., Cobb, N.S., Hanna, D., Delph, R., Ford, P., Turner, D., 2009. Relationship of stand characteristics to drought-induced mortality in three Southwestern pinon–juniper woodlands. Ecol. Appl. 19 (5), 1223–1230.

47. Frigg, R., 2003. Self-organised criticality – what it is and what it isn't. Stud. Hist. Philos. Sci. A 34 (3), 613–632.

48. Gallant, J.C.,Moore, I.D., Hutchinson,M.F., Gessler, P., 1994. Estimating fractal dimension of profiles: a comparison of methods. Math. Geol. 26 (4), 455–481.

49. Galle, S., Ehrmann, M., Peugeot, C., 1999. Water balance in a banded vegetation pattern: a case study of tiger bush in western Niger. Catena 37 (1), 197–216.

50. Gilman, K., Newson, M.D., 1980. Soil Pipes and Pipeflow. A Hydrological Study in Upland Wales.

51. Geobooks, Norwich, UK. Gooding, D.C., Addis, T.R., 2008. Modelling experiments as mediating models. Found. Sci. 13 (1), 17–35.

52. Harvey, A.M., 2001. Coupling between hillslopes and channels in upland fluvial systems: implications for landscape sensitivity, illustrated from the Howgill Fells, northwest England. Catena 42 (2), 225–250.

53. Horton, R.E., 1933. The role of infiltration in the hydrologic cycle. Trans. Am. Geophys. Union 14, 446–460.

54. Horton, R.E., 1942. Remarks on hydrologic terminology. Trans. Am. Geophys. Union 23 (2), 479–482.

55. Huffman, D.W., Crouse, J.E., Walker Chancellor, W., Fulé, P.Z., 2012. Influence of time since fire on pinyon–juniper woodland structure. For. Ecol. Manage. 274, 29–37.

56. Hunt, J.C.R., Eames, I.,Westerweel, J., 2006. Mechanics of inhomogeneous turbulence and interfacial layers. J. Fluid Mech. 554, 499–519.

57. Jayasuriya, M.D.A., Dunn, G., Benyon, R., O'Shaughnessy, P.J., 1993. Some factors affecting water yield from mountain ash (Eucalyptus regnans) dominated forests in south-east Australia. J. Hydrol. 150 (2), 345–367.

58. Jones, J.A.A., 2010. Soil piping and catchment response. Hydrol. Process. 24 (12), 1548–1566.

59. Kleidon, A., Schymanski, S., 2008. Thermodynamics and optimality of the water budget on land: a review. Geophys. Res. Lett. 35, L20404, http://dx.doi.org/10.1029/2008G L035393.

60. Kleidon, A., Zehe, E., Lin, H., 2012. Thermodynamic limits of the critical zone and their relevance to hydropedology. In: Lin, H. (Ed.), Hydropedology. Elsevier, New York, pp. 243–281.

61. Lane, S.N., 2014. Acting, predicting and intervening in a socio-hydrological world. Hydrol. Earth Syst. Sci. (in press).

62. Li, J., Gou, X., Cook, E.R., Chen, F., 2006. Tree-ring based drought reconstruction for the central Tien Shan area in northwest China. Geophys. Res. Lett. 33 (7), L07715.

63. Liang, E., Liu, X., Yuan, Y., Qin, N., Fang, X., Huang, L., Zhu, H., Wang, L., Shao, X., 2006. The 1920s drought recorded by tree rings and historical documents in the semi-arid and arid areas of northern China. Climate Change 79, 403–432.

64. Limousin, J., Bickford, C.P., Dickman, L.T., Pangle, R.E., Hudson, P.J., Boutz, A.L., Gehres, N., Osuna, J.L., Pockman, W.T., McDowell, N.G., 2013. Regulation and acclimation of leaf gas exchange in a pinon–juniper

woodland exposed to three different ˜ precipitation regimes. Plant Cell Environ. 36 (10), 1812–1825.

65. McBride, J.R., 1983. Analysis of tree rings and fire scars to establish fire history. Tree-Ring Bull. 43, 39–49.
66. Meneveau, C., Sreenivasan, K.R., 1987. Simple multifractal cascade model for fully developed turbulence. Phys. Rev. Lett. 59 (13), 1424.
67. Montanari, A., Young, G., Savenije, H.H.G., Hughes, D., Wagener, T., Ren, L.L., Koutsoyiannis, D., Cudennec, C., Toth, E., Grimaldi, S., Bloeschl, G., Sivapalan, M., Beven, K., Gupta, H., Hipsey, M., Schaefli, B., Arheimer, B., Boegh, E., Schymanski, S.J., Di Baldassarre, G., Yu, B., Hubert, P., Huang, Y., Schumann, A., Post, D.A., Srinivasan, V., Harman, C., Thompson, S., Rogger, M., Viglione, A., McMillan, H., Characklis, G., Pang, Z., Belyaev, V., 2013. "Panta Rhei–Everything Flows": change in hydrology and society – The IAHS Scientific Decade 2013–2022. Hydrol. Sci. J. 58 (6), 1256–1275, Chicago.
68. Montgomery, D.R., Dietrich,W.E., 1994. A physically basedmodel for the topographic control on shallow landsliding. Water Res. Res. 30 (4), 1153–1171.
69. Montgomery, D.R., Schmidt, K.M., Greenberg, H.M., Dietrich,W.E., 2000. Forest clearing and regional landsliding. Geology 28 (4), 311–314.
70. Montgomery, D.R., Dietrich, W.E., Heffner, J.T., 2002. Piezometric response in shallow bedrock at CB1: implications for runoff generation and landsliding. Water Resour. Res. 38 (12), 1274, http://dx.doi.org/10.1029/2002WR001429.
71. Moody, J.A., Kinner, D.A., 2006. Spatial structures of stream and hillslope drainage networks following gully erosion after wildfire. Earth Surf. Process. Landf. 31 (3), 319–337.
72. Morgan, R.P.C., 2001. A simple approach to soil loss prediction: a revised Morgan–Morgan–Finney model. Catena 44 (4), 305–322.
73. Morgan, R.P.C., Duzant, J.H., 2008. Modified MMF (Morgan–Morgan–Finney) model for evaluating effects of crops and vegetation cover on soil erosion. Earth Surf. Process. Landf. 33 (1), 90–106.
74. Morton, A., 1993. Mathematical models: questions of trustworthiness. Br. J. Philos. Sci. 44, 659–674.
75. Mueller, R.C., Scudder, C.M., Porter, M.E., Talbot Trotter, R., Gehring, C.A., Whitham, T.G., 2005. Differential tree mortality in response to severe drought: evidence for long-term vegetation shifts. J. Ecol. 93 (6), 1085–1093.
76. Newson, M.D., 1980. The geomorphological effectiveness of floods: a contribution stimulated by two recent events in mid-Wales. Earth Surf. Process. Landf. 5, 1–16.
77. Ogle, K., Whitham, T.G., Cobb, N.S., 2000. Tree-ring variation in pinyon predicts likelihood of death following severe drought. Ecology 81 (11), 3237–3243.
78. Pelletier, J.D., Quade, J., Goble, R.J., Aldenderfer, M.S., 2011. widespread hillslope gulleying on the southeastern Tibetan plateau: human or climate-change induced. Geol. Soc. Am. Bull. 123, 1926–1938.
79. Pelletier, J.D., DeLong, S.B., Orem, C.A., Becerra, P., Compton, K., Gressett, K., Lyons-Baral, J., McGuire, L.A., Molaro, J.L., Spinler, J.C., 2012. How do vegetation bands form in dry lands? Insights from numerical modeling and field studies in southern Nevada, USA. J. Geophys. Res. 117, F04026, http://dx.doi.org/10.1029/2012JF002465.
80. Peterman, W., Waring, R.H., Seager, T., Pollock, W.L., 2012. Soil properties affect pinyon pine – juniper response to drought. Ecohydrology 6, 455–463.
81. Phillips, J.D., 1997. Simplexity and the reinvention of equifinality. Geogr. Anal. 29 (1), 1–15. Phillips, J.D., 2007. Perfection and complexity in the lower Brazos River. Geomorphology 91, 364–377.
82. Phillips, J.D., 2009. Landscape evolution space and the relative importance of geomorphic processes and controls. Geomorphology 109, 79–85.
83. Phillips, J.D., 2011. Predicting modes of spatial change from state-and-transition models. Ecol. Model. 222, 475–484.
84. Phillips, J.D., 2012. Log-jams and avulsions in the San Antonio River Delta, Texas. Earth Surf. Process. Landf. 37 (9), 936–950.
85. Philllips, J.D., 2013. Networks of historical contingency in earth surface systems. J. Geol. 121, 1–16.
86. Phillips, J.D., Park, L., 2009. Forest blowdown impacts of Hurricane Rita on fluvial systems. Earth Surf. Process. Landf. 34, 1069–1081.
87. Phillips, J.D., Marion, D.A., Turkington, A.V., 2008. Pedologic and geomorphic impacts of a tornado blowdown event in a mixed pine-hardwood forest. Catena 75 (3), 278–287.
88. Porada, P., Kleidon, A., Schymanski, S.J., 2011. Entropy production of soil hydrological processes and its maximisation. Earth Syst. Dyn. Discuss. 2 (1), 105–132.
89. Reggiani, P., Sivapalan, M., Majid Hassanizadeh, S., 1998. A unifying framework for watershed thermodynamics: balance equations for mass, momentum, energy and entropy, and the second law of thermodynamics. Adv. Water Resour. 22 (4), 367–398.
90. Reggiani, P., Sivapalan,M., Hassanizadeh, S., 2000. Conservation equations governing hillslope responses: exploring the physical basis of water balance.Water Resour. Res. 36, 1845–1863.
91. Renwick, W.H., 1992. Equilibrium, disequilibrium and nonequilibrium landforms in the landscape. In: Phillips, J.D., Renwick, W.H. (Eds.), Geomorphic Systems. Geomorphology 5, 265–276.
92. Rodriguez-Iturbe, I., Rinaldo, A., 2001. Fractal River Basins: Chance and SelfOrganization. Cambridge University Press, Cambridge, UK.
93. Schaefli, B., Harmon, C.J., Sivapalan, M., Schymanski, S.J., 2011. HESS Opinions: hydrologic predictions in a changing environment: behavioralmodeling. Hydrol. Earth Syst. Sci. 15, 635–646.

94. Schymanski, S.J., Roderick, M.L., Sivapalan, M., Hutley, L.B., Beringer, J., 2007. A test of the optimality approach to modelling canopy properties and CO2 uptake by natural vegetation. Plant Cell Environ. 30 (12), 1586–1598.

95. Schymanski, S.J., Sivapalan, M., Roderick, M.L., Hutley, L.B., Beringer, J., 2009. An optimality-based model of the dynamic feedbacks between natural vegetation and the water balance. Water Resour. Res. 45 (1), W01412, http://dx.doi.org/10.1029/2008WR006841.

96. Schymanski, S.J., Kleidon, A., Stieglitz, M., Narula, J., 2010. Maximum entropy production allows a simple representation of heterogeneity in semiarid ecosystems. Philos. Trans. R. Soc. B: Biol. Sci. 365 (1545), 1449–1455.

97. Sidle, R.C., Tsuboyama, Y., Noguchi, S., Hosoda, I., Fujieda, M., Shimizu, T., 2000. Stormflow generation in steep forested headwaters: a linked hydrogeomorphic paradigm. Hydrol. Process. 14 (3), 369–385.

98. Sidle, R.C., Noguchi, S., Tsuboyama, Y., Laursen, K., 2001. A conceptual model of preferential flow systems in forested hillslopes: evidence of self-organization. Hydrol. Process. 15 (10), 1675–1692.

99. Sivapalan, M., Savenije, H.H., Blöschl, G., 2012. Socio-hydrology: a new science of people and water. Hydrol. Process. 26 (8), 1270–1276.

100. Swetnam, T.W., 1993. Fire history and climate change in giant sequoia groves. Science 262 (5135), 885–889. Stock, J.D., Dietrich, W.E., 2006. Erosion of steepland valleys by debris flows. Geol. Soc. Am. Bull. 118 (9–10), 1125–1148.

101. Suckling, E.B., Smith, L.A., 2013. An evaluation of decadal probability forecasts from state-of-the-art climate models. J. Climate 26 (23), 9334–9347.

102. Theiler, J., 1990. Estimating fractal dimension. J. Opt. Soc. Am. A 7, 1055–1073.

103. Tuan, Y.-F., 1966. New Mexican gullies: a critical review and some recent observations. Ann. Assoc. Am. Geogr. 56.4, 573–597.

104. Valentin, C., d'Herbès, J.M., Poesen, J., 1999. Soil and water components of banded vegetation patterns. Catena 37 (1), 1–24.

105. Van Noordwijk, M., Spek, L.Y., De Willingen, P., 1994. Proximal root diameters as predictors of total root system size for fractal branching models. I. Theory. Plant Soil 164, 107–118.

106. Vigiak, O., Sterk, G., Romanowicz, R.J., Beven, K.J., 2006. A semi-empirical model to assess uncertainty of spatial patterns of erosion. Catena 66 (3), 198–210.

107. Watson, F.G., Vertessy, R.A., Grayson, R.B., 1999. Large-scale modelling of forest hydrological processes and their long-term effect on water yield. Hydrol. Process. 13 (5), 689–700.

108. Weiler, M., McDonnell, J.J., 2007. Conceptualizing lateral preferential flow and flow networks and simulating the effects on gauged and ungauged hillslopes. Water Resour. Res. 43 (3), W03403.

109. Zehe, E., Blume, T., Blöschl, G., 2010. The principle of "maximum energy dissipation": a novel thermodynamic perspective on rapid water flow in connected soil structures. Phil. Trans. R. Soc. B 365, 1377–1386, http://dx.doi.org/10.1098/rstb.2009.0308.

CHAPTER 9

FRESHWATER ECOSYSTEMS: WATER QUALITY GUIDELINES AND MANAGEMENT

Magdalena K. Grove, Gary S. Bilotta, Robert R. Woockman, John S. Schwartz

Keywords: Suspended sediment, Suspended solids, Suspended particulate matter, Reference-condition, Water quality, Guidelines

INTRODUCTION

Managing global water resources is one of the greatest challenges of the 21st century (Garrido and Dinar, 2009, Poff, 2009 and Staddon, 2012). Water is a resource that is under growing pressure as the global population rises, and the natural supply, in the form of precipitation, is becoming increasingly variable and uncertain with climate change (Giorgi et al., 2004, Räisänen et al., 2004 and Trenberth et al., 2003). It is therefore essential that water resources are managed sustainably in terms of both their quantity and quality. One of the most commonly attributed causes for the impairment of water quality globally is the presence of excess suspended sediments, ranging from nano-scale particles and colloids to sand-sized sediments (Gray, 2008 and Richter et al., 1997). Suspended sediments (SS) can have a range of detrimental effects on water resources, from aesthetic issues and higher costs of water treatment, to a decline in the fisheries resource and serious ecological degradation (Alabaster and Lloyd, 1982, Bilotta and Brazier, 2008, Cordone and Kelley, 1961, Gammon, 1970, Newcombe and MacDonald, 1991, Owens et al., 2005, Peddicord, 1979, Ryan, 1991 and Wood and Armitage, 1997). Ultimately this can lead to a significant decline in the associated freshwater ecosystem services, estimated to have a global value in excess of $1.7 trillion per annum (Costanza et al., 1998).

Citation: Magdalena K. Grove, Gary S. Bilotta, Robert R. Woockman, John S. Schwartz, Freshwater ecosystems: water quality guidelines and management, Science of The Total Environment, Volume 502, 1 January 2015, Pages 481-492, ISSN 0048-9697, http://dx.doi.org/10.1016/j.scitotenv.2014.09.054.

In recognition of the potential for SS to cause aquatic degradation, and in an effort to minimise this degradation, government-led environmental organisations from around the world have established water quality guidelines and standards, which state recommended targets for SS (sometimes referred to as suspended solids, and occasionally assessed through proxy measurements such as turbidity) (Bilotta and Brazier, 2008 and Collins et al., 2011). However, at present these guidelines are often blanket values that do not recognise the natural spatial and temporal variations of SS in streams/rivers, and are not well-linked to the biological/ecological impact evidence; therefore ultimately these guidelines may not reflect the specific requirements of the biological communities that they are designed to protect (Bilotta et al., 2012, Collins et al., 2011, Schwartz et al., 2008 and Schwartz et al., 2011). For example, in Europe the Freshwater Fisheries Directive (78/659/EEC) (2006/44/EC) guideline for SS stated that concentrations should not exceed 25 mg L^{-1} in salmonid and cyprinid waters except in exceptional weather conditions or exceptional geographic circumstances (Bilotta and Brazier, 2008 and Collins et al., 2011). This Directive has now been repealed and replaced by the Water Framework Directive (2000/60/EC) (2008/105/EC), but there are currently no new guidelines for SS under this Directive.

It is important to recognise that unlike some aquatic pollutants (e.g. pesticides, pharmaceuticals, veterinary medicines), SS is a natural component of freshwater ecosystems, critical to habitat heterogeneity and ecological functioning (Maitland, 2003, Swietlik et al., 2003, Vannote et al., 1980 and Yarnell et al., 2006). There is natural spatial variation in SS conditions. For example, Bilotta et al. (2012) found background concentrations of SS varied by more than 15-fold between 42 temperate ecosystem-types that were in reference condition.[1] Furthermore, these differences could be predicted using data on a number of the natural environmental characteristics associated with each river's catchment, including metrics of climate, catchment geology and topography, and channel hydromorphology. The interpretation of these findings was that differences in the natural environment create unique SS conditions that support unique freshwater communities. If water quality managers wish to protect these unique freshwater communities through establishing and implementing water quality guidelines, then these guidelines should be environment-specific.

There are also natural temporal variations in SS conditions, which may be critical to the ontogeny of certain organisms (e.g. Maitland, 2003), and which should be expected given the temporally variable contributions of SS from channel and non-channel sources. Wood and Armitage (1997) suggest that the principal sources of SS available to a river from channel sources are: (i) river banks; (ii) mid-channel and point bars; (iii) fine bed material; (iv) natural backwaters; (v) fine particles associated with aquatic vegetation; and (vi) other biotic particles including phytoplankton and zooplankton. In addition there may be in-channel generation of SS due to the decomposition of aquatic macrophytes, biofilms and invertebrates. Benthic invertebrate faecal material has been shown to constitute a significant source of fine SS (Ladle and Griffiths, 1980 and Ward et al., 1994). Finally, some aquatic organisms (invertebrates and vertebrates) may also act to re-suspend material stored in the bed and banks of a water body through burrowing, feeding, and breeding behaviours (Harvey et al., 2011, Hassan et al., 2008 and Montgomery et al., 1996). The main non-channel sources of SS supplied to a river are: (i) exposed soils subject to erosion — this material is transported to the channel via surface and subsurface runoff; (ii) mass failures within the catchment, such as landslides and soil creep; (iii) litter fall, principally leaf material from vegetation adjacent to the channel; and (iv) atmospheric deposition, due to aeolian processes and precipitation (Wood and Armitage, 1997).

In catchments where these SS sources have been modified through anthropogenic activities, the resultant modified SS regimes can affect the biological community. However, at present the SS regime required to maintain or restore biological integrity in a given environment has not been defined. It is known that some aquatic organisms are sensitive to changes in SS, in particular, changes to the duration, frequency and timing that a given concentration is experienced (Diehl and Wolfe, 2010, Kerr, 1995, Newcombe and Jensen, 1996, Newcombe and MacDonald, 1991, Reid and Anderson, 1999, Waters, 1995 and Yount and Niemi, 1990), which collectively are referred to as the SS regime (concentration, duration, frequency). The current water quality guidelines for SS, however, through their use of mean annual concentration values or total mean daily loads, fail to recognise the importance of the SS regime, because an observed mean concentration, even if regarded as being compliant with guidelines, could be achieved through an infinite number of concentration scenarios (from highly variable regimes with concentration extremes to relatively stable regimes with little variability about the mean), each scenario potentially having very different biological effects. There is therefore a need for more advanced water quality guidelines that can take into account these natural spatial and temporal variations, so that water quality managers can identify where and when SS pollution is taking place and to what extent. The first step to achieving this is to understand what the natural background SS regimes are in contrasting ecosystems that are in reference-condition.

The aims of this study are to (1) monitor the SS regimes of contrasting reference-condition river ecosystems and to examine the spatial and temporal variations in SS regimes, and based on these findings (2) consider the appropriateness of the current regulatory water quality guidelines and monitoring.

MATERIALS AND METHODS

Field sites

Field sites were selected from the RIVPACS IV database (May 2011 version) (River Invertebrate Prediction and Classification System — © NERC [CEH] 2006. Database rights NERC [CEH] 2006 all rights reserved). The RIVPACS IV model and database are described in details by Wright et al. (2000) and Clarke et al. (2011), and are only summarised here. The database contains invertebrate, water quality and catchment characteristics data, recorded between 1978 and 2004, from 795 streams and river sites in the UK. All of the sites are in reference condition; defined as having no, or only very minor, anthropogenic alterations to the values of the hydrochemistry and hydromorphology, with biota usually associated with such undisturbed or minimally disturbed conditions. In the development of the RIVPACS model, the sites were divided into similar biological communities, referred to as end groups, based on the invertebrate community composition. Forty-three end group communities were identified in RIVPACS IV. These end groups are a proxy for the wider ecosystems, i.e. specific invertebrate communities occur in specific environments and are associated with specific floral and faunal (vertebrate) communities. In the RIVPACS IV model these end groups were characterised, using Multiple Discriminant Analysis (MDA), in terms of the temporally invariant properties of the environments that they inhabit. The original purpose of the RIVPACS model was for use in assessments of ecological status. The model can be applied to any site wherever these environmental properties can be measured, allowing the user to make a prediction of the expected community composition. The magnitude of deviation between observed and expected community compositions at a site is used as an indication of the ecological status and therefore anthropogenic disturbance.

In this article, the authors used the RIVPACS IV database to identify ecosystems that occur in the most contrasting environments. In order to achieve this, the environmental properties of the 43 end groups were plotted for eight of the twelve environmental variables identified through MDA as being discriminant variables for invertebrate community composition. Latitude and longitude were excluded because they have no direct physical basis and are therefore likely to produce models that are only applicable locally. Alkalinity was excluded because it is a water quality parameter which can be influenced by pollution which could result in an incorrect reference condition model prediction. Air temperature range was excluded because it is highly correlated with mean annual air temperature.

The end groups that had the lowest and highest mean values for each of the remaining eight environmental variables were selected as the most contrasting ecosystem types; subsequently the field sites from each of the identified end groups were selected. This was decided by choosing the sites that had the value closest to the mean for that particular environmental variable for that end group. In a few instances, for logistical reasons (proximity/access), the sites that had the second closest value to the mean for the environmental parameter variable were chosen. Using this process, ten field sites were selected to encompass a wide range of environments, varying in their: (1) Climate — mean annual precipitation totals of 450 to 1400 mm and mean annual temperatures ranging from 8.42 to 11.33 °C, (2) Geology — varying from catchments with almost 100% hard igneous rocks to catchments with almost 100% clay or sandstone, and (3) Topography — altitudes at the river's source varying from 128 to 1216 m above sea level. The stream and river sites also vary in their morphometry with widths ranging from 0.60 to 50 m and depths ranging from 0.08 to 1.90 m (mean of 3 seasonal measurements). As such the findings and model developed from this study will be applicable to a wide range of streams and rivers located in mid-latitude zones. The approximate location of each site can be seen in Fig. 1, which also describes the range of environmental characteristics of each site.

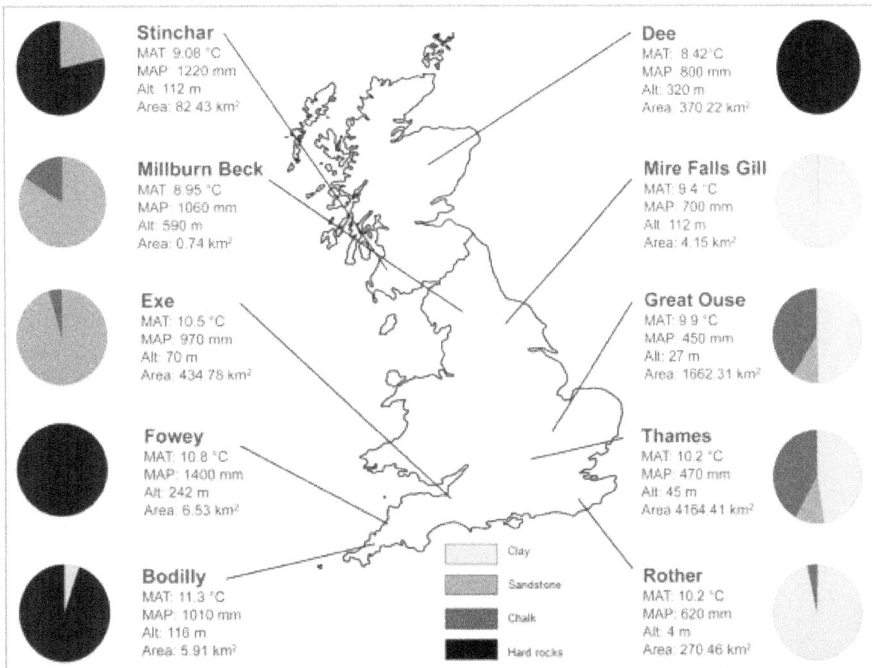

Figure 1: Location of the ten reference-condition stream/river sites, including information on their catchment characteristics: mean annual temperature (MAT), mean annual precipitation (MAP), mean altitude of upstream catchment (Alt), and catchment area (Area). Dominant geology classes, shown in the pie-charts, represent an amalgamation of the 116 Solid Geology classes of the British Geological Survey into 8 River Habitat Survey super-classes. The environmental characteristics of each site, excluding MAP, were sourced from the RIVPACS IV (May 2011 version) database, (Clarke et al., 2011). Mean annual precipitation data were sourced from the UK Meteorological Office (http://www.metoffice.gov.uk/climatechange/science/monitoring/ukcp09/) and derived from 1 km gridded data for the monitoring location during the period 1961–1990.

Sampling, monitoring and laboratory analysis

Each field site was instrumented with a multi-parameter water quality probe (*Aquaread Aquaprobe*) and logger (*Aquaread Aqualogger*), which recorded turbidity — a surrogate measure of SS; every 15 min. These turbidity data were converted to estimated SS concentrations using site-specific turbidity-SS regression equations derived from:

(1) SS data measured through laboratory analysis of water samples that were collected from the sites on a 4–8-week temporal frequency, between May 2011 and May 2013. Water samples were collected manually in 1 L plastic bottles in a depth- and width-integrating manner across each river site. Samples were refrigerated at < 4 °C and stored in the dark prior to analysis in order to avoid degradation. Following Clesceri et al. (1998), samples were analysed in the laboratory for concentrations of SS and volatile organic matter (VOM). This involved filtration of a known volume of sample through a dried, pre-combusted (550 °C for 30 min) and pre-weighed glass fibre filter paper (*Whatman GFF*), with a pore size of 0.70 μm. The filters and their residue were then dried at 105 °C for 60 min and re-weighed. Following this the filters and their residue were ignited in a muffle furnace at 550 °C for 30 min and then re-weighed to determine VOM (VOM equal to the dried particle mass minus the combusted particle mass). This is the standard technique for the determination of VOM (APHA, 1995 and APHA, 1998).

(2) Laboratory-based turbidity-SS calibrations using SS collected from time-integrated samplers between May 2011 and May 2013, that were combined with filtered river water from the site to make up a wide range of known SS concentrations against which to measure turbidity. The time-integrated SS samples were collected using the time-integrating passive trap sampler method described by Phillips et al. (2000). The time-integrated SS samplers were one metre

long and consisted of commercially available PE pipes with an outer diameter of 110 mm and a wall thickness of 4.2 mm. They were sealed with a plugged polythene funnel at the inlet and a cap at the outlet. A plastic tube with an inner diameter of 4 mm was passed through the funnel and the cap as inlet and outlet. The samplers were mounted parallel to the riverbed at two upright steel bars driven into the channel bed, with the inlet tube pointing upstream. The greater cross-sectional area of the main cylinder compared to that of the inlet tube reduces the flow velocity within the samplers by a factor of 600 relative to that of the ambient flow. This reduction in flow velocity induces sedimentation of the SS as the water moves through the cylinder towards the outlet tube (Phillips et al., 2000). The time-integrated SS samplers are designed to collect a statistically representative sample under field conditions (Perks et al., 2013, Phillips et al., 2000 and Wildhaber et al., 2012). The SS was extracted from these samplers on a 4–8-week temporal frequency.

The best-fitting regression equations for all of the sites are shown in Fig. 2.

Science of the Total Environment

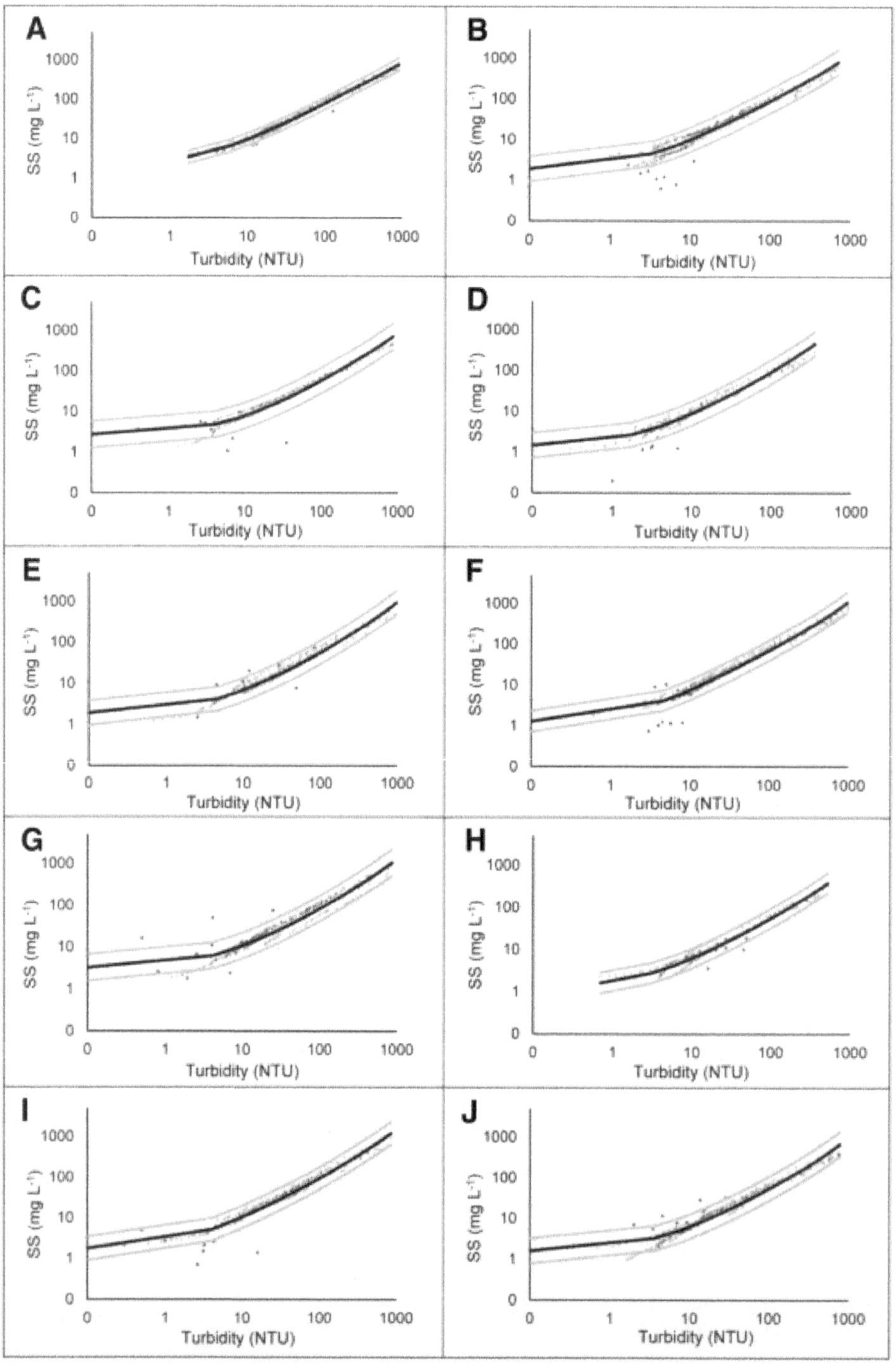

Figure 2: Observed turbidity-SS values (grey crosses for time integrated samples; black diamonds for grab samples) and the predicted values based on regression analysis at each of the study sites (black lines represent the fitted regression value, grey lines represent the upper and lower 95% prediction intervals). Regression equations used to make predictions were carried out using quadratic equations on log-transformed turbidity (NTU) (X-axis) and SS concentration data (Y-axis). All regression equations are statistically significant (p-values < 0.001). (A) Bodilly, $R^2 = 0.98$. (B) Dee, $R^2 = 0.94$. (C) Exe, $R^2 = 0.93$. (D) Fowey, $R^2 = 0.94$. (E) Great Ouse, $R^2 = 0.95$. (F) Millburn Beck, $R^2 = 0.96$. (G) Mire Falls Gill, $R^2 = 0.93$. (H) Rother, $R^2 = 0.95$. (I) Stinchar, $R^2 = 0.96$. (J) Thames, $R^2 = 0.94$.

Data processing and analysis

All field data were checked for consistency in turbidity signals by visual inspection as well as against discharge data gathered by the Environment Agency and the Scottish Environment Protection Agency or from the discharge estimations computed from the stage-discharge data collected during the study. In cases where the stream debris or biota covered the turbidity sensors, or if sensors did not respond to changes in stream discharge for prolonged periods of time suggesting equipment failure, the data were removed.

All statistical analysis was conducted using SPSS statistical software (IBM® SPSS® Statistics 20). Prior to the statistical analysis, the SS concentration data sets for all the ten study sites were investigated for normality and homogeneity of variance. In all cases the Kolmogorov–Smirnov test for normality ($p < 0.001$) suggests that the SS concentration data are not normally distributed. The test of homogeneity of variance (Levene Statistic; F (9, 536530) = 7192.1, $p = < 0.001$) also indicated that the variances are significantly different. Tests of normality and of the homogeneity of variance were also carried out on the transformed SS data (common and natural logarithm, square, square root) with the same outcomes. An examination of the skewness and kurtosis values for the sites also suggests that the SS concentration data are not normally distributed. Based on those findings, non-parametric tests were applied for comparative data analysis.

RESULTS

Spatial variability

Table 1 presents the summary statistics for the estimated (from herein 'estimated' refers to turbidity-derived) SS concentrations in the ten river sites during the 2011–2013 monitoring period. The estimated mean SS concentrations range from 3 mg L^{-1} (Fowey) to 29 mg L^{-1} (Bodilly and Great Ouse). A Kruskal–Wallis and post-hoc tests were conducted to determine if the differences in estimated mean SS concentrations between the sites are statistically significant. Pairwise comparisons were performed using Dunn›s (1964) procedure with a Bonferroni correction for multiple comparisons. Statistical significance was accepted at the $p < 0.05$ level for the omnibus test and $p < 0.05$ for the multiple comparisons. Estimated SS concentrations were significantly different between the sites, 2 (9) = 115315.17, $p < 0.001$. Post-hoc analysis revealed statistically significant differences in all comparisons ($p < 0.001$), except between the following pairs of sites: the River Dee (median = 3.2 mg L^{-1}) and Milburn Beck (median = 3.0 mg L^{-1}) ($p = 0.555$); the River Rother (median = 6.1 mg L^{-1}) and River Stinchar (median = 6.2 mg L^{-1}) ($p = 0.048$); and the River Mire Falls Gill (median = 7.7 mg L^{-1}) and River Exe (median = 6.8 mg L^{-1}) ($p = 0.027$).

Table 1: Estimated mean SS concentrations with lower and upper prediction intervals (based on 95% confidence intervals derived from the turbidity-SS regression), estimated median SS, estimated minimum and maximum SS, and 5th, 30th, 50th and 90th percentiles for estimated SS (mg L^{-1}) in the ten study sites monitored between May 2011 and May 2013. n refers to the number of turbidity readings upon which the estimate is based

Site	n	Mean	Lower–upper prediction intervals	5th Percentile	30th Percentile	50th Percentile	90th Percentile	Min	Max
				C95	C70	C50 (median)	C10		
Bodilly	59,976	29	20–42	0	11	22	66	0	357
Dee	42,052	6	3–13	0	1	3	14	0	304
Exe	57,840	18	9–39	3	3	7	35	2	591
Fowey	55,489	3	1–6	1	1	1	4	1	460
Great Ouse	51,544	29	15–57	3	3	6	79	2	857
Milburn Beck	62,157	6	3–10	1	1	3	14	1	380
Mire Falls Gill	58,169	15	8–32	3	3	8	27	2	492
Rother	43,331	11	6–19	1	3	6	25	1	407
Stinchar	54,120	15	8–28	1	1	6	32	1	701
Thames	51,862	11	6–22	2	4	7	22	1	234

Table 2 presents the low-resolution, grab sample-derived mean SS concentrations observed in the ten river sites monitored between 2011 and 2013. Observed mean SS concentrations measured between May 2011 and 2013 ranged from 1 mg L^{-1} (Dee) to 17 mg L^{-1} (Bodilly and Mire Falls Gill). Table 2 also compares these observations with the historical low-resolution (monthly grab sample-derived) mean SS concentrations observed in other reference-condition (RIVPACS) sites belonging to the same ecosystem-types or 'end groups'. All of the river sites monitored between 2011 and 2013, except for the River Bodilly, had mean SS concentrations that were within the minimum and maximum mean SS concentrations observed in the sites belonging to the same end-groups. The River Bodilly›s mean concentration exceeded the range of mean concentrations observed in the sites belonging to the same end-group. A comparison of the observed (grab-sample) mean SS concentration for each site against the equivalent predicted² SS ranges (see Fig. 3), reveals that the model predicted the correct SS range for seven of the sites, and predicted the correct or next closest SS range for the remaining three sites.

Table 2: Comparison of the observed mean SS concentration (mg L^{-1}, based on low-resolution grab sampling) against the mean, minimum and maximum SS concentrations (mg L^{-1}) observed in the sites (n = 638) belonging to the same RIVPACS IV end groups

River	Observed (grab) mean SS 2011–2013	Mean average SS of sites in the same end group	Min average SS of sites in the same end group	Max average SS of sites in the same end group
Bodilly	17	8	6	9
Dee	1	3	1	6
Exe	13	9	1	24
Fowey	2	4	1	10
Great Ouse	16	19	8	45
Millburn Beck	5	4	1	10
Mire Falls Gill	17	13	5	26
Rother	14	12	4	26
Stinchar	4	4	1	9
Thames	14	19	8	45

Science of the Total Environment

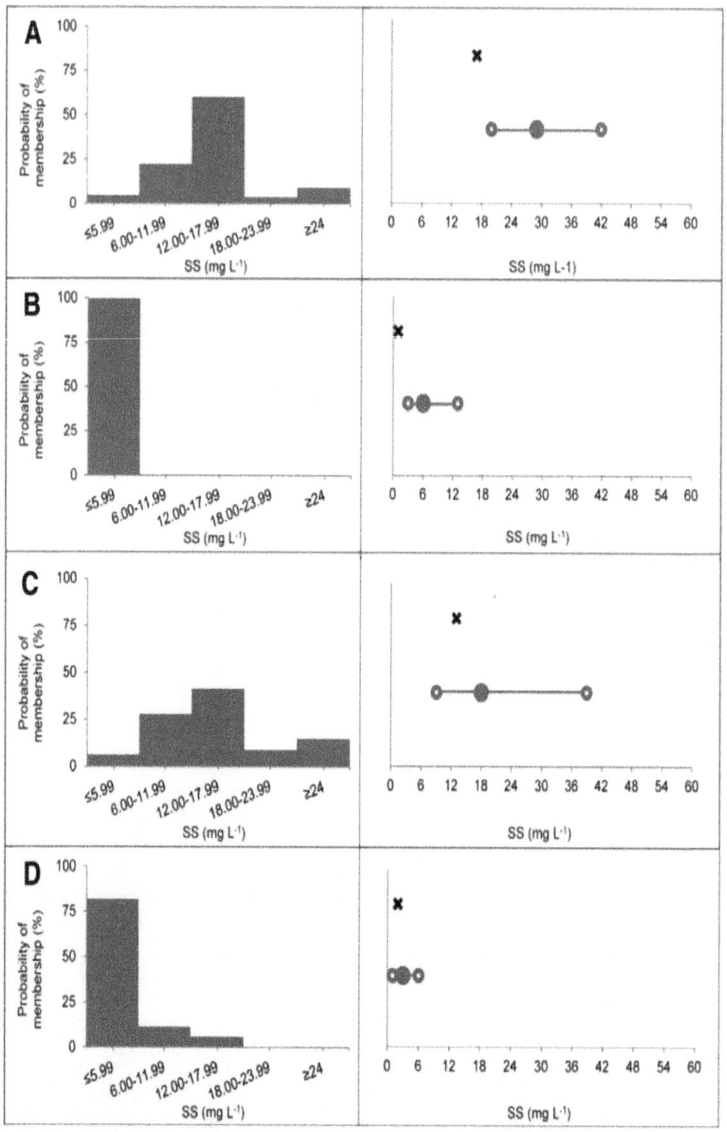

Science of the Total Environment

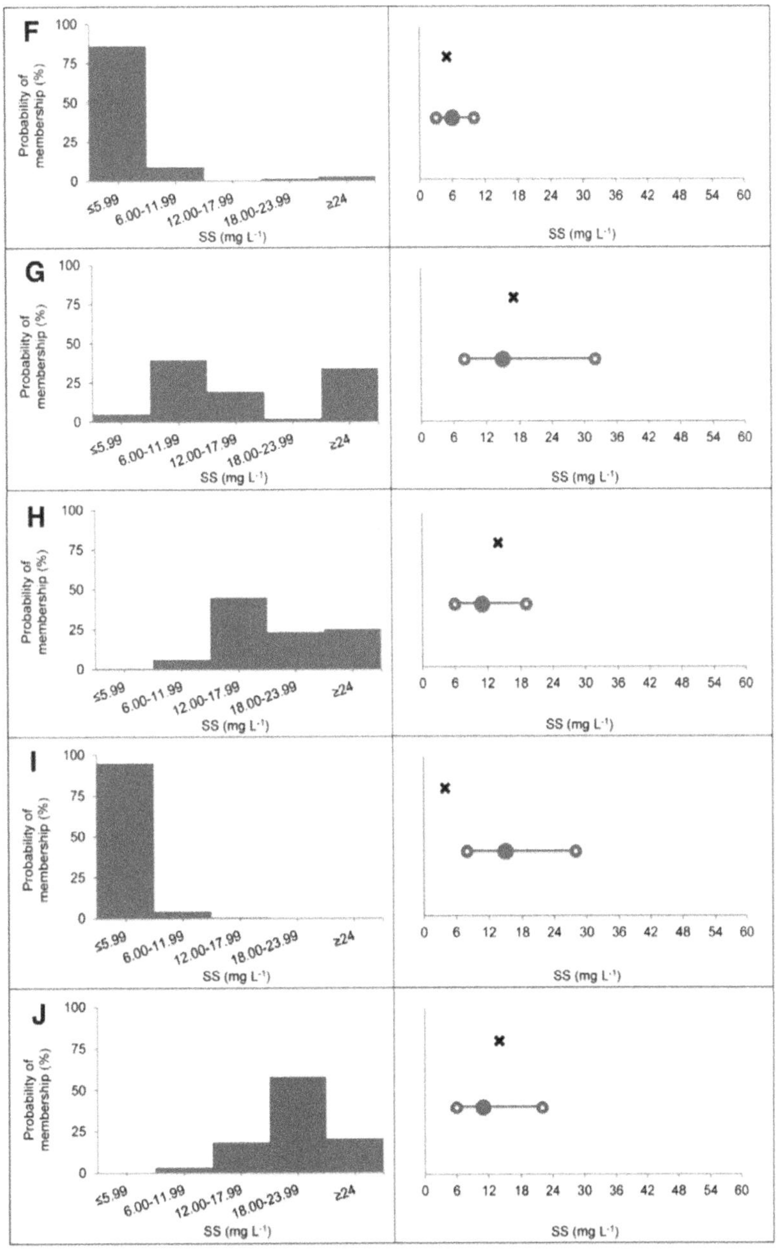

Figure 3: A comparison of the model-predicted SS ranges (bar charts on the left panel) versus the observed (grab sample-derived; illustrated by crosses on the right panel) and estimated (turbidity-derived; illustrated by circles: a closed circle for the central estimate and open circles for the lower and upper prediction intervals) SS concentrations for 2011–2013 monitoring period. (A) Bodilly, (B) Dee, (C) Exe, (D) Fowey, (E) Great Ouse, (F) Millburn Beck, (G) Mire Falls Gill, (H) Rother, (I) Stinchar, (J) Thames.

Temporal variability

Table 3 presents the estimated mean annual SS concentration observed in each river site for year one (May 2011 to May 2012) and year two (May 2012 to May 2013) of monitoring. These data are also illustrated in Fig. 4 alongside the estimated mean SS concentration for the total monitoring period. A Mann–Whitney *U* test showed that there are statistically significant differences in the distribution of SS concentrations observed in year one and year two for all the sites ($p < 0.001$ in all cases). All the river sites, except the Dee, Stinchar and Thames, experienced higher estimated mean annual SS concentrations in year two of the study. This difference appears to be partially related to the mean annual discharge. The largest difference in mean annual SS concentration was observed in the River Great Ouse, which experienced a three-fold increase between year one (14 mg L^{-1}) and year two (42 mg L^{-1}) concomitant with roughly three-fold increases in mean annual discharge observed at this site from year one (6.37 $m^3 s^{-1}$) to year two (19.56 $m^3 s^{-1}$) (Table 4). The Mann–Whitney *U* test showed significant differences ($p < 0.05$ in all cases) in annual average discharge observed in year one and year two for all the sites. The relationship between estimated mean annual SS and mean annual discharge is not evident for the Thames site, which experienced remarkably similar estimated mean annual SS concentrations in year one (10.94 mg L^{-1}) and year two (10.86 mg L^{-1}), despite experiencing contrasting mean annual discharges (16.34 $m^3 s^{-1}$and 55.90 $m^3 s^{-1}$).

Table 3: Estimated mean annual SS concentrations in the ten river sites and the percentage differences between the seasons

SS concentration (mg L^{-1})			
	2011–2012	2012–2013	Percentage difference between 2011–2012 and 2012–2013
	Mean	Mean	
Bodilly	26.70	32.60	22.1
Dee	6.53	6.33	− 3.1
Exe	15.48	21.25	37.3
Fowey	2.21	3.57	61.5
Great Ouse	13.94	42.08	201.9
Milburn Beck	5.64	5.79	2.7
Mire Falls Gill	13.14	17.26	31.4
Rother	8.17	13.02	59.4
Stinchar	17.46	11.76	− 32.6
Thames	10.94	10.86	− 0.7

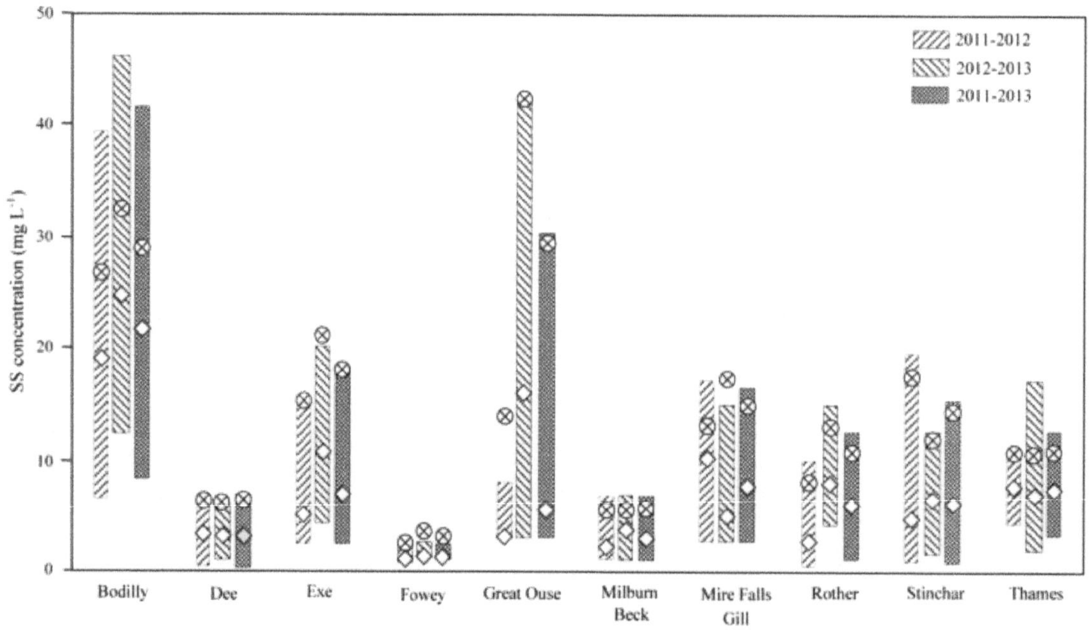

Figure 4: Estimated (turbidity-derived) SS concentrations for the ten study sites based on high-resolution (15-min frequency) monitoring. Data displayed for 2011–2012, 2012–2013 and 2011–2013. Circle symbols represent the mean SS values. Diamond symbols represent the median SS values.

Table 4: Average river discharge for study sites for 2011–2012, 2012–2013, 2011–2013 and the long-term historical average (CEH, 2014)

River	Mean discharge (m^3 s^{-1})			Percentage difference between 2011–2012 and 2012–2013	Historical mean discharge[a] (m^3 s^{-1}) and period of record (in parenthesis)
	2011–2012	2012–2013	2011–2013		
Bodilly	0.14	0.30	0.22	114.1	N/A (N/A)
Dee	17.27	13.91	15.61	− 19.4	15.7 (1982–2012)
Exe	8.89	11.85	10.37	33.3	12.7 (1960–2012)

Fowey	0.20	0.34	0.28	68.2	N/A (N/A)
Great Ouse	6.37	19.56	13.18	207.0	11.5 (1972–2011)
Milburn Beck	0.10	0.14	0.12	40.9	N/A (N/A)
Mire Falls Gill	0.05	0.11	0.08	119.3	0.1 (1974–2011)
Rother	1.42	4.42	2.83	211.0	3 (1962–2012)
Stinchar	3.22	2.77	2.99	− 13.9	2.6 (1973–2012)
Thames	16.34	55.90	36.08	242.0	34.5 (1938–2012)

[a]Data are area-corrected from the National River Flow Archive's supplied by the Centre for Ecology and Hydrology where available (2013).

These temporal variations in discharge and SS are natural phenomena that shape biological communities. Therefore ideally (a) water quality guidelines should acknowledge this, and (b) the sampling frequencies used should enable accurate assessments of compliance with these guidelines. In order to assess whether the environment-specific SS prediction model developed by Bilotta et al. (2012) could predict this inter-annual variation, the model input data on mean annual temperature (1961–1990 average) and mean annual precipitation (1961–1990) was replaced with mean annual temperature (°C) and total annual precipitation (mm) measured in the year for which the prediction is to be made at UK Meteorological Office stations closest to each of the ten study sites.[3]

In order to assess the accuracy of the sampling frequency that is currently used in routine regulatory monitoring (Bowes et al., 2009 and Cooper et al., 2003), the high-resolution (15 min frequency) SS datasets were systematically subsampled, selecting one sample per month (selected at 08:00 am on the first day of each month), to calculate the low-resolution mean SS concentrations over the 2011–2013 monitoring period for each site, simulating routine regulatory water quality monitoring. Percentage errors as well as absolute differences were calculated between the high-resolution (15 min frequency) mean SS concentrations and the simulated low-resolution (monthly frequency) mean SS concentrations. The results, presented in Table 5, demonstrate that mean SS concentrations derived from monthly sampling frequencies suffered from errors of up to 53% of values derived from 15-min sampling frequencies, with an average error of 24% for all the ten sites.

Table 5: Estimated mean SS concentrations in the study sites based on 15 min resolution and simulated monthly resolution sampling regimes

		Mean SS concentration (mg L^{-1})		% Errors	Absolute difference
		15 min resolution	Monthly resolution	Between mean SS concentration (mg L^{-1}) sampled every 15 min and monthly	
Bodilly	Mean	29.00	23.07	− 20.5	5.9
	St. Dev.	28.17	18.48		
Dee	Mean	6.42	7.15	11.4	0.7
	St. Dev.	13.17	10.38		
Exe	Mean	18.28	19.17	4.8	0.9
	St. Dev.	40.49	38.32		
Fowey	Mean	3.00	2.99	− 0.3	0.0
	St. Dev.	10.21	3.57		

Great Ouse	Mean	29.38	27.17	− 7.5	2.2
	St. Dev.	61.44	39.14		
Milburn Beck	Mean	5.71	3.25	− 43.0	2.5
	St. Dev.	9.20	3.78		
Mire Falls Gill	Mean	15.34	23.50	53.1	8.2
	St. Dev.	28.00	43.26		
Rother	Mean	10.85	5.66	− 47.8	5.2
	St. Dev.	17.66	4.69		
Stinchar	Mean	14.56	8.34	− 42.7	6.2
	St. Dev.	28.52	10.17		
Thames	Mean	10.90	12.43	14.1	1.5
	St. Dev.	15.01	11.87		

DISCUSSION

Spatial variability

The results presented in Table 1 show that there were over 9-fold differences in estimated mean SS concentrations between the ten sites. Two of the river sites, Bodilly and Great Ouse, had estimated mean SS concentrations (29 mg L^{-1}) that would have exceeded the Freshwater Fisheries Directive (FFD) guideline despite being in reference-condition. The implications of this finding are that the FFD's blanket water quality guideline is likely to have been inappropriate for these specific ecosystem-types as they may be non-compliant regardless of the intensity of land-use. Five of the river sites studied had estimated mean SS concentrations that were less than half of the FFD guideline, with the lowest mean SS observed in the River Fowey (3 mg L^{-1}). Although these river sites would be compliant with the FFD guideline, it is questionable whether the guideline was appropriate for them, as previous research has demonstrated that detrimental impacts can be experienced by some freshwater organisms, of all trophic levels, when exposed to concentrations below 25 mg L^{-1}. For example, Quinn et al. (1992) observed a 40% reduction in the biomass of phytoplankton exposed to SS concentrations of 10 mg L^{-1}. Lloyd (1987) observed a 3–13% reduction in the primary productivity of macrophytes and periphyton exposed to SS concentrations of 8 mg L^{-1}. Rosenberg and Wiens (1978) observed an increase in the amount of drift of benthic invertebrates exposed to SS concentrations of 8 mg L^{-1}. Robertson et al. (2007)observed increased foraging activity in Atlantic Salmon when exposed to SS concentrations of 20 mg L^{-1}. Therefore, it is suggested here that some ecosystems may require much lower guideline values in order to be effectively protected.

The degree of spatial variation in estimated mean SS concentrations (3–29 mg L^{-1}) and observed (grab sample-derived) mean SS concentrations (1–17 mg L^{-1}), supports earlier research by Bilotta et al. (2012) that analysed the mean SS concentrations derived from low-resolution (monthly grab sample) monitoring programmes in 638 reference-condition sites in the UK (grouped into 42 ecosystem-types or 'end groups'). Bilotta et al. (2012) reported that the mean SS concentrations for the 42 end groups varied between 1.7 and 26 mg L^{-1}. All of the river sites monitored between 2011 and 2013, except for the River Bodilly, had observed (grab sample-derived) mean SS concentrations that were within the minimum and maximum mean SS concentrations observed in the sites belonging to the same end-groups. Bilotta et al. (2012) proposed a model capable of predicting mean (monthly grab sample-derived) SS concentrations to the correct range (6 mg L^{-1} ranges) or the next closest range (in 90% of cases), as a function of a range of site characteristics: mean annual air temperature, mean annual precipitation, mean altitude of upstream catchment, distance from source, slope to source, channel width and depth, the percentage of catchment area comprised of clay, chalk, and hard rock solid geology, and the percentage of the catchment area comprised of blown sand as the surface (drift) material. This model was used to make predictions for the sites monitored in this study. A comparison of the observed (grab-sample) mean SS concentration for each site against the equivalent predicted SS ranges (see Fig. 3), reveals that the model predicted the correct SS range for the seven sites, and predicted the correct or next closest SS range for the remaining three sites.

Temporal variability

There were statistically significant differences in the distribution of SS concentrations observed in year one and year two of the study for all the sites ($p < 0.001$ in all cases). All river sites, except the Dee, Stinchar and Thames, experienced higher estimated mean annual SS concentrations in year two of the study. This difference appears to be partially related to the mean annual discharge. The largest difference in mean annual SS concentration was observed in the River Great Ouse, which experienced a three-fold increase between year one (14 mg L^{-1}) and year two (42 mg L^{-1}) concomitant with roughly three-fold increases in mean annual discharge observed at this site from year one (6.37 $m^3 s^{-1}$) to year two (19.56 $m^3 s^{-1}$) (Table 4). This inter-annual variability could be predicted reasonably well for all the sites except the River Rother, using the environment-specific model described above, but with modified input data to take into account the mean annual temperature (°C) and total annual precipitation (mm) in the year for which the mean SS prediction was to be made.

The average discharge for the 2011–2013 monitoring period is within 20% of the historical average annual discharge for each site (where data are available). Therefore, although the individual estimated mean annual SS concentrations reported in this paper may represent observations from a drier than average year and a wetter than average year for many sites, the combined mean SS concentration for the whole monitoring period (2011–2013) may be considered to be representative of average hydrological conditions for each site and perhaps therefore average SS conditions for each site. Of course mean values can conceal significant inter-annual differences in the dynamics of both SS and discharge, distorting any relationships between these parameters. For example, although the mean annual SS concentrations for the Thames were similar in year one and year two, the Mann–Whitney *U* test showed that there was a significant difference between the distribution of estimated SS concentrations across the two years.

Sampling frequency considerations

The results presented in Table 5 demonstrate that mean SS concentrations derived from monthly sampling frequencies suffer from errors of up to 53% of values derived from 15-min sampling frequencies, with an average error of 24% for all the ten sites. Similar errors have been reported by other researchers (Coynel et al., 2004, Skarbøvik et al., 2012 and Thompson et al., 2014). As the errors vary in both directions there is no simple assumption that less intense sampling underestimates the 'true' values as is sometimes inferred in literature (Coynel et al., 2004 and Thompson et al., 2014). Based on the results of the current study and previous research (Coynel et al., 2004, Skarbøvik et al., 2012 and Thompson et al., 2014), we suggest that water quality managers should place lower confidence in the assessments of compliance with water quality guidelines, when mean annual concentrations are based on monthly-resolution sampling; higher-resolution monitoring of SS would be recommended in order to make more accurate assessments of compliance with water quality guidelines. Thompson et al. (2014)suggest that the optimum monitoring frequency for SS, in terms of accuracy and practicality (e.g. feasible using an autosampler), is likely to be achieved through a 24/7 sampling regime (i.e. a SS sample taken every 7 h to give 24 samples per week). In order to determine the accuracy of this sampling regime in the ten sites studied here, the high-resolution (15-min frequency) SS datasets were systematically subsampled (selecting one sample every 7 h) to calculate a mean SS concentration for each site based on simulated 24/7 sampling. The average error compared to the high-resolution means was just 2%. In addition to enabling a more accurate quantification of the mean SS concentration, high-resolution monitoring of SS also enables water quality managers to gain a better insight into the temporal dynamics of SS (concentration–duration–frequencies) which can be used to predict the impacts on the ecosystem. High-resolution monitoring is more likely to capture the occasional low-frequency high concentration events such as runoff from agricultural land during storms, which may otherwise be missed by low-resolution monitoring but nevertheless can have significant impacts on the aquatic biota. High-resolution monitoring may also help to identify the sources of the pollution based on temporal patterns in concentrations. It must be recognised, however, that: (1) At present there is no environment-specific model that is capable of predicting the high-resolution reference-condition dynamics of SS at a given site in order to aid assessment. At best, the environment-specific model developed by Bilotta et al. (2012)can only predict mean annual concentration ranges. The development of an environment-specific model that is capable of predicting high-resolution dynamics or alternative metrics of exposure (e.g. concentration–duration–frequencies), would require high-resolution monitoring data from studies, similar to this one, collected from a significant number of reference-condition sites. (2) If high resolution monitoring is achieved using a surrogate measure of SS (e.g. turbidity monitoring rather than measuring SS through sampling and analysis), the resultant estimated concentrations will suffer from uncertainties that arise from the turbidity-SS regression analysis. As illustrated in Fig. 3, these uncertainties can be substantial in some cases.

The EU WFD recommends that sampling for the purposes of surveillance monitoring of surface waters should be conducted a minimum of four times a year, but it allows the Member States an option 'to tailor their monitoring frequencies according to the conditions and variability within their own waters' (EC, 2003). The WFD also recognises that sampling programmes will differ greatly between WQ elements, water body types, areas and countries. It also acknowledges that if the quality elements that are to be monitored are temporally variable or spatially discontinuous, more intensive sampling may be required. The observations of this study suggest that SS may be an example of one such temporally variable or spatially discontinuous water quality parameter. In this study the low-resolution sampling error of up to 53% was translated into a maximum absolute error of 8.2 mg L^{-1}, which some might consider to be relatively small. However, it is important to note that this study monitored reference-condition rivers. The same percentage error on a polluted river with higher mean annual SS concentrations would result in larger absolute errors and potentially significant uncertainties in the assessments of compliance (Skarbøvik et al., 2012).

CONCLUSIONS

Mean SS concentrations, measured between May 2011 and May 2013, in the ten contrasting reference-condition river sites varied up to 9-fold. The degree of spatial variation in estimated mean SS concentrations (3–29 mg L^{-1}), supports earlier research by Bilotta et al. (2012) that analysed the mean SS concentrations derived from low-resolution (monthly grab sample) monitoring programmes in 638 reference-condition sites in the UK (grouped into 42 ecosystem-types or 'end groups'). Bilotta et al. (2012)reported that the mean SS concentrations for the 42 end groups varied between 1.7 and 26 mg L^{-1}. The spatial variation in mean SS could be predicted using the environment-specific SS prediction model developed by Bilotta et al. (2012); which, when applied to the ten study sites, predicted the correct or next closest SS range for all of the sites.

Analysis of temporal variation of SS showed that there are statistically significant differences in the mean annual SS concentrations observed between the two study years. The greatest difference was observed in the Great Ouse river site, which experienced a three-fold increase between year one (14 mg L^{-1}) and year two (42 mg L^{-1}). This difference corresponds well with increases in mean annual discharge observed at this site in year one (6.37 m^3 s^{-1}) and two (19.56 m^3 s^{-1}). These temporal variations in discharge and SS are natural phenomena that shape biological communities and therefore it is important that water quality guidelines acknowledge this. This inter-annual variability in mean annual SS concentration could be predicted reasonably well for all the sites except the River Rother, using the Bilotta et al. (2012) environment-specific SS prediction model, but with modified input data to take into account the mean annual temperature (°C) and total annual precipitation (mm) in the year for which the mean SS prediction is to be made.

This study assessed the accuracy of the sampling frequency that is currently used as part of routine regulatory water quality monitoring. The results demonstrate that mean SS concentrations derived from simulated monthly sampling frequencies suffered from errors of up to 53% compared to 15-min resolution monitoring, with an average error of 24% for all the ten sites. These errors vary in both directions and there is no simple correction that can be applied. Water quality managers should place lower confidence in the assessments of compliance with water quality guidelines, when mean annual concentrations are based on monthly-resolution sampling. The observations of this study suggest that SS is a temporally variable and spatially discontinuous water quality parameter that requires higher-resolution monitoring in order to make an accurate assessment of SS exposure. A 24/7 sampling regime (one sample every 7 h) may be an optimum sampling strategy for SS in terms of accuracy (mean SS concentrations calculated from this regime are, on average, within 2% of the mean SS concentrations calculated from 15-min frequency data) and practicality (samples can be collected by an automated sampler).

The authors are not suggesting that high-resolution SS monitoring should be conducted on a routine basis in all rivers in order to assess compliance with water quality guidelines. This would be prohibitively expensive. Rather, a better system for regulating SS would be to use low-resolution (lower-cost) biological monitoring to determine the ecological status of river sites on a periodic basis (as is current practice across the EU for a number of other water quality parameters) (Turley et al., 2014). At sites where the biological community composition deviates significantly from the reference-condition community composition for that ecosystem-type, high-resolution water quality monitoring could be targeted and used to confirm the likely cause of the biological deviation. If the SS regime differs significantly from that observed in equivalent ecosystems that are in reference condition, then it is likely that SS pollution is occurring and mitigation may be needed. In terms of improving the water quality guidelines for SS, the results of this study demonstrate the need for the more advanced guidelines that take into account natural spatial and temporal variations. One way to achieve this might be to develop environment-specific concentration–duration–frequency guidelines, using high-resolution data from ecosystem-scale observations in reference-condition sites, combined with existing data from microcosm to ecosystem-scale biological/ecological experiments. For the assessments of compliance against modelled guidelines, the resolution of SS monitoring at a given site, must be similar to the resolution of SS data used to develop these guidelines.

ACKNOWLEDGMENTS

The authors would like to acknowledge the funding for this work which was provided by an Engineering and Physical Sciences Research Council CASE Studentship (Grant Reference: EP/1501274/1) which was co-funded by Aquaread Ltd. The authors would also like to acknowledge the following organisations for their contribution to the RIVPACS IV Database (© NERC [CEH] 2006. Database rights NERC [CEH] 2006 all rights reserved) and the WFD119 Project's extension to this database: Centre for Ecology and Hydrology and other Stakeholders/Centre for Ecology and Hydrology, Countryside Council for Wales, Department for Environment, Food and Rural Affairs, English Nature, Environment Agency, Environment and Heritage Service, Freshwater Biological Association, Scotland and Northern Ireland Forum for Environmental Research, Scottish Environment Protection Agency, Scottish Executive, Scottish Natural Heritage, South West Water, Welsh Assembly Government. The authors are extremely grateful for the work of John Davey-Bowker and Michael Dunbar, for their work in compiling the RIVPACS IV database.

REFERENCES

1. Alabaster JS, Lloyd R. Finely divided solids. In: Alabaster JS, Lloyd R, editors. Water Quality Criteria for Freshwater Fish. London: Butterworth Scientific; 1982. p. 1–20.
2. APHA. Standard methods for the examination of water and wastewater. Washington DC, USA: American Public Health Association/American Water Works Association/ Water Environment Federation; 1995.
3. APHA. Standard methods for the examination of water and wastewater. Washington, DC: American Public Health Association; 1998.
4. Bilotta G, Brazier R. Understanding the influence of suspended solids on water quality and aquatic biota. Water Res 2008;42:2849–61.
5. Bilotta GS, Burnside NG, Cheek L, Dunbar MJ, Grove MK, Harrison C, et al. Developing environment-specific water quality guidelines for suspended particulate matter. Water Res 2012;46:2324–32.
6. Bowes MJ, Smith JT, Neal C. The value of high-resolution nutrient monitoring: a case study of the River Frome, Dorset, UK. J Hydrol 2009;378:82–96.
7. Clarke R, Davy-Bowker J, Dunbar M, Laize C, Scarlett P, Murphy J. Enhancement of the river invertebrate classification tool. Final Report, Project WFD119; 2011.
8. Clesceri LS, Greenberg AE, Eaton AD. Standard methods for the examination of water and wastewater. Washington, DC.: American Public Health Association; 1998

9. Collins A, Naden P, Sear D, Jones J, Foster ID, Morrow K. Sediment targets for informing river catchment management: international experience and prospects. Hydrol Process 2011;25:2112–29.

10. Cooper D, Naden P, Smith B, Gannon B. Siltation in rivers. 2: a minimum monitoring strategy for two cSAC rivers. Conserving Natura 2000 Rivers Conservation Techniques Series. Peterborough: English Nature; 2003.

11. Cordone AJ, Kelley DW. The influences of inorganic sediment on the aquatic life of streams. California Department of Fish and Game; 1961.

12. Costanza R, d'Arge R, De Groot R, Farber S, Grasso M, Hannon B, et al. The value of the world's ecosystem services and natural capital. Ecol Econ 1998;25:3–15.

13. Coynel A, Schäfer J, Hurtrez J-E, Dumas J, Etcheber H, Blanc G. Sampling frequency and accuracy of SPM flux estimates in two contrasted drainage basins. Sci Total Environ 2004;330:233–47.

14. Dunn OJ. Multiple comparisons using rank sums. Technometrics 1964;6:241–52.

15. Diehl TH, Wolfe WJ. Suspended-sediment concentration regimes for two biological reference streams in Middle Tennessee. J Am Water Res Assoc 2010;46:824–37.

16. EC. EC Directive 2000/60/EC of the European Parliament and of the Council of 23 October 2000 of establishing a framework for community action in the field of water policy; 2003. p. 1–160.

17. Gammon JR. Effect of inorganic sediment on stream biota. Vol 18050 DWC 12/70. Washington, DC: USGPO, . USEPA: Water Pollution Control Research Series; 1970.

18. Garrido A, Dinar A. Managing water resources in a time of global change: mountains, valleys and flood plains. London: Routledge; 2009.

19. Giorgi F, Bi X, Pal J. Mean, interannual variability and trends in a regional climate change experiment over Europe. II: climate change scenarios (2071–2100). Climate Dynam 2004;23:839–58.

20. Gray NF. Drinking water quality. Cambridge University Press; 2008.

21. Harvey GL, Moorhouse TP, Clifford NJ, Henshaw AJ, Johnson MF, Macdonald DW, et al. Evaluating the role of invasive aquatic species as drivers of fine sediment-related river management problems: the case of the signal crayfish (Pacifastacus leniusculus). Prog Phys Geogr 2011;35:517–33.

22. Hassan MA, Gottesfeld AS, Montgomery DR, Tunnicliffe JF, Clarke GK, Wynn G, et al. Salmon-driven bed load transport and bed morphology in mountain streams. Geophys Res Lett 2008;35.

23. Kerr SJ. Silt, turbidity and suspended sediments in the aquatic environment: an annotated bibliography and literature review: Ontario Ministry of Natural Resources. Science and Technology Transfer Unit Technical Report TR-008; 1995.

24. Ladle M, Griffiths B. A study on the faeces of some chalk stream invertebrates. Hydrobiologia 1980;74:161–71.

25. Lloyd DS. Turbidity as a water quality standard for salmonid habitats in Alaska. N Am J Fish Manag 1987;7:34–5.

26. Maitland PS. Ecology of the river, brook and Sea Lamprey. Conserving Natura 2000 Rivers Ecology Series No. 5. English Nature, Peterborough. Peterborough: English Nature; 2003.

27. Montgomery DR, Buffington JM, Peterson NP, Schuett-Hames D, Quinn TP. Stream-bed scour, egg burial depths, and the influence of salmonid spawning on bed surface mobility and embryo survival. Can J Fish Aquat Sci 1996;53:1061–70.

28. Newcombe CP, Jensen JO. Channel suspended sediment and fisheries: a synthesis for quantitative assessment of risk and impact. N Am J Fish Manag 1996;16:693–727.

29. Newcombe CP, MacDonald DD. Effects of suspended sediments on aquatic ecosystems. N Am J Fish Manag 1991;11:72–82.

30. Owens P, Batalla R, Collins A, Gomez B, Hicks D, Horowitz A, et al. Fine-grained sediment in river systems: environmental significance and management issues. River Res Appl 2005;21:693–717.

31. Peddicord R. Direct effects of suspended sediments on aquatic organisms. Abstracts of papers of the American Chemical Society. 1155 16th Street NW, Washington DC 20036: American Chemical Society; 1979. p. 38.

32. Perks MT, Warburton J, Bracken L. Critical assessment and validation of a time- integrating fluvial suspended sediment sampler. Hydrol Process 1979.

33. Perks MT, Warburton J, Bracken L. Critical assessment and validation of a time-integrating fluvial suspended sediment sampler. Hydrol Process 2013;28(17):4795–807.

34. Phillips J, Russell M, Walling D. Time-integrated sampling of fluvial suspended sediment: a simple methodology for small catchments. Hydrol Process 2000;14: 2589–602.

35. Poff NL. Managing for variability to sustain freshwater ecosystems. J Water Res Plan Manag 2009;135:1–4.

36. Quinn JM, Davies-Colley RJ, Hickey CW, Vickers ML, Ryan PA. Effects of clay discharge on streams: 2. Benthic invertebrates. Hydrobiologia 1992;248:235–47.

37. Räisänen J, Hansson U, Ullerstig A, Döscher R, Graham L, Jones C, et al. European climate in the late twenty-first century: regional simulations with two driving global models and two forcing scenarios. Climate Dynam 2004;22:13–31.

38. Reid SM, Anderson PG. Effects of sediment released during open-cut pipeline water crossings. Can Water Res J 1999;24:235–51.

39. Richter BD, Braun DP, Mendelson MA, Master LL. Threats to imperiled freshwater fauna. Conserv Biol 1997;11:1081–93.

40. Robertson MJ, Scruton DA, Clarke KD. Seasonal effects of suspended sediment on the behaviours of juvenile Atlantic salmon. Trans Am Fish Soc 2007;136(3):822–8.

41. Rosenberg DM, Wiens AP. Effects of sedimentation on macrobenthic invertebrates on a northern Canadian river. Water Res 1978;12:753–63.

42. Ryan PA. Environmental effects of sediment on New Zealand streams: a review. N Z J Mar Freshw Res 1991;25:207–21.

43. Schwartz JS, Dahle M, Robinson RB. Concentration–frequency–duration curves for stream turbidity: possibilities for use assessing biological impairment. J Am Water Resour Assoc 2008;44(4):879–86.

44. Schwartz JS, Simon A, Klimetz L. Use of fish functional traits to associate in-stream suspended sediment transport metrics with biological impairment. Environ Monit Assess 2011;179:347–69.

45. Skarbøvik E, Stålnacke P, Bogen J, Bønsnes TE. Impact of sampling frequency on mean concentrations and estimated loads of suspended sediment in a Norwegian river: implications for water management. Sci Total Environ 2012;433:462–71.

46. Staddon C. Managing Europe's water resources: twenty-first century challenges. Ashgate Publishing, Ltd.; 2012

47. Swietlik W, Berry W, Gardner T, Hill B, Jha M, Kaufmann P. Developing water quality criteria for suspended and bedded sediments (SABS): potential approaches. Washington, DC.: US EPA, Office of Water; 2003

48. Thompson J, Cassidy R, Doody DG, Flynn R. Assessing suspended sediment dynamics in relation to ecological thresholds and sampling strategies in two Irish headwater catchments. Sci Total Environ 2014;468–469:345–57.

49. Trenberth KE, Dai A, Rasmussen RM, Parsons DB. The changing character of precipitation. Bull Am Meteorol Soc 2003;84.

50. Turley M, Bilotta GS, Brazier RE, Extence C. Evaluation of a fine sediment biomonitoring tool across a wide range of temperate rivers and streams. Freshw Biol 2014. http://dx.doi.org/10.1111/fwb.12429.

51. Vannote RL, Minshall GW, Cummins KW, Sedell JR, Cushing CE. The river continuum concept. Can J Fish Aquat Sci 1980;37:130–7.

52. Ward G, Ward A, Dahm C, Aumen N. Origin and formation of organic and inorganic particles in aquatic systems. The biology of particles in aquatic systems; 1994. p. 45–73.

53. Waters TF. Sediment in streams: sources, biological effects and control. Am Fish Soc Monogr, Bethesda, Maryland 1995;7:251.

54. Wildhaber YS, Michel C, Burkhardt-Holm P, Banninger D, Alewell C, Van Meerveld I. Measurement of spatial and temporal fine sediment dynamics in a small river. Hydrol Earth Syst Sci 2012;16:1501–15.

55. Wood PJ, Armitage PD. Biological effects of fine sediment in the lotic environment. Environ Manage 1997;21:203–17.

56. Wright JF, Sutcliffe DW, Furse MT. Assessing the biological quality of freshwaters. RIVPACS and other techniques. Ambleside, England: Freshwater Biological Association; 2000.

57. Yarnell SM, Mount JF, Larsen EW. The influence of relative sediment supply on riverine habitat heterogeneity. Geomorphology 2006;80:310–24.

58. Yount JD, Niemi GJ. Recovery of lotic communities and ecosystems from disturbance — a narrative review of case studies. Environ Manage 1990;14:547–69.

CHAPTER 10

DEVELOPING A MULTI-POLLUTANT CONCEPTUAL FRAMEWORK FOR THE SELECTION AND TARGETING OF INTERVENTIONS IN WATER INDUSTRY CATCHMENT MANAGEMENT SCHEMES

J.W. Bloodworth, I.P. Holman, P.J. Burgess, S. Gillman, Z. Frogbrook, P. Brown

Keywords: Catchment management, Water industry, Drinking water protection, Conceptual framework

INTRODUCTION

Variability in catchment water quality is an issue that impacts on many aspects of the environment and society. Improvements to water quality have resulted from the improved regulation of industries that discharge effluent into the water environment from an individual source (e.g. EU Urban Waste Water Directive 91/271/EEC). However the improved control of point sources means that increasing attention is being placed on diffuse sources of pollutants in catchments (Edwards and Withers, 2008). The spatially-diverse nature of diffuse pollution makes it difficult to pinpoint areas for regulation and investment and therefore an integrated catchment based approach has commonly been adopted for its control (Harris, 2013), e.g. the Watershed Approach Framework (USA) (EPA, 1996), Catchment Management Authorities (New South Wales, Australia) (NSW Government, 2003).

In Europe, this integrated approach forms the underlying management structure of the EU Water Framework Directive (WFD) which aims to achieve a 'good' ecological and chemical status for all EU water bodies (Council Directive 2000/60/EC of the European Parliament and the Council of 23, 2000 and Holzwarth, 2002). Achieving the required water body status relies on the designated Competent Body (such as the national environmental regulator) outlining a programme of measures, within their river basin management plans, to tackle a range of pollutants which may be implemented by multiple organisations or stakeholders, which may include water companies.

Water companies as recipients of poor water quality and with their own regulatory issues have an increasing interest in controlling pollution at source using catchment management rather than relying on increased treatment. At the EU level this is consistent with Article 7 of the WFD which aims at "*avoiding deterioration in their* (Water Companies) *quality to reduce the level of purification treatment required in the production of drinking water*" (2000/60/EC). This, combined with encouragement from the water industry regulatory bodies and the perceived benefits of catchment management within the water industry (such as reducing costs of treatment, promoting sustainability and reducing greenhouse gas emissions) (UKWIR, 2012), have led to all of the major water companies in the UK adopting some form of integrated catchment management (Spiller et al., 2013). However despite national concerns about drinking water limits being breached

Citation: J.W. Bloodworth, I.P. Holman, P.J. Burgess, S. Gillman, Z. Frogbrook, P. Brown, Developing a multi-pollutant conceptual framework for the selection and targeting of interventions in water industry catchment management schemes, Journal of Environmental Management, Volume 161, 15 September 2015, Pages 153-162, ISSN 0301-4797, http://dx.doi.org/10.1016/j.jenvman.2015.06.050.

by multiple pollutants, catchment management investigations have tended to be reactionary and to focus on a single pollutant issue in isolation. This potentially means that companies are not realising the full benefit of an integrated approach to improve raw drinking water quality.

Drinking water catchments often present an inherent mosaic of different land uses, soil types, geology and anthropogenic influences that promote heterogeneity in catchment pollutant processes. For catchment management to be effective and sustainable it is vital that interventions are carefully selected and targeted at Critical Source Areas (CSAs) that pose disproportionately higher risk than others (Strauss et al., 2007, White et al., 2009 and Doody et al., 2012). Aside from improved effectiveness and reduced implementation costs, disruption to other catchment stakeholders will be reduced when compared to widespread implementation of interventions (Beharry-Borg et al., 2013).

The selection and targeting of measures has been supported by the development of a number of conceptual frameworks and models to aid stakeholder decisions. However, they have largely been developed with single pollutant issues in mind e.g. the Nutrient Export Risk Matrix (Hewett et al., 2004 and Hewett et al., 2009) or the CatchIS modelling framework (Brown et al., 2002). Frameworks have also been developed which highlight certain components of pollutant risk such as the SciMap modelling framework which makes an assessment of spatial risk based on hydrological connectivity in a catchment (Lane et al., 2009). Frameworks that concentrate on singular pollutants or risk components however do not always allow for the assessment of the range of pollutants that need to be considered by water companies. Where multiple pollutant frameworks have been produced, they have either been developed for specific land use and soil types (Granger et al., 2010) or tend to focus on a single component of risk (e.g. source comparison in Dawson and Smith, 2010).

The aim of this paper therefore is to develop and demonstrate a generic conceptual framework that allows comparison of the spatial and temporal drinking water quality risks associated with multiple pollutants. This will allow water utilities and their partners to proactively identify critical source areas for multiple pollutants and subsequently better select and target a programme of interventions.

This paper (i) identifies catchment process similarities between different pollutants as a basis for integrating multiple pollutants within a single framework; (ii) proposes a new conceptual framework (CaRPoW – Catchment Risk to Potable Water) to facilitate the selection and targeting of catchment interventions to address multiple pollutants, that meets the needs of water companies based on criteria developed with water company professionals and (iii) discusses the merits and drawbacks of using such a framework to select and target interventions.

CATCHMENT HETEROGENEITY AND MULTIPLE POLLUTANT PROCESSES

The development of this generic framework is based on the presumption that different pollutants sometimes show similarities in either their source, mobilisation or delivery and that this will result in common critical source areas. Building on work by Haygarth et al. (2005) and Granger et al. (2010) we took the key pollutants of concern to drinking water source protection and reviewed their processes within the Source-Mobilisation-Delivery continuum framework that describes the cascade of groupings of pollutant processes that lead to the contamination of drinking water sources (Haygarth et al., 2005). Source processes concern whether a pollutant occurs naturally or as a result of human intervention (Granger et al., 2010). Mobilisation relates to the mechanism(s) by which a pollutant moves from its source either in solution and/or attached to particulate matter. Finally the delivery component refers to the pathway that a mobilised pollutant takes to reach the receptor (water body). Pollutants reviewed include pesticides and nitrate, which have regulated limits under the EU Drinking Water Directive (98/83/EC), and Dissolved Organic Carbon (DOC), sediment and phosphorus which cause issues relating to disinfection by-products, turbidity and reservoir algal blooms, respectively.

Pesticides

The importance and strength of the source term for most pesticides depends primarily on the agricultural land use, which in turn determines the rate, frequency and timing of active ingredient application. The mobilisation and delivery of the active ingredients are then often primarily determined by hydrological events which can affect runoff, leaching and drain-flow (Leu et al., 2004 and Reichenberger et al., 2007). Although less common, spray-drift and overspraying into water bodies can also result from poor pesticide spraying practices (Reichenberger et al., 2007).

Mobilisation can be both in particulate and soluble forms, and is influenced by both soil properties and the sorption strength and solubility of the pesticide (Wauchope et al., 2002 and Gavrilescu, 2005). Soil organic matter and clay content determine sorption sites and soil physical properties such as porosity determine the water storage capacity and thus propensity of pesticide sorption (Spark and Swift, 2002 and Arias-Estévez et al., 2008). Soil texture and topographical features such as slope, which affect erosion rates, influence mobilisation in particulate forms (Arias-Estévez et al., 2008). Rainfall intensity, duration and timing can influence the onset of soil detachment (particulate mobilisation), the mobilisation of freshly applied pesticides on the soil surface and pesticides in soil solution when soil moisture content is above field capacity (Kladivco et al., 2001, Nolan et al., 2008 and Lewan et al., 2009).

Pesticides can be delivered to water bodies by high energy, low energy and non-hydrological delivery pathways. Again pesticide properties are key determinands, with low sorbing and soluble pesticides more likely to be delivered in low energy pathways such as throughflow and leaching (Kördel et al., 2008) and stronger sorbing and less soluble pesticides delivered through higher energy processes such as surface runoff and preferential flow (Riise et al., 2004 and Reichenberger et al., 2007). Throughflow and leaching processes are likely to be more prevalent in lighter, sandier soils (Leu et al., 2004) and the higher energy runoff and preferential pathways associated with heavier clay soils that may be subject to artificial drainage (Akay and Fox, 2007 and Brown and van Beinum, 2009). Precipitation characteristics, especially the timing and magnitude of the first rainfall event after application (Louchart et al., 2001 and Guo et al., 2004), are important in the delivery of pesticides. The significance of non-hydrological processes, such as spray drift and volatilisation, are dependent on the proximity of application to surface water, spraying technique and the properties of the pesticide (Gil and Sinfort, 2005 and FOCUS., 2008).

Nitrogen

Sources of nitrogen on agricultural land include inorganic fertiliser (Domburg et al., 1998), cycled sources such as livestock waste either directly applied by livestock or in the form of manure or slurry applications (Hooda et al., 2000), and internal sources such as nitrate from soil mineralisation (Di and Cameron, 2002). These sources exist in a number of forms including nitrate and ammonium (Hooda et al., 2000, Granger et al., 2010 and Jarvie et al., 2010). Agricultural sources are generally diffuse with some point sources associated with farm yards (Edwards et al., 2008) and other non-agricultural sources, such as Waste Water Treatment Works (WWTW) discharges and septic tanks. The main determinands of nitrate sources in a catchment therefore are the type of agriculture (although it varies with cropping and livestock system, and fertiliser and stocking rates) and the presence of any point sources. Nitrate concentrations in rivers can be higher in winter months when plants are not actively taking nitrate from the soil and leaching rates are higher (Jarvie et al., 2010).

Nitrate is mostly mobilised via solubilisation processes (Granger et al., 2010) due to its high solubility (Di and Cameron, 2002). However, the solubilisation of nitrate is closely linked to the soil moisture content which in turn is affected by rainfall and the soil properties such as texture and porosity (Torbert et al., 1999). Incidental mobilisation of ammonium straight after organic fertiliser application can locally occur, although this is dependent on significant rainfall shortly after application (Smith et al., 2001) and reduced soil infiltration capacity (e.g. from livestock poaching) (Butler et al., 2008).

Nitrate is generally delivered via lower energy leaching and throughflow, with approximately 5 times higher losses of nitrate via leaching than surface runoff in both intensive and extensive agriculture (Parn et al. 2012). Nitrate losses via leaching and throughflow are increased by a soil moisture content high enough to drive vertical and lateral throughflow; soils with a high hydraulic conductivity (Bergström and Johansson, 1991); land management activities such as artificial drainage (Singleton et al., 2001) and the lack of cover crops (MacDonald et al., 2005). Although these processes dominate, ammoniacal-N mobilised incidentally can also be delivered in high energy surface runoff and preferential flow processes (e.g. Ming-kui et al., 2007).

Phosphorus

Similar to nitrogen, phosphorus can be supplied from external sources such as the application of inorganic fertilisers; cycled sources from the direct and indirect application of livestock waste; and internal sources from the P surplus in many soils (Edwards and Withers, 1998, Withers et al., 2003 and Hodgkinson and Withers, 2007). Diffuse sources occur from fields receiving fertiliser and slurry whereas point sources can either be losses from livestock waste in farmyards or effluent discharges from waste water treatment works and septic tanks (Macintosh et al., 2011). The strength of the source depends on the type of livestock (Smith et al., 1998), stocking rate (Withers et al., 2001), fertiliser application rate which can be related to crop type and the P status of the soil, and the presence of non-agricultural point sources in the catchment (Macintosh et al., 2011).

P mobilisation has been demonstrated to occur in particulate form, in solution and incidentally (Granger et al., 2010). Due to the high sorption capabilities of P, particulate detachment processes are generally considered the dominant P mobilisation mechanism (Kleinman et al., 2011). Detachment is dependent on significant erosional rainfall inputs, but can be exacerbated by other factors such as soil texture, tillage, soil compaction and livestock poaching. Solubilisation of P is also possible via the dissolution of P compounds or via desorption when the P sorption equilibrium of the soil is exceeded (Styles et al., 2006). Variables influencing the likelihood of P solubilisation include the soil type (sandy soils have low P sorption coefficients), the organic matter content and the soil moisture content of the soil (Hooda et al., 2000, McDowell et al., 2001 and Djodjic et al., 2004). Incidental mobilisation occurs when a slurry or fertiliser application precedes or coincides with a hydrologically effective rainfall event, with inorganic fertilisers generally causing the highest stream P concentrations after incidental mobilisation (Preedy et al., 2001). This can locally be the dominant mobilisation form at the field scale but effects are diluted at the catchment scale (Withers et al., 2003).

High particle associated and soluble P loads can be delivered via surface runoff in both arable and grassland catchments (McDowell et al., 2001, Haygarth et al., 2006 and Bilotta et al., 2008). Factors controlling the likelihood of surface runoff P delivery are again related to effective rainfall inputs (Shigaki et al., 2007), topography, soil properties such as texture, and land management activities such as cultivation and soil compaction (Silgram et al., 2010). The presence of preferential delivery pathways (macropores and drainflow) have also been shown to deliver P in both soluble and particulate formats (Heathwaite and Dils, 2000 and Gächter et al., 2004). The likelihood of preferential flow pathways as a P delivery mechanism is increased by the presence of artificial drainage (Hodgkinson et al., 2002) and clay soils (due to the presence of macropores) (Van Es et al., 2004). Although the dominant delivery pathways for P are generally thought to be high energy processes such as surface runoff and preferential flow, P in a soluble form can be delivered via lower energy pathways, especially in soils with a low P sorption coefficient, a porous structure and a shallow water table (Börling, 2003).

Sediment

The water industry considers sediment to be an important pollutant as it causes turbidity (regulated under the Drinking Water Directive (98/83/EC)) and it provides a conduit for other pollutants such as pesticides and phosphorus (Bilotta and Brazier, 2008). 'Naturally' derived sediment is produced from erosive processes on the land and from bank and bed erosion in streams and rivers (Walling, 2005). Vegetation cover plays an important role in controlling erosion and sediment loss from land, and thus arable rotations that include periods of minimal vegetation cover often have a high disproportional sediment load contribution (Collins et al., 2009). In lowland river systems inputs from both urban areas and sewage treatment effluent can also provide a significant source of sediment, for example Carter et al. (2003) report annual percentage sediment load values of 19–22% from roads and 14–18% from sewage effluent.

The mobilisation of sediment sources depends on significant energy inputs that enable detachment processes (Granger et al., 2010). The detachment of land based sources is usually via rain drop action or hydraulic detachment from runoff processes (Morgan, 2005). Where livestock is present, poaching can also detach sediment both from land based and river bank sources (Skinner et al., 1997).

The dominant form of sediment delivery is associated with high energy runoff processes. The proportion of mobilised sediment delivered to surface waters is therefore reliant upon runoff processes having enough energy to maintain sediment suspension or saltation. Since energy along runoff pathways is often not maintained, sediment can be deposited before reaching surface waters (Morgan, 2005). Therefore a disproportionate amount of sediment is delivered in the highest energy storm events. Although runoff is the dominant delivery process, studies such as Deasy et al. (2009), have demonstrated that significant macro-pore connections in the soil subsurface can result in fine sediment delivery in sub-surface drains.

DOC/colour

The majority of DOC (Dissolved Organic Carbon) in surface waters is sourced from the dissolution of organic matter in soils, with some inputs from atmospherically deposited carbon (Dawson and Smith, 2007). Hence, there is a relationship between the organic carbon content of a soil and the DOC concentration/load in the surface waters that drain the soil (Clark et al., 2004, Holden, 2005 and Buckingham et al., 2008). The largest source of DOC in British surface waters is from catchments with a high fraction of peat soils (Holden et al., 2007 and Billett et al., 2010).

By its nature DOC is mobilised via solubilisation processes with the rate of solubilisation dependent on the rate of organic matter decomposition (Dawson and Smith, 2007). In peatland catchments, decomposition is regulated by temperature, pH and water table depth (Bonnett et al., 2006). The temperature driven nature of DOC production and mobilisation means there is often a distinct observed seasonality with higher loads and concentrations in soil and surface water in the summer-autumn when compared to the winter–spring (Dawson et al., 2011). Whilst these controls explain intra-annual variations in surface water DOC concentrations they do not explain the increase in DOC that has been observed in surface waters over the last 40 years (Evans et al., 2005). Explanations for this include increases in temperature (e.g. Freeman et al., 2001); a reduction in acid deposition due to controls over sulphur emissions (e.g. Evans et al., 2006); CO_2enrichment from increased emissions (e.g. Freeman et al., 2004); and changes to upland land management such as increases in drainage and moorland burning that influence local hydrology (e.g. Yallop and Clutterbuck, 2009), although no single driver fully explains trends in all catchments.

In peatlands DOC is transported by both low energy throughflow processes and high energy surface runoff and preferential flow (Clark et al., 2008 and Holden, 2005) including through macropores or soil pipes in peatlands (Holden et al., 2012). The perceived dominance of each process in the literature is variable but a few studies have quantified the importance of runoff events for intra-annual DOC flux with Clark et al. (2007) showing that 50% of annual DOC exports for a peatland catchment were in the highest 10% of runoff events.

Pollutant process links

If the same processes within the Source-Mobilisation-Delivery modules control the behaviour of different pollutants then it is reasonable to assume that there is a potential for shared critical source areas. In turn this highlights possibilities for the implementation of interventions within a catchment to have a positive impact on more than one pollutant. Conversely if pollutant processes are different then it acts as a warning for potential pollutant swapping with certain interventions (Stevens and Quinton, 2009). Fig. 1highlights the dominant process similarities between the different pollutants of direct (or indirect) interest to water companies. Source processes are characterised according to the dominant land uses they are associated with; mobilisation processes to whether pollutants are mobilised in solution, attached to particles and/or incidentally after application; and delivery is characterised by the important pathways to drinking water bodies.

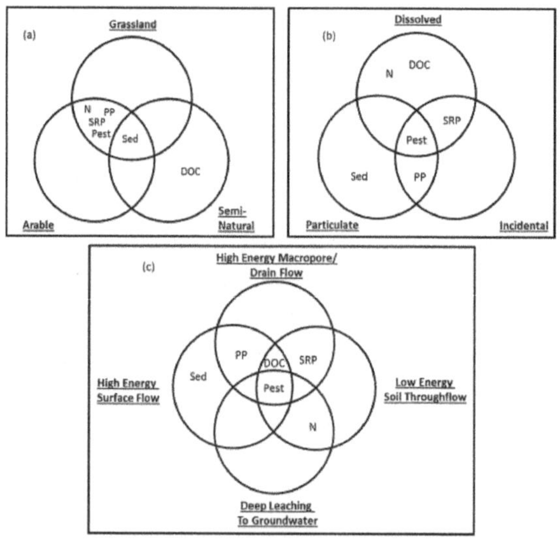

Figure 1: Process linkages between the water industry priority pollutants framed within the (a) Source, (b) Mobilisation and (c) Delivery continuum (adapted from Granger et al., 2010). Pest – Pesticides, PP – Particulate Phosphorus, SRP – Soluble Reactive Phosphorus, N – Nitrate, Sed – Sediment, DOC – Dissolved Organic Carbon.

Analysis of Fig. 1 highlights similarities in the dominant process characterisations between pollutants. For example sources of nutrients, pesticides and sediment are likely to be spatially similar as they are dominated by both grassland and arable land uses. Regarding mobilisation of these sources PP, sediment and some pesticides are inherently linked and SRP and N could potentially be mobilised in similar circumstances (mobilisation of soil pore water). Pesticides have the potential to be delivered in the same pathways as any of the other pollutants, whilst sediment and PP are associated with high energy delivery pathways and N and SRP with sub surface delivery pathways.

The process similarities therefore suggest that there is potential for pollutants to be considered within a single generic framework for the purposes of defining catchment risk. Furthermore process similarity could also be used as the conceptual basis for selecting and targeting interventions in an attempt to abate catchment pollutant risk for priority pollutants.

A CONCEPTUAL FRAMEWORK OF MULTIPLE POLLUTANT PROCESSES

Any new framework developed for water company catchment management projects clearly requires input from the end users of the framework within its development. Hence a set of criteria, informed by Graves et al. (2005), were developed in collaboration with water company professionals involved in catchment management, water quality regulation and the delivery of the Water Framework Directive.

The criteria are divided into nine key sub-sections (Table 1). End users were presented with guidance information and the list of sub-sections in Table 1 and asked to each devise criteria. Following further discussion the final criteria were collated (the second column in Table 1) which formed the basis of the conceptual framework development.

Table 1: Criteria framework used to define conceptual framework with input from water industry professionals (adapted from Graves et al., 2005)

Criteria sub-sections	Water company defined criteria
1.Background – General information on framework/model	1.1Operate in English
	1.2Supporting methodology for drinking water source protection decisions
2.Systems Modelled – Components of the system represented by framework/model	2.1Represents lowland and upland systems; arable, grassland and moorland dominated systems
	2.2Focus on surface water systems although consideration of groundwater made in some capacity
3.Objectives	3.1Characterise dominant diffuse pollution processes from source to delivery in drinking water catchments
	3.2Assess spatial and temporal variation in process characterisations
	3.3Classify risk of pollutant characterisations
	3.4Compare risk classifications between different pollutants
	3.5Select and target interventions according to the high risk areas
4.Viewpoint of analysis – Who the methodology is being developed for	4.1Modelled from the viewpoint of a water company with a focus on abstracted raw water quality
5.Spatial scale and arrangement	5.1Field/land unit scale
6.Temporal scale	6.1Monthly for some model components but output to be seasonal or annual risk
7.Generation and use of data – How the framework/model is used	7.1GIS methodology with potential to derive information from other models if necessary
	7.2Potential for framework to be used in a qualitative assessment
8.Platform and interface	8.1Initial development in spatial modelling platform
9.Inputs and outputs	9.1Inputs are spatial-temporal datasets and parameters defined by user
	9.2First output is modular (source, mobilisation and delivery) process risk
	9.3Second output combined total risk output with all three modules
	9.4Third output is risk comparison between pollutants
	9.5Intervention options selected according to process characterisation in post-processing of outputs 9.1–9.3

Assessing current modelling frameworks

Models and frameworks previously conceptualised for understanding the linkages of risk between different pollutant processes were selected for assessment against the criteria in Table 1 on the basis that they represented at least one of the key pollutants of concern, accounted for spatial risk, included some form of consideration for interventions (even if just implied) and have been applied to UK or Northern European countries.

Various limitations in the associated pollutants, represented scales and modelling typologies mean that none of the models and frameworks assessed in Table 2 fully matches the criteria. In some respects this is unsurprising as the outline criteria are bespoke to the needs of the water industry which many of the models and frameworks were not developed for, with the exception of SaGIS, CatchIS, Foster and MacDonald (2000) and Grayson et al. (2012). These models that were developed from the viewpoint of a water utility are either focused on single pollutant issues (CatchIS, Grayson et al., 2012) or do not represent the necessary processes within the Source-Mobilisation-Delivery continuum required to select and target specific interventions (SaGIS, Foster and MacDonald, 2000). The Territ'eau framework (Gascuel-Odoux et al., 2009) considers the 'fate' and 'transfer' of phosphorus, nitrates and pesticides to the river network, but has modules for individual pollutants rather than process modules (that are applied to all pollutants) which makes it difficult to compare risks for different pollutants. This limits the identification of potentially appropriate interventions or pollutant swapping.

Table 2: Evaluation of existing frameworks and models against the water company criteria components

Related criteria sub-sections		1, 4	2	2	2	3	3	3	3	5, 6	7, 8, 9
Framework/ model	Reference	Drinking water specific?	Land uses represented	Pollutants represented	Hydrological systems represented?	Separate source-Mobilisation-Delivery?	Pollutant comparison?	Intervention targeting?	Transferable, generic methodology?	Spatio-temporal scale?	Platform and outputs
Phosphorus Indicators Tool	Heathwaite et al. (2003)	✗	Upland and lowland – grassland, arable, semi-natural,	Soluble and particulate phosphorus	Surface water	S-M-D	✗	✗	✗ (principles could possibly be generic)	1 km2 – Annual	GIS risk maps – total and component risk
FARMSCOPER	Gooday et al. (2014)	✗	Lowland – grassland and arable	Soluble and particulate phosphorus, nitrate, sediment, pesticides	Surface water and groundwater	S-M-D	✓	✓	✓	Farm scale – annual	Numerical assessment
Granger et al. (2010)	Granger et al. (2010)	✗	Upland and lowland – grassland	Soluble phosphorus, particulate phosphorus, nitrate, nitrite, ammonia, fine sediment	Surface water and groundwater	S-M-D	✓	✗	✓	n/a	Qualitative classification
NERM	Hewett et al. (2004)	✗	Upland and lowland – grassland and arable	Nitrate and phosphorus	Surface water and groundwater	✗	✓ (Potentially Implicit)	✓	✓	Farm scale – n/a	Qualitative classification
SNIFFER – Diffuse Pollution Screening Tool	Sniffer (2006)	✗	Upland and lowland – grassland, arable, semi-natural	Phosphorus, nitrate, sediment, pesticides, metals	Surface water and groundwater	✗	✗	✗	✓	1 km2 – annual	GIS risk maps
SCiMap	Lane et al. (2009)	✗	Upland and lowland – grassland, arable, semi-natural	Potential for all pollutants	Surface water	D	✗	✗	✓	User defined – n/a	GIS risk maps – total risk

SAGIS	Comber et al. (2013)	✓(but not exclusively)	Upland and lowland – grassland, arable, semi-natural	Phosphorus, nitrate, sediment, metals	Surface water	X	X	X	✓	Catchment scale – Annual	GIS risk maps – total risk
CatchIS	Brown et al. (2002)	✓	Lowland – grassland, arable, semi-natural	Pesticides, nitrate	Surface water and groundwater	X	X	X	X	Catchment scale – Daily (time series), Annual (spatial risk)	GIS risk maps – total risk
Foster and MacDonald (2000)	Foster and MacDonald (2000)	✓	Upland and lowland – grassland, arable,semi-natural	Cryptospridium, pesticides, oil and grease, colour, trace metals, faecal bacteria, lead, phosphorus, nitrate	Surface water	X	X	X	✓	Catchment scale – annual	GIS risk maps – total risk
Grayson et al. (2012)	Grayson et al. (2012)	✓	Upland – grassland and semi-natural	DOC and water colour	Surface water	X	X	X	X	Catchment scale – annual	GIS risk maps
PSYCHIC model	Davison et al. (2008)	X	Upland and lowland – grassland, arable, semi-natural	Soluble phosphorus, particulate phosphorus and sediment	Surface water	S-M-D	X	X	X	1 km2 grid (Tier 1), farm scale (Tier 2) – monthly	GIS risk maps – total and component risk
The Territ'eau framework	Gascuel-Odoux et al. (2009)	X	Upland and lowland – grassland, arable, semi-natural	Phosphorus, nitrate, sediment and pesticides	Surface water	X	X	✓	✓	Field to catchment scale – annual	GIS risk maps – total risk

This review has highlighted the need for a framework that meets the water utility criteria outlined in Table 1, notwithstanding the greater need for generic frameworks at the catchment scale to assess multiple diffuse pollutants for the purposes of risk identification.

The CaRPoW framework

The Catchment Risk to Potable Water (CaRPoW) Framework (Fig. 2) has therefore been developed in response to the need for a generic framework applicable to key drinking water protection concerns. The core basis of the framework is centred on the Source-Mobilisation-Delivery continuum (Haygarth et al., 2005) which has previously been developed for phosphorus and applied to other agricultural pollutants in heavy soil grassland systems by Granger et al. (2010). The selection of the continuum as the key basis for the framework is centred on the need to disaggregate pollutant processes from how pollutants are derived (Source) to how they reach a water body (Mobilisation and Delivery), so that intervention decisions are informed by the main component of risk.

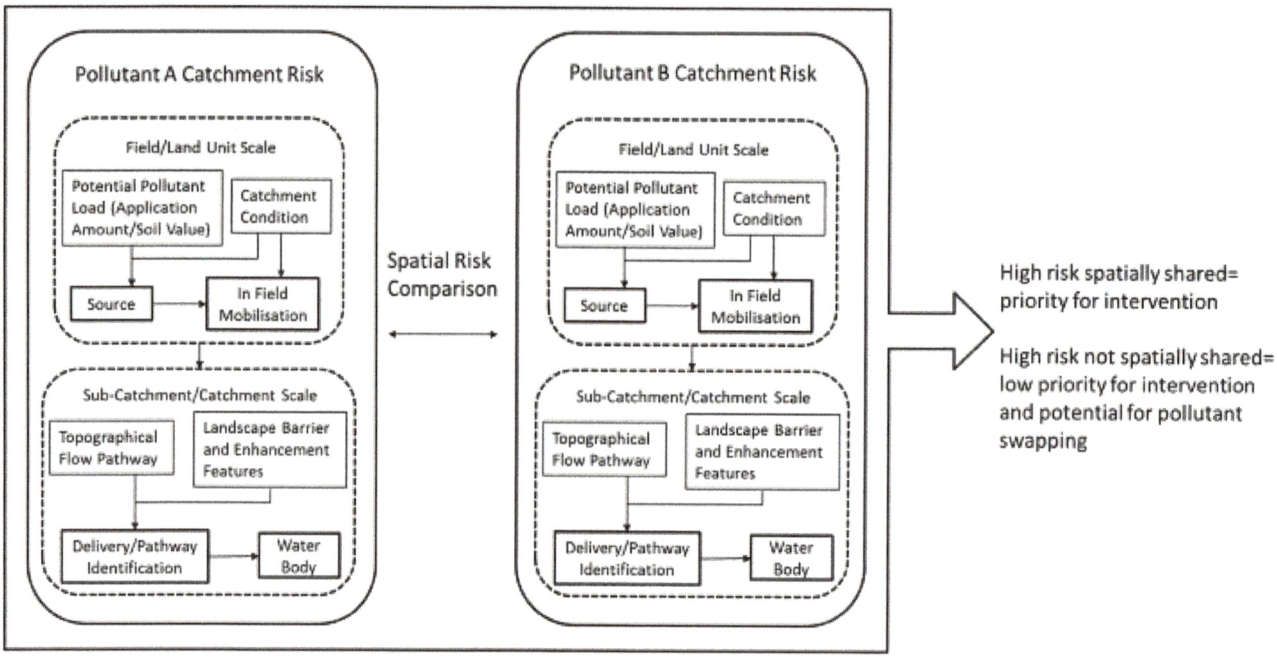

Figure 2: The CaRPoW Framework. Risk is defined within the three modules for pollutant A and Pollutant B so that risks can be compared overall and between components. Areas that have shared high risk between different pollutants are prioritised for intervention.

Defining pollutant risk

The CaRPoW framework takes a novel approach to the spatial application of the three modules of the Source-Mobilisation-Delivery continuum in order to determine pollutant risk. The source and mobilisation modules are both implemented at the 'field scale' as the key scale for agronomic intervention. Within CaRPoW 'field scale' relates to areas of land that share common land use, soil and drainage characteristics i.e. the smallest scale at which differences in substance source strength and mobility might exist using available spatial datasets. However these are also still separated by 'real world' field delineations to give a mosaic of spatial units across the catchment each with a unique source and mobilisation risk.

As with the traditional Source-Mobilisation-Delivery continuum the source module of CaRPoW refers to the mass of pollutant potentially available for mobilisation, as not all of the source may be available for mobilisation (e.g. pesticides bound to the soil, nitrate taken up by crops). This can consist of a single source (e.g. pesticides applied to fields) or multiple sources (e.g. phosphorus from manure, inorganic fertiliser and/or the soil) depending on the pollutant.

The mobilisation module represents the proportion of the available source module output that is mobilised and delivered to the edge of the field. It is therefore dependent on the initiation of runoff and drainflow processes and hence the field's soil water balance. For some pollutants such as pesticides the time factor between source availability and the initialisation of mobilisation processes is also important for source degradation and is therefore considered within the mobilisation module.

The delivery module accounts for the movement of mobilised pollutants from field edge to the water body (such as a river or reservoir) from which drinking water is abstracted, and represents the scale of opportunity for boundary feature and water course management to influence pollutant loads. It therefore incorporates principles of hydrological connectivity, i.e. the better connected a field is to the water body the more likely mobilised pollutants will be delivered. This gives a coefficient of delivery to each field within the catchment i.e. the fraction of mobilised pollutant that is delivered to the water body.

The combination of these three modules provides the overall risk as the mass of pollutant delivered to the water body per annum (Equation (1)).

$$Risk_t = Source_t \cdot KMob_t \cdot KDelivery_t \quad (1)$$

where $Risk_t$ is the mass of pollutant delivered to the water body (kg ha yr^{-1}), $Source_t$ is the mass of pollutant potentially available for mobilisation (kg ha yr^{-1}), $KMob_t$ is the coefficient of mobilisation (dimensionless proportion between 0 and 1) and $KDelivery_t$ is the coefficient of delivery (dimensionless proportion between 0 and 1).

CaRPoWs main utility is as a generic modelling framework in which modelling methodologies are implemented in the source and mobilisation components for each pollutant. The delivery component however is based on the same principles of hydrological connectivity for all pollutants. This largely stems from the fact that the movement of different pollutants beyond the field scale is much less understood (Haygarth et al., 2005), along with uncertainties over the accepted processes and complexities of hydrological connectivity within the literature (Bracken et al., 2013). One of the benefits of a generic modular modelling framework such as CaRPoW is that, as understanding of pollutant processes improves and better modelling methodologies are developed, the framework components can be updated.

Pollutant comparison and measure selection

Once risk has been assessed for each of a range of pollutants, comparisons can be made to determine if the risks are spatially concurrent between pollutants. The root cause of risk can be assessed for individual high risk fields by determining which of the three modular components is the most dominant in the high risk classification. Once determined, interventions can be referenced from an inventory (e.g. Newell Price et al., 2011) that is classified by risk component and the pollutants mitigated.

It is also important to be aware of pollutants that do not have spatially consistent risk as there may be potential for pollutant swapping. Although CaRPoW does not explicitly incorporate intervention effectiveness i.e. how risk changes after intervention implementation, it is important for the end user to make assessments of where potential pollutant swapping might occur when selecting interventions for one pollutant. For example if a change in land use is the intervention deemed best for a certain field it is important for the end user to judge if this will increase the risk for another pollutant. This is particularly prevalent for pesticides where changing from one crop to another may lead to product substitution. Overall the framework provides the end user information on the potential pre-conditions of multiple pollutant risk, upon which decisions about the most appropriate intervention in a particular area can be made.

DISCUSSION – IS CARPOW FIT FOR PURPOSE?

Benefits and potential uses of CaRPoW

A key advantage of the CaRPoW approach is that it provides the end user (a water company), more insight into their drinking water supply areas. Where catchments are large, heterogeneous and water is abstracted from a river system it is often difficult for water companies to ascertain the catchment processes that dominate pollutant risks (Spiller et al., 2013).

For example Scottish Water is currently implementing an agri-envrionment scheme in the River Ugie catchment in the North East of Scotland in order to reduce pesticide concentrations in abstracted river water. The catchment covers an area of approximately 330 km^2, has a mixture of arable and grassland farming and approximately 400 different farm holdings which poses challenges for water company staff administering the scheme. The catchment size and number of farmers makes it extremely difficult for Scottish Water staff to conduct whole-catchment walkovers to gain the comprehensive spatial understanding of risks needed to prioritise interventions from the many land owners applying for finance from the scheme. This is further highlighted by Spiller et al. (2013) who found that water companies in England and Wales were more likely to implement catchment management schemes if catchments were small, relatively homogenous and the risks are "easy" to identify. By using CaRPoW, water companies can gain a better understanding of how and where catchment management may contribute to reducing water quality risks in the larger more heterogeneous catchments that they previously may have overlooked.

Although CaRPoWs main purpose is to be used as a modelling framework, it can also be used in a purely qualitative manner by industry experts to conceptually frame pollutant risks into the three components when initiating catchment management schemes. The benefits of disaggregating risk into its constituent components in a holistic, systems-based way cannot be gained from other frameworks and models that just define overall risk.

The framework has potential uses in addressing other key catchment management based questions. A big uncertainty in catchment management is where both short to longer term future risks may arise (Pal et al., 2010). This is particularly relevant to pollutants such as pesticides where the location of sources is likely to vary spatially and temporally depending on crop rotations (Balderacci et al., 2008) or changes in crop profitability associated with climate, agronomic or economic change. The framework enables the user to make assumptions, incorporate knowledge gathered from farmers and agronomists, or implement land use/cropping scenarios to predict risk. By having a better understanding of future catchment risks, water companies should be able to plan water abstraction regimes more effectively.

Potential limitations of CaRPoW

As with any framework or approach, the limitations of CaRPoW should be clearly addressed and communicated. One such limitation is the uncertainty associated with the lack of an accepted and cohesive theory of hydrological connectivity which forms the basis of the delivery component. Bracken et al. (2013) highlight this succinctly in their review of five different approaches to investigating hydrological connectivity (soil moisture connectivity, flow process connectivity, terrain connectivity, connectivity modelling and indices of connectivity), that are all influenced by different conceptual understandings. There is also much to be said about the difference between structural connectivity which is more easily represented in spatial modelling approaches and the more dynamic functional or process-based connectivity which looks at temporal variance in connectivity which is less understood and thus more difficult to model (Bracken and Croke, 2007 and Bracken et al., 2013).

The capabilities of CaRPoW to identify where sub-catchment scale interventions should be prioritised in small, homogenous catchments, such as in parts of Scotland and Northern England where drinking water is abstracted from upland reservoir systems with water colour issues, are likely to be limited. Such catchments are often fed by comparatively small hydrologically 'flashy' catchments, with homogenised moorland land uses, peat soils and extensive agricultural drainage ditch networks. Previous modelling studies have demonstrated that these catchments are high risk to water colour as a whole when compared against other catchments (e.g. Aitkenhead-Peterson et al., 2007 and Grayson et al., 2012), but there is a dearth of studies that investigate spatial risk within these catchments.

CONCLUSIONS

In the face of deteriorating raw water quality from a number of diffuse pollution pressures, water companies have started to adopt catchment management interventions for the provision of clean drinking water. Catchment heterogeneity dictates that supply catchments are a mosaic of different land uses, soil typologies and hydrology making it difficult for water companies to ascertain the components and spatial nature of pollutants risks. Thus for interventions to be effectively targeted, conceptual frameworks are required that are able to assess the risks of the multitude of pollutants that must be removed from raw drinking water. We demonstrate that catchment risk from pollutants of concern to the water industry can be split into three constituent components of risk (Source-Mobilisation-Delivery). Highlighted key linkages between different pollutants present opportunities for shared spatial risk and therefore multiple benefits from interventions.

Criteria were developed with water industry catchment management professionals for a new conceptual framework that can uniquely be used to identify pollutant risk, compare risks between different pollutants of water-industry concern and select interventions according to the main components of risk. The Catchment Risk to Potable Water (CaRPoW) framework was specifically created for the water industry to split the components of pollutant catchment risk into Source-Mobilisation-Delivery at field-to-catchment scales. Although modular modelling methodologies are unique for each pollutant their outputs can be compared to assess where spatial risk is shared, under current or future conditions. Once the highest risk fields for multiple pollutants are defined, the dominant components of risk can be ascertained and cross referenced against an inventory of interventions to determine the potentially most suitable intervention. CaRPoW provides a valuable, transferable framework for water utilities to better understand the linkages between a range of pollutant risks in their water supply catchments and hence make more informed decisions on financing interventions.

ACKNOWLEDGEMENTS

This research was supported by the Engineering and Physical Sciences Research Council (EPSRC) through the STREAM Industrial Doctorate Centre (EPSRC grant number EP/G037094/1), with co-funding from Scottish Water. The views presented in the paper represent those of the authors and do not necessarily reflect those of Scottish Water. No new data were created in this study.

REFERENCES

1. Aitkenhead-Peterson, J.A., Smart, R.P., Aitkenhead, M.J., Cresser, M.S., McDowell, W.H., 2007. Spatial and temporal variation of dissolved organic carbon export from gauged and ungauged watersheds of Dee Valley, Scotland: effect of land cover and C: N. Water Resour. Res. 43, 1e11.
2. Akay, O., Fox, G.A., 2007. Experimental investigation of direct connectivity between macropores and subsurface drains during infiltration. Soil Sci. Soc. Am. J. 71, 1600.
3. Arias-Estevez, M., L opez-Periago, E., Martínez-Carballo, E., Simal-G andara, J., Mejuto, J.-C., García-Río, L., 2008. The mobility and degradation of pesticides in soils and the pollution of groundwater resources. Agric. Ecosyst. Environ. 123, 247e260.
4. Balderacci, M., Di Guardo, A., Costantino, V., Trevisan, M., 2008. The effect of crop rotation on pesticide leaching in a regional pesticide risk assessment. Environ. Sci. Technol. Lett. 42 (21), 8000e8006.
5. Beharry-Borg, N., Smart, N.B.J.C.R., Termansen, M., Hubacek, K., 2013. Evaluating farmers ' likely participation in a payment programme for water quality protection in the UK uplands. Reg. Environ. Change 13, 633e647.
6. Bergstrom, L.F., Johansson, R., 1991. Leaching of nitrate from monolith lysimiters of € different types of agricultural soils. J. Environ. Qual. 20 (4), 801e807.
7. Billett, M., Charman, D., Clark, J., Evans, C., Evans, M., Ostle, N., Worrall, F., Burden, A., Dinsmore, K., Jones, T., McNamara, N., Parry, L., Rowson, J., Rose, R., 2010. Carbon balance of UK peatlands: current state of knowledge and future research challenges. Clim. Res. 45, 13e29.
8. Bilotta, G.S., Brazier, R.E., 2008. Understanding the influence of suspended solids on water quality and aquatic biota. Water Res. 42, 2849e2861.
9. Bilotta, G.S., Brazier, R.E., Haygarth, P.M., Macleod, C.J.A., Butler, P., Granger, S., Kreuger, T., Freer, J., Quinton, J., 2008. Rethinking the contribution of drained and undrained grassland to sediment related water quality problems. J. Environ. Qual. 37, 906e914.
10. Bonnett, S.A.F., Ostle, N., Freeman, C., 2006. Seasonal variations in decomposition processes in a valley-bottom riparian peatland. Sci. Total Environ. 370 (2e3), 561e573.
11. Borling, A., 2003. Phosphorus Sorption, Accumulation and Leaching: Effects of € Long-term Inorganic Fertilization of Cultivated Soils. Uppsala University (Unpublished PhD thesis).
12. Bracken, L.J., Croke, J., 2007. The concept of hydrological connectivity and its contribution to understanding runoff-dominated geomorphic systems. Hydrol. Process. 1763, 1749e1763.
13. Bracken, L.J., Wainwright, J., Ali, G.A., Tetzlaff, D., Smith, M.W., Reaney, S.M., Roy, a. G., 2013. Concepts of hydrological connectivity: research approaches, pathways and future agendas. Earth Sci. Rev. 119, 17e34.
14. Brown, C.D., Bellamy, P.H., Dubus, I.G., 2002. Prediction of pesticide concentrations found in rivers in the UK. Pest Manage. Sci. 58, 363e373.
15. Brown, C.D., van Beinum, W., 2009. Pesticide transport via sub-surface drains in Europe. Environ. Pollut. 157, 3314e3324.
16. Buckingham, S., Tipping, E., Hamilton-Taylor, J., 2008. Concentrations and fluxes of dissolved organic carbon in UK topsoils. Sci. Total Environ. 407, 460e470.
17. Butler, D.M., Ranells, N.M., Franklin, D.H., Poore, M.H., Green, J.T., 2008. Runoff water quality from manured riparian grasslands with contrasting drainage and simulated grazing pressure. Agric. Ecosyst. Environ. 126, 250e260.
18. Carter, J., Owens, P.N., Walling, D.E., Leeks, G.J.L., 2003. Fingerprinting suspended sediment sources in a large urban river system. Sci. Total Environ. 314e316, 513e534.
19. Clark, M.J., Cresser, M.S., Smart, R., Chapman, P.J., Edwards, a. C., 2004. The influence of catchment characteristics on the seasonality of carbon and nitrogen species concentrations in upland rivers of Northern Scotland. Biogeochemistry 68, 1e19.

20. Clark, J.M., Lane, S.N., Chapman, P.J., Adamson, J.K., 2007. Export of dissolved organic carbon from an upland peatland during storm events: implications for flux estimates. J. Hydrol. 347, 438e447.

21. Clark, J.M., Lane, S.N., Chapman, P.J., Adamson, J.K., 2008. Link between DOC in near surface peat and stream water in an upland catchment. Sci. Total Environ. 404, 308e315.

22. Collins, A.L., Anthony, S.G., Hawley, J., Turner, T., 2009. Catena the potential impact of projected change in farming by 2015 on the importance of the agricultural sector as a sediment source in England and Wales. Catena 79, 243e250.

23. Comber, S.D., Smith, R., Daldorph, P., Gardner, M.J., Constantino, C., Ellor, B., 2013. Development of a chemical source apportionment decision support framework for catchment management. Environ. Sci. Technol. 47 (17), 9824e9832.

24. Council Directive 91/271/EEC of 21 May 1991 concerning urban waste-water treatment.

25. Council Directive 98/83/EC of 3 November 1998 on the quality of water intended for human consumption as amended by regulations 1882/2003/EC and 596/2009/EC.

26. Council Directive 2000/60/EC of the European Parliament and the Council of 23 October 2000 establishing a framework for community action in the field of water policy.

27. Davison, P.S., Withers, P.J.A., Lord, E.I., Betson, M.J., Stromqvist, J., 2008. PSYCHIC € e a process-based model of phosphorus and sediment mobilisation and delivery within agricultural catchments. Part 1: model description and parameterisation. J. Hydrol. 350, 290e302.

28. Dawson, J.J.C., Smith, P., 2007. Carbon losses from soil and its consequences for landuse management. Sci. Total Environ. 382, 165e190.

29. Dawson, J.J.C., Smith, P., 2010. Integrative management to mitigate diffuse pollution in multi-functional landscapes. Curr. Opin. Environ. Sustainable 2, 375e382.

30. Dawson, J.J.C., Tetzlaff, D., Speed, M., Hrachowitz, M., Soulsby, C., 2011. Seasonal controls on DOC dynamics in nested upland catchments in NE Scotland. Hydrol. Process 25, 1647e1658.

31. Deasy, C., Brazier, R.E., Heathwaite, A.L., Hodgkinson, R., 2009. Pathways of runoff and sediment transfer in small agricultural catchments. Hydrol. Process 1358, 1349e1358.

32. Di, H.J., Cameron, K.C., 2002. Nitrate leaching in temperate agroecosystems: sources, factors and mitigating strategies. Nutrient Cycl. Agroecosyst. 46, 237e256.

33. Djodjic, F., Borling, K., Bergstr € om, L., 2004. Phosphorus leaching in relation to soil € type and soil phosphorus content. J. Environ. Qual. 33, 678e684.

34. Domburg, P., Edwards, A.C., Sinclair, A.H., Wright, G.G., Ferrier, R.C., 1998. Changes in fertilizer and manurial practices during 1960 e 1990: implications for N and P inputs to the Ythan catchment, N. E. Scotland. Nutr. Cycl. Agroecosyst. 19e29.

35. Doody, D.G., Archbold, M., Foy, R.H., Flynn, R., 2012. Approaches to the implementation of the Water Framework Directive: targeting mitigation measures at critical source areas of diffuse phosphorus in Irish catchments. J. Environ. Manage 93, 225e234

36. Edwards, A.C., Withers, P.J.A., 1998. Soil phosphorus management and water quality: a UK perspective. Soil Use Manage. 14, 124e130.

37. Edwards, a. C., Withers, P.J.A., 2008. Transport and delivery of suspended solids, nitrogen and phosphorus from various sources to freshwaters in the UK. J. Hydrol. 350, 144e153.

38. Edwards, A.C., Kay, D., McDonald, A.T., Francis, C., Watkins, J., Wilkinson, J.R., Wyer, M.D., 2008. Farmyards, an overlooked source for highly contaminated runoff. J. Environ. Manage. 87, 551e559.

39. EPA., 1996. The Watershed Approach. EPA 840-S-96e001.

40. Evans, C.D., Monteith, D.T., Cooper, D.M., 2005. Long-term increases in surface water dissolved organic carbon: observations, possible causes and environmental impacts. Environ. Pollut. 137, 55e71.

41. Evans, C.D., Chapman, P.J., Clark, J.M., Monteith, D.T., Cresser, M.S., 2006. Alternative explanations for rising dissolved organic carbon export from organic soils. Glob. Chang. Biol. 12, 2044e2053.

42. FOCUS., 2008. Pesticides in Air: Considerations for Exposure Assessment, p. 327. Report of the FOCUS Working Group on Pesticides in Air, EC Document Reference SANCO/10553/2006 Rev 2 June 2008.

43. Foster, J.A., McDonald, A.T., 2000. Assessing pollution risks to water supply intakes using geographical information systems (GIS). Environ. Model. Softw. 15, 225e234.

44. Freeman, C., Evans, C.D., Monteith, D.T., Reynolds, B., Fenner, N., 2001. Export of organic carbon from peat soils. Nature 412, 785.

45. Freeman, C., Fenner, N., Ostle, N.J., Kang, H., Dowrick, D.J., Reynolds, B., Lock, M.A., Sleep, D., Hughes, S., Hudson, J., 2004. Export of dissolved organic carbon from peatlands under elevated carbon dioxide levels. Nature 430, 195e198.

46. Gachter, R., Steingruber, S.M., Reinhardt, M., Wehrli, B., 2004. Nutrient transfer from € soil to surface waters: differences between nitrate and phosphate. Aquat. Sci. 66, 117e122.

47. Gascuel-Odoux, C., Massa, F., Durand, P., Merot, P., Troccaz, O., Baudry, J., Thenail, C., 2009. Framework and tools for agricultural landscape assessment relating to water quality protection. Environ. Manage 43, 921e935.

48. Gavrilescu, M., 2005. Fate of Pesticides in the Environment and its Bioremediation. Eng. Life Sci. 5, 497e526.

49. Gil, Y., Sinfort, C., 2005. Emission of pesticides to the

air during sprayer application: a bibliographic review. Atmos. Environ. 39, 5183e5193.

50. Gooday, R.D., Anthony, S.G., Chadwick, D.R., Newell-Price, P., Harris, D., Duethman, D., Fish, R., Collins, A.L., Winter, M., 2014. Modelling the costeffectiveness of mitigation measures for multiple pollutants at the farm scale. Sci. Total Environ. 468e469, 1198e1209

51. Granger, S.J., Bol, R., Anthony, S., Owens, P.N., White, S.M., Haygarth, P.M., 2010. Towards a holistic classification of diffuse agricultural water pollution from intensively managed grasslands on heavy soils. Adv. Agron. 105, 83e115.

52. Graves, H.R., Burgess, P.J., Liagre, F., Terreaux, J.-P., Dupraz, C., 2005. Agrofor. Syst. 65, 53e65.

53. Grayson, R., Kay, P., Foulger, M., Gledhill, S., 2012. A GIS based MCE model for identifying water colour generation potential in UK upland drinking water supply catchments. J. Hydrol. 420e421, 37e45

54. Guo, L., Nordmark, C.E., Spurlock, F.C., Johnson, B.R., Li, L., Lee, J.M., Goh, K.S., 2004. Characterizing dependence of pesticide load in surface water on precipitation and pesticide use for the Sacramento River watershed. Environ. Sci. Technol. 38, 3842e3852.

55. Harris, B., 2013. The catchment based approach. IAH (Irish Group) conference "Groundwater & catchment management". Tullamore 1e8.

56. Haygarth, P.M., Condron, L.M., Heathwaite, A.L., Turner, B.L., Harris, G.P., 2005. The phosphorus transfer continuum: linking source to impact with an interdisciplinary and multi-scaled approach. Sci. Total Environ. 344, 5e14.

57. Haygarth, P.M., Bilotta, G.S., Bol, R., Brazier, R.E., Butler, P.J., Freer, J., Gimbert, L.J., Granger, S.J., Krueger, T., Macleod, C.J.A., Naden, P., Old, G., Quinton, J.N., Smith, B., Worsfold, P., 2006. Processes affecting transfer of sediment and colloids, with associated phosphorus, from intensively farmed grasslands: an overview of key issues. Hydrol. Process 4413, 4407e4413.

58. Heathwaite, A., Dils, R., 2000. Characterising phosphorus loss in surface and subsurface hydrological pathways. Sci. Total Environ. 251e252, 523e538.

59. Heathwaite, A., Fraser, A.I., Johnes, P.J., Hutchins, M., Lord, E., Butterfield, D., 2003. The phosphorus indicators tool: a simple model of diffuse P loss from agricultural land to water. Soil Use Manage. 19, 1e11.

60. Hewett, C.J.M., Quinn, P.F., Whitehead, P.G., Heathwaite, A.L., Flynn, N.J., 2004. Towards a nutrient export risk matrix approach to managing agricultural pollution at source. Hydrol. Earth Syst. Sci. 8, 834e845.

61. Hewett, C.J.M., Quinn, P.F., Heathwaite, A.L., Doyle, A., Burke, S., Whitehead, P.G., Lerner, D.N., 2009. A multi-scale framework for strategic management of diffuse pollution. Environ. Model. Softw. 24, 74e85.

62. Hodgkinson, R.A., Chambers, B.J., Withers, P.J.A., Cross, R., 2002. Phosphorus loss to surface waters following organic manure application to a drained clay soil. Agric. Water Manage. 57, 155e173.

63. Hodgkinson, R.A., Withers, P.J.A., 2007. Sourcing, transport and control of phosphorus loss in two English headwater catchments. Soil Use Manage. 23, 92e103.

64. Holden, J., 2005. Peatland hydrology and carbon release: why small-scale process matters. Philos. Trans. A. Math. Phys. Eng. Sci. 363, 2891e2913.

65. Holden, J., Shotbolt, L., Bonn, A., Burt, T.P., Chapman, P.J., Dougill, A.J., Fraser, E.D.G., Hubacek, K., Irvine, B., Kirkby, M.J., Reed, M.S., Prell, C., Stagl, S., Stringer, L.C., Turner, A., Worrall, F., 2007. Environmental change in moorland landscapes. Earth Sci. Rev. 82 (1e2), 75e100.

66. Holden, J., Chapman, P.J., Palmer, S.M., Kay, P., Grayson, R., 2012. The impacts of prescribed moorland burning on water colour and dissolved organic carbon: a critical synthesis. J. Environ. Manage 101, 92e103.

67. Holzwarth, F., 2002. The EU Water Framework directive e a key to catchment-based governance. Water Sci. Technol. 45 (8), 105e112.

68. Hooda, P.S., Edwards, A.C., Anderson, H.A., Miller, A., 2000. A review of water quality concerns in livestock farming areas. Sci. Total Environ. 250, 143e167.

69. Jarvie, H.P., Withers, P.J.A., Bowes, M.J., Palmer-Felgate, E.J., Harper, D.M., Wasiak, K., Wasiak, P., Hodgkinson, R.A., Bates, A., Stoate, C., Neal, M., Wickham, H.D., Harman, S.A., Armstrong, L.K., 2010. Streamwater phosphorus and nitrogen across a gradient in ruraleagricultural land use intensity. Agric. Ecosyst. Environ. 135, 238e252.

70. Kladivco, E.J., Brown, L.C., Baker, J.L., 2001. Pesticide transport to subsurface tile drains in humid regions of North America. Crit. Rev. Environ. Sci. Technol. 31 (1), 1e62.

71. Kleinman, P.J.A., Sharpley, A.N., McDowell, R.W., Flaten, D.N., Buda, A.R., Tao, L., Bergstrom, L., Zhu, Q., 2011. Managing agricultural phosphorus for water quality protection: principles for progress. Plant Soil 349, 169e182.

72. Kordel, W., Egli, H., Klein, M., 2008. Transport of pesticides via macropores (IUPAC € Technical Report). Pure Appl. Chem. 80, 105e160.

73. Lane, S.N., Reaney, S.M., Heathwaite, A.L., 2009. Representation of landscape hydrological connectivity using a topographically driven surface flow index. Water Resour. Res. 45, 1e10.

74. Leu, C., Singer, H., Stamm, C., Müller, S.R., Schwarzenbach, R.P., 2004. Simultaneous assessment of sources, processes, and factors influencing herbicide losses to surface waters in a small agricultural catchment. Environ. Sci. Technol. 38, 3827e3834.

75. Lewan, E., Kreuger, J., Jarvis, N., 2009. Implications of precipitation patterns and antecedent soil water content for leaching of pesticides from arable land. Agric. Water Manage. 96, 1633e1640.

76. Louchart, X., Voltz, M., Andrieux, P., Moussa, R., 2001. Herbicide transport to surface waters at field and watershed scales in a Mediterranean vineyard area. J. Environ. Qual. 30, 982e991.

77. Macdonald, A.J., Poulton, P.R., Howe, M.T., Goulding, K.W.T., Powlson, D.S., 2005. The use of cover crops in cereal-based cropping systems to control nitrate leaching in SE England. Plant Soil 273, 355e373.
78. Macintosh, K.A., Jordan, P., Cassidy, R., Arnscheidt, J., Ward, C., 2011. Low flow water quality in rivers; septic tank systems and high-resolution phosphorus signals. Sci. Total Environ. 412e413, 58e65.
79. McDowell, R.W., Sharpley, A.N., Condron, L.M., Haygarth, P.M., Brookes, P.C., 2001. Processes Controlling Soil Phosphorus Release to Runoff and Implications for Agricultural Management, pp. 269e284.
80. Ming-kui, Z., Li-ping, W., Zhen-li, H.E., 2007. Spatial and temporal variation of nitrogen exported by runoff from sandy agricultural soils. J. Environ. Sci. 19, 1086e1092.
81. Morgan, R.P.C., 2005. Soil Erosion and Conservation. Blackwell, Oxford, pp. 1e316.
82. Newell Price, J.P., Harris, D., Taylor, M., Williams, J.R., Anthony, S.G., Duethmann, D., Gooday, R.D., Lord, E.I., Chambers, B.J., Chadwick, D.R., Misselbrook, T.H., 2011. An Inventory of Mitigation Methods and Guide to Their Effects on Diffuse Water Pollution, Greenhouse Gas Emissions and Ammonia Emissions from Agriculture. Defra report WQ0106.
83. New South Wales Government, 2003. Catchment Management Authorities Act 2003 No 104.
84. Nolan, B.T., Dubus, I.G., Surdyk, N., Fowler, H.J., Burton, A., Hollis, J.M., Reichenberger, S., Jarvis, N.J., 2008. Identification of key climatic factors regulating the transport of pesticides in leaching and to tile drains. Pest Manage. Sci. 944, 933e944.
85. Pal, A., Gin, K.Y.-W., Lin, A.Y.-C., Reinhard, M., 2010. Impacts of emerging organic contaminants on freshwater resources: review of recent occurrences, sources, fate and effects. Sci. Total Environ. 408, 6062e6069.
86. Parn, J., Pinay, G., Mander, Ü., 2012. Indicators of nutrients transport from agricul- € tural catchments under temperate climate: a review. Ecol. Indic. 22, 4e15.
87. Preedy, N., McTiernan, K., Matthews, R., Heathwaite, L., Haygarth, P., 2001. Rapid incidental phosphorus transfers from grassland. J. Environ. Qual. 30, 2105e2112.
88. Reichenberger, S., Bach, M., Skitschak, A., Frede, H.-G., 2007. Mitigation strategies to reduce pesticide inputs into ground- and surface water and their effectiveness; a review. Sci. Total Environ. 384, 1e35.
89. Riise, G., Lundekvam, H., Wu, Q.L., Haugen, L.E., Mulder, J., 2004. Loss of pesticides from agricultural fields in SE Norwayerunoff through surface and drainage water. Environ. Geochem. Health 26, 269e276.
90. Shigaki, F., Sharpley, A., Prochnow, L.I., 2007. Rainfall intensity and phosphorus source effects on phosphorus transport in surface runoff from soil trays. Sci. Total Environ. 373, 334e343.
91. Silgram, M., Jackson, D., Bailey, A., Quinton, J., Stevens, C., 2010. Hillslope scale surface runoff, sediment and nutrient losses associated with tramline wheelings. Earth Surf. Process. Landforms 35, 699e706.
92. Singleton, P.L., McLay, C.D.A., Barkle, G.F., 2001. Nitrogen leaching from soil lysimeters irrigated with dairy shed effluent and having managed drainage. Aust. J. Soil Res. 39 (2), 385e396.
93. Skinner, J.A., Lewis, K.A., Bardon, K.S., Tucker, P., Catt, J.A., Chambers, B.J., 1997. An overview of the environmental impact of agriculture in the U.K. J. Environ. Manage 50, 111e128.
94. Smith, K.A., Chalmers, A.G., Chambers, B.J., Christie, P., 1998. Organic manure phosphorus accumulation, mobility and management. Soil Sci. 154e159.
95. Smith, K.A., Jackson, D.R., Withers, P.J., 2001. Nutrient losses by surface run-off following the application of organic manures to arable land. 2. Phosphorus. Environ. Pollut. 112, 53e60.
96. SNIFFER, 2006. Diffuse Pollution Screening Tool: Stage 3. SNIFFER project WFD 277. Spark, K.M., Swift, R.S., 2002. Effect of soil composition and dissolved organic matter on pesticide sorption. Sci. Total Environ. 298, 147e161.
97. Spiller, M., McIntosh, B.S., Seaton, R.A.F., Jeffrey, P., 2013. Implementing pollution source control e learning from the innovation process in English and Welsh water companies. Water Resour. Manage 27, 75e94.
98. Stevens, C.J., Quinton, J.N., 2009. Diffuse pollution swapping in arable agricultural systems. Crit. Rev. Environ. Sci. Technol. 39, 478e520.
99. Strauss, P., Leone, A., Ripa, M.N., Turpin, N., Lescot, J.-M., Laplana, R., 2007. Using critical source areas for targeting cost-effective best management practices to mitigate phosphorus and sediment transfer at the watershed scale. Soil Use Manage. 23, 144e153.
100. Styles, D., Donohue, I., Coxon, C., Irvine, K., 2006. Linking soil phosphorus to water quality in the Mask catchment of western Ireland through the analysis of moist soil samples. Agric. Ecosyst. Environ. 112, 300e312.
101. Torbert, H.A., Potter, K.M., Hoffman, D.W., Gerik, T.J., Richardson, C.W., 1999. Surface residue and soil moisture affect fertilizer loss in simulated runoff on a heavy clay soil. Agron. J. 91 (4), 606e612.
102. UKWIR., 2012. Quantifying the Benefits of Catchment Management. UKWIR report 12/WR/26/10.
103. Van Es, H.M., Schindelbeck, R.R., Jokela, W.E., 2004. Effect of manure application timing, crop, and soil type on phosphorus leaching. J. Environ. Qual. 33, 1070e1080.
104. Walling, D.E.T., 2005. Tracing suspended sediment sources in catchments and river systems. Sci. Total Environ. 344, 159e184.
105. Wauchope, R.D., Yeh, S., Linders, J.B.H.J., Kloskowski, R., Tanaka, K., Rubin, B., Katayama, A., Kordel, W., Gerstl, Z., Lane, M., Unsworth, J.B., 2002. Pesticide soil € sorption parameters: theory, measurement, uses, limitations and reliability. Pest Manage. Sci. 58, 419e445.
106. Withers, P.J.A., Edwards, A.C., Foy, R.H., 2001. Phosphorus cycling in UK agriculture and implications

for phosphorus loss from soil. Soil Use Manage. 17, 139e149.

107. Withers, P.J.A., Ulen, B., Stamm, C., Bechmann, M., 2003. Incidental phosphorus lossese are they significant and can they be predicted? J. Plant Nutr. Soil Sci. 166, 459e468.

108. White, M.J., Storm, D.E., Busteed, P.R., Stoodley, S.H., Phillips, S.J., 2009. Evaluating nonpoint source critical source area contributions at the watershed scale. J. Environ. Qual. 38 (4), 1654e1663.

109. Yallop, A.R., Clutterbuck, B., 2009. Land management as a factor controlling dissolved organic carbon release from upland peat soils 1: spatial variation in DOC productivity. Sci. Total Environ. 407, 3803e3813.

CHAPTER 11

REHABILITATING MANGROVE ECOSYSTEM SERVICES: A CASE STUDY ON THE RELATIVE BENEFITS OF ABANDONED POND REVERSION FROM PANAY ISLAND, PHILIPPINES

Clare Duncan, Jurgenne H. Primavera, Nathalie Pettorelli, Julian R. Thompson, Rona Joy A. Loma, Heather J. Koldewey

Keywords: Mangroves, Ecosystem services, Rehabilitation, Abandoned aquaculture ponds, Carbon stocks, Coastal protection

INTRODUCTION

Environmental management is placing increasing emphasis on the services provided by the world's ecosystems (Cardinale et al., 2012). Mangrove forests deliver numerous important ecosystem services (ES) to humans, valued at \$194,000 ha^{-1} yr^{-1} (Costanza et al., 2014): food and fuel, nursery habitat, recreation (Barbier et al., 2008 and Barbier et al., 2011). Mangroves are of particular significance in the context of climate change (Duarte et al., 2013), affording among the largest per hectare global carbon stores and coastal protection from regular waves and frequent tropical storms (Dahdouh-Guebas et al., 2005 and Donato et al., 2011). Growing global policy emphasis on both emissions reduction and climate impact mitigation in vulnerable countries (UNFCCC, 2015) places ever higher significance on the climate change mitigation and adaptation (CCMA) properties of mangroves. High susceptibility to anthropogenic activities and climate change impacts (Primavera, 2005, Duke et al., 2007 and Lovelock et al., 2015) has, however, led to mangrove areal declines of 30–50% globally (Field et al., 1998 and Valiela et al., 2001), with continued losses of 0.16–0.39% per annum (Hamilton and Casey, 2016 and Richards and Friess, 2016). 16% of mangrove species are now threatened with global extinction (Polidoro et al., 2010). Extensive loss has left degraded and highly fragmented mangroves in many parts of their global distribution (Giri et al., 2011 and Hamilton and Casey, 2016) that may have limited potential to deliver CCMA services into the future (Koch et al., 2009, Barbier et al., 2011 and Lee et al., 2014).

To combat mangrove losses, and to enhance CCMA efforts in the tropics, rehabilitation is an essential management tool (Ellison, 2000, Kairo et al., 2001, Lewis, 2005, Primavera and Esteban, 2008 and Primavera et al., 2012a). Rehabilitated mangrove blue carbon-based Payments for Ecosystem Services (PES) projects are emerging (Wylie et al., 2016), and governments are increasingly recognising the significance of mangrove coastal protection (Marois and Mitsch, 2015), with national coastal greenbelt replanting programmes now widespread following recent natural disasters (Dahdouh-Guebas et al., 2005 and Primavera et al., 2014). Recent studies on potential blue carbon PES schemes have concluded that projects would benefit from inclusion of "bundled services" to offset low voluntary carbon market prices (Locatelli et al., 2014 and Thompson et al., 2014), with particular reference to coastal protection (Kairo et al., 2009 and Locatelli et al., 2014). However, mangrove rehabilitation efforts have historically seen low successes (e.g. Primavera and Esteban, 2008, Dale et al., 2014 and Bayraktarov et al., 2015; but see Arnaud-Haond et al., 2009 and Goessens et al., 2014), and where

Citation: Clare Duncan, Jurgenne H. Primavera, Nathalie Pettorelli, Julian R. Thompson, Rona Joy A. Loma, Heather J. Koldewey, Rehabilitating mangrove ecosystem services: A case study on the relative benefits of abandoned pond reversion from Panay Island, Philippines, Marine Pollution Bulletin, Volume 109, Issue 2, 30 August 2016, Pages 772-782, ISSN 0025-326X, http://dx.doi.org/10.1016/j.marpolbul.2016.05.049.

established, longer-term monitoring of functionality has been minimal (Bosire et al., 2008). Where this has been monitored, the structure and specific functionality of rehabilitated mangroves can be comparable to adjacent natural stands (Kairo et al., 2001, Bosire et al., 2008, Ren et al., 2010, Salmo et al., 2013 and Nam et al., 2016); however, their relative potential for high multiple ES delivery is mostly unknown (but see Rönnbäck et al., 2007 and Nam et al., 2016). We thus currently lack quantitative information on the combined CCMA potential of current mangrove rehabilitation efforts.

There are two major potential sources of variation in the ability of rehabilitated mangroves to deliver high multiple CCMA ES. First, ignorance of or noncompliance to scientific guidelines has driven many rehabilitation efforts to take place in low-intertidal seafront areas where sub-optimal hydrological conditions limit survival and growth of replanted mangroves (Iftekhar, 2008, Primavera and Esteban, 2008, Primavera et al., 2012a, Primavera et al., 2012b and Primavera et al., 2014). Rehabilitation in such areas may result in low relative mangrove biomass and density, and associated carbon stocks and coastal protection potential, particularly where rehabilitation failure has historically been high. Second, site areal configuration may heavily impact the potential ES delivery of rehabilitated mangroves. Rehabilitated mangrove carbon stocks may be expected to increase linearly with site area, while coastal protection rapidly increases with mangrove greenbelt width (Koch et al., 2009). Indeed, low-intertidal rehabilitated mangroves exist primarily in monospecific narrow-fringing stands (Ellison, 2000, Iftekhar, 2008, Primavera et al., 2012a and Primavera et al., 2012b), with potentially severely limited ability to deliver effective coastal protection (Ewel et al., 1998, Barbier et al., 2008, Barbier et al., 2011 and Koch et al., 2009). Larger rehabilitation sites in the mid- to upper-intertidal zone may thus be expected to deliver much higher multiple CCMA benefits than narrow, low-intertidal rehabilitated mangroves. However, the spatial configuration and area of suitable land for mangrove rehabilitation is often constrained by land tenure conflicts and complexities in the coastal zone (e.g. agri- and aquaculture); often the major driver for prioritizing low-intertidal zone rehabilitation (Iftekhar, 2008, Primavera and Esteban, 2008, Primavera et al., 2012b and Primavera et al., 2014). CCMA arguments for rehabilitation actions may be key in future decision-making and spatial planning. To guide effective coastal zone management in the face of climate change, there is thus a need to identify and prioritise rehabilitation locations in which high CCMA gains may co-occur with minimal tenure issues (Locatelli et al., 2014, Primavera et al., 2014 and Thompson et al., 2014).

This study examines the CCMA potential of mangrove rehabilitation in abandoned aquaculture ponds relative to low-intertidal, seafront areas across Panay Island, Philippines. We first quantify the relative vegetation and sediment carbon stocks, and coastal protection potential of rehabilitated mangrove areas (mid- to upper-intertidal abandoned fishpond and low-intertidal seafront areas), against mature natural reference mangrove stands, to explore the ES potential of these different rehabilitation strategies. We then conduct a municipality-specific case study to model the potential CCMA benefits of targeted abandoned fishpond reversion, with specific reference to current coastal greenbelt rehabilitation efforts. We conclude by examining the feasibility of prioritising abandoned fishpond reversion for CCMA goals under current fishpond tenure status across the case study.

MATERIALS & METHODS

Study areas: Panay Island, Philippines

The Philippines is among the most typhoon-ravaged countries in the world (Peduzzi et al., 2012 and United Nations University (UNU) – Institute for Environment). High typhoon-exposure of coastal areas, and the infrastructural and institutional vulnerability to typhoon events (UNU, 2014), has been recently evidenced by devastating impacts suffered during super-typhoon Haiyan (Soria et al., 2015). The Philippines has experienced substantial mangrove loss: approximately 50% of the former 500,000 ha (Spalding et al., 2010) disappeared over the last century, due primarily to shallow brackish-water fishpond aquaculture development in former estuarine, basin and riverine mangroves (Primavera, 2005). Some of the highest fishpond densities occur in the West Visayas Region; e.g. on Panay Island (Primavera and Esteban, 2008). Development is largely unregulated, and despite laws mandating 50–100 m of mangrove greenbelt (Primavera et al., 2012b), fishponds are often built to the shoreline. Abandonment is high (estimates in the thousands of hectares; see Samson and Rollon, 2011 and Primavera et al., 2012b), due primarily to bank breaches in sea-facing fishponds over low productivity (Primavera et al., 2014). Fishponds are tenured by among the wealthiest in society, and operated by the poorest. Reversion of abandoned fishponds to former mangroves for greenbelt resurrection could thus benefit coastal community livelihoods through associated fisheries enhancement (Walton et al., 2006).

Philippines' public mangrove land is released by the Department of Environment and Natural Resources (DENR) for aquaculture under multiple tenure arrangements: from titled ownership, to temporary leaseholds under Fishpond Lease Agreements (FLAs) granted under the jurisdiction of the Bureau of Fisheries and Aquatic Resources of the Department of Agriculture (DA-BFAR). Under Philippine law, failure to adhere to FLA terms should preclude FLA cancellation by DA-BFAR, and reversion of jurisdiction to the Forest Management Bureau of DENR for subsequent mangrove rehabilitation. This includes Abandoned (no operational activity, subleasing, or neglect of payments), Underutilised (no commercial production within three years), and Undeveloped (pond infrastructure absent) (AUU) FLA fishponds (see Primavera et al., 2014). Herein, the term 'abandoned fishpond' refers to all AUU fishponds. However, non-coordination between government departments (DA-BFAR and DENR), low institutional capacity, exclusion of local government units (LGUs) and coastal communities from decision-making, and a lack of political will means FLA monitoring is minimal, and cancellation and reversion rarely occurs: large areas of former mangrove lie fallow. Furthermore, cancelled abandoned FLAs are often absorbed and re-tenured under new FLA leases or operated illegally (Primavera et al., 2014).

Due to the challenges of abandoned fishpond reversion, national greenbelt rehabilitation programmes continue to focus on low- and sub-intertidal planting seaward of coastal infrastructure and fishponds ('seafront rehabilitation'). High mortality in plantations of inappropriate species wastes public and international funds, while threatening other intertidal systems (seagrasses, mudflats; Primavera and Esteban, 2008, Samson and Rollon, 2008 and Samson and Rollon, 2011). Surviving seafront rehabilitated mangroves are often small areas growing at the limits of their physiological tolerance ranges (Tomlinson, 1986). In contrast, some Non-Governmental Organisation-led projects, in partnership with specific LGUs, have begun to target rehabilitation of abandoned fishponds in the mid-upper intertidal zone where more natural hydrological conditions largely remain (Primavera et al., 2012b and Primavera et al., 2014).

This study investigated the relative CCMA ES delivery by rehabilitated low-intertidal seafront and abandoned fishpond areas across Panay Island, with reference to adjacent natural stands. Six mangrove stands from four sites in Iloilo and Aklan Provinces were used (Fig. 1):

1. Bakhawan ecopark, Buswang, Kalibo, a remnant area of a former deltaic mangrove at the mouth of Aklan River (Cadaweng and Aguirre, 2005 and Walton et al., 2006). Following over-exploitation of mangrove timber, large portions of the seaward area have been replanted with *Rhizophora* spp. since the early 1990s. A wide band of mature natural *Avicennia marina* and *Sonneratia* spp.-dominated mangrove remains behind the rehabilitated areas. This study focused on (1) a seafront area replanted in 2006 with *Rhizophora apiculata*, and subsequently naturally recolonised by *A. marina*, *Nypa fruticans* and *Sonneratia alba* individuals ("Bakhawan rehab"); and (2) the inland natural mangrove area ("Bakhawan natural").
2. Ermita, Dumangas. A remnant now-fringing area of a former deltaic mangrove cleared inland for fishpond aquaculture, bordered in the landward direction by active fishponds and a coastal road. The site contains a narrow band of remaining natural mangrove (*A. marina* dominant; "Ermita natural"). Seaward of the natural mangrove is a low-intertidal stand planted in 2007 ("Ermita rehab"). The area was originally planted with *S. alba*, *A. marina* and *Rhizophora* spp. seedlings; however, only *S. alba*survived algal (*A. marina*) and barnacle (*Rhizophora* spp.) infestation.
3. A sea-facing abandoned fishpond (FP) in Nabitasan, Leganes ("Nabitasan FP"), which was reverted and replanted under a partnership between the Zoological Society of London (ZSL) and Leganes LGU in 2009 (Primavera et al., 2012a). *A. marina* is dominant, with *Rhizophora* spp., *S. alba* and *A. rumphiana* also present. Prior to assisted rehabilitation, very low mangrove cover and seaward bank destruction drove erosion of former fishpond sediments at the seaward edge. The area drains directly into the sea and is flushed twice-daily by the tide.
4. A sea-facing abandoned fishpond in Dumangas municipality ("Dumangas FP") which was abandoned following a seaward bank breach in 2005–2006. The site has subsequently been naturally recolonised. Vegetation is dense and dominated by *A. marina*, with *S. alba* and *Rhizophora* spp. also present. The area drains directly into the sea; however, much of the seaward bank remains. The site is not presently flushed completely with the tide; in areas organic matter is trapped in waterlogged sediments. The area is currently leased under an FLA, and the leaseholder is in breach of terms following abandonment (Primavera et al., 2014).

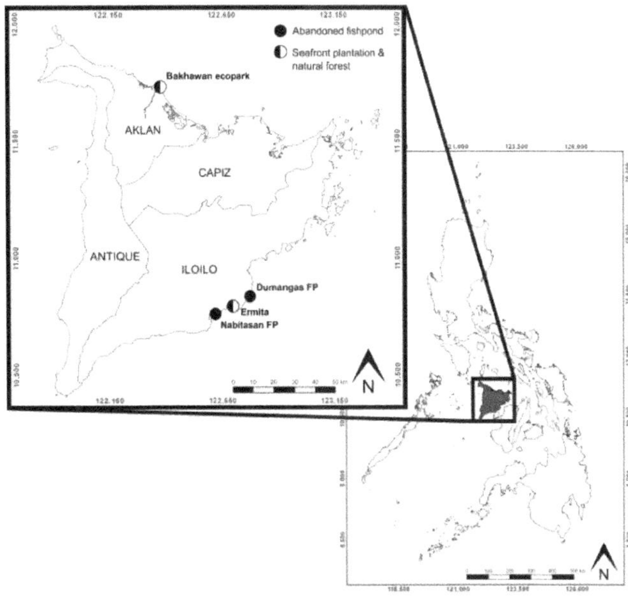

Figure 1: Location of rehabilitated and natural mangrove sites considered in this study, and location of Panay Island in West Visayas (Region VI), Philippines. Filled circles donate rehabilitated abandoned fishpond sites, while half-filled circles donate replanted seafront sites and adjacent natural areas. Details of all sites are provided in Table 1.

Field data collection

Field data collection was conducted from 2014–2015. Temporary circular plots (radius = 7 m) were established via stratified sampling with variation in distance from the shoreline (Kauffman and Donato, 2012) within a 15 × 15 m grid (N = 8 per site). For each plot, tree species, diameter at breast height (DBH; at 1.37 m height, or above the highest prop root for *Rhizophora* spp.; Kauffman and Donato, 2012), height, and maximum canopy width (m; Kauffman and Donato, 2012) of all trees (> 1.37 m height) were measured. Small trees were defined as those with ≤ 1.5 cm DBH, in order to avoid underestimation of biomass in areas with a high abundance of small trees (e.g. rehabilitated sites; Kauffman and Donato, 2012), and were measured within a 3 m radius sub-plot (from the plot centre). All larger trees (> 1.5 cm DBH) were measured throughout the plot. Low, shrubby, heterogeneous canopies in younger rehabilitated sites restricted traditional methods of rapid canopy closure estimation (e.g. spherical densitometer or vertical position digital photography; Korhonen et al., 2006). To avoid error from ocular estimates (Korhonen et al., 2006), we estimated average plot-level canopy closure (%) at all sites from horizontal position digital photographs (N = 4 per plot; at the plot centre facing toward the plot 'corners'). Average plot-level canopy closure was estimated as the average percentage of non-sky, -water or -sediment image pixels, classified for all plot photographs from the ratio of green to red light. These estimates were then averaged across all plots for each site (N = 8) to estimate site-specific average canopy closure (%).

This approach enabled us to index both large canopy gaps (where present) and canopy penetration across plots.

Sediment cores were taken in a triangular configuration within the 3 m sub-plot ($N = 3$ per plot) with an Eijkelkamp gouge auger (30 mm diameter, 50 cm sampling length). Within each core, six 5 cm samples were taken at specified depths and aggregated to represent the 0–50 cm ($N = 3$ depth subsamples × 3 cores = 9) and 50–150 cm ($N = 3$ depth subsamples × 3 cores = 9) sediment horizons (5–15 cm, 15–30 cm and 30–50 cm, and 50–100 cm, 100–150 cm and 150 cm respectively; Kauffman and Donato, 2012). Sediments were sampled to 150 cm depth due to laboratory constraints (rather than the entire sediment profile; Donato et al., 2011 and Kauffman and Donato, 2012). Average sediment profile depth was measured ($N = 3$ per plot) in a triangular configuration within 3.5 m of the plot centre, avoiding areas close to large trees (aerial roots), by inserting an iron rod (diameter = 1.5 cm) vertically into the sediment by hand until it could no longer be pushed.

Ecosystem services quantification

Vegetation carbon stocks

Vegetation biomass was calculated via allometric equations. For single-stemmed trees, biomass was estimated via the general mangrove allometric equations derived by Komiyama et al. (2005) (Eqs. (1) and (2)), in order to ensure continuity in biomass estimation across plots and sites. Individual tree above- (B_{AG}; kg) and belowground biomass (B_{TB}; kg) were estimated by:

$$B_{AG} = 0.251 \times p \times D^{2.46} \tag{1}$$

$$B_{TB} = 0.199 \times p^{0.899} \times D^{2.22} \tag{2}$$

where p is species-specific wood density (g cm^{-3}) (Komiyama et al., 2005). Two individual trees in the natural area at Ermita, Dumangas were beyond the DBH limit of these equations and their biomass was overestimated (Komiyama et al., 2005 and Kauffman and Donato, 2012). In the absence of guidelines to accommodate this situation (see Kauffman and Donato, 2012 and Thompson et al., 2014), these individuals were here treated as multiple trees. Their DBH was split to create two separate 'individuals' (e.g. DBH1 = 49 cm, DBH2 = 12 cm) and their biomass calculated. While this method likely still overestimates biomass, the resulting overestimates were substantially smaller than those caused by calculation via actual DBH. Estimates of p were taken from the Global Wood Density Database (Chave et al., 2009 and Zanne et al., 2009). Species-specific averages of p were taken across all Southeast Asian estimates, in the absence of Philippines-specific estimates for all species in this study.

For multi-stemmed trees, aboveground biomass was estimated via the method of Fu and Wu (2011). B_{AG} (kg) was calculated as:

$$B_{AG} = CD^2 \times \mathrm{H} \tag{3}$$

where CD is the maximum canopy diameter (m) and H is tree height (m). This method has been found to be applicable to species with variable growth form (e.g. *A. marina*;Fu and Wu, 2011). Belowground biomass of multi-stemmed trees was calculated via the general equation of Komiyama et al. (2005) (Eq. (2)), using an artificial DBH calculated from the DBH of the largest stem and the number of stems per individual.

All sites were highly heterogeneous in biomass distribution; remote sensing methods were thus employed to index spatial variability. We employed a technique similar to that derived by Simard et al. (2006) and Fatoyinbo et al. (2008), using Shuttle Radar Topography Mission (SRTM) near-global Digital Elevation Model (DEM) data (30 m resolution; Rodriguez et al., 2006). All analyses were conducted in R v. 3.2.1 (R Development Core Team, 2015). SRTM pixel values (SRTM-derived height; m) were first extracted for each temporary field plot location. C- and X-band SRTM radar scattering is influenced by vegetation density, biomass and canopy closure, and error exists between observed SRTM DEM height and true canopy height (Simard et al., 2006 and Fatoyinbo et al., 2008). We thus established a linear regression to predict field plot mean height (m) from corresponding SRTM DEM pixel height (m). Plot-level mean height data was used to derive (1) a mean height-aboveground biomass relationship, and (2) a mean height-belowground biomass (plot level Mg [tonnes], extrapolated to Mg 900 m^2) relationship with our allometrically-estimated biomass values (Komiyama et al., 2005), using linear regression. These equations were then applied across all SRTM DEM pixels at each site, and summed (biomass value × proportion of pixel within the site boundary) to estimate site-level biomass. Site-level above- and belowground biomass estimates were then multiplied by carbon concentration values of 0.464 (Donato et al., 2011) and 0.39 (Kauffman and Donato, 2012) respectively to obtain vegetation carbon stock estimates. Mean per hectare above- and belowground vegetation carbon stocks were calculated across all pixels with ≥ 50% of their area within the site boundary.

Sediment carbon stocks

Sediment samples were analysed at the Bureau of Soils and Water Management, Cebu. Bulk density (BD; g cm^{-3}) was determined against the sample volume as:

$$BD = \frac{DM}{SV} \tag{4}$$

where DM is oven-dried mass (g) and SV is the sample volume (cm^3). Air-dried subsamples were homogenised and sieved (2 mm) (Donato et al., 2011 and Thompson et al., 2014). Organic carbon content (OC; %) was determined gravimetrically for each subsample (0.25 g) via the dry combustion method (Schumacher, 2002). Sediment horizon carbon stock was calculated by summing the carbon stock for each of the two sampled depth intervals (Kauffman and Donato, 2012). The 50–150 cm depth samples were used to extrapolate sediment carbon to the entire sediment profile depth for each plot, as OC does not vary greatly below 30 cm (Thompson et al., 2014). Sediment carbon (SC; Mg ha^{-1}) was calculated for each depth interval as:

$$SC_i = \mathrm{BD} \times D \times OC \tag{5}$$

where D is the sediment depth interval (cm; e.g. here 50 cm and maximum depth–50 cm for the 0–50 cm and 50–sediment profile depth intervals respectively) and OC is the sample OC (%; Kauffman and Donato, 2012). Linear regressions (lms) were applied to model differences in vegetation and sediment carbon estimates (Mg ha^{-1}) across sites (with Site Name as a predictor variable). Site-level sediment carbon stock estimates were calculated from the plot-level estimates from each site (Mg 153.94 m^2). The area of each site was divided by plot area (153.94 m^2) to create 'sub-site' areas, and an estimate of sediment carbon stock for each 'sub-site' was drawn from a normal distribution created from the mean and standard deviation of plot-level estimates for each depth interval. These 'sub-site' estimates were then summed to create total site sediment carbon storage estimates for each depth interval at each site (mean over 1,000,000 runs).

Coastal protection potential

The model of wave attenuation derived by Bao (2011) was employed to assess the coastal protection potential of each site. This model estimates the width of mangrove greenbelt required to attenuate a regular wave of three metres height to a 'safe height' of 0.3 m behind the forest (Bao, 2011). A Forest Structure Index (FSI), calculated from field-derived mangrove structural parameters, is calculated and characterises different forests into protection level categories (I-V). FSI was calculated from mean plot-level data for each site as:

$$FSI = -0.048 + 0.0016\,H + 0.00178\log(SD) + 0.0077\log(CC) \quad (6)$$

where H is mean mangrove height (m), SD is mean stem density (N ha^{-1}), and CC is mean canopy closure (%). The required greenbelt width (BW; m) required at each site was calculated as:

$$B_W = \frac{\log(H_{safe}) - \log(0.9899\,H_0 + 0.3526)}{0.048 - 0.0016\,H - 0.00178\log(SD) - \log(CC)} \quad (7)$$

where H_{safe} is the 'safe wave height' (30 cm), and H_0 is initial wave height (300 cm). Required mangrove greenbelt width (BW) was compared to the median actual landward width (B_{Actual}; m) to determine the current coastal protection potential of each site. Landward and seaward boundaries of the polygon outline of each site were extracted in QGIS (QGIS Development Team, 2015) and the median distance between seaward and landward boundary vertices (N ≈ 2000) were calculated for each site using the function 'gDistance' (Bivand et al., 2015).

Case study: Dumangas municipality

Fishpond density in Dumangas, Iloilo is among the highest in West Visayas (4282.7 ha; Primavera et al., 2014). We mapped the distribution of abandoned fishponds (mangrove vegetation present) from high-resolution satellite imagery (Google Earth, 2015). SRTM-derived vegetation biomass estimates (Mg ha^{-1}) from the two abandoned fishpond sites (Nabitasan and Dumangas) were used to project potential vegetation carbon stock accumulation across Dumangas' abandoned fishponds over a period of 6.5 years (mean abandoned fishpond site age) via random sampling from the observed distribution of vegetation carbon stock (Mg ha^{-1}) estimates (mean over 1,000,000 runs). This method was employed to produce a conservative estimate, as some abandoned fishponds are currently in relatively advanced stages of mangrove re-colonisation (Primavera et al., 2014). Potential sediment carbon accretion was estimated using a conservative accretion rate of 25 mm y^{-1} for the Indo-Pacific Region (Lovelock et al., 2015; 1.63 cm surface accretion over 6.5 years) and random sampling from the observed distribution of sediment carbon stock (Mg ha^{-1}) for the 0–50 cm depth interval (mean over 1,000,000 runs). We compared the mapped abandoned fishponds against a recent survey of tenure status in Dumangas (e.g. breached FLA leases; Primavera et al., 2014).

We then subset mapped sea-facing abandoned fishponds and calculated median distance between landward and seaward boundary vertices as above. Vertices with median distance to landward boundaries greater than estimated required greenbelt widths to attenuate three metre initial waves (BW; estimated for abandoned fishpond sites as above; Bao, 2011) were retained to calculate the length of Dumangas' coastline with potential for effective coastal protection from regular wind waves following abandoned fishpond reversion. We then compared this against tenure status for identified sea-facing abandoned fishponds (Primavera et al., 2014).

RESULTS

Vegetation structure

Mean vegetation basal area, DBH, height and biomass were greatest in the natural areas, followed by the rehabilitated seafront areas, and abandoned fishponds (Table 1). Stem density was highly variable across natural and rehabilitated sites, being greatest at the rehabilitated seafront area at Bakhawan (11,839.18 stems ha^{-1}) and lowest at the rehabilitated seafront area at Ermita (1916.36 ha^{-1}) (Table 1).

Table 1: Mean field-derived structural parameters across plots (N = 8) in all surveyed sites. Biomass refers to mean total (above- and below-ground) biomass

Site	Basal area (m^2 ha^{-1})	Stem density (N ha^{-1})	Canopy closure (%)	DBH (cm)	Mean height (m)	Biomass (Mg ha^{-1})
Bakhawan Natural	27.68	6496.12	88.10	5.76	6.16	177.88
Bakhawan Rehab	16.52	11,839.18	84.45	2.71	3.78	79.27
Ermita Natural	33.16	2151.84	84.85	18.0	6.67	241.62

Ermita Rehab	10.75	1916.36	87.19	6.86	4.63	52.88
Nabitasan FP	3.38	2379.21	63.60	3.27	1.99	15.25
Dumangas FP	8.17	6950.85	84.00	3.98	2.89	37.96

SRTM vegetation biomass modelling

A positive linear relationship was found between mean plot-level vegetation height (m; log-transformed) and SRTM-derived height (m; square root-transformed): $R^2 = 0.43$; root-mean-square error (RMSE) = 0.41 (Fig. 2). Allometrically-estimated (Section 2.3) plot-level above- (linear; $R^2 = 0.73$; RMSE = 0.50; Fig. 3a) and belowground biomass (quadratic; $R^2 = 0.89$; RMSE = 0.44; Fig. 3b), both extrapolated to Mg 900 m^2, showed a positive relationship with mean plot-level vegetation height (m; log-transformed). We applied these equations across SRTM tiles for Panay to predict above- (Fig. 4) and belowground biomass across sites.

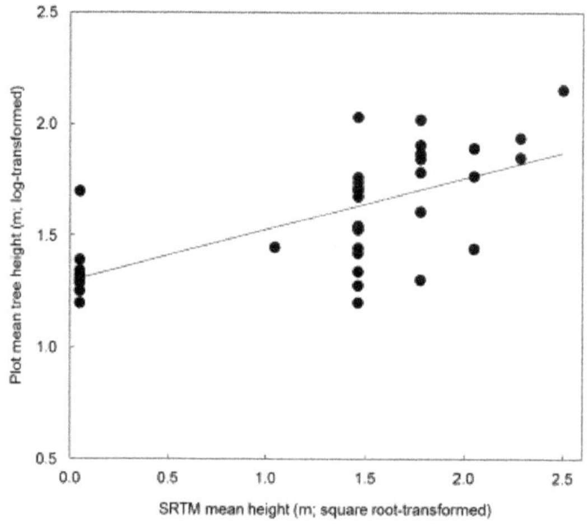

Figure 2: Regression of observed plot-level mean mangrove height (m; log-transformed) against SRTM DEM-derived elevation (m; square root-transformed). $\log(\textit{mean height}) = 1.31 + 0.23\sqrt{\textit{SRTMelevation}}$; $R^2 = 0.43$; RMSE = 0.41.

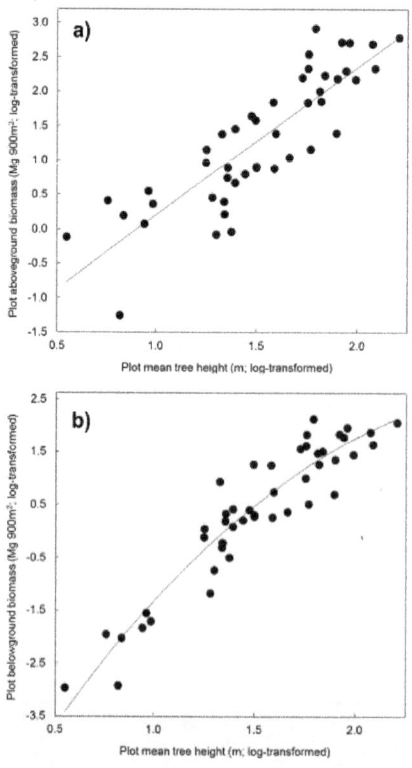

Figure 3: Best-fitting regressions (highest variance explained; R^2) of (a) above- and (b) belowground biomass (allometric estimation from field DBH data; Komiyama et al., 2005) against observed mean mangrove height (m; log-transformed). Allometrically-estimated above- and belowground biomass has been scaled from plot- to pixel-level estimates (900m^2; log-transformed). The relationship between allometrically-estimated aboveground biomass and observed mean mangrove height was linear in form ($\log(\textit{aboveground biomass}) = 2.14\log(\textit{mean height}) - 1.84$; $R^2 = 0.73$; RMSE = 0.50), while the relationship for allometrically-estimated below-ground biomass was a quadratic relationship ($\log(\textit{belowground biomass}) = -1.05\log(\textit{mean height})^2 + 6.13\log(\textit{mean height}) - 6.25$; $R^2 = 0.89$; RMSE = 0.44).

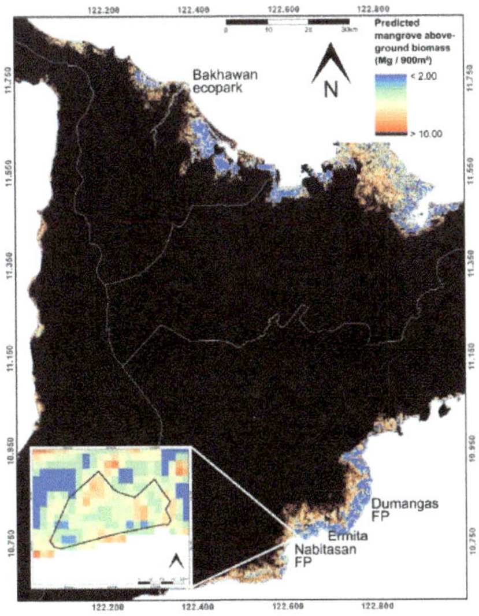

Figure 4: Predicted aboveground mangrove biomass (Mg 900m^2) across the two SRTM DEM tiles on Panay Island. Blue pixels denote areas of low biomass, while red areas denote higher biomass areas. Dark blue pixels indicate active aquaculture pond areas, and black pixels denote areas with biomass > 10 Mg 900 m^2. N.B. This figure illustrates predictions of areas outside of the distribution of mangroves on Panay Island (e.g. beach forest and terrestrial forest and plantation areas), which were not included in the analyses of this study.

Carbon stocks

Mean above- and belowground vegetation carbon stocks (Fig. 5) were greatest at the two natural areas (Ermita: 50.41 Mg ha^{-1}; Bakhawan: 46.95 Mg ha^{-1}), followed by the rehabilitated seafront areas (Ermita: 41.01 Mg ha^{-1}; Bakhawan: 30.42 Mg ha^{-1}), and were lowest in the abandoned fishponds (Dumangas: 25.68 Mg ha^{-1}; Nabitasan: 5.17 Mg ha^{-1}). Both rehabilitated seafront sites had significantly lower SRTM-derived vegetation carbon stocks (Mg ha^{-1}) than adjacent natural areas (log(vegetation carbon); intercept = 4.46 ± 0.10 (1 s.e.); Bakhawan rehab: − 1.08 ± 0.18 (1 s.e.), $p < 0.001$; Ermita rehab: − 1.35 ± 0.18 (1 s.e.), $p < 0.001$). Both abandoned fishponds had significantly lower SRTM-derived vegetation carbon stocks (Mg ha^{-1}) than rehabilitated seafront areas (log(vegetation carbon); intercept = 3.25 ± 0.13 (1 s.e.); Dumangas: − 0.58 ± 0.23 (1 s.e.), $p = 0.02$; Nabitasan: − 1.42 ± 0.23 (1 s.e.), $p < 0.001$).

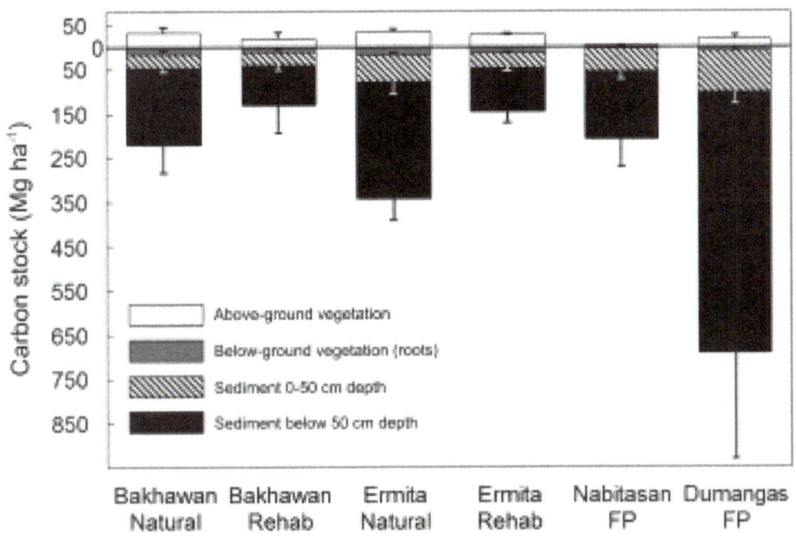

Figure 5: Mean mangrove vegetation and sediment carbon stocks across all study sites (Mg ha^{-1}). Mean plot-level sediment carbon stock to 30 cm depth is depicted in hashed bars, and from 30 cm to soil profile depth in filled black bars. Mean aboveground vegetation carbon stock is depicted in white bars, and below-ground (roots) carbon stock in grey bars. Mean above- and below-ground vegetation carbon stocks (Mg ha^{-1}) were calculated as the mean SRTM DEM-predicted above- and below-ground carbon storage across all pixels with > 50% of their area within each given study site (see Methods). Mean sediment carbon stock estimates (Mg ha^{-1}) were calculated across all plot-level measurements ($N = 8$ per site). Error bars depict one standard deviation from the mean of each estimate at each site.

Sediment OC estimates for the 0–50 cm depth interval ranged from 0.71% (Ermita Rehab) to 4.88% (Dumangas FP), and for the 50–150 cm depth interval ranged from 0.82% (Bakhawan Rehab) to 4.59% (Dumangas FP) (Table 2). Bulk density ranged from 0.37–0.38 g cm^{-3} at the Dumangas abandoned fishpond to 0.85–0.91 g cm^{-3} at the Ermita seafront rehabilitation site (Table 2). Mean total sediment carbon stocks were similarly variable (Fig. 5), being lowest at the two seafront rehabilitated areas (Bakhawan: 120.85; Ermita: 131.14 Mg ha^{-1}), followed by the natural area at Bakhawan (204.47 Mg ha^{-1}), the Nabitasan abandoned fishpond (207.04 Mg ha^{-1}), and the natural area at Ermita

(324.85 Mg ha^{-1}). The largest mean total sediment carbon stock occurred at the Dumangas abandoned fishpond (684.67 Mg ha^{-1}) (Fig. 5). Both rehabilitated seafront sites had significantly lower plot-level total sediment carbon stocks (Mg ha^{-1}) than adjacent natural areas (log(total sediment carbon); intercept = 5.52 ± 0.10 (1 s.e.); Bakhawan rehab: − 0.86 ± 0.17 (1 s.e.), $p < 0.001$; Ermita rehab: − 0.67 ± 0.17 (1 s.e.), $p < 0.001$). Plot-level total sediment carbon stocks at the Nabitasan abandoned fishpond were not significantly different than at the two natural areas (log(total sediment carbon); intercept = 5.52 ± 0.09 (1 s.e.); Nabitasan: − 0.23 ± 0.16 (1 s.e.), $p = 0.16$), while those at the Dumangas abandoned fishpond were significantly larger (Dumangas: 0.92 ± 0.16 (1 s.e.), $p < 0.001$).

Table 2: Data summary from soil analysis, organised by site and soil depth interval. Values represent mean values across all plots at each site ($N = 8$) ± one standard deviation. Sediment depth values denote the range of depths across all plots

Site	Depth interval (cm)	Bulk density (g cm^{-3})	OC (%)	Sediment depth (cm)
Bakhawan Natural	0–50	0.60±0.07	1.01±0.29	207–238
	50–150	0.67±0.08	1.61±0.80	
Bakhawan Rehab	0–50	0.65±0.16	0.85±0.33	181–226
	50–150	0.68±0.16	0.82±0.55	
Ermita Natural	0–50	0.70±0.11	1.84±1.11	163–391
	50–150	0.63±0.07	2.33±0.83	
Ermita Rehab	0–50	0.91±0.07	0.71±0.24	159–172
	50–150	0.85±0.06	1.02±0.32	
Nabitasan FP	0–50	0.62±0.04	1.72±0.60	146–250
	50–150	0.63±0.03	1.70±0.51	
Dumangas FP	0–50	0.38±0.08	4.88±0.97	380–400
	50–150	0.37±0.12	4.59±1.41	

Site-level carbon storage

The largest site-level total carbon stock occurred at the Dumangas abandoned fishpond (vegetation carbon = 1176.12 Mg; sediment carbon = 31,482.69 Mg), followed by the natural area at Bakhawan (vegetation carbon = 1813.37 Mg; sediment carbon = 7888.11 Mg), the seafront rehabilitated area at Bakhawan (vegetation carbon = 599.06 Mg; sediment carbon = 2358.92 Mg), and the Nabitasan abandoned fishpond (vegetation carbon = 46.82 Mg; sediment carbon = 1870.91 Mg). Owing to small areal coverage (2.34 ha and 0.50 ha respectively), total site carbon stock was low at both the natural (vegetation carbon = 114.03 Mg; sediment carbon = 760.10 Mg) and seafront rehabilitated areas at Ermita (vegetation carbon = 7.19 Mg; sediment carbon = 64.60 Mg) (Table 3).

Table 3: Results of site-level carbon stock analysis

Site	Aboveground vegetation C (Mg)	Belowground vegetation C (Mg)	Sediment C 0–50 cm (Mg)	Sediment C below 50 cm (Mg)	Total (Mg)	Site area (ha)
Bakhawan Natural	1267.27	546.10	1176.94	6711.17	9701.48	38.59
Bakhawan Rehab	430.18	168.88	563.18	1795.74	2957.98	19.52
Ermita Natural	79.30	34.73	140.31	619.79	908.86	2.34
Ermita Rehab	5.05	2.14	15.56	49.04	71.79	0.50
Nabitasan FP	41.35	5.47	477.97	1392.94	1917.73	9.04
Dumangas FP	856.90	319.22	4315.52	27,167.17	32,658.81	45.99

Site-level coastal protection

Variable vegetation structure (Table 1) produced variable potential coastal protection and required greenbelt width (*BW*) across sites (Table 4). Both natural and rehabilitated areas at the Bakhawan site provide relatively strong protection (protection categories III and II respectively) and require narrow greenbelt widths (*BW* = 201 and 270 m respectively). This is within the median actual greenbelt width (*BActual* = 842 m) (Table 4). At the Ermita site, despite high FSI and protection category

(III and II respectively), the required *BW* for adequate coastal protection potential (229–331 m) is not achieved, due to a narrow fringing area available for rehabilitation (median *BActual* = 81 m; Table 4). Both abandoned fishpond sites had high median actual greenbelt width (*BActual*; Nabitatasan = 268 m and Dumangas = 827 m). High stem density and vertical growth (Table 1) translated to a higher protection category for the Dumangas abandoned fishpond site (II) and adequate current coastal protection potential (*BW* = 370 m). However, low vegetation density and height (Table 1) at the Nabitasan abandoned fishpond site resulted in a low protection category (I) and high required greenbelt width (*BW* = 2405 m).

Table 4: Results of coastal greenbelt protection analysis. FSI = Forest Structure Index and the associated protection category (min = I; max = V), *BW* = required greenbelt width (Bao, 2011), *BActual* = median actual greenbelt width calculated from GIS distance analysis (Bivand et al., 2015)

Site	FSI	Protection category	*BW* (m)	*BActual* (m)	Age (years)
Bakhawan Natural	0.012	III	201	842	–
Bakhawan Rehab	0.009	II	270	842	8
Ermita Natural	0.011	III	229	81	–
Ermita Rehab	0.007	II	331	81	7
Nabitasan FP	0.001	I	2405	268	5
Dumangas FP	0.006	II	370	827	8

Case study: Dumangas municipality

We delimited an area of 377.25 ha of abandoned fishponds in Dumangas (8.8% of total current aquaculture area; Primavera et al., 2014). Potential total vegetation carbon stock accumulation across this area over 6.5 years was estimated at 4691.73 Mg. We estimated potential sediment carbon accretion (1.63 cm; 0.25 cm yr^{-1}) over 6.5 years at 911.32 Mg. We further undertook this analysis based on only vegetation and sediment carbon stock estimates (Mg ha^{-1}) from the partially-banked Dumangas abandoned fishpond site in order to model potential abandoned fishpond reversion carbon gains under seaward dike management for high propagule and organic matter retention (see Section 2.1). Under this scenario, we estimated potential total vegetation carbon accumulation of 6564.49 Mg and sediment carbon accretion of 1166.41 Mg in Dumangas municipality's abandoned fishponds over 6.5 years. Of the 377.25 ha of abandoned fishponds identified in Dumangas, 167.59 ha (44.4%) are currently leased under FLAs (Primavera et al., 2014).

We estimated a length of 8.59 km of Dumangas' coastline (30.5% of total coastline) currently fringed by abandoned fishponds. All abandoned fishponds within this 8.59 km had an estimated median landward width > 100 m (current greenbelt mandate). 3.66 km of coastline had an estimated median landward width > 370 m (estimated required bandwidth (*BW*) based on the Dumangas abandoned fishpond site), translating to 13.0% of the Dumangas municipality coastline with potential adequate coastal protection from abandoned fishpond mangroves over eight years from reversion (Dumanagas abandoned fishpond site age; Table 4). Of this, 3.45 km (96.7%) is fringed by abandoned fishponds in breach of FLAs (Primavera et al., 2014).

DISCUSSION

This study provides a quantitative analysis of relative CCMA ES delivery by different mangrove rehabilitation areas and adjacent natural stands. While per hectare carbon stocks were variable across both rehabilitated and natural areas, rehabilitation for enhanced CCMA goals appears more promising in abandoned fishponds. Despite currently lower per hectare biomass production, carbon-rich sediments and large areal coverage enhanced the overall carbon stocks and coastal protection potential of rehabilitated abandoned fishponds. Our municipality-specific case study revealed that overlap may exist between areas of high rehabilitation potential for CCMA goals and low competing opportunity costs, with 96.7% of the identified wide sea-facing abandoned fishponds currently in breach of lease agreements (FLAs) on public lands. Our results may also have implications regarding reverted abandoned fishpond management for high carbon stocks.

All rehabilitated stands exhibited structural parameters (stem density, DBH, biomass) within observed trajectories to maturity (15–20 years; Bosire et al., 2008 and Alongi, 2011; Table 1), suggesting good rehabilitation status to date. However, contrary to expectations according to their mid- to upper-intertidal position, both abandoned fishponds had comparatively low per hectare vegetation carbon stocks (Table 1; Fig. 5). This may in part reflect methodological limitations that could have underestimated true biomass and canopy height: first, shrubby *A. marina* was dominant at both sites, and biomass calculation for multi-stemmed individuals (Fu and Wu, 2011) may have underestimated biomass; second, SRTM-DEM data were acquired prior to rehabilitation of these areas (Rodriguez et al., 2006). Application of new high resolution TanDEM-X global DEM data (2014) to be released for scientific use in late 2016-2017 will enhance the application of SAR-derived DEM data to recent mangrove rehabilitation monitoring (Zink et al., 2015). However, the SRTM-DEM related limitation also applied to the seafront rehabilitated areas, where estimated vegetation carbon stocks remained high (Fig. 5). High heterogeneity across both abandoned fishponds, due to on-going natural recolonization at the Dumangas site and high pre-rehabilitation erosion at the Nabitasan site, likely contributed to low average per hectare vegetation carbon stocks. However, lower relative biomass production may also indicate possible hydrological or fishpond effluent constraints to biomass production in abandoned fishponds (Lewis,

2005, Matsui et al., 2010 and Primavera et al., 2014). These results suggest further active rehabilitation may be necessary to enhance mangrove functioning in reverted abandoned fishponds. Indeed, the Bakhawan seafront site had the highest per hectare vegetation carbon stocks of all rehabilitated sites (Fig. 5), which may reflect a combination of active replanting and high natural re-colonisation as an effective strategy in low-mid intertidal areas (sensu Matsui et al., 2010), and a possible positive influence of multi-species rehabilitation (Lang›at et al., 2011).

As observed in other studies from the region (Thompson et al., 2014), carbon stocks in natural areas (Fig. 5) were at the lower end of estimates for Indo-Pacific mangroves (Donato et al., 2011). The abandoned fishponds were either not different (Nabitasan) or had significantly greater (Dumangas) plot-level sediment carbon stocks than natural areas, and greater than seafront rehabilitated areas (Fig. 5). This reflects their position on former mangrove sediments, and highlights their greater potential over seafront rehabilitation sites in reforestation PES schemes that recognise existing stocks (Locatelli et al., 2014). The large sediment carbon stocks at the Dumangas fishpond are notable (mean 684.67 ± 263.39 Mg ha^{-1}), as these are among the highest recorded in Panay. This includes a large ancient basin mangrove (mean sediment carbon stock 372.60 ± 128.61 Mg ha^{-1}; C. Duncan unpublished data). A possible driver is site configuration: much of the seaward bank is retained and reduces complete tidal flushing, trapping organic matter in partially-waterlogged sediments. Conversely, near-complete loss of the Nabitasan fishpond seaward bank before recolonization has resulted in extensive erosion at the seaward margin. While our observations are limited to only two abandoned fishponds, these results suggest that retaining partial seaward banks may be important for preventing erosion, retaining sediment carbon stocks and improving propagule establishment in recovering abandoned fishponds (similar to breakwater interventions: Hashim et al., 2010 and Primavera et al., 2012a). Relatively high bulk densities and low OC at seafront rehabilitation sites compared to adjacent natural areas (Table 2) suggest low rates of soft sediment accretion, and slow gains in potential sequestration-oriented PES schemes (Locatelli et al., 2014). Further surveying of abandoned fishpond areas, and study of accretion rates in abandoned fishponds and seafront plantations, will be required to investigate these findings further and establish relative sediment carbon sequestration rates.

Overall, the comparatively large size, mid- to upper-intertidal position and high sediment carbon of the abandoned fishponds translated to high relative total ecosystem carbon stocks (Table 3). We moreover predicted high vegetation and sediment carbon sequestration gains over 6.5 years of potential future rehabilitation of abandoned fishponds in Dumangas municipality (5809.95–7995.75 Mg). It is important to note that these estimates are preliminary and likely underestimates; they employ basic (but conservative) accretion rates and do not account for important processes such as carbon burial (Lee et al., 2014). Large area translates to landward widths available for rehabilitation at abandoned fishpond sites (Table 4) that are considerably greater than at rehabilitated seafront areas (Bakhawan: 186 m; Ermita: 41 m). Seafront rehabilitation sites were bordered at their landward margins by existing natural mangrove, increasing effective greenbelt width (Table 4). In the unique case of Bakhawan ecopark (see Section 2.1), the combination of rehabilitated and natural forest make the total greenbelt well above that estimated for effective coastal protection potential (Table 4). However, for Ermita, a site more typical of remaining seafront mangroves in the Philippines, the combined greenbelt width of both rehabilitated and natural areas is far below that required (Table 4). Repeated LGU planting in adjacent areas to the Ermita site since the early 2000s have seen no survival. The existence of a coastal road between these now fringing areas and the former deltaic mangrove (now active fishponds) inland further prohibits development of an effective greenbelt in the area.

In contrast, both abandoned fishponds had large median landward widths (Nabitasan: 268 m; Dumangas: 827 m). Low vegetation density caused very large estimates of required greenbelt width for Nabitasan abandoned fishpond mangroves (Table 4). However, saplings (high density at Nabitasan) were not considered in these analyses, and can contribute to protection from regular wind waves (Mazda et al., 1997). Furthermore, rehabilitated mangroves at this site are comparatively young (Table 4), and assisted rehabilitation by the Leganes LGU is on-going. The landward greenbelt width of the Dumangas abandoned fishpond may provide ample coastal protection, even while in relatively young stages of development (Table 4; Alongi, 2011). In direct conflict with current rehabilitation programmes in Dumangas, we found a large length of coastline (3.66 km) with adequate future protection potential from abandoned fishpond mangroves (> 370 m wide) eight years post-rehabilitation. While this wide greenbelt requirement will reduce as rehabilitated abandoned fishpond mangroves mature (Alongi, 2008, Bao, 2011, McIvor et al., 2012 and Lee et al., 2014), it is important to note that: (1) these estimates are derived for protection from regular waves (Bao, 2011), while greenbelt requirements for storm surge attenuation may be substantially greater (Koch et al., 2009 and McIvor et al., 2012), and (2) natural stands were estimated here to have required greenbelt widths of > 200 m (Table 4). This suggests that current Philippines greenbelt laws (50–100 m) may be inadequate for coastal protection in typhoon-prone areas. Reallocation of Philippines' National Greening Program (NGP) funds and capacity away from seafront planting toward targeted reversion and assisted rehabilitation of large sea-facing abandoned fishponds is thus advisable if effective mangrove greenbelt reestablishment and integrated CCMA goals are to be realised (Blankespoor et al., 2016). Our municipality-specific case study revealed that 96.7% of coastline identified as having rehabilitation potential for effective coastal protection within eight years is currently tenured under and in breach of FLA terms. As such, reversion (DA-BFAR) and rehabilitation (DENR) of these areas is a legal requirement, is not in direct conflict with opportunity costs of fishpond leaseholders, and could provide substantial benefits to inland coastal communities from coastal protection and fisheries enhancement (Walton et al., 2006 and Primavera et al., 2012a).

Integrated spatial planning and coastal zone management approaches incorporating ES goals are increasingly on the agenda (Lester et al., 2013 and Arkema et al., 2015) to prioritise areas for management to reduce ecosystem degradation (e.g. Atkinson et al., 2016). Similarly, such approaches, including multiple CCMA ES and considering their value to the full cohort of stakeholders, could be an important tool for identifying key areas for whole ecosystem rehabilitation and restoration in converted lands. However, existing formally-recognised and local tenure structures provide a major challenge to spatial planning and prioritisation approaches (e.g. Adger et al., 2005 and Brown et al., 2014), and may impact the effectiveness of management decisions (Weeks et al., 2010). In contrast, our case study identified substantial areas with minimal tenure conflict within priority areas for CCMA mangrove rehabilitation (mid- to upper-intertidal zone and large areal coverage). Spatial planning exercises to prioritise multiple CCMA ES greenbelt rehabilitation (based on e.g. coastal vulnerability assessment) should first aim to evaluate the existence of formally-recognised tenure gaps (e.g. tenure breaches or unproductive lands) to evaluate the rehabilitation potential of currently-tenured areas. In cases such as that of mangrove rehabilitation in the Philippines, this may maximise benefit:cost ratios of CCMA efforts for coastal communities and avoid ineffective and wasteful allocation of limited conservation funds. Such spatial planning approaches for mangrove rehabilitation may have particular relevance elsewhere in South and Southeast Asia where fishpond abandonment is similarly high: e.g. Malaysia (60%; Choo, 1996), Thailand (50–80%; Hossain and Lin, 2001) and Sri Lanka (60–90%; Jayakody et al., 2012 and Bournazel et al., 2015). In other, more typhoon-prone countries believed to have high rates of fishpond abandonment (e.g. Vietnam, Taiwan; Stevenson, 1997), the potential CCMA benefits of abandoned pond identification and rehabilitation may be of particular consequence. In many parts of Asia, much unproductive abandoned fishpond area is rapidly being converted to alternative uses, cementing mangrove loss (e.g. salt pans, agriculture: Stevenson, 1997, Hossain and Lin, 2001 and Jayakody et al., 2012). Monitoring and evaluation of aquaculture productivity within these coastal areas and their inclusion within spatial planning for mangrove rehabilitation may provide a means to halt such loss. However, the relevance of such approaches to wider regional areas may be comparatively limited where (1) wider continental shelves enhance the suitability of low-intertidal seafront rehabilitation, or (2) titled ownership is the predominant form of coastal zone land tenure.

In recent decades, evidence has been mounting on the relative potential of abandoned fishpond rehabilitation for conserving mangrove forests (e.g. Lewis, 2005, Matsui et al., 2010, Primavera et al., 2012a, Primavera et al., 2012b, Primavera et al., 2014 and Brown et al., 2014). Our case study highlights a high contribution of ES strings to the bow of fishpond reversion, with a high relative potential for coastal greenbelt rehabilitation for integrated CCMA goals in the Philippines. While our case study revealed favourable tenure status for most potential greenbelt abandoned fishpond areas, strong political will and capacity shifts will be required to cancel leaseholds and rehabilitate former mangrove areas (Primavera et al., 2014). However, mid-upper intertidal mangrove rehabilitation in abandoned fishpond areas may hold one of the many keys to safeguarding mangrove forests and their CCMA ES delivery, particularly important in the context of the increasing severity of storms and sea level rise associated with climate change (Intergovernmental Panel on Climate Change (IPCC), 2013, Lovelock et al., 2015 and Blankespoor et al., 2016).

ACKNOWLEDGEMENTS

The authors thank the communities and people's organisations in all barangays studied here, as well as Bugtong-Bato, Ibajay, Aklan, for field assistance and logistical support. We are very grateful to Mayor Jaen of Leganes, Mayor Distura of Dumangas, and Mayor Lachica and former-Mayor and -Congressman Quimpo of Kalibo municipalities, as well as all LGU staff, for access, support and assistance. We also thank the soils laboratory of the Bureau of Soils and Water Management, Department of Agriculture, Cebu, and in particular Esther Caligacion, for sediment analysis. Huge thanks to Venus Sadio-Calanda for field assistance and support, and to all of the ZSL Philippines team; in particular to Jofel Coching, Myrtle Arias and Jude Baguio. In addition, we thank members of the Kalibo Save the Mangroves Association (KASAMA) and the managers and operators of Bakhawan ecopark, Kalibo; in particular Elizabeth Ramos. This study was generously supported by a UCL BEAMS Impact Award, the Darwin Initiative (21-010), and the Rufford Foundation (15169-1).

REFERENCES

1. Adger, W.N., Hughes, T.P., Folke, C., Carpenter, S.R., Rockström, J., 2005. Social-ecological resilience to coastal disasters. Science 12, 1036–1039.
2. Alongi, D.M., 2008. Mangrove forests: resilience, protection from tsunamis, and responses to global climate change. Estuar. Coast. Shelf Sci. 76, 1–13.
3. Alongi, D.M., 2011. Carbon payments for mangrove conservation: ecosystem constraints and uncertainties of sequestration potential. Environ. Sci. Pol. 14, 462–470.
4. Arkema, K.K., Verutes, G.M., Wood, S.A., Clarke-Samuels, C., Rosado, S., Canto, M., Rosenthal, A., Ruckelshaus, M., et al., 2015. Embedding ecosystem services in coastal planning leads to better outcomes for people and nature. PNAS 112, 7390–7395.
5. Arnaud-Haond, S., Duarte, C.M., Teixeira, S., Massa, S.I., Terrados, J., Tri, N.H., Hong, P.N., Serrão, E.A., 2009. Genetic recolonization of mangrove: genetic diversity still increasing in the Mekong Delta 30 years after Agent Orange. Mar. Ecol. Prog. Ser. 390, 129–135.
6. Atkinson, S.C., Jupiter, S.D., Adams, V.M., Ingram, J.C., Narayan, S., Klein, C.J., Possingham, H.P., 2016. Prioritising mangrove ecosystem services results in spatially variable management priorities. PLoS One 11, e0151992.
7. Bao, T.Q., 2011. Effect of mangrove forest structures on wave attenuation in coastal Vietnam. Oceanologia 53, 807–818.
8. Barbier, E.B., Koch, E.W., Silliman, B.R., Hacker, S.D., Wolanski, E., Primavera, J.H., Granek, E.F., Polasky, S., et al., 2008. Coastal ecosystem-based management with nonlinear ecological functions and values. Science 319, 321–323.
9. Barbier, E.B., Hacker, S.D., Kennedy, C., Koch, E.W., Stier, A.C., Silliman, B.R., 2011. The value of estuarine and coastal ecosystem services. Ecol. Monogr. 81, 169–193.
10. Bayraktarov, E., Saunders, M.I., Abdullah, S., Mills,

M., Beher, J., Possingham, H.P., Mumby, P.J., Lovelock, C.E., 2015. The cost and feasibility of marine coastal restoration. Ecol. Appl. (Early Online View).

11. Bivand, R., Rundel, C., Pebesma, E., Hufthammer, K.O., 2015. rgeos: Interface to Geometry Engine – Open Source (GEOS). V. 0.3–15. R Foundation for Statistical Computing, Vienna.
12. Blankespoor, B., Dasgupta, S., Lange, G.-M., 2016. Mangroves as Protection From Storm Surges in a Changing Climate. World Bank Policy Research Working Paper (WPS7596). World Bank Group, Washington DC.
13. Bosire, J.O., Dahdouh-Guebas, F., Walton, M.E.M., Crona, B.I., Lewis, R.R., Field, C., Kairo, J.G., Koedam, N., 2008. Functionality of restored mangroves: a review. Aquat. Bot. 89, 251–259.
14. Bournazel, J., Kumara, M.P., Javatissa, L.P., Viergever, K., Morel, V., Huxham, M., 2015. The impacts of shrimp farming on land-use and carbon storage around Puttalam lagoon, Sri Lanka. Ocean Coast. Manag. 113, 18–28.
15. Brown, B., Fadillah, R., Nurdin, Y., Soulsby, I., Ahmad, R., 2014. Community based ecological mangrove rehabilitation (CBEMR) in Indonesia. Sapiens 7, 2.
16. Cadaweng, E.A., Aguirre, J.A.N., 2005. Forest from the Mud: The Kalibo Experience. In: Durst, P.B., Brown, C., Tacio, H.D., Ishikawa, M. (Eds.), In Search of Excellence: Exemplary Forest Management in Asia and the Pacific. FAO, Bangkok, pp. 39–48.
17. Cardinale, B.J., Duffy, J.E., Gonzalez, A., Hooper, D.U., Perrings, C., Venail, P., Narwani, A., Mace, G.M., et al., 2012. Biodiversity loss and its impact on humanity. Nature 486, 59–67.
18. Chave, J., Coomes, D.A., Jansen, S., Lewis, S.L., Swenson, N.G., Zanne, A.E., 2009. Towards a worldwide wood economics spectrum. Ecol. Lett. 12, 351–366.
19. Choo, P.S., 1996. Aquaculture Development in the Mangrove. In: Suzuko, S., Hayase, S., Kawahasa, S. (Eds.), Sustainable Utilisation of Coastal Ecosystems. Proceedings of the Seminar on Sustainable Utilisation of Coastal Ecosystems for Agriculture, Forestry and Fisheries in Developing Countries, p. 63.
20. Costanza, R., de Groot, R., Sutton, P., van der Ploeg, S., Anderson, S.J., Kubiszewski, I., Farber, S., Turner, R.K., 2014. Changes in the global value of ecosystem services. Glob. Environ. Chang. 26 (152-158.-71).
21. Dahdouh-Guebas, F., Jayatissa, L.P., Di Nitto, D., Bosire, J.O., Lo Seen, D., Koedam, N., 2005. How effective were mangroves as a defence against the recent tsunami? Curr. Biol. 15, 443–447.
22. Dale, P.E.R., Knight, J.M., Dwyer, P.G., 2014. Mangrove rehabilitation: a review focusing on ecological and institutional issues. Wetl. Ecol. Manag. 22, 587–604.
23. Donato, D.C., Kauffman, J.B., Murdiyarso, D., Kurniato, S., Stidham, S., Kannien, M., 2011. Mangroves among the most carbon-rich forests in the tropics. Nat. Geosci. 4, 293–297.
24. Duarte, C.M., Losada, I.K., Hendriks, I.E., Mazarrasa, I., Marbà, N., 2013. The role of coastal plant communities for climate change mitigation and adaptation. Nat. Clim. Chang. 3, 961–968.
25. Duke, N.C., Meyneche, J.O., Dittmann, S., Ellison, A.M., Anger, K., Berger, U., Cannicci, S., Diele, K., et al., 2007. A world without mangroves? Science 317, 41–42.
26. Ellison, A.M., 2000. Mangrove restoration: do we know enough? Restor. Ecol. 8, 219–229.
27. Ewel, K.C., Twilley, R.R., Ong, J.E., 1998. Different kinds of mangrove forests provide different goods and servies. Glob. Ecol. Biogeogr. Lett. 7, 83–94.
28. Fatoyinbo, T.E., Simard, M., Washington-allen, R.A., Shugart, H.H., 2008. Landscape-scale extent, height, biomass and carbon estimation of Mozambique's mangrove forests with Landsat ETM+ and Shuttle Radar Topography Mission elevation data. J. Geophys. Res. 113, G02–S06.
29. Field, C., Osborn, J., Hoffman, L., Polsenberg, J., Ackerly, D., Berry, J., Björkman, O., Helf, A., et al., 1998. Mangrove biodiversity and ecosystem function. Glob. Ecol. Biogeogr. Lett. 7, 3–14.
30. Fu, W., Wu, Y., 2011. Estimation of aboveground biomass of different mangrove trees based on canopy diameter and tree height. Procedia Environ. Sci. 10, 2189–2194.
31. Giri, C., Ochieng, E., Tieszen, L.L., Zhu, Z., Singh, A., Loveland, T., Masek, J., Duke, N.C., 2011. Status and distribution of mangrove forests of the world using earth observation satellite data. Glob. Ecol. Biogeogr. 20, 154–159.
32. Goessens, A., Satyanarayana, B., Van der Stocken, T., Zuniga, M.Q., Mohd-Lokman, H., Sulong, I., Dahdouh-Guebas, F., 2014. Is Matang mangrove forest in Malaysia sustainably rejuvenating after more than a century of conservation and harvesting management? PLoS One 9, e105069.
33. Google Earth, 2015. Google Earth V. 7.1.5. Google, Mountain View (Available at: http:// www.google.com/earth).
34. Hamilton, S.E., Casey, D., 2016. Creation of a high spatio-temporal resolution global database of continuous mangrove forest cover for the 21st century. Glob. Ecol. Biogeogr. 25, 729–738.
35. Hashim, R., Kamali, B., Tamin, N.M., Zakaria, R., 2010. An integrated approach to coastal rehabilitation: mangrove restoration in Sungai Haji Dorani, Malaysia. Estuar. Coast. Shelf Sci. 86, 118–124.
36. Hossain, Z., Lin, C.K., 2001. Diversified uses of abandoned shrimp ponds – a case study in the Upper Gulf of Thailand. ITCZM Monographs No. 5 (23 pp).
37. Iftekhar, M.S., 2008. Functions and development of reforested mangrove areas: a review. IJBSM 4, 1–14.
38. Intergovernmental Panel on Climate Change (IPCC), 2013. Climate Change 2013: The Physical Science Basis. In: Stocker, T.F., Qin, G.-K., Plattner, M., Tignor, M., Allen, S.K., Boschung, J., Nauels, A., Xia, Y., Bex, V., Midgley, P.M. (Eds.), Contribution of Working Group I to the Fifth Assessment Report of the Intergovernmental

Panel on Climate Change. Cambridge University Press, Cambridge.

39. Jayakody, S., Jayasinghe, J.M.P.K., Wijesundara, A., 2012. Active vs passive restoration of mangroves: developing models for sustainable rejuvenation of mangrove ecosystems used for shrimp farming in North-Western Province of Sri Lanka. In: Macintosh, D.J., Mahindapala, R., Markopoulos, M. (Eds.), Sharing Lessons on Mangrove Restoration. IUCN, Gland, pp. 179–189.

40. Kairo, J.G., Dahdouh-Guebas, F., Bosire, J.O., Koedam, N., 2001. Restoration and management of mangrove systems – a lesson for and from the East African region. S. Afr. J. Bot. 67, 383–389.

41. Kairo, J.G., Wanjiru, C., Ochiewo, J., 2009. Net pay: economic analysis of a replanted mangrove plantation in Kenya. J. Sustain. For. 28, 395–414.

42. Kauffman, J.B., Donato, D.C., 2012. Protocols for the Measurement, Monitoring and Reporting of Structure, Biomass and Carbon Stocks in Mangrove Forests. CIFOR, Bogor.

43. Koch, E.W., Barbier, E.B., Silliman, B.R., Reed, D.J., Perillo, G.M.E., Hacker, S.D., Granek, E.F., Primavera, J.H., et al., 2009. Non-linearity in ecosystem services: temporal and spatial variability in coastal protection. Front. Ecol. Environ. 7, 29–37.

44. Komiyama, A., Poungparn, S., Kato, S., 2005. Common allometric equations for estimating the tree weight of mangroves. J. Trop. Ecol. 21, 471–477.

45. Korhonen, L., Korhonen, K.T., Rautiainen, M., Stenberg, P., 2006. Estimation of forest canopy cover: a comparison of field measurement techniques. Silva Fenn. 40, 577–588.

46. Lang'at, J.K.S., Kirui, B.K.Y., Skov, M.W., Kairo, J.G., Mencuccini, M., Huxham, M., 2011. Species mixing boosts root production of mangrove trees. Oecologia 172, 271–278.

47. Lee, S.Y., Primavera, J.H., Dahdouh-Guebas, F., McKee, K., Bosire, J.O., Cannicci, S., Diele, K., Fromard, F., et al., 2014. Ecological role and services of tropical mangrove ecosystems: a reassessment. Glob. Ecol. Biogeogr. 23, 726–743.

48. Lester, S.E., Costello, C., Halpern, B.S., Gaines, S.D., White, C., Barth, J.A., 2013. Evaluating tradeoffs among ecosystem services to inform marine spatial planning. Mar. Policy 38, 80–89.

49. Lewis, R.R., 2005. Ecological engineering for successful management and restoration of mangrove forests. Ecol. Eng. 24, 403–418.

50. Locatelli, T., Binet, T., Kairo, J.G., King, L., Madden, S., Patenaude, G., Upton, C., Huxham, M., 2014. Turning the tide: how blue carbon and payments for ecosystem services (PES) might help save mangrove forests. Ambio 43, 981–995.

51. Lovelock, C.E., Cahoon, D.R., Friess, D.A., Guntenspergen, G.R., Krauss, K.W., et al., 2015. The vulnerability of Indo-Pacific mangrove forests to sea-level rise. Nature 526, 559–563.

52. Marois, D.E., Mitsch, W.J., 2015. Coastal protection from tsumanis and cyclones provided by mangrove wetlands – a review. Int. J. Biodivers. Sci. Ecosyst. Serv. Manage. 11, 71–83.

53. Matsui, N., Suekuni, J., Nogami, M., Havanond, S., Salikul, P., 2010. Mangrove rehabilitation dynamics and soil organic carbon changes as a result of full hydraulic restoration and re-grading of a previously intensively managed shrimp pond. Wetl. Ecol. Manag. 18, 233–242.

54. Mazda, Y., Wolanski, E., King, B., Sase, A., Ohtsuka, D., Michimasa, M., 1997. Drag force due to vegetation in mangrove swamps. Mangrove Salt Marshes 1, 193–199.

55. McIvor, A.L., Möller, I., Spencer, T., Spalding, M., 2012. Reduction of wind and swell waves by mangroves. Natural Coastal Protection Series: Report 1Cambridge Coastal Research Unit Working Paper Vol. 40. Nature Conservancy and Wetlands International, Cambridge.

56. Nam, V.N., Sasmito, S.D., Murdiyarso, D., Purbopusito, J., MacKenzie, R.A., 2016. Carbon stocks in artificially regenerated mangrove ecosystems in the Mekong Delta. Wetl. Ecol. Manag. 24, 231–244.

57. Peduzzi, P., Chatenoux, B., Dao, H., De Bono, A., Herold, C., Kossin, J., Mouton, F., Nordbeck, O., 2012. Global trends in tropical cyclone risk. Nat. Clim. Chang. 2, 289–294.

58. Polidoro, B.A., Carpenter, K.E., Collins, L., Duke, N.C., Ellison, A.M., Ellison, J.C., Farnsworth, E.J., Fernando, E.S., et al., 2010. The loss of species: mangrove extinction risk and geographic areas of global concern. PLoS One 5, e10095.

59. Primavera, J.H., 2005. Mangroves, fishponds and the quest for sustainability. Science 310, 57–59.

60. Primavera, J.H., Esteban, J.M.A., 2008. A review of mangrove rehabilitation in the Philippines: successes, failures and future prospects. Wetl. Ecol. Manag. 16, 345–358.

61. Primavera, J.H., Savaris, J.P., Bajoyo, B.E., Coching, J.D., Curnick, D.J., Golbeque, R.L., Gizman, A.T., Henderin, J.Q., et al., 2012a. Manual on community-based mangrove rehabilitation. Mangrove Manual Series Vol. No. 1. Zoological Society of London, London.

62. Primavera, J.H., Rollon, R.N., Samson, M.S., 2012b. The pressing challenges of mangrove rehabilitation: pond reversion and coastal protection. In: Chicharo, L., Zalewski, M. (Eds.), Treatise on Estuarine and Coastal Science. Volume 10: Ecohydrology and Restoration. Elsevier, Amsterdam, pp. 217–244.

63. Primavera, J.H., Yap, W.G., Savaris, J.P., Loma, R.J.A., Moscoso, A.D.E., Coching, J.D., Montilijao, C.L., Poingan, R.P., et al., 2014. Manual on mangrove reversion of abandoned and illegal brackishwater fishponds. Mangrove Manual Series Vol. No. 2. Zoological Society of London, London.

64. QGIS Development Team, 2015. QGIS Geographic Information System V. 2.8.1. Open Source Geospatial Foundation Project (Available at: http://qgis.osgeo.org).

65. R Development Core Team, 2015. R: A Language and

Environment for Statistical Computing. R Foundation for Statistical Computing, Vienna.

66. Ren, H., Chen, H., Li, Z., Han, W., 2010. Biomass accumulation and carbon storage of four different ages Sonneratia apetala plantations in Southern China. Plant Soil 327, 279–291.

67. Richards, D.R., Friess, D.A., 2016. Rates and drivers of mangrove deforestation in Southeast Asia, 2000–2012. PNAS 113, 344–349.

68. Rodriguez, E., Morris, E., Belz, J.E., 2006. A global assessment of the SRTM assessment. Photogramm. Eng. Remote. Sens. 19, 1873–1886.

69. Rönnbäck, P., Crona, I., Ingwall, L., 2007. The return of ecosystem goods and services in replanted mangrove forests: perspectives from local communities. Environ. Conserv. 4, 313–324.

70. Salmo, S.G., Lovelock, C., Duke, N.C., 2013. Vegetation and soil characteristics as indicators of restoration trajectories in restored mangroves. Hydrobiologia 720, 1–18.

71. Samson, M.S., Rollon, R.N., 2008. Growth performance of planted mangroves in the Philippines: revisiting forest management strategies. Ambio 37, 234–240.

72. Samson, M.S., Rollon, R.N., 2011. Mangrove revegetation potentials of brackish-water pond areas in the Philippines. In: Sladonja, B. (Ed.), Aquaculture and the Environment – A Shared Destiny. InTech, Riejeka, pp. 31–50.

73. Schumacher, B., 2002. Methods for the Determination of Total Organic Carbon (TOC) in Soils and Sediments. United States Environmental Protection Agency, Los Angeles.

74. Simard, M., Zhang, K., Rivera-Monroy, V.H., Ross, M.S., Ruiz, P.L., Castañeda-Moya, E., Twilley, R.R., Rodriguez, E., 2006. Mapping height and biomass of mangrove forests in Everglades National Park with SRTM elevation data. Photogramm. Eng. Remote. Sens. 72, 299–311.

75. Soria, J., Switzer, A., Villanoy, C., Fritz, H., Bilgera, P., Cabrera, O., Siringan, F., Sta. Marina, Y., et al., 2015. Repeat storm surge disasters of Typhoon Haiyan and its 1897 predecessor in the Philippines. Bull. Am. Meteorol. Soc. 97, 31–48.

76. Spalding, M., Kainuma, M., Collins, L., 2010. World Atlas of Mangroves. Earthscan, London. Stevenson, N.J., 1997. Disused shrimp ponds: options for redevelopment of mangroves. Coast. Manag. 25, 425–435.

77. Thompson, B.S., Clubbe, C.P., Primavera, J.H., Curnick, D.J., Koldewey, H.J., 2014. Locally assessing the economic viability of blue carbon: a case study from Panay Island, the Philippines. Ecosyst. Serv. 8, 128–140.

78. Tomlinson, P.B., 1986. The Botany of Mangroves. Cambridge University Press, Cambridge.

79. United Nations Framework Convention on Climate Change (UNFCCC), 2015C. Adoption of the Paris Agreement. 21st Conference of the Parties. United Nations, Paris. United Nations University (UNU) – Institute for Environment and Security, 2014t. World Risk Report 2014. (Available at: http://ehs.unu.edu/news/news/world-risk-report- 2014.html).

80. Valiela, I., Bowen, J.L., York, J.K., 2001. Mangrove forests: one of the world's threatened major tropical environments. Bioscience 51, 807–815.

81. Walton, M.E.M., Samonte-Tan, G.P.B., Primavera, J.H., Edwards-Jones, G., Le Vay, L., 2006. Are mangroves worth replanting? The direct benefits of a community-based reforestation project. Environ. Conserv. 33, 335–343.

82. Weeks, R., Russ, G.R., Bucol, A.A., Alcala, A.C., 2010. Incorporating local tenure in the systematic design of marine protected area networks. Conserv. Lett. 3, 445–453.

83. Wylie, L., Sutton-Grier, A.E., Moore, A., 2016. Keys to successful blue carbon projects: lessons learned from global case studies. Mar. Policy 65, 76–84.

84. Zanne, A.E., Lopez-Gonzalez, G., Coomes, D.A., Ilic, J., Jansen, S., Lewis, S.L., Miller, R.B., Swenson, N.G., et al., 2009. Data From: Towards a Worldwide Wood Economics Spectrum. Dryad Digital Repository (available at: http://dx.doi.org/10.5061/dryad.234).

85. Zink, M., Bachmann, M., Brautigam, B., Fritz, T., Hajnsek, I., Krieger, G., Moreira, A., Wessel, B., 2015. TanDEM-X: a single-pass SAR interferometer for global DEM generation and demonstration of new SAR techniques. Geoscience and Remote Sensing Symposium (IGARSS), IEEE International, pp. 2888–2891.

CHAPTER 12 IMPACTS OF AQUACULTURE ON SOCIAL NETWORKS IN THE MANGROVE SYSTEMS

Steven E. Orchard, Lindsay C. Stringer, Claire H. Quinn

Keywords: Social network analysis, Social capital, Resilience, Social-ecological systems, Natural resource management, Wetland, Sustainable livelihoods, Vulnerability, Adaptive capacity

INTRODUCTION

Mangrove systems are among the most productive and biologically important ecosystems in the world (Giri et al., 2011). They form a crucial component of the livelihoods of coastal communities in developing countries (Joffre and Schmitt, 2010), providing: fish, crustaceans and other sea life for food and income; wood for fuel and energy; protection of shorelines from erosion, flooding and storm damage; and a filter for pollutants to help maintain water quality (Spalding et al., 2010). However, mangrove systems have experienced rapid change in the form of degradation and loss due to increased human activity (cf. Ellis, 2000), particularly from intensive and extensive commercial aquaculture (Gopal, 2013; Barbier, 2006). These changes undermine their ability to support coastal livelihoods (Primavera, 2006). Aquaculture has also contributed to changes in mangrove system access and livelihood opportunities within coastal communities, which can cause conflict in mangrove system use (van Hue and Scott, 2008). This impacts social capital through alterations in the networks that coastal communities draw upon to access the resources necessary to pursue their livelihoods (i.e. financial, human, physical, and natural), and which facilitate cooperation, trust, and the ability of communities to self-organise in response to change (Misra et al., 2014, Speranza et al., 2014 and Nath et al., 2010). In order to increase resilience to change, communities require sufficient capacity within their social networks to self-organise and generate the necessary social capital to support livelihoods (Djalante et al., 2013).

The rapid rise of large scale, intensive aquaculture in Vietnam has been facilitated by the transition from a centrally planned to a socialist-orientated market economy (Adger, 2000 and Tri et al., 1998). Subsequently, Vietnam has lost 69% of its 269,000 ha of mangrove forests held in 1980, with an estimated 77% of this loss due to aquaculture (Hamilton, 2013). Social networks have long been central to household responses to change in Vietnam, being used to pool risk and promote security and stability (Luong, 2003). However, the political and economic reforms experienced have altered relations within mangrove system dependent communities (MSDC) (van Hue and Scott, 2008) as they become increasingly influenced by market processes, particularly for aquaculture goods, altering the ways in which communities interact with one another and the wider world. Growing income inequalities, changing livelihood opportunities, and alterations in access to mangrove system resources have the potential to significantly impact the social network structure of coastal communities. The changing structure of social networks will influence the resilience of MSDC through alterations in social capital and the ability to self-organise.

Citation: Steven E. Orchard, Lindsay C. Stringer, Claire H. Quinn, Impacts of aquaculture on social networks in the mangrove systems, Ocean & Coastal Management, Volume 114, September 2015, Pages 1-10, ISSN 0964-5691, http://dx.doi.org/10.1016/j.ocecoaman.2015.05.019.

As the complexity of coastal livelihoods increases due to increased pressure on mangrove systems from aquaculture, understanding the structure of the social networks that sustain livelihoods is critically important in order to understand the resilience of MSDC to change. However, despite the importance of social networks in fostering resilience through supporting rural livelihoods and self-organisation, social network analysis has gained little attention in livelihood research (Misra et al., 2014), particularly in Vietnam. This paper addresses this gap. It aims to assess the impact of differing degrees of aquaculture on the livelihoods and social networks of three MSDC in northern Vietnam to provide insights into the resilience of these communities to change. The objectives are to: (1) assess and compare livelihood contexts at each degree of aquaculture; (2) assess and compare social network structures at each degree of aquaculture; and (3) examine the relationship between social network structures and livelihood characteristics at each degree of aquaculture. We find that a greater degree of aquaculture weakens community networks as livelihoods are more market orientated and social networks are larger and expand beyond the local community. This can lead to divergent understandings of mangrove functions and processes that impact the ability of communities to self-organise. Supporting network diversity through a balance of internal bonding and external bridging ties is required in order to facilitate self-organisation through the sharing and increased understanding of divergent perceptions of mangrove systems, and to foster resilience for future mangrove system planning.

The next section outlines the central concepts related to resilience and social network analysis, followed by details of the research process. Quantitative analysis then offers insights into the influence of aquaculture and resulting livelihood alterations on social networks between and within communities. Lessons from these insights are discussed, drawing out key conclusions that could be used to inform future mangrove planning.

Central concepts

Livelihoods, vulnerability and resilience

A livelihood comprises the capabilities, assets and activities necessary for a means of living (Chambers and Conway, 1992), and is sustainable when it is able to cope with perturbations and maintain its capabilities while not undermining the natural resource base (Scoones, 1998). Vulnerability is defined as "*... the degree to which a person, system or unit is likely to experience harm due to exposure to perturbations or stresses*" (Kasperson et al., 2002: p253), and is a function of exposure, sensitivity and capacity to respond (Gallopin, 2006). Perturbations and stresses encompass disruptions to household livelihoods in response to changing circumstances (Adger, 2000), such as the change observed in Vietnams mangrove systems in the shape of degradation and loss due to increased aquaculture (Tri et al., 1998). For vulnerable households, change is often intractable and related to underlying socio-economic factors such as income (Fisher, 2001), livelihood diversity (Ellis, 2000), and dependency on and access to natural resources such as mangroves (Chambers, 1989 and Shackleton and Shackleton, 2004).

Although much debate surrounds the link between vulnerability and resilience (i.e. that the concepts are not simply linear antonyms of each other), they do generally have an inverse relationship (Cassidy and Barnes, 2012). Research suggests that vulnerability to stress can be reduced through strengthening social resilience (Nelson et al., 2007), defined as "*... the ability of individuals and communities to absorb external changes and stresses ... while maintaining the sustainability of their livelihoods*" (Adger et al., 2002: p358). In Vietnam, a large portion of the population is rural and living in coastal areas with highly variable environmental conditions (Adger, 2000). Households living within these communities are engaged in a range of primarily natural resource based livelihoods which are being severely impacted by mangrove system change related to the rapid growth of aquaculture (van Hue and Scott, 2008). This in turn is leading to considerable divergence in the vulnerability context within communities due to changes in the livelihood opportunities available to households (Orchard et al., 2015). The resilience of communities to such change is embedded within and available to them through the social networks, which households use to gain access to the necessary livelihood resources required to respond to change (Cassidy and Barnes, 2012, Nath et al., 2010 and Bodin et al., 2006).

Social capital, social networks and resilience

Historically, natural resource dependent communities have self-organised to manage changes to the resource base on which their livelihoods depend, with social networks shaping access to the necessary resources to respond (Adger, 2003). Social networks are constituent of two or more actors (e.g. individuals, households or organisations) that are connected through one or more relations (Abbasi et al., 2012). The structure and function of social networks shapes the trust, norms and reciprocity that forms a crucial part of social capital and enables people to act together to pursue shared goals (Putnam, 1995; Pretty and Ward, 2001). Social networks are also an attribute of self-organisation, which refers to the autonomy, freedom, and power of actors to coordinate themselves (Pelling, 2011). Self-organisation relates to resilience through the ability of communities to reorganize in the face of change (Whaley and Weatherhead, 2014). Hence, communities endowed with social capital will have greater resilience through their ability to self-organize in order to understand and tackle environmental challenges (Bunting et al., 2010, Fleischman et al., 2010, Sanginga et al., 2007 and Folke et al., 2005; Pelling and High, 2005). Hence, aspects of resilience reside in the social networks of natural resource dependent communities through: facilitating access to livelihood resources in order to respond to change (Smit and Wandel, 2006); social connectedness that increases the ability of communities to self-organise and build social capital (Crona and Bodin, 2010). In Vietnam, social networks have long been central to household responses to change, being used to pool risk and promote security and stability (Luong, 2003). Analysing how aquaculture impacts social network structure and the ability of MRDC to self-organise is crucial for improving our understanding of the factors that shape resilience. Vietnam provides a highly relevant context for this research, having experienced rapid social, economic, political and mangrove system change due to aquaculture expansion following transition.

This paper uses an egocentric approach (i.e. analysis of individual household networks rather than whole networks) to assess social network density, degree and betweenness centrality, efficiency, effectiveness and constraint. These measures provide indicators of how information and resources may flow through particular types of network (e.g. large or small, dense or sparse, open or closed). They also help analyse the opportunities and constraints that actors experience, and the potential for self-organisation, as a result of social network structures.

Degree centrality is simply the number of contacts a household has and is an important indicator of how integrated a household is within the network (Valente and Foreman, 1998). Network density is the number of existing contacts divided by the number of possible contacts. Network density relates to bonding social capital in that it involves strong social linkages within localised networks (Barnes-Mauthe et al., 2013) which can lead to the creation of trust and the promotion of norms for acceptable resource use (Pretty and Ward, 2001). High degree centrality and network density can increase resilience by providing a number of redundant contacts (Magsino, 2009). However, too much degree centrality and network density may constrain a household's behaviour due to homogenisation of knowledge and perspectives and reduced flexibility, and hence reduce resilience (Frank and Yasumoto, 1998 and Bodin and Crona, 2009). Betweenness centrality refers to households that connect other households who would otherwise not be linked (Burt, 2004). Betweenness centrality has similarities to bridging social capital, although it does not differentiate between households within or outside a community. High betweenness centrality provides the potential for a household to control the flow of and access to a variety of resources between the households it connects (Bodin and Crona, 2009). However, bridging capital is characterised by weaker linkages, and networks with high levels of betweenness are vulnerable to fragmentation should bridging households leave the network (Borgatti, 2003).

Burt (1992) states that, in terms of productive capacity, using the least number of ties to access as wide a range of resources as possible increases the effectiveness and efficiency of social networks. Fig. 1 illustrates an inefficient network (A) comprising a large number of redundant contacts (i.e. duplicated ties) accessing the same cluster of resources, compared to an efficient network (B) with low levels of redundancy as only one tie is used to access each cluster. Burt (1992) uses the term effective size to denote the number of clusters that an actor can receive benefits from, so network A and network B both have an effective size of 4. Network efficiency is calculated by dividing the effective size of the network by the total number of ties: in the case of network A giving a low score of 0.25 (i.e. 4/16 = 0.25), whilst in network B we observe perfect efficiency of 1 (i.e. 4/4 = 1). Network constraint measures the degree to which a household's contacts are connected to each other. In terms of network productivity, if a household's potential trading partners are all connected and have one another as potential trading partners, that household is highly constrained (Hanneman and Riddle, 2005). Research on network productivity demonstrates that high efficiency and low constraint are useful indicators of an individual's ability to 'get ahead' in terms of performance and ideas (Burt, 2004 and Podolny and Baron, 1997). However, a more in-depth analysis of the nature of the relationships in the social networks of natural resource dependent communities is required to assess how they affect resilience (Brockhaus et al., 2012).

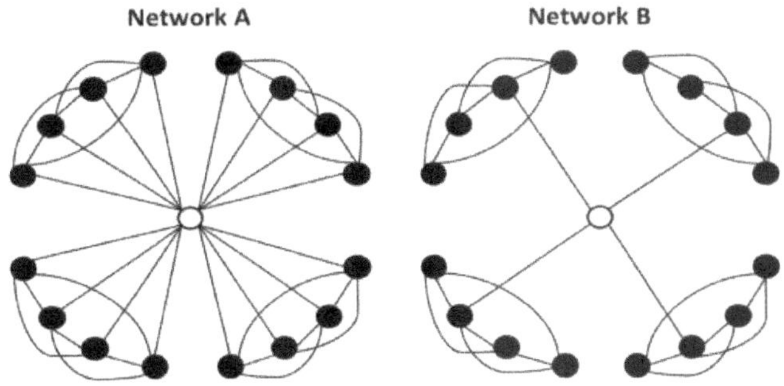

Figure 1: Inefficient (A) and efficient (B) networks. Adapted from Burt (1992).

By focussing on productive capacity rather than resilience in social networks, resilience can be compromised by a narrow focus on efficiencies (Walker et al., 2006). Increasing network efficiency results in a loss of redundancy, which represents buffering capacity in the case of loss (i.e. if one or more actors are weakened or lost, others can fill the position and continue to perform the necessary functions (Janssen et al., 2006)). Hence, social networks assessed purely from a productive capacity standpoint might be considered efficient by having low or no redundancy, but in terms of resilience the system requires redundancy so that the network does not fragment if a household leaves. There is growing recognition that governance of natural resources such as mangroves requires a deeper understanding of social structures, and the ways in which relationships among different actors facilitate or constrain the way natural resources are managed (Henry and Dietz, 2011 and Crona and Hubacek, 2010). Management efforts often fail due to inadequate attention to the role of social relationships in shaping environmental outcomes (Pretty and Ward, 2001).

Social network analysis is employed here to analyse three MRDC in northern Vietnam with different degrees of aquaculture, providing a snap shot of the social structures at the local scale. This approach does not provide a time dimension (e.g. as with social-ecological system approaches (Folke et al., 2005)), consideration of network links to higher levels of governance (e.g. as with network governance approaches (Carlsson and Sandström, 2008)), or the power dynamics related to each respective social network (e.g. as with political ecology approaches (Escobar, 2006)), as these are beyond the scope of this research. Rather, by analysing the current structure of social networks in MRDC with differing degrees of aquaculture, it is expected that crucial insights will be gained into the impact of aquaculture on resilience in order to support future mangrove system planning and to guide appropriate policy development.

MATERIALS AND METHODS

Three coastal communities in northern Vietnam were selected for data collection during February–August 2012: Giao Xuan (Nam Dinh province); Da Loc (Thanh Hoa province); and Dong Rui (Quang Ninh province) (Fig. 2). A community is distinguished here as a sub-set of the lowest administrative level of Vietnam's government, the commune, and defined as the socio-economic impact area of a mangrove system (Glaser,

2003). Communities were all situated on the northern coast of Vietnam, yet they represented distinct mangrove systems, geographically separate and with different histories. All three communities were located on the brackish shoreline in river estuaries where mangrove systems comprise mangrove trees, intertidal wetlands and mudflats that provide provisioning goods for local livelihoods. Research participants in all study communities had some degree of access to surrounding mangrove systems. For the purpose of this research, mangrove system provisioning goods (MSPG) refer to wild fish, clam, shrimp, crab and other shoreline animals collected from mangrove system areas held in common.

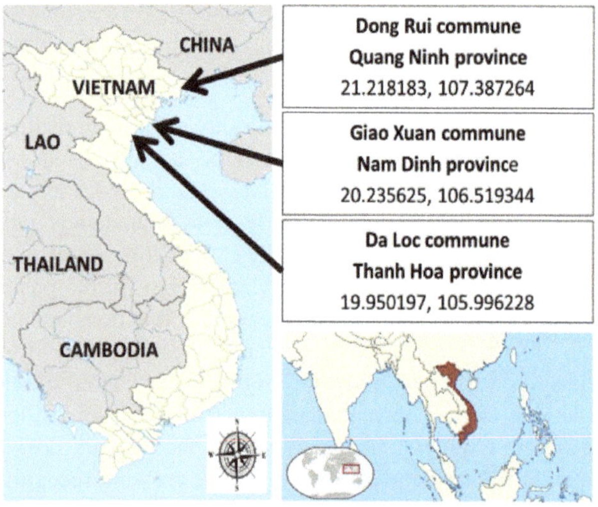

Figure 2: Study site locations and coordinates (*source*: Wikipedia, 2014).

Each community exhibits different degrees of aquaculture: Giao Xuan has a greater degree of aquaculture involvement; Da Loc has a moderate degree of aquaculture; and Dong Rui has a lower degree of aquaculture. Degree is used here to indicate "*the amount, level, or extent to which something happens or is present*" (Oxford English Dictionary, 2011: 142), and involvement indicates "*the fact or condition of being involved with or participating in something*" (Oxford English Dictionary, 2011: 296). Giao Xuan has a highly developed clam aquaculture sector which was established in the early 1990s. The sector was facilitated and supported by emerging trading connections between local households and aquaculture investors from China following market liberalisation (1986). Since then, the number of aquaculture farms developed by people external to the community has increased, and Giao Xuan is now one of the largest producers of clams in Vietnam, supplying domestic and international markets. Aquaculture farms are situated in the mudflat area located beyond the mangrove forest and covering the full extent of the coastline adjacent to the community. In Da Loc the clam aquaculture sector is in the initial stages of development, having been started by local households in 2010 who observed the success of a neighbouring community's clam aquaculture ventures. Aquaculture farms have been developed by locals who have little experience, knowledge or support to guide them but productivity is rising quickly and markets are rapidly growing domestically. As with Giao Xuan, the aquaculture farms are situated in the mudflat area located beyond the mangrove forest, but do not cover the full extent of coastline adjacent to the community.

Dong Rui experienced a short, intense and highly productive period of shrimp and fish aquaculture during the late 1980s/early 1990s. Large scale commercial aquaculture enterprises were initiated and implemented by external investors from a neighbouring province that had experienced recent success in the industry. These powerful and influential investors were able to use their connections in local authorities to acquire land for aquaculture farms, often illegally, and exclude the local community. Dong Rui is situated on an island archipelago, and aquaculture enterprises were located in adjacent areas surrounding the community on all sides. After an initial 2–3 years of high productivity the sector collapsed due to mismanagement, which resulted in severe environmental damage and the abandonment of the farms by investors. Hence, considering the minor impact on community livelihood opportunities and social networks, and the time elapsed since the collapse of the sector, aquaculture can be deemed as having limited influence on the community. Throughout this research each community is used to examine the impacts of different degrees of aquaculture on aspects of resilience related to the research objectives.

A quantitative approach to data collection was taken. Household surveys (n = 248: Giao Xuan, n = 79; Da Loc, n = 70; Dong Rui, n = 99) were conducted with household heads to identify: (i) livelihood characteristics regarding income, mangrove dependency (i.e. per cent of total income coming from sale of MSPG), and livelihood activity diversity; and (ii) social connectivity through name-generator questions. Name-generator questions were designed to produce a list of individuals with whom that particular household communicated regarding issues relating to mangrove systems. Each individual listed in the name-generator represented a communication tie of that household, and the full set of individuals on the name-generator list comprised that household's full ego-network. Local partners and key informants helped identify an initial set of key households engaged in mangrove system related livelihood activities. These key households then identified further respondents for the subsequent snowball sampling (Luttrell, 2006 and Pereira et al., 2005). Sampling continued in a respondent-driven way until saturation of target areas had been reached (i.e. until the same names started to reoccur in the list of names provided by respondents). Although it was recognised that this approach may miss households unconnected to the network of the initial respondents, respondent-driven sampling did permit less well-known households to be identified, as those best able to access members of hidden populations are their own peers (Heckathorn, 1997). Reaching saturation also meant that the configuration of the total sample was fully independent from the initial key respondents, hence yielding an unbiased sample (Heckathorn, 1997). In addition, this approach also prevented time being wasted talking to respondents that were not engaged in mangrove system use, thus permitting more focussed data collection. Household names were removed and replaced with a number to maintain confidentiality throughout this research.

Data for social network analysis were collated using quantitative methods to produce numerical data on the presence or absence of ties (Edwards, 2010). Such an approach enabled the measurement of network structure properties of density, degree and betweenness centrality, efficiency, effective size, and constraint. Although quantitative methods can overlook culture, agency and the processes through which relationships are created, maintained or reconfigured over time (Emirbayer and Goodwin, 1994), employing a quantitative approach in this study permitted the analysis of large sets of data using statistical techniques in order to identify patterns and connections in the data, which would not have been possible with qualitative information. Furthermore, the structure of networks was able to be analysed from the perspective of all actors in the network at the same time, and not just one individual perspective (Scott, 1976).

Data analysis first involved the identification of livelihood variables, i.e. total household annual income, mangrove dependency and livelihood diversity. Name-generator data, representing connectivity, were entered into UCINET 6 software to produce degree, density and betweenness scores for each household. For objectives 1 and 2 (assess the difference in livelihood diversity and social network measures across communities), once livelihood and connectivity measures and scores were obtained for each household, descriptive statistics were produced using SPSS v19 software. Objective 3 (identify differences within communities in livelihoods and social connectivity) involved the categorisation of livelihood diversity variables using two-step cluster analysis (Table 1). Inferential statistics, with livelihood diversity categorisations being the independent variables tested against the dependent connectivity measures, were produced using Kruskal–Wallis and Mann–Whitney tests (Ahenkan and Boon, 2011 and Cox et al., 2010).

Table 1: Categorisation of livelihood diversity variables

	Dong Rui	Da Loc	Giao Xuan
Income ($ per capita)	Low: 0–572 (n = 32)	Low: <350 (n = 23)	Low: 0–730 (n = 17)
	Middle: 573–1156 (n = 34)	Middle: 350–800 (n = 24)	Middle: >730–<1330 (n = 28)
	High: >1156 (n = 33)	High: >800 (n = 23)	High: >1330 (n = 34)
MSPG dependency	Low = 0–<33 (n = 44)	Low = 0–<33 (n = 47)	No = 0 (n = 53)
	Medium = 33–<66 (n = 27)	Medium = 33–>34 (n = 16)	Low = <32 (n = 18)
	High = >66 (n = 28)	High = >66 (n = 7)	High = >33 (n = 8)
Livelihood diversity	Low: <3 activities (n = 5)	Low: <5 (n = 22)	Low: <2 activities (n = 15)
	Med: 3–4 activities (n = 47)	Med: 5 (n = 28)	Med: 3 activities (n = 31)
	High: >4 activities (n = 47)	High: >5 (n = 20)	High: >3 activities (n = 33)

RESULTS

Comparing livelihood diversity characteristics across communities

In relation to objective 1, establishing a set of livelihood characteristics and values for each household allowed us to explore the similarities and differences in the relationship between livelihoods and aquaculture in each of the study communities (Table 2). A significant difference was observed in the mean income values between Giao Xuan and both Da Loc and Dong Rui, indicating that higher incomes are associated with greater degrees of aquaculture. The link between income inequality and degree of aquaculture is illustrated by observing the distribution of data in the descriptive statistics tables, whereby the 5% trimmed mean and income range are both considerably higher in Giao Xuan, followed by Da Loc, then Dong Rui.

Table 2: Livelihood characteristic measures between communities (*GX = Giao Xuan, DL = Da Loc, DR = Dong Rui*)

	Livelihood characteristic measures		
	Test statistic	Degrees of freedom	Post-hoc r score
Income	22.97*	2	0.148 (GX > DR)
Mangrove dependency	89.25*	2	0.485 (DR > GX)
Livelihood diversity	72.2*	2	0.405 (DL > FGX)

* Kruskal–Wallis test score significant at p = 0.025.

A significant disparity occurred in the mean mangrove dependency values, suggesting an inverse relationship between aquaculture prevalence and mangrove dependency. The 5% trimmed mean indicates that omitting the lowest 5% and highest 5% values from the data had less influence in Dong Rui than in Giao Xuan and Da Loc. However, the lower level of mangrove dependency observed in Giao Xuan, with a well-established aquaculture industry, combined with greater skewdness and kurtosis values, suggests that greater degrees of aquaculture were associated with lower mangrove dependency.

A significant variation was noted in the mean livelihood diversification values between Giao Xuan and both Da Loc and Dong Rui, suggesting that degrees of aquaculture has an inverse relationship with household livelihood diversification. Although there is no notable deviation from the mean value observed in the 5% trimmed mean and range values in all three communities, a greater range of livelihood activities available to households in Giao Xuan but a lower mean value of livelihood activities undertaken suggests that households are specialising their livelihoods in response to greater degrees of aquaculture. This is supported by the kurtosis values which suggest that households cluster around the mean in Giao Xuan, with a flatter distribution found in Da Loc and Dong Rui. Hence, in communities with high degrees of aquaculture, household livelihood activities are lower and concentrated into that industry.

Comparing social network characteristics across communities

Establishing a set of social network measures and values for each household allowed us to explore the similarities and differences between community networks (objective 2). Analysis of social network measures (Table 3) indicated a statistically significant difference was observed in mean degree centrality values between Dong Rui and both Giao Xuan and Da Loc, with larger network sizes observed in communities with high degrees of aquaculture. However, a significant disparity was also observed in mean density values, suggesting that higher degrees of aquaculture are associated with networks of lower connectivity. This could be due to the changing structure of social networks, from close-knit and family based networks associated with low degrees of aquaculture, and wider reaching commercial networks associated with high degrees of aquaculture. Although there was no significant difference between mean betweenness values, the lower value observed in Dong Rui supports the previous finding from network density values, i.e. Dong Rui networks are more closely connected with a household's contacts all having many connections to each other.

Table 3: Difference in social network measures between communities

	Social network measures		
	Test statistic	Degrees of freedom	Post-hoc r score
Degree	32.64**	2	0.178 (DL-DR)
Density	8.64*	2	0.076 (GX-DL)
Betweenness	0.241	2	– Ω
Effective size	31.08**	2	0.172 (DL-DR)
Efficiency	9.45**	2	0.081 (DL-GX)
Constraint	15.15**	2	0 (GX-DR)

*Kruskal–Wallis test score p = 0.026, **p = 0.05.

Ω = no significant relationship observed, therefore no score provided.

A significant difference was observed in the mean value of the effective size of networks between communities. The lower value observed in Dong Rui suggests that low degrees of aquaculture are associated with a greater redundancy among a household›s contacts, which is supported by the greater network density observed in the community. Although this may be detrimental to the productivity of a household›s social network, the increased redundancy can increase resilience. A significant disparity was detected in the mean value of network efficiency. The greater network efficiency observed in Giao Xuan than in either Da Loc or Dong Rui suggests that households in Giao Xuan obtain greater productivity from their social network for each unit of effort invested in their contacts, but the observed lower levels of redundancy (i.e. network density) may translate into lower resilience. A significant difference was also observed for the mean value of network constraint, with a greater constraint observed in Dong Rui suggesting that the high density of networks can constrain household›s behaviour in terms of productivity, but increase resilience in terms of the greater number of redundant contacts.

Comparing livelihood diversity and social network characteristics within communities

Having established a set of livelihood and social network characteristics and values, we can determine whether there is a relationship between livelihood characteristics within communities and their social connectivity (objective 3) (Table 4). In Dong Rui there were no significant differences in mean degree, density or betweenness values according to livelihood diversity. However, there was a significant difference in scores according to network density, with higher income households having greater network density than lower income households. Although not statistically significant, a difference was noted in betweenness scores according to mangrove dependency. There was no significant difference found in mean social network values of effective size, efficiency or constraint according to mangrove dependency or livelihood diversity. However, a significant difference was observed in efficiency values according to income, with lower income groups having more efficient networks, suggesting that they have more non-redundant ties in their networks.

Table 4: Results of Kruskal–Wallis tests for significant differences in social network scores according to livelihood measures in the three study communities

	Density	Degree	Betweenness	Effective size	Efficiency	Constraint
Dong Rui						
Income	9.15**	3.26	0.91	1.11	10.78*	0.66
Mangrove dependency	1.88	1.82	4.66*	2.46	1.85	4.55
Livelihood diversity	0.7	0.49	1.48	0.71	0.35	1.1
Da Loc						
Income	0.51	0.54	1.67	0.21	0.431	0.59
Mangrove dependency	2.92	3.71	5.92*	4.54	3.13	3.04
Livelihood diversity	4.45	2.33	0.02	1.69	0.94	1.68
Giao Xuan						
Income	1.63	0.31	0.96	0.46	1.22	2.24
Mangrove dependency	4.66*	0.85	0.28	0.26	3.88	0.19
Livelihood diversity	4.49	0.75	3.01	0.54	4.36	5.93*

**p = 0.05, * = 0.1 (nb: ** indicates statistical significance, while * is apparent but not statistically significant).

In Da Loc, there were no significant differences in degree, density or betweenness values according to all livelihood diversity measures. However, there was a difference in betweenness according to mangrove dependency, with more dependent groups scoring lower in betweeness. There were no significant differences in effective size, efficiency or constraint according to any of the livelihood measures. In Giao Xuan, there were no significant differences in mean values of degree, density or betweenness values according to livelihood income and diversity. There was a small but not significant difference in network density values according to mangrove dependency, between groups with no dependency and those with low and high dependency. No significant differences were observed in mean effective size, efficiency and constraint values according to livelihood income and mangrove dependency measures. There was a small but not significant difference in constraint according to livelihood diversity, with lower livelihood diversity groups possessing less constrained networks.

DISCUSSION

The impact of aquaculture on livelihoods

This study found high degrees of aquaculture are associated with lower resilience in coastal communities through greater income and inequality, lower mangrove system dependency and lower livelihood diversity. Adger et al. (2006) show how income inequality can negatively impact resilience as the concentration of resources among a small number of individuals reduces resource access for the rest of the community. Allenby and Fink (2005) suggest that reductions in the regenerative ability of ecosystems are undermining the resilience of those with greater levels of natural resource dependency by reducing their ability to effectively respond to change in order to sustain their livelihoods. Turning to livelihood diversity, findings correspond with those from Cinner and Bodin (2010) in their study of fishing communities in Kenya. The divergence observed in aggregate livelihood diversity measures between communities is important because livelihoods are the means by which households in MSDC interact with one another and give value and meaning to the changing environment around them (cf. Frost et al., 2006 and Wiesmann, 1998), which greatly influences resilience (Vincent, 2007). Cinner and Bodin (2010) argue that increased levels of development are associated with changes in livelihoods which influence the structure of social networks.

The impact of aquaculture on bonding social capital and resilience

High degrees of aquaculture are associated with low community bonding social capital, with negative impacts on resilience through lower levels of network redundancy and potential for self-organisation. Results indicate that different degrees of aquaculture involvement are associated with distinct livelihood contexts, and subsequent variation in bonding social capital with regard to the size and density of social networks. Communities with low degrees of aquaculture are associated with smaller and denser networks than communities with high degrees of aquaculture. This suggests that such communities are characterised by a larger stock of bonding social capital and higher degrees of resilience due to a greater number of redundant network ties. Furthermore, Djalante et al. (2013) argues that high levels of network density can foster greater interaction and trust between individuals and groups in natural resource dependent communities. Previous research suggests that greater levels of trust have the potential to increase the resilience of MRDC in a number ways: first, it reduces the risk and cost of collaborating with others which is crucial for self-organisation to occur (Bodin et al., 2006); second, it fosters shared identification and understandings of environmental issues necessary for self-organisation to occur (Pretty and Ward, 2001); third, it facilitates the creation of and compliance with mutual norms with regards to acceptable behaviour in resource use (Barnes-Mauthe et al., 2013); finally, it reduces the potential for network fragmentation (Coleman, 1988).

Communities with high degrees of aquaculture were found to be associated with larger and less dense networks, with a greater number of non-redundant ties suggesting lower levels of resilience. Sandstrom and Rova (2010) argue that less dense networks can exhibit conflicting interests and perceptions, lowering resilience through a lack of common understanding and problem identification, such as resource condition, quantity/quality of stocks and rules of use, that are necessary for self-organisation to occur. Results presented here concur with that of Baird and Gray (2014) in their study of the influence of economic transition on Maasai communities in Tanzania, which indicate that: livelihood opportunities are low and social network interactions are high prior to transition; livelihood opportunities increase with development, which prompts changes in the traditional use of social networks; subsequently, households reduce their engagement with traditional social networks.

Research from our study suggest that communities differentiate subsequent to the transition process, leading to reduced levels of resilience in MRDC by hindering the potential for self-organisation (cf. Cumming, 2011). King (2000) suggests that actors who are successful in furthering their goals will actively seek ties with others to continue the pursuit of their goals. For example, in communities with high degrees of aquaculture, wealthy households with little mangrove system dependency and large and expansive market oriented networks are typically more concerned with market demands and less aware of the degradation of local mangrove systems. This could act as a barrier to self-organisation within MRDC as it could reduce shared understandings of natural resource issues and the support of wealthy and influential households (cf. Bodin and Crona, 2009). In light of this, Sandstrom and Rova (2010) argue that denser community networks made up of heterogeneous actors and groups within MRDC can promote the reconciling of conflicting perspectives, and facilitate the development of common understandings of natural resource issues and dilemmas.

The impact of aquaculture on bridging social capital and resilience

The large and expansive social networks associated with high degrees of aquaculture can reduce resilience through less redundant ties and potential for self-organisation. Communities with higher degrees of aquaculture are associated with larger and less dense social networks, indicating greater access to external sources of capital, skills and knowledge, market opportunities and social networks. The social networks of these communities were found to exhibit greater effectiveness and efficiency and less constraint, indicating a high number of non-redundant ties. Previous research on organisational networks advocates this type of network with regard to increasing productivity and gaining competitive advantage in market settings (Burt, 2004). However, resilience can be compromised by a narrow focus on efficiencies (Walker et al., 2006). The increased reliance on a small number of weak market based bridging ties can lower resilience by reducing the number of redundant ties required to buffer networks against the loss of any particular actor (Prell et al., 2012). Furthermore, in line with findings from Baird and Gray (2014) in their study of traditional social networks and household livelihood strategies in northern Tanzania, market forces may increase household economic independence in communities with high degrees of aquaculture involvement. This can reduce resilience through lower capacity to self-organise in response to change.

Previous research indicates that only a small number of households in communities with high aquaculture have bridging social capital due to their engagement in aquaculture market networks (Orchard et al., 2015). These households are able to maintain their advantageous network position through bridging ties that facilitate their access to and command over external resources (cf. Isaac et al., 2007). Research suggests that as local networks become increasingly integrated into market orientated networks, it is the local social networks that largely determine who benefits (Frank et al., 2007). Furthermore, results show that households with high dependence on MSPG that are unable to access new market opportunities maintain their traditional bonding networks as a resource to respond to change (cf. Busby and Albers, 2010; Cassidy and Barnes, 2012 and Baird and Gray, 2014). Hence, it is possible that bonding capital within successful socio-economic groups is reinforced over time, and the resources attained through bridging ties become captured within these groups (cf. Isaac et al., 2007). Reduced community cohesion, in this case through disengagement in community-level networks due to large and expansive aquaculture networks, can reduce the ability of communities to self-organise in response to change (Adger, 2000).

Whilst communities with high degrees of aquaculture have built bridging ties to other communities to develop and maintain aquaculture, this could be at the expense of bridging among different socio-economic groups within communities. Although self-organisation may occur within distinct socio-economic groups, either among high income groups seeking to maximise their power and wealth, or among marginalised groups pooling resources in order to respond to change, the subsequent homogenisation has fractionalised communities with high degrees of aquaculture. A lack of bridging among socio-economic groups obstructs the opportunity for self-organisation and prevents the creation of shared understanding of environmental issues (cf. Crona and Bodin, 2006). This highlights the need to balance the bonding and bridging ties of communities to help build trust across diverse groups, encourage a diversity of ideas, increase network flexibility and foster resilience (Baird and Gray, 2014). This should link marginalised groups who have rich knowledge of mangrove systems due to their high dependence, with those from higher socio-economic groups that are integrated into external networks of diverse actors and resources other than those based solely on market relations (e.g. NGOs, governments) (Bodin and Crona, 2009).

CONCLUSION

By analysing the impact of aquaculture on livelihoods and social networks, our findings illustrate the importance of considering how these interacting elements have shaped resilience in three mangrove system dependent communities in northern Vietnam. By employing an approach that provides insights into social capital in communities with differing degrees of aquaculture, we have discussed how the livelihood context and the structure of social networks shape the ability of communities to self-organise in response to change. Whilst efforts to increase social capital in natural resource dependent communities in order to increase resilience are welcomed, the various ways in which aquaculture impacts the structure of social networks and the ability to self-organise must be acknowledged.

Findings demonstrate how economic transition alters mangrove system governance through the increasing influence of market mechanisms on the structure of social networks. For example, small and dense social networks based on kinship have traditionally played a crucial role in rural Vietnam, representing a component of social capital used as an asset and coping strategy for households with few alternative assets. However, our findings show that communities with a greater degree of aquaculture are associated with larger and less dense networks that are shaped by market relations for aquaculture goods that extend beyond the immediate community. We have demonstrated how market relations can negatively impacted resilience by: (1) lowering the level of redundancy in social networks, reducing buffering capacity in the event that ties are lost; and (2) reducing the level of connectedness within communities as networks become less dense, compromising the ability of communities to self-organise.

Hence, it is crucial that the impact of aquaculture on livelihoods and social networks is acknowledged if policies to sustainably manage mangrove systems are to be successful. In particular, understanding how aquaculture impacts the structure of social networks is vital in order to provide targeted support to community networks to increase resilience. Building and supporting social networks among MSDC can help foster self-organisation to effectively manage and respond to external shocks through shared understanding, not only of the resource itself, but of the perspectives of divergent mangrove system stakeholders. This is crucial in transition economies as the state is rolled back and traditional community networks (i.e. bonding capital) are replaced by external networks (bridging capital) oriented towards markets and commerce. Further research will be necessary to identify the specific kinds of support communities will need, and also to understand the structure and role of networks that extend beyond the community and across governance levels.

Findings presented here highlight important features of communities that should be considered within environmental governance more widely. For example, the increasing influence of external market relations means that community networks risk becoming fractionalised among groups with differing needs regarding networks of productivity or adaptability, and priorities for responding to market or mangrove system changes. While networks that extend beyond the immediate community present an opportunity to access external resources, they are fragile and have the potential to extract resources away from communities. How mangrove system change affects resilience depends on the impact of aquaculture on livelihood contexts and social network structures, and will manifest differently depending on the diversity and balance of social networks necessary for the sustainable governance of mangrove systems.

ACKNOWLEDGEMENTS

This work was funded through the Centre for Climate Change Economics and Policy(RES-599-28-0001) (CCCEP), an Economic and Social Research Council programme. We are grateful to the communities of Giao Xuan, Da Loc and Dong Rui, Marinelife Conservation and Community Development (MCD), the Centre for Natural Resource and Environment Studies (CRES), and CARE Vietnam.

REFERENCES

1. Abbasi, A., Chung, K.S.K., Hossain, L., 2012. Egocentric analysis of co-authorship network structure, position and performance. Inf. Process. Manag. 48, 671e679.
2. Adger, N.W., Kelly, N.P., Winkels, A., Huy, L.Q., Locke, C., 2002. Migration, remittances, livelihood trajectories, and social resilience. Ambio 31, 358e366.
3. Adger, W.N., 2000. Institutional adaptation to environmental risk under the transition in Vietnam. Ann. Assoc. Am. Geogr. 90, 738e758.
4. Adger, W.N., 2003. Social capital, collective action, and adaptation to climate change. Econ. Geogr. 79, 387e404.
5. Adger, W.N., Paavola, J., Huq, S., Mace, M.J. (Eds.), 2006. Fairness in Adaptation to Climate Change. MIT Press, Cambridge.
6. Ahenkan, A., Boon, E., 2011. Improving Nutrition and Health through non-timber Forest products in Ghana. J. Health Popul. Nutr. 29, 141e148.
7. Allenby, B., Fink, J., 2005. Towards inherently secure and resilient societies. Science 309 (5737), 1034e1036.
8. Baird, T.D., Gray, C.L., 2014. Livelihood diversification and shifting social networks of exchange: a social network transition? World Dev. 60, 14e30
9. Barbier, E.B., 2006. Mangrove dependency and the livelihoods of coastal communities in Thailand. In: Hoanh, C.T., Tuong, T.P., Gowing, J.W., Hardy, B. (Eds.), Environment and Livelihoods in Tropical Coastal Zones: Managing Agriculture- fishery-aquaculture Conflicts. CAB International, Wallingford, UK, pp. 126e139.
10. Barnes-Mauthe, M., Arita, S., Allen, S.D., Gray, S.A., Leung, P., 2013. The influence of ethnic diversity on social network structure in a common-pool resource System: Implications for collaborative management. Ecol. Soc. 18.
11. Bodin, O., Crona, B., Ernstson, H., 2006. Social networks in natural resource management: what is there to learn from a structural perspective? Ecol. Soc. 11.
12. Bodin, O., Crona, B.I., 2009. The role of social networks in natural resource governance: what relational patterns make a difference? Glob. Environ. ChangeHum. Policy Dimens. 19, 366e374.
13. Borgatti, S.P., 2003. The key player problem. In: Breiger, R., Carley, K., Pattison, P. (Eds.), Dynamic Social Network Modeling and Analysis. National Academy of Sciences Press, pp. 241e252.
14. Brockhaus, M., Obidzinski, K., Dermawan, A., Laumonier, Y., Luttrell, C., 2012. An overview of forest and land allocation policies in Indonesia: is the current framework sufficient to meet the needs of REDDþ? For. Policy Econ. 18, 30e37.
15. Bunting, S.W., Pretty, J., Edwards, P., 2010. Wastewater-fed aquaculture in the East Kolkata Wetlands, India: anachronism or archetype for resilient ecocultures? Rev. Aquac. 2 (3), 138e153.

16. Burt, R.S., 1992. Structural Holes: the Social Structure of Competition. Harvard University Press. Burt, R.S., 2004. Structural holes and good ideas. Am. J. Sociol. 110, 349e399.
17. Busby, G., Albers, H.J., 2010. Wildfire risk management on a landscape with public and private ownership: who pays for Protection? Environ. Manag. 45 (2), 296e310.
18. Carlsson, L., Sandstrom, A., 2008. Network governance of the commons. Int. J. € Commons 2 (1), 33e54.
19. Cassidy, L., Barnes, G.D., 2012. Understanding household connectivity and resilience in marginal rural communities through social network analysis in the village of Habu, Botswana. Ecol. Soc. 17.
20. Chambers, R., 1989. Editorial Introduction: Vulnerability, coping and policy. IDS Bull. 20, 1e7.
21. Chambers, R., Conway, G., 1992. In: PAPER, I.D. (Ed.), Sustainable Rural Livelihoods: Practical Concepts for the 21st Century. IDS, Brighton.
22. Cinner, J.E., Bodin, O., 2010. Livelihood diversification in tropical coastal communities: a network-based approach to analyzing 'Livelihood landscapes'. PloS One 5.
23. Coleman, J.S., 1988. Social capital in the creation of human-capital. Am. J. Sociol. 94, S95eS120.
24. Cox, M., Arnold, G., Villamayor Tom as, S., 2010. A review of design principles for community-based natural resource management. Ecol. Soc. 15, 38.
25. Crona, B., Bodin, O., 2006. What you know is who you know? communication patterns among resource users as a prerequisite for co-management. Ecol. Soc. 11.
26. Crona, B., Bodin, O., 2010. Power asymmetries in small-scale fisheries: a barrier to governance transformability? Ecol. Soc. 15 (4), 32.
27. Crona, B., Hubacek, K., 2010. The right connections: how do social networks lubricate the machinery of natural resource governance? Ecol. Soc. 15 (4), p18.
28. Cumming, G.S., 2011. Spatial resilience: integrating landscape ecology, resilience, and sustainability. Landsc. Ecol. 26 (7), 899e909.
29. Djalante, R., Holley, C., Thomalla, F., Carnegie, M., 2013. Pathways for adaptive and integrated disaster resilience. Nat. Hazards 69 (3), 2105e2135.
30. Edwards, G., 2010. Mixed-method Approaches to Social Network Analysis. ESRC National Centre for Research Methods Review paper, NCRM/015.
31. Ellis, F., 2000. The determinants of rural livelihood diversification in developing countries. J. Agric. Econ. 51, 289e302.
32. Emirbayer, M., Goodwin, J., 1994. Network analysis, culture, and the problem of Agency. Am. J. Sociol. 99, 1411e1454.
33. Escobar, E., 2006. Difference and conflict in the struggle over Natural Resources: a political ecology framework. Development 49 (3), 6e13.
34. Fisher, D.R., 2001. Resource dependency and rural poverty: rural areas in the United States and Japan. Rural. Sociol. 66, 181e202.
35. Folke, C., Hahn, T., Olsson, P., Norberg, J., 2005. Adaptive governance of socialecological systems. Annu. Rev. Environ. Resour. 30 (1), 441e473.
36. Fleischman, F.D., Boenning, K., Garcia-Lopez, G.A., Mincey, S., Schmitt-Harsh, M., Daedlow, K., Lopez, M.C., Basurto, X., Fischer, B., Ostrom, E., 2010. Disturbance, response, and persistence in self-organized Forested communities: analysis of robustness and resilience in Five communities in southern Indiana. Ecol. Soc. 15 (4).
37. Frank, K.A., Mueller, K., Krause, A., Taylor, W.W., Leonard, N.J., 2007. The intersection of global trade, social networks, and fisheries. In: Taylor, W.W., Schecter, M.G., Wolfson, L.G. (Eds.), Globalization: Effects on Fisheries Resources. Cambridge University Press, New York.
38. Frank, K.A., Yasumoto, J.Y., 1998. Linking action to social structure within a system: social capital within and between subgroups. Am. J. Sociol. 104, 642e686.
39. Frost, P., Campbell, B., Medina, G., Usongo, L., 2006. Landscape-scale approaches for integrated natural resource management in tropical forest landscapes. Ecol. Soc. 11.
40. Gallopin, G.C., 2006. Linkages between vulnerability, resilience, and adaptive capacity. Global Environ. Change 16, 293e303.
41. Giri, C., Ochieng, E., Tieszen, L.L., Zhu, Z., Singh, A., Loveland, T., Masek, J., Duke, N., 2011. Status and distribution of mangrove forests of the world using earth observation satellite data. Glob. Ecol. Biogeogr. 20 (1), 154e159.
42. Glaser, M., 2003. Interrelations between mangrove ecosystem, local economy and social sustainability in Caete Estuary, North Brazil. Wetl. Ecol. Manag. 11, 265e272.
43. Gopal, B., 2013. Future of wetlands in tropical and subtropical Asia, especially in the face of climate change. Aquat. Sci. 75 (1), 39e61.
44. Hamilton, S., 2013. Assessing the role of commercial aquaculture in displacing mangrove forest. Bull. Mar. Sci. 89, 585e601.
45. Hanneman, R.A., Riddle, M., 2005. Introduction to Social Network Methods. University of California, Riverside.
46. Heckathorn, D.D., 1997. Respondent-driven sampling: a new approach to the study of hidden populations. Soc. Probl. 44, 174e199.
47. Henry, A.D., Dietz, T., 2011. Information, networks, and the complexity of trust in commons governance. Int. J. Commons 5 (2), 188e212.
48. Isaac, M.E., Erickson, B.H., Quashie-Sam, S.J., Timmer, V.R., 2007. Transfer of knowledge on agroforestry management practices: the structure of farmer advice networks. Ecol. Soc. 12.
49. Janssen, M.A., Schoon, M.L., Ke, W., Borner, K., 2006. Scholarly networks on resil- € ience, vulnerability and adaptation within the human dimensions of global

environmental change. Glob. Environ. Change 16, 240e252.

50. Joffre, O.M., Schmitt, K., 2010. Community livelihood and patterns of natural resources uses in the shrimp-farm impacted Mekong Delta. Aquac. Res. 41, 1855e1866.
51. Kasperson, J.X., Kasperson, R.E., Turner, B.L., Hsieh, W., Schiller, A., 2002. Vulnerability to global environmental change. In: Diekmann, A., Dietz, T., Jaeger, C.C., Rosa, E.A. (Eds.), The Human Dimensions of Global Environmental Change. MIT presss, Cambridge.
52. King, A., 2000. Managing without Institutions : the Role of Communication Networks in Governing Resource Access and Control (PhD thesis). University of Warwick.
53. Luong, H.V., 2003. Postwar Vietnam: Dynamics of a Transforming Society. Institute of Southeast Asian Studies, Singapore.
54. Luttrell, C., 2006. Adapting to aquaculture in Vietnam: securing livelihoods in a context of change in two coastal communities. Environ. Livelihoods Trop. Coast. Zones Manag. Agric. Fish. Aquac. Confl. 2, 17e29.
55. Magsino, S.L., 2009. Applications of Social Network Analysis for Building Community Disaster Resilience. Board on Earth Sciences and Resources: Division on Earth and Life Studies. The National Academies Press, Washington, D. C.
56. Misra, S., Goswami, R., Basu, D., Jana, R., 2014. Application of social network analysis in livelihood system study. Space Cult. India 2 (3), 24e46.
57. Nath, T.K., Inoe, M., Pretty, J., 2010. Formation and function of social capital for Forest resource management and the improved livelihoods of indigenous people in Bangladesh. J. Rural Community Dev. 5 (3), 104e122.
58. Nelson, D.R., Adger, W.N., Brown, K., 2007. Adaptation to environmental change: contributions of a resilience framework. Annu. Rev. Environ. Resour. 32, 395e419.
59. Orchard, S.E., Stringer, C.L., Quinn, C.H., 2015. Mangrove system dynamics in Southeast Asia: linking livelihoods and ecosystem services in Vietnam. Reg. Environ. Change. Open access (Available online from: http://link.springer. com/article/10.1007%2 Fs10113-015-0802-5 http://dx.doi.org/10.1007/s10113-015-0802-5.
60. Oxford English Dictionary, twelfth ed., 2011. Oxford University Press, Oxford.
61. Pelling, M., 2011. Adaptation to Climate Change: from Resilience to Transformation. Routledge, Abingdon, UK.
62. Pelling, M., High, C., 2005. Understanding adaptation: what can social capital offer assessments of adaptive capacity? Global Environ. Change A 15 (4), 308e319.
63. Pereira, E.,C., Queiroz, H., Pereira, H., Vicente, L., 2005. Ecosystem services and human well-being: a participatory study in a mountain community in portugal. Ecol. Soc. 10, 14.
64. Podolny, J.M., Baron, J.N., 1997. Resources and relationships: social networks and mobility in the workplace. Am. Sociol. Rev. 62, 673e693.
65. Prell, C., 2012. Social network analysis: history, theory, and methodology. Sage, Thousand Oaks, CA.
66. Pretty, J., Ward, H., 2001. Social capital and the environment. World Dev. 29, 209e227.
67. Primavera, J.H., 2006. Overcoming the impacts of aquaculture on the coastal zone. Ocean Coast. Manag. 49 (9e10), 531e545.
68. Putnam, R.D., 1995. Bowling alone: America's declining social capital. J. Democr 6 (1), 64e78.
69. Sandstrom, A., Rova, C., 2010. Adaptive co-management networks: a comparative analysis of two fishery conservation areas in Sweden. Ecol. Soc. 15.
70. Sanginga, P.C., Kamugisha, R.N., Martin, A.M., 2007. Conflicts management, social capital and adoption of agroforestry technologies: empirical findings from the highlands of southwestern Uganda. Agrofor. Syst. 69 (1), 67e76.
71. Scoones, I., 1998. Sustainable Rural Livelihoods: a Framework. IDS Working Paper, 72.
72. Scott, J., 1976. The Moral Economy of the Peasant: Rebellion and Subsistence in Southeast Asia. Yale University Press.
73. Shackleton, C., Shackleton, S., 2004. The importance of non-timber forest products in rural livelihood security and as safety nets: a review of evidence from South Africa. South Afr. J. Sci. 100, 658e664.
74. Smit, B., Wandel, J., 2006. Adaptation, adaptive capacity and vulnerability. Global Environ. Change Hum. Policy Dimens 16 (3), 282e292.
75. Spalding, M., Kainuma, M., Collins, L., 2010. World Atlas of Mangroves. A Collaborative Project of ITTO, ISME, FAO, UNEP-WCMC, UNESCO-MAB, UNU-INWEH and TNC, London
76. Speranza, C.I., Wiesmann, U., Rist, S., 2014. An indicator framework for assessing livelihood resilience in the context of social-ecological dynamics. Glob. Environ. Change-Hum. Policy Dimens. 28, 109e119.
77. Tri, N.H., Adger, N.W., Kelly, N.P., 1998. Natural resource management in mitigating climate impacts: the example of mangrove restoration in Vietnam. Glob. Environ. Change 8 (1), 49e61.
78. Valente, T.W., Foreman, R.K., 1998. Integration and radiality: measuring the extent of an individual's connectedness and reachability in a network. Soc. Netw. 20, 89e105.
79. van Hue, L.T., Scott, S., 2008. Coastal livelihood transitions: socio-economic consequences of changing mangrove forest management and land allocation in a commune of Central Vietnam. Geogr. Res. 46 (1), 62e73.
80. Vincent, K., 2007. Uncertainty in adaptive capacity and the importance of scale. Glob. Environ. Change 17, 12e24.
81. Walker, B.H., Anderies, J.M., Kinzig, A.P., Ryan, P., 2006. Exploring resilience in social-ecological systems through comparative studies and theory development: introduction to the special issue. Ecol. Soc. 11.

82. Whaley, L., Weatherhead, E.K., 2014. An integrated approach to analyzing (Adaptive) comanagement using the "Politicized" IAD framework. Ecol. Soc. 19 (1).

83. Wiesmann, U., 1998. Sustainable Regional Development in Rural Africa: Conceptual Framework and Case Studies from Kenya. In: African Studies Series, vol. A14. University of Berne, Switzerland.

CHAPTER 13 ECOLOGICAL ENGINEERING ON WATER BALANCE

Fei Tian, Yi He Lü, Bo Jie Fu, Yong Hui Yang, GuoYu Qiu, Chuanfu Zang, Lu Zhang

Keywords: SWAT, Water balance, Ecosystem restoration, Qilian Mountain

INTRODUCTION

As the originate district of freshwater resources, mountain accounting for 24% of the earth's land area, but providing more than 80% of terrestrial freshwater resources, while in arid and semi-arid areas, this ratio can even achieve 90% (FAO, 2000), thus mountain ecosystems play a significant role in the subsistence of the global ecosystem (Chase et al., 1999 and Huss et al., 2008). Hydrological processes are one of the most important elements in interaction between different ecosystems (Buttle et al., 2000 and Hanjra and Qureshi, 2010), mountain hydrology has been intensively studied (Huber et al., 2005 and Moran-Tejeda et al., 2014).

Several studies have been conducted to assess the impact of climate change on mountain hydrology with specific regard to runoff (Wang and Li, 2006), water yield (Huss et al., 2008 and Sun et al., 2006), and rarely evapotranspiration (Calanca et al., 2006). Hydrological cycle not only subject to weather conditions, but also closely associated with vegetation type (Wang et al., 2011a and Wang et al., 2011b), thus the relationship between vegetation and water budgets in catchments have been widely studied and became the subject of intense debate from hydrological as well as ecological point of view (Yang et al., 2005). For example, Yu et al. (2010) stated runoff reduction rate of 25.6 mm per 10% forest increase at watershed scale by employing 23 scenarios, that runoff at the site scale will be reduced to zero when grass- or shrub-lands are converted to forest in arid Qilian Mountain. Farley et al. (2005) and Sun et al. (2006) reported that this runoff reduction would be more severe in humid regions. To our knowledge, very few studies have addressed the effects of terrestrial ecosystem restoration on water balance in mountain areas, especially with a focus on soil moisture content. Due to the limitation of scale, and research methods, it was considered as unresolved central issues in the related fields (Mu et al., 2007).

Qilian Mountains is an important ecological functioning zone nourishing the oasis in northwest China (Yu et al., 2010). Due to the unsustainable use of the forest, the oasis ecosystem was seriously damaged and experienced deterioration of its ecological environment (Chen and Xia, 1999). To alleviate desertification and alleviate soil erosion, government has adopted ecological engineering such as "Grain for Green" program (GFG), "Natural Forest Protection Project", and "The Three-North Shelter Forest Program" since 2005 in this region.

In arid and semiarid region, some researchers indicated that increasing forest and grassland cover will reduce the total runoff volume and strengthen drought (Huber et al., 2008 and Sun et al., 2006), Qiu et al. (2011) showed that the ecological engineering have increased soil water retention capacity, a greater proportion of precipitation reaching the catchment is absorbed by the soil rather than flowing out of the region as runoff, water resources were more efficiently used and the ecological environment was protected. Water supply in the up mountain area limits ecological water use in the middle and lower reaches of the river (Jin et al., 2005 and Li et al., 2012), however, they have not determine whether ecosystem restoration reduce or increase regional runoff volume in the mountains of arid inland river basins, the effects of ecosystem restoration on water balance is unclear.

While answer this question will play a key role in sustainable development of water resources, and guide the recent massive forestation and ecological restoration campaigns. In this study, Landsat TM remote sensing images (30m - resolution)

Citation: Fei Tian, Yi He Lü, Bo Jie Fu, Yong Hui Yang, GuoYu Qiu, Chuanfu Zang, Lu Zhang, Ecological engineering on water balance, Journal of Hydrology: Regional Studies, Volume 5, March 2016, Pages 324-335, ISSN 2214-5818, http://dx.doi.org/10.1016/j.ejrh.2015.11.015.

obtained in 2000 and 2010 were used to monitor the vegetation changes and identify the magnitude of ecosystem restoration. In an effort to fully address the scarcity of this information, the Soil and Water Assessment Tool (SWAT) model (Arnold et al., 1998) was used to quantify the effects of terrestrial ecosystem restoration on water balance in Qilian Mountains.

MATERIALS AND METHODS

Study area

The up Qilian Mountain in the arid areas of northwest China, is the headwater of the Heihe river, covering an area of 9319 km^2. It lies between longitudes 97°23′–101°19′E, and latitudes 37°42′–39°41′N at elevations ranging from 1985 to 4723 m above sea level (Fig. 1). Subject to a temperate semi-arid and semi-humid continental monsoon climate, Fig. 2(a) vividly showed that 60–70% of the precipitation occurred in the rainy summer season from June to September. Its monthly mean temperature was about minus 11 °C in winter, and 12 °C in summer. Annual precipitation is relatively stable during past four decades (Fig. 2(b)). Temperature, on the other hand, has experience drastic increase. From the 1970s to the1980s, temperature was relatively stable, while during the following decade, an abrupt change occurred, and it increased by 64.66% (0.49 °C). From the 1990s to the 2000s, it risen 50.10% (0.63 °C), the warming trend was dramatically in the Qilian Mountain.

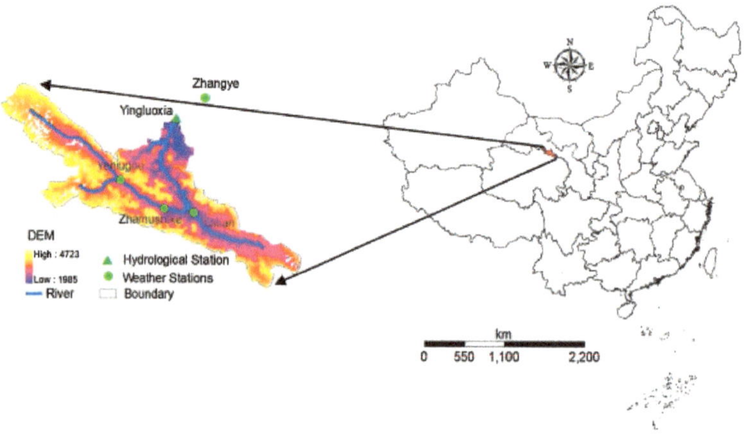

Figure 1: Location of the up Qilian Mountains with its hydro-meteorological stations.

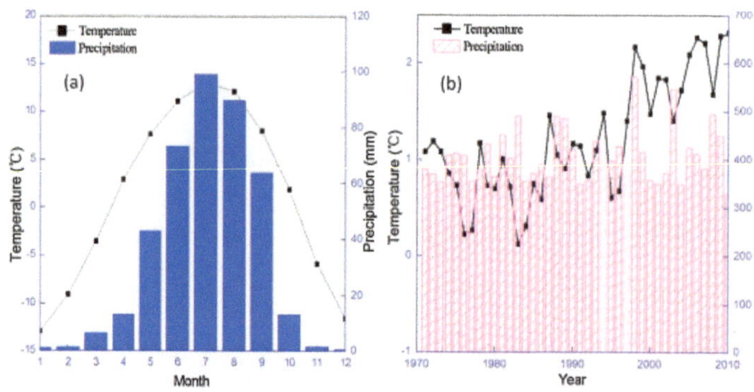

Figure 2: Climate diagrams of air temperature (a) and precipitation (b) at the up Qilian Mountain in monthly and annual scale during1970–2010.

To alleviate eco-environmental degraded, with a budget of 7.9 billion (over 1.2 billion), government has adopted ecological engineering since 2005 such as "Grain for Green" program (GFG), "Natural Forest Protection Project", and "The Three-North Shelter Forest Program". According to the objective of the project, after successful pending completion ecological engineering, vegetation coverage will be achieved to 30–50%, and forest coverage rate to 7% by 2020.

In order to test the effect of ecological engineering at regional scale, 30 m resolution Landsat TM remote sensing images in 2000 and 2010 were collected. Meanwhile the pre-processing of satellite images such as radiometric correction and geometric rectification should be carried out (Coppin and Bauer, 1994), the classification was based on the land cover and land use classification system developed by Liu (1996) for interpretation of remote sensor data at various scales and resolutions, with overall classification accuracies of 80%. In this typical region, that the main land use types are farmland, forest, and grassland—collectively accounting for about 90% of study area as depicted in Table 1. Farmland covered about 0.04% of the land area in the 2000s, as opposed to 0.19% in the 2010s. Forest expansion was most evidently from 584 km^2 to 1608 km^2, increased by 176% during past decades. The high coverage grassland (vegetation coverage >50%) expansion was not as obvious as forest, it also increased by 200 km^2 (5%). Compared above mentioned, the medium (20% < vegetation coverage < 50%) and low coverage grassland (5% < vegetation coverage < 20%) has experienced great decrease, which achieved to about 25%, and 21%, respectively. Furthermore, decline of glacier area was relatively drastic, from 457 km^2 to 38 km^2, declined by nearly 92%.

Table 1: The land type area variability of the up Qilian Mountain from 2000 to 2010

	2010	2000	Vegetation change (km^2)	Vegetation change (%)
Farmland	17.55	3.5	14.05	401.43
Forest	1607.7	583.49	1024.21	175.53
High coverage grassland	4453.34	4252.92	200.42	4.71
Medium coverage grassland	1962.83	2601.12	−638.29	−24.54
Low coverage grassland	1022.13	1291.72	−269.59	−20.87
Water	110.32	63.65	46.67	73.32
Ice coverage	37.74	456.83	−419.09	−91.74
Residential area	6.91	7.53	−0.62	−8.23
Bare land	100.67	58.39	42.28	72.41

Model set up

SWAT model is one of the most widely used distributed Hydrological Models for clarifying effects of land management practices on watershed. The SWAT-simulated hydrologic cycle is based on the following water balance equation (Arnold et al., 1998):

$$SW_t = SW_0 + \sum_{i=1}^{t}(R_{day} - Q_{surf} - E_a - W_{seep} - Q_{gw}) \quad (1)$$

where SW_t is the final soil water content (mm); SW_0 is the initial soil water content on day i (mm); t is the time (days); R_{day} is the amount of precipitation on day i (mm); $Q_{surf}$is the amount of surface runoff on day i (mm); E_a is the amount of evapotranspiration on day i (mm); W_{seep} is the amount of water entering the vadose zone from the soil profile on day i (mm); and Q_{gw} is the amount of return flow on day i (mm). Since SWAT model has been widely used and described in detail (Singh et al., 2005, Moran-Tejeda et al., 2014 and Rahman et al., 2015), additional details on the SWAT model have been documented by Neitsch et al. (2002), a further description of the model is not given here.

Data input and scenario design

In this study, the following sets of data were used: Landsat TM images with a spatial resolution of 30 m for 2000, and 2010. A DEM of 90m resolution was used to delineate the boundaries, river networks, and hydrologic response units (HRUs) (Li et al., 2009) in the study area. In this study, thresholds for defining HRUs in the SWAT model were set at 20% for land use and 10% for soils within a subbasin. Subbasins resulted from the overlay of land use and soil data were showed in Fig. 3. Soil type imaging (1:1,000,000) was obtained from the Heihe Watershed Allied Telemetry Experimental Research (HiWATER) (Li et al., 2013). Hydrological data included annual/monthly/daily data collected from the catchment's hydrological control station Yingluoxia station from 1978 to 2004 (Fig. 1), one of the runoff gauges situated at main outlets of Qilian Mountain, was used in the calibration and validation analysis of the SWAT model. Daily weather data from 1971 to 2010 were collected from four meteorological stations, which included precipitation, maximum and minimum temperatures, wind speed, hours of sunshine, and relative humidity. Glacier data is poor in this research, nevertheless, a number of adjustments were applied and documented in order to simulate the glacier processes in the mountainous region, glacier cover data was incorporated into SWAT model according to its contribution to the region (Pereira, 2011), and included into the snow-ice melt routing of SWAT according to Rahman et al. (2013).

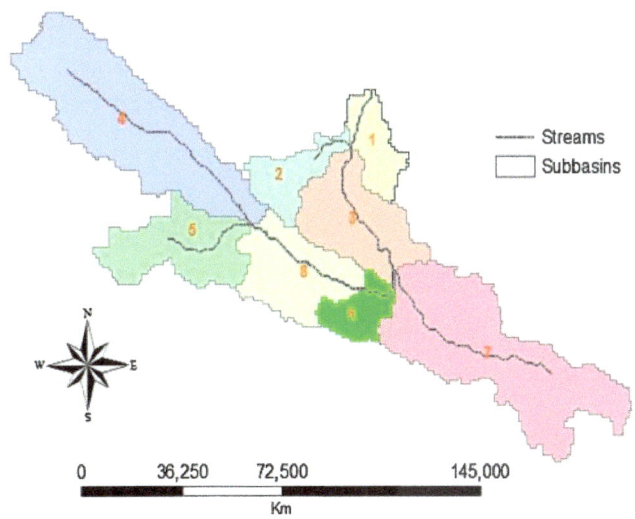

Figure 3: Subbasins divided via SWAT modeling.

Over the last decade, numerous ecological programs have been launched and the area of vegetation has continued to change in the arid regions of northwestern China. Thus, effects of ecological restoration on water balance were evaluated by fixed climate and change land use and land cover scenario into the model. This was done by replacing the TM imaging of 2000 with 2010 in the same climate scenario (1971–2010) with available data, running the SWAT model, and then simulating the changes of water yield, runoff, evapotranspiration, and soil moisture content from 1971 to 2010. Comparison of simulated values under different vegetation scenarios reveals the effects of vegetation restoration on water balance.

Calibration and validation

First, the model was calibrated to match with monthly runoff for the period 1979–1987. This period was chosen because runoff started measured since 1979, while in 1988 and 1989, no data traces, as no measurement conducted in the two years. Second, runoff for 1990 and 2004 were used for validation analysis. The first two years were used as warm-up period to mitigate the effect of unknown initial conditions, therefore were excluded from the analysis. The simulation period was from 1971 to 2010. 11 most sensitive parameters were the same as that identified in Zang et al. (2012). The performance of the model was evaluated by correlation coefficient (R^2), and Nash–Sutcliffe coefficients (Nash and Sutcliffe, 1970), and relative error (*RE*).

The Nash–Sutcliffe coefficients are calculated as:

$$E = 1 - \frac{\sum_{i=1}^{n}(Q_m - Q_o)^2}{\sum_{i=1}^{n}(Q_m - Q_{avg})^2} \quad (2)$$

where E is the Nash–Sutcliffe determinacy coefficient; Q_m is simulated runoff (m^3/s); Q_o is observed runoff (m^3/s); Q_{avg} is the average observed runoff (m^3/s); n is number of observed values; and i is the index of the observed values. E is used to estimate the agreement between the observed and simulated values. The range for E is from negative infinity to 1; and the nearer E is to 1, the better the result.

Relative error was calculated as:

$$RE = \left[\frac{Q_o - Q_m}{Q_o}\right] \times 100\% \quad (3)$$

where RE is the relative error, Q_m is simulated runoff (m^3/s); Q_o is observed runoff (m^3/s). The smaller the RE, the better the result.

RESULTS

Results of model calibration and validation

The calibration and validation with SWAT model at the monthly scale was satisfactory, as indicated by high values of R^2 and most of data points were closely distributed around the 1:1 line (Fig. 4). At monthly time steps, the model performance during the validation period (Fig. 4(b)) was better than calibration period (Fig. 4(a)), with high values of R^2($R^2 = 0.96$) and E ($E = 0.93$), and low values of relative error (3.51%). R^2 and Nash–Sutcliffe coefficients for calibration period reach 0.93, and 0.88, respectively. And the average value of relative error was only 5.68%, indicating that the SWAT model generally exhibits a good performance in the study area.

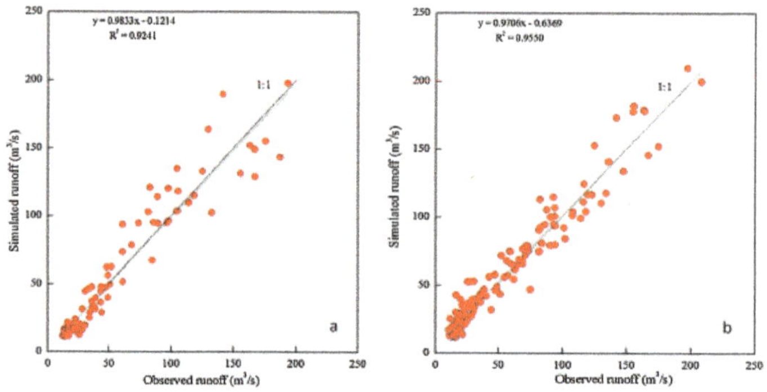

Figure 4: Observed and simulated monthly runoff at Yingluoxia station for the calibration (from 1979 to 1987, a) and validation period (from 1990 to 2004, b).

At yearly scale, the model performance ($R^2 = 0.88$) (Fig. 5(b)) was not as good as that in monthly steps ($R^2 = 0.95$) (Fig. 5(a)), however, the evaluation result was acceptable. More detailed information related E, RE, and R^2 at yearly scale were shown in Table 2. The good agreement between the observed and simulated results demonstrated that the SWAT model was acceptable and could be used to simulate the ecohydrological processes at up Qilian Mountain.

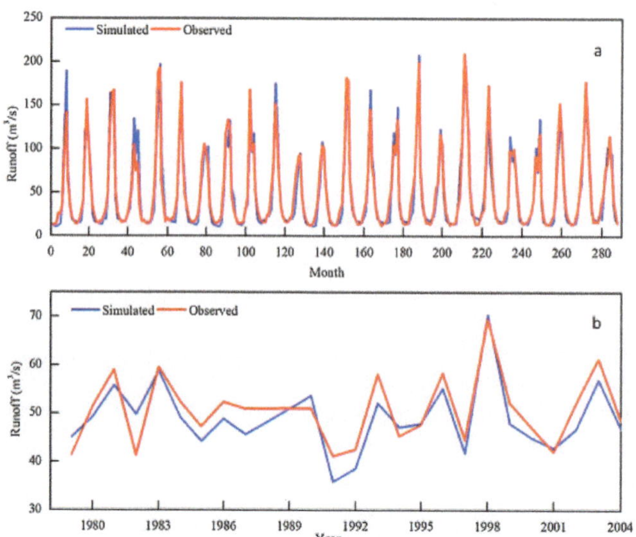

Figure 5: Observed and simulated monthly (a) and yearly runoff (b) at Yingluoxia station for the period of 1979–2004 (excluded 1988 and 1989).

Table 2: The Nash–Sutcliffe coefficients (E), relative error (RE), and correlation coefficient (R^2) during the calibration and validation periods at yearly scale

	Year	E	RE	R^2
Calibrationperiod	1979	0.79	−9.02	0.93
	1980	0.85	4.05	
	1981	0.82	5.37	
	1982	0.60	−20.49	
	1983	0.93	1.41	
	1984	0.81	6.09	
	1985	0.80	6.30	
	1986	0.79	6.57	
	1987	0.75	10.28	
	Average	0.80	1.17	
Validationperiod	1990	0.82	−5.05	0.95
	1991	0.73	12.70	
	1992	0.80	9.15	
	1993	0.76	10.18	
	1994	0.85	−4.17	
	1995	0.98	−0.55	
	1996	0.82	5.33	
	1997	0.81	6.11	
	1998	0.93	−1.42	
	1999	0.80	0.08	
	2000	0.85	4.47	
	2001	0.93	−1.82	
	2001	0.77	10.24	
	2003	0.82	6.96	
	2004	0.87	3.45	
	Average	0.85	4.23	

Water balance response to ecological engineering

Fig. 6 showed the comparisons of the runoff-precipitation correlations for the natural and ecosystem restoration periods. The natural period is defined as the period from 1971 to 2010 with TM imaging of 2000, while the ecosystem restoration period is defined as the period from 1971 to 2010 with TM imaging of 2010. Although a decrease in precipitation was accompanied by a decrease in runoff, the regression line for the natural period lied above that for the ecosystem restoration period, this implies that under similar precipitation, runoff under the ecosystem restoration period is much weaker. It is reasonable to state that ecological engineering, rather than climate or precipitation change, could be the major driving factor of runoff decline in the basin.

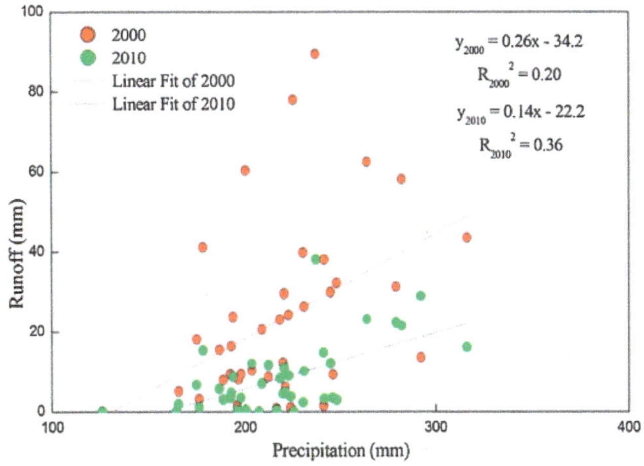

Figure 6: Runoff and precipitation correlations comparison for the natural and ecological restoration periods at the up Qilian Mountain. Also plotted on the figures is runoff decline under similar precipitation for the ecological restoration period.

We mapped the water balance impact of ecological engineering across the up Qilian Mountain to show its sensitivity to terrestrial ecosystem restoration (Fig. 7). Temporal hydrological parameters change such as runoff change was computed as the runoff amount after ecosystem restoration period (2010 land use scenario) minus that under the baseline condition (i.e. before ecosystem restoration period, 2000 land use scenario). As forest expansion evidently 176%, and the medium and low coverage grassland decreased 25%, and 21%, respectively, from 2000 to 2010, water yield increased by 32% (Fig. 7(a)), and soil water content experienced an increment of 46% (Fig. 7(d)). Instead on the other hand, severe runoff reduction of 48% occurred (Fig. 7(b)) over the same period, and evapotranspiration exhibited similar decreasing trend with runoff, with a magnitude of 4% (Fig. 7(c)). Therefore, conversion of grass or shrub to forest eventually led to a variety of environmental benefits, such as intensifying soil water storage, it also could cause a complete loss of runoff at the regional scale.

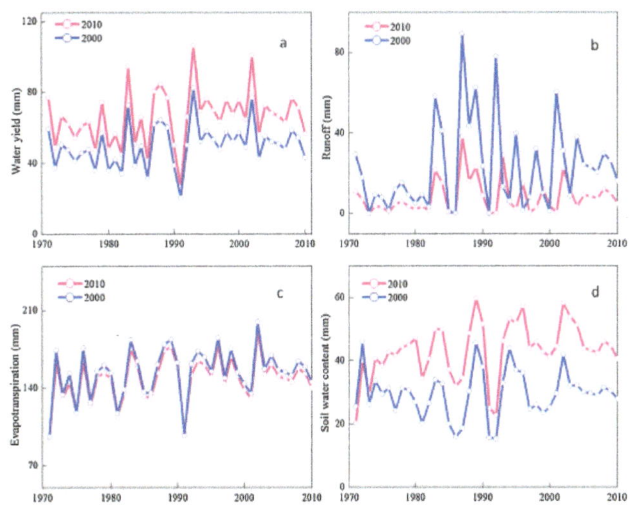

Figure 7: Comparison of annual water yield (a), runoff (b), evapotranspiration (c), and soil water content (d) before and after ecological engineering at the up Qilian Mountain.

Water balance comparison between different vegetation types

All the hydrological parameters such as runoff, and evapotranspiration were computed for individual land cover types in SWAT model (Cristea et al., 2014). And the landscape patterns impact water yield, runoff, evapotranspiration, and soil water content at the up Qilian Mountain under ecological engineering scenario were presented in Fig. 8. Temporal hydrological parameters change such as runoff change was defined as the difference between runoff distribution for grassland and forest lands. The water yield changing trend was almost the same for different vegetation types. The annual water yield from forest was 70.7 mm on average, and then followed by low and high coverage grassland, with values of 66.9 mm, and 56.1 mm, respectively. Medium coverage grassland has the lowest values of 27.4 mm, about 61.2% lower than forest. The evapotranspiration change case noted inversely with water yield. Medium coverage grassland presented the highest evapotranspiration with 175. 6 mm, that was 24.9% higher than forest. Generally, the evapotranspiration of high coverage grassland is much greater than that of low coverage grassland, which exceeds that of forest. In contrast to the water yield and evapotranspiration, the annual runoff from high and medium coverage grassland varied from 49.8 mm to 20.6 mm. The forest and low coverage grassland absorbed part of the lateral water flow as well as generating no runoff from precipitation. This is clearly demonstrated by the changing trend of soil water content. Generally, the soil water content was relatively stable, except the wet years. Forest has the highest values of soil water content (43.1 mm), which was 65.3%, and 31.5% higher than high and medium coverage grassland, respectively. This indicated that the conversion of medium and low coverage grassland to forest could cause a complete loss of runoff at the regional scale. A greater proportion of precipitation reaching the region is absorbed by the soil and used for forest growth, rather than flowing out of the region as runoff, water resources were more efficiently used and the ecological environment was protected.

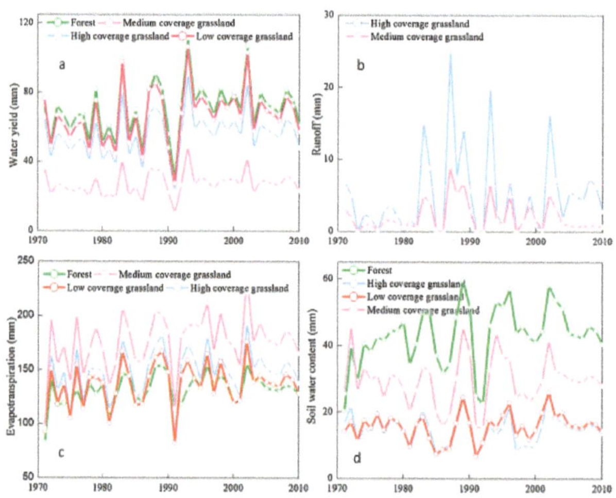

Figure 8: Comparison of annual water yield (a), runoff (b), evapotranspiration (c), and soil water content (d) between different vegetation types under ecological engineering scenario at the up Qilian Mountain. (FRSE: Forest; PAST: High coverage grassland; SWRN: Low coverage grassland; WPAS: medium coverage grassland).

DISCUSSION

Relationship between vegetation cover and water balance

Sun et al. (2006) indicated that the average water yield reduction vary from 50% in the semi-arid region in northern China to 30% in the tropical southern region. Feng et al. (2012) also detected 38% of the Loess Plateau experienced water yield decrease (1–48 mm per year) as a result of afforestation. However, our results indicated that water yield in the Qilian Mountain in northwest China increased by 32% during post-vegetation restoration period (Fig. 7(a)), this may comes from Qilian Mountain is a complex ecosystem consisting of grassland, forest, and glaciers. The relationship between vegetation cover and water yield are relatively complex (Wang et al., 2011a and Wang et al., 2011b). According to Cristea et al. (2014), higher air temperature declined glacier area and produced less snowmelt water under warmer temperatures than under historic temperatures, when trees are not present then accumulation is higher due to the lack of canopy interception. Therefore we argue that ecological engineering campaigns may have obvious local effects on water yield reduction, however, regional effects are of less concern since it vary greatly as determined by eco-environmental conditions (i.e. soil depth, climate, and plant physiological characteristics, etc.) and socioeconomics (Sun et al., 2005), relationship between water yield and vegetation cover can be inconsistent.

Farley et al. (2005) reported that global averaged runoff reduction in humid regions was 44%, while in semi-humid and semi-arid regions of China this reduction was about 50% (Sun et al., 2006). Our modeled result indicated that ecological engineering at the up Qilian Mountain may cause 48% runoff decrease (Fig. 7(b)), when grassland was converted to forest in this arid region. This is in accordance with Yu et al. (2010), who indicated that runoff at the site scale will be reduced to zero when grasslands or shrublands are converted to forest. At the mountainous region runoff could not reduce to zero, as only part of the area in this mountainous region can support forest. Observations at Qilian Mountain also found that forest vegetation might reduce surface runoff, minimizing flooding, and postpone runoff period (Wang & Che, 1998). However, such a runoff reduction my cause severely consequence in semi-arid region that already suffering from serious water crisis. As expected, small change in runoff amount resulted in a large relative change in the dry portion due to their low background (Feng et al., 2012 and Bi et al., 2013). It is thus important to policy makers in arid region, especially in the semi-arid mountains which is the headwaters of inland rivers.

Soil water content experienced an increment of 46%, therefore conversion of grass or shrub to forest eventually led to a variety of environmental benefits, such as intensifying soil water storage. For Qilian Mountain it is important for maintaining the "soil water pool" function. With limited precipitation and groundwater, soil moisture becomes the main water source and the limiting factor of plant growth (Qiu et al., 2011), great differences lie in the low and high flow period (Zang and Liu, 2015), it directly affected the vegetation distribution and community structure, and sometimes restricting the ecological engineering construction. Ecological engineering programs such as "Natural Forest Protection Project" and "The Three-North Shelter Forest Programme" are the direct ecological effects of vegetation change better and prevention of soil erosion. Although water conservation function of ecological engineering programs has been the community's consensus, soil water content increment was seldom clearly quantified before. This Soil water change represents a small fraction of available freshwater, however, it affects evapotranspiration which is the central process in the climate system and a nexus of water, energy, and carbon cycles, it also influence climate and weather (Dobriyal et al., 2012). Therefore soil moisture and plants, and there coupling are at the heart of ecohydrology and the soil water balance, this suggests that the interaction between vegetation development and soil moisture needs to be studied extensively. However, very few studies have addressed this issue in mountain areas, although soil moisture estimation can help identify win–win association between ecosystem services and establish payment for watershed services. In this research, evapotranspiration decreased 4% after ecological engineering, while due to the lack of evapotranspiration data hampers the validation of the model. Evapotranspiration has a large impact on water balance, thus the precise estimation of ET is still a priority for hydrological research (Wang et al., 2011a and Wang et al., 2011b).

Policy problems and future prospects

China has been suffering from some of the world's worst land degradation since 1960s (Zhou and Tan, 2011), and the degraded area increased by 15% each decades (Wang and Han, 2000), government has responded by developing laws, policies, and national programs to arrest the rate of land degradation, where possible, reverse it. Ecological engineering such as "Grain for Green" was the first program launched in (Long, 2014). It is a milestone in Chinese forest policy. Although there is a growing body of studies documenting the significance of restoration (Qiu et al., 2011), over planting, on the contrary, in water-limited region leaded to many policy failures, thus "small age trees" with heights of 3–5 m appeared widely in Loess Plateau-the largest highly erodible area on earth, where the pilot phased of GFG were implemented (Chen et al., 2008), and large-scale forest restoration in arid areas will undoubtedly resulted in water consumption, reducing the availability of runoff water, and aggravated water scarcity.

Different hydrological effects are caused by different landscape patterns. Based on the quantitative understanding of ecological restoration effects on water balance, a trade-off between the numerous benefits of vegetation coverage increase and its negative impact on water balance should be carefully addressed in semi-arid region with inherently severe water shortage when considering ecological engineering.

Assessing the long term effects of vegetation based campaigns on water balance is important for the success of forestation projects, therefore such mountain hydrology studies are essential to guide the recent ecological restoration campaigns. This may provide implications for eco-environmental restoration program in China, and also be applicable to other arid and semi-arid region of the world.

Uncertainties and limitations of this study

This research represents a great effort to fully examine the hydrological balances across a typical mountainous region by employing a modelling approach. However, there are still some uncertainties and limitations. Firstly, uncertainties consistently come from the compiled datasets and throughout the analyses. For example, the Penman–Monteith equation, which is used to estimate evapotranspiration, requires significant data including, but not limited to, solar radiation, wind speed, and canopy cover characteristics. Considerable uncertainty exists in weather generation. Secondly, the limited and uneven distribution of meteorological stations may influence the accuracy of simulation. Only one of the runoff gauges situated at main outlets of Qilian Mountain and four meteorological stations were used in this case study, this will influence the simulation accuracy. The evaluation result was acceptable, however, at yearly scale, the model performance ($R^2 = 0.88$) was not as good as other swat model simulation (Li et al., 2009 and Qiu et al., 2011). Thirdly, the lack of soil water content and evapotranspiration data hampers the validation of the model. Last but not least, human activities such as hydropower station and dam construction may influence the water cycling, however, they were not addressed quantitatively. According to Rahman et al. (2013), uncertainties related to future energy demand needs to address as the decision on energy consumption is linked with type of energy generation sources (i.e. hydropower) and their dependency based upon the future demand. Therefore the model used in this study still has room for further improvement.

CONCLUSION

This research quantitatively evaluated the effects of ecological engineering on the water balance via distributed hydrological model—SWAT model. The SWAT model showed good performance in up Qilian Mountain with high R^2 values over 0.90 on monthly scales, and low relative error values of 3.51%, and 5.68%, respectively, for both calibration and validation periods (Fig. 5). Simulation result indicated that as forest expansion evidently 176%, and the medium and low coverage grassland decreased 25%, and 21%, respectively, water yield and soil water content increased by 32% (Fig. 7(a)), and 46% (Fig. 7(d)), severe runoff reduction of 48% occurred (Fig. 7(b)). The landscape patterns impact water balance indicated that the annual water yield from forest was the highest (70.7 mm on average), then followed by low (66.9 mm), high (56.1 mm), and medium (27.4 mm) coverage grassland (Fig. 8). However, the forest and low coverage grassland absorbed part of the lateral water flow as well as generating no runoff from precipitation. This is clearly demonstrated by the changing trend of soil water content. Forest has the highest values of soil water content (43.1 mm), which was 65.3%, and 31.5% higher than high and medium coverage grassland, respectively. This indicated that the conversion of medium and low coverage grassland to forest could cause a complete loss of runoff at the regional scale. A greater proportion of precipitation reaching the region is absorbed by the soil and used for forest growth, rather than flowing out of the region as runoff, water resources were more efficiently used and the ecological environment was protected.

ACKNOWLEDGEMENTS

We acknowledge, with gratitude, the financial support from the 12th five year science and technology development program (2012BAC08B01), the international postdoctoral exchange fellowship program (20130002), the China Postdoctoral Science Foundation(2012M520403), and the Key Project of Chinese National Programs for Fundamental Research and Development (2013CB429905-01).

REFERENCES

1. Arnold, J.G., Srinivasan, R., Muttra, R.S., Williams, J.R., 1998. Large area hydrologic modelling and assessment part I: Model development. J. Am. Water Resour. Assoc. 34, 73–89.
2. Bi, L.M., Hou, H.Y., Yang, L.F., 2013. Analysis on change trend of runoff in yingluoxia station of Heihe River. Yellow River 35, 23–25.
3. Buttle, J.M., Creed, I.F., Pomeroy, J.W., 2000. Advances in Canadian forest hydrology, 1995–1998. Hydrol. Processes 14 (9), 1551–1578.
4. Calanca, P., Roesch, A., Jasper, K., Wild, M., 2006. Global warming and the summertime evapotranspiration regime of the Alpine Region. Clim. Change 79, 65–78.
5. Chase, T.N., Pielke, R.A., Kittel, T.G.F., Nemani, R.R., Running, S.W., 1999. Simulated impacts of historical land cover changes on global climate in northern winter. Clim. Dyn. 16, 93–105.
6. Chen, J.Q., Xia, J., 1999. Facing the challenge: barriers to sustainable water resources development in China. Hydrol. Sci. J. 44, 507–516.
7. Chen, H.S., Shao, M.A., Li, Y.Y., 2008. Soil desiccation in the Loess Plateau of China. Geoderma 143, 91–100.
8. Coppin, P.R., Bauer, M.E., 1994. Processing of multitemporal Landsat TM imagery to optimize extraction of forest cover change features. IEEE Trans. Geosci. Remote Sens. 32 (4), 918–927.
9. Cristea, N.C., Lundquist, J.D., Loheide, S.P., Lowry, C.S., Moore, C.E., 2014. Modelling how vegetation cover affects climate change impacts on streamflow timing and magnitude in the snowmelt-dominated upper Tuolumne Basin, Sierra Nevada. Hydrol. Processes 28, 3896–3918.
10. Dobriyal, P., Qureshi, A., Badola, R., Hussain, S.A., 2012. A review of the methods available for estimating soil moisture and its implications for water resource management. J. Hydrol. 458–459, 110–117.
11. FAO. International Year of Mountains: Concept paper[C]. Rome, 2000.
12. Farley, K.A., Jobbagy, E.G., Jackson, R.B., 2005. Effects of afforestation on water yield: a global synthesis with implications for policy. Global Change Biol. 11, 1565–1576.
13. Feng, X.M., Sun, G., Fu, B.J., Su, C.H., Liu, Y., Lamparski, H., 2012. Regional effects of vegetation restoration on water yield across the Loess Plateau, China. Hydrol. Earth Syst. Sci. 16, 2617–2628.
14. Hanjra, M.A., Qureshi, M.E., 2010. Global water crisis and future food security in an era of climate change. Food Policy 35, 365–377.
15. Huber, U.M., Bugmann, H.K.M., Reasoner, M.A., 2005. Global Change and Mountain Regions [M]. Springer, Netherlands.
16. Huber, A., Iroume, A., Bathurst, J., 2008. Effect of Pinus radiata plantations on water balance in Chile. Hydrol. Processes 22, 142–148.
17. Huss, M.D., Farinotti, A.B., Funk, M., 2008. Modeling runoff from highly glacierized alpine drainage basins in a changing climate. Hydrol. Processes 22, 3888–3902.
18. Jin, F., Zhang, Z.M., Yu, X.X., Rao, L.Y., Niu, J.Z., Lu, S.W., Xie, Y.Y., 2005. Value evaluation of forest ecosystem services of Qilian mountain in Gansu province. Sci. Soil Water Conserv. 3, 53–57.
19. Li, X., Li, X.W., Li, Z.Y., 2012. Progresses on the watershed allied telemetry experimental research (WATER). Remote Sens. Technol. App. 27, 637–649.
20. Li, X., Cheng, G.D., Liu, S.M., Ma, M.G., Jin, R., et al., 2013. Heihe Watershed Allied Telemetry Experimental Research (HiWATER): Scientific Objectives and Experimental Design. Ame. Meteorol. Soc. 94, 1145–1160.
21. Liu, J.Y., 1996. Study on National Resources & Environment Survey and Dynamic Monitoring Using Remote Sensing. China Science and Technology Press, Beijing.
22. Li, Z.L., Xu, Z.X., Shao, Q.X., 2009. Parameter estimation and uncertainty analysis of SWAT model in upper reaches of the Heihe river basin. Hydrol. Processes 23, 2744–2753.
23. Long, H.L., 2014. Land use policy in China: introduction. Land Use Policy 40, 1–5.
24. Moran-Tejeda, E., Zabalza, J., Rahman, K., Gago-Silva, A., Lopez-Moreno, J.I., Vicente-Serrano, S., Lehmann, A., Tague, C.L., Beniston, M., 2014. Hydrological impacts of climate and land-use changes in a mountain watershed: uncertainty estimation based on model comparison. Ecohydrology, http://dx.doi.org/10.1002/eco.1590.
25. Mu, Q., Heinsch, F.A., Zhao, M., Running, S.W., 2007. Development of a global evapotranspiration algorithm

based on MODIS and global meteorology data. Remote Sens. Environ. 111, 519–536.

26. Nash, J.E., Sutcliffe, J.V., 1970. Riverflow forecasting through conceptual model. J. Hydrol. 10, 282–290.
27. Neitsch, S.L., Arnold, J.R.K., Srinivasan, J.R., Williams, J.R., 2002. Soil and Water Assessment Tool User's Manual: Version 2000. TR-192. Texas Water Resources Inst., College Station.
28. Pereira, G., 2011. Water management in the Mendoza River Watershed. In: Interview (Ed.). Departamento de hidrologia del DGI, Mendoza.
29. Qiu, G.Y., Yin, J., Tian, F., Geng, S., 2011. Effects of the conversion of cropland to forest and grassland program on the water budget of the Jinghe River Catchment in China. J. Environ. Qual. 40, 1745–1755.
30. Rahman, K., Maringanti, C., Beniston, M., Widmer, F., Abbaspour, K., Lehmann, A., 2013. Streamflow modeling in a highly managed mountainous glacier watershed using SWAT: the upper Rhone River Watershed case in Switzerland. Water Resour. Manag. 27, 323–339.
31. Rahman, K., Silva, A.G.D., Tejeda, E.M., Gobiet, A., Beniston, M., Lehmann, A., 2015. An independent and combined effect analysis of land use and climate change in the Upper Rhone River Watershed, Switzerland. Appl. Geogr. 63, 264–272.
32. Singh, J., Knapp, H.V., Arnold, J.G., Demissie, M., 2005. Hydrological modelling of the Iroquois River watershed using HSPF and SWAT. J. Am. Water Resourc. Assoc. 14, 343–360.
33. Sun, G., Mcnulty, S.G., Lu, J., Amatya, D.M., Liang, Y., Kolka, R.K., 2005. Regional annual water yield from forest lands and its response to potential deforestation across the southeastern United States. J. Hydrol. 308, 258–268.
34. Sun, G., Zhou, G.Y., Zhang, Z.Q., Wei, X.H., Mcnulty, S.G., Vose, J.M., 2006. Potential water yield reduction due to forestation across China. J. Hydrol. 328, 548–558.
35. Wang, G.X., Deng, W., Yang, Y., Cheng, G.W., 2011a. The advances, priority and developing trend of alpine ecology. J. Mt. Sci. 29, 129–140.
36. Wang, J., Li, S., 2006. Effect of climatic change on snowmelt runoffs in mountainous regions of inland rivers in Northwestern China. Sci. China: Ser. D Earth Sci. 49, 881–888.
37. Wang, K., Han, J.G., 2000. The restoration and reconstruction for degraded grassland in the west of China. Sci. Silvae Sin. 36, 7–8, In Chinese.
38. Wang, S., Fu, B.J., He, C.S., Sun, G., Gao, G.Y., 2011b. A comparative analysis of forest cover and catchment water yield relationships in northern China. For. Ecol. Manag. 262, 1189–1198.
39. Yang, J.G., Xiao, D.N., Zhou, L.H., Tang, C.W., 2005. Hydrological effects of forest landscape patterns in the Qilian Mountains. Mt. Res. Dev. 25 (3), 262–268.
40. Yu, P.T., Wang, Y.H., Wu, X.D., Dong, X.H., Xiong, W., Bu, G.W., Wang, S.L., Wang, J.Y., Liu, X.D., Xi, L.H., 2010. Water yield reduction due to forestation in arid mountainous regions, northwest China. Int. J. Sediment Res. 25, 423–430.
41. Zang, C.F., Liu, J.G., Velde, M.V.D., Kraxner, F., 2012. Assessment of spatial and temporal patterns of green and blue water flows under natural conditions in inland river basins in Northwest China. Hydrol. Earth Syst. Sci. 16, 2859–2870.
42. Zang, C.F., Liu, J.G., 2015. l and temporal pattern difference of blue-green water flows in typical reference years in the Heihe River Basin, northwestern China. J. Beijing For. Univ. 35 (3), 1–10.
43. Zhou, K., Tan, B., 2011. Capacity of legislation and policy to control land degradation. In: Du, Q., Hannam, I. (Eds.), Law, Policy and Dryland Ecosystems in the People'sRepublic of China. IUCN, Environmental Policy and Law. IUCN, Gland, pp. 73–83.

CHAPTER 14

EXPLORING EFFECTS OF HYPOXIA ON FISH AND FISHERIES

Kim de Mutsert, Jeroen Steenbeek, Kristy Lewis, Joe Buszowski, James H. Cowan Jr., Villy Christensen

Keywords: Hypoxia, Gulf of Mexico, Ecosystem model, Ecopath with Ecosim, Ecospace, Fisheries

INTRODUCTION

Nutrient rich waters flowing from the Mississippi River into the Gulf of Mexico result in high primary productivity in this coastal area (Turner et al., 2006). Bacterial decomposition of this organic matter in combination with summer stratification has led to the occurrence of an extensive area of low bottom oxygen since at least the early 1970s (Rabalais and Turner, 2006). While often referred to as the 'dead zone', the effect on living marine resources of this annually reoccurring area of hypoxic bottom waters off the coast of Louisiana is not necessarily lethal.

Hypoxia refers to oxygen levels of 2 mg/l or lower, which can lead to decreased feeding and growth rates, changes in activity level, avoidance behavior, and death in fish and shellfish (Bell and Eggleston, 2005, Robert et al., 2011 and Goodman and Campbell, 2007). The exact level of dissolved oxygen that results in effects on physiology or behavior is species-specific, which can results in community structure shifts and changes in species interactions (Essington and Paulsen, 2010). Indirect effects occur through predator–prey relationships; fish could be affected not by hypoxia, but by the response of their prey or predators to hypoxia, and the effects could be either positive or negative (Altieri, 2008, Pierson et al., 2009 and Eby et al., 2005). Effects on fisheries may be even more complicated, as catch per unit effort (CPUE) could decrease when the abundance of target species is reduced by hypoxia, or could increase due to aggregation of target species at the edge of the hypoxic zone, which may enhance their susceptibility to be caught (Craig, 2012).

A significantly obscuring mechanism is the fact that the same nutrient enriched waters that are the main cause of bottom hypoxia (Rabalais and Turner, 2001), are responsible for the high primary and secondary production in this region (Gunter, 1963, Nixon and Buckley, 2002 and Chesney et al., 2000). It is likely due to these complications, that holistic effects of hypoxia on the fisheries ecosystem of the northern Gulf of Mexico have remained elusive (Rose, 2000 and Rose et al., 2009).

The purpose of this study is to analyze effects of hypoxia on fish and fisheries through ecosystem model simulations, and to provide a tool that can be used in management scenario analyses pertaining to Mississippi River nutrient load reductions and coastal fisheries management. To this purpose an Ecospace model was developed using Ecopath with Ecosim (EwE) software that was enabled to receive spatio-temporal primary productivity and dissolved oxygen output from a coupled physical–biological hypoxia model developed by Fennel et al. (2011). Since a reduction in hypoxia would entail a reduction in nutrients that enter the Gulf of Mexico, it is important to incorporate the effects of nutrient enrichment on phytoplankton (and changes therein) in an ecosystem model that studies effects of hypoxia and scenarios that may reduce this hypoxia. Output of the Fennel et al. (2011) model of dissolved oxygen (DO) as well as Chl *a* was used as forcing functions in the Ecospace model to account for both effects. Similar approaches to incorporate effects of biogeochemistry on foodweb models, often referred to as End-to-End modeling, have been used in other studies (see e.g. Libralato and Solidoro, 2009).

The ecosystem model developed for this study takes a holistic approach by simulating species interactions, while accounting for changes in biomass as well as spatial distribution

Citation: Kim de Mutsert, Jeroen Steenbeek, Kristy Lewis, Joe Buszowski, James H. Cowan Jr., Villy Christensen, Exploring effects of hypoxia on fish and fisheries, Ecological Modelling, Volume 331, 10 July 2016, Pages 142-150, ISSN 0304-3800, http://dx.doi.org/10.1016/j.ecolmodel.2015.10.013.

changes, and by explicitly simulating fisheries with dynamic fleets. The model allows for simulations of all direct and indirect effects on fish and fisheries, in an environment where hypoxia and primary productivity fueling can be evaluated together and separately. While this ecosystem model contains sixty groups to provide a representative simulation of the ecosystem, the main focus of this paper is on a select group of species that are of economic or ecological significance. These species are Gulf menhaden (*Brevoortia patronus*), which is largest fishery in Louisiana by weight; brown, white and pink shrimp (*Farfantepenaeus aztecus*, *Litopenaeus setiferus*, and *Farfantepenaeus duorarum*), together comprising the largest fishery by value; red snapper (*Lutjanus campechanus*), a popular sportfish; Atlantic Croaker (*Micropogonias undulatus*), the most dominant forage fish in the model area; and jellyfish, a group of organisms of interest because of previous documented responses to hypoxia in other areas.

METHODS

Data preparation

Fisheries independent survey data from the SEAMAP program of the Gulf States Marine Fisheries Commission (seamap.gsmfc.org) was used to determine which species were representative of the area, and to determine the biomass of each species present in the model area. Initial biomass in the base model was based on the average biomass of each group (species or functional group) from 2005 to 2008. Fishing was represented by including shrimp trawls, recreational fishing, snapper/grouper fishery, crab pots, menhaden fishery, squid fishery, and longlines as 'fleets' in the model. Annual landings of model groups by these fleets were based on NOAA Fisheries Annual Commercial Landings Statistics (st.nmfs.noaa.gov), and trip ticket data from the Louisiana Department of Wildlife and Fisheries. These data were used to develop the Ecopath model.

Landings data from 1950 to 2010, and SEAMAP data collected in the model area from 1982 to 2010 were used to calculate annual landings and biomass (t/km^2) respectively for each group in the model for which these data were available. In addition, an oxygen forcing function was developed from data collected during Lumcon cruises from 1998 to 2007 (D. Obenauer, personal communication), and a nutrient forcing function from NO*x*data collected in the Mississippi River by USGS from 1950 to 2010 (toxics.usgs.gov) to simulate nitrogen load into the coastal area from the Mississippi River. These time series and forcing functions were used for model calibration in Ecosim.

In EwE, a nutrient forcing function serves as a multiplier on primary production. In order for groups to respond to the level of dissolved oxygen, empirically derived sigmoidal oxygen response curves were developed. These curves were developed by determining catch rates at each level of dissolved oxygen, using all SEAMAP data where dissolved oxygen was measured during collections. The tolerance curves were then used as a multiplier on effective search rate in Ecosim (and Ecospace, using a plug-in described in Section 2.5) as described in Christensen et al. (2008) and de Mutsert et al. (2012), to affect biomass of each specific group (Fig. 1).

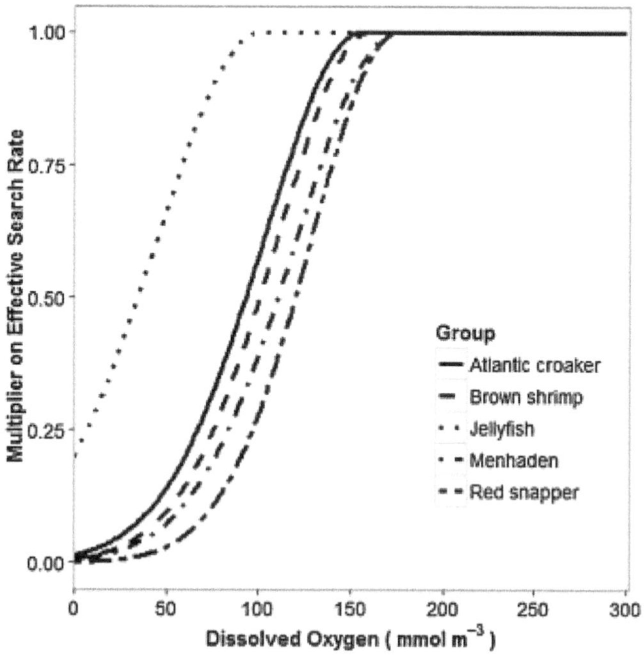

Figure 1: Oxygen response curves of selected species.

Model preparation

The EwE modeling suite was used to build the model (www.ecopath.org). The virtual representation of the ecosystem was developed in Ecopath, the static model of the EwE modeling suite. Groups in the model represent single species as well as species aggregated in functional groups. Where deemed necessary to represent ontogenetic diet changes or size-selective fisheries, species were split into multiple life stages. For those species, the initial biomass of only one life stage was derived from empirical data, and the biomass of other stages were determined using a *von Bertalanffy* growth model. Some functional groups were represented with multiple life stages as well. This resulted in 60 groups (Table 1). Parameters included for each group to develop a mass-balanced Ecopath model in addition to biomass (*B*), were the *P*/*B* (production/biomass) ratio, *Q*/*B*(consumption/biomass) ratio, and the total fisheries catch rate (*Y*) for the groups that are fished. Parameters were derived from other Gulf of Mexico food web models (Walters et al., 2008 and de Mutsert et al., 2012) or fishbase (fishbase.org).

Table 1: Initial conditions of mass-balanced Ecopath model. *B* = biomass, *Z* = total mortality, *P*/*B* = production to biomass ratio, *Q*/*B* = consumption to biomass ratio, EE = ecotrophic efficiency

Nr.	Group name	*B* (t km^{-2})	*Z* (yr^{-1})	*P*/*B*	*Q*/*B*	EE
1	marine mammals	0.069		0.02	11.97	0.623
2	tunas	0.024		0.9	13.00	0.811
3	jacks	0.018		0.8	3.30	0.693
4	birds	0.011		0.25	35.00	0.722
5	juv Atlantic cutlassfish	0.003	2		8.48	0.011
6	Atlantic cutlassfish	0.228	0.41		2.05	0.745
7	lizardfish	0.384		0.6	5.00	0.806
8	juv coastal sharks	1.2E−04	2		5.52	0.625
9	coastal sharks	0.148	0.08		1.00	0.646
10	mackerel	0.300		0.7	2.00	0.591
11	0–3 seatrout	2.5E−04	6		23.96	0.056
12	3–18 seatrout	0.072	1.4		4.11	0.279
13	18+ seatrout	0.647	0.7		1.60	0.478
14	0–6 red snapper	0.001	3		9.20	0.065
15	6–24 red snapper	0.032	2		2.91	0.659
16	24+ red snapper	0.090	0.6		1.20	0.222
17	0–1 groupers	0.008	2		5.13	0.011
18	1–3 groupers	0.090	0.6		2.07	0.027
19	3+ groupers	0.226	0.45		1.30	0.452
20	other snappers	0.141		1.3	13.70	0.405
21	0–3 red drum	4.4E−06	2		30.83	0.065
22	3–8 red drum	1.2E−04	3.5		11.16	0.451
23	8–18 red drum	0.001	1.1		5.10	0.298
24	18–36 red drum	0.003	0.6		3.03	0.810
25	36+ red drum	0.029	0.15		1.86	0.084
26	juv rays & skates	0.001	2		4.49	0.577
27	rays & skates	0.082	0.3		1.00	0.319
28	flounders	0.202		0.42	6.36	0.274
29	pompano	0.002		1	8.00	0.450
30	Atlantic bumper	0.434		1.2	6.00	0.632
31	scad	0.182		1.65	5.00	0.526
32	juv Atlantic croaker	1.303	2		4.01	0.014
33	Atlantic croaker	4.344	1.5		2.00	0.263
34	catfish	0.582		0.8	7.60	0.340
35	spot	0.690		1.1	12.00	0.909
36	squid	0.168		1	3.90	0.986
37	pinfish	0.094		2	5.00	0.744
38	porgies	1.223		2.52	8.00	0.468
39	anchovy	2.032		2.53	14.00	0.322
40	juv menhaden	1.891	2.3		14.53	0.008
41	menhaden	6.240	1.9		6.00	0.614

42	clupeids	4.448		1.8	12.11	0.219
43	mullets	0.100		0.8	8.00	0.309
44	sea turtles	0.030		0.11	6.76	0.082
45	small forage fish	3.715		2.53	12.00	0.851
46	jellyfish	0.360		22	67.00	0.727
47	blue crab	0.244		2.4	8.50	0.960
48	juv brown shrimp	0.007	3		17.36	0.027
49	brown shrimp	0.558	2.4		5.00	0.680
50	juv white shrimp	0.004	3		17.36	0.019
51	white shrimp	0.300	2.4		5.00	0.236
52	juv pink shrimp	2.6E−04	3		17.36	0.037
53	pink shrimp	0.020	2.4		5.00	0.208
54	other shrimp	0.369		2.4	19.20	0.551
55	benthic crabs	0.045		2	7.00	0.948
56	benthic invertebrates	12.08		4.5	22.00	0.800
57	zooplankton	7.642		36	89.00	0.387
58	benthic algae/weeds	29.8		25		0.072
59	phytoplankton	25		182		0.203
60	Detritus	100				0.046

Two master equations must be satisfied to correctly parameterize the Ecopath model. The first equation describes the production of each functional group as a set of n linear equations for n groups:

$$\left(\frac{P_i}{B_i}\right) \cdot B_i \cdot EE_i - \sum_{j=1}^{n} B_j \cdot \left(\frac{Q_j}{B_j}\right) \cdot DC_{ji} - Y_i - E_i - BA_i = 0 \qquad (1)$$

where (*Pi*/*Bi*) is the production to biomass ratio for group *i*, *EEi* is the ecotrophic efficiency (the proportion of production used in the system), *Bi* and *Bj* are the biomasses of the prey and predators respectively, (*Qj*/*Bj*) is the consumption to biomass ratio, *DCji* is the fraction of prey *i* in predator *j*'s diet, *Yi* is catch rate for the fishery for group *i*, *Ei* is the net migration rate, and *BAi* is the biomass accumulation for group *i*.

The Ecopath model assumes conservation of mass over a year. Energy balance within each group is ensured with the second master equation:

$$\text{Consumption} = \text{production} + \text{respiration} + \text{unassimilated energy} \qquad (2)$$

where production can be described as:

$$\text{Production} = \text{predation mortality} + \text{catches} + \text{net migration} + \text{biomass accumulation} + \text{other mortality} \qquad (3)$$

More succinctly, production can be described by the following equation:

$$P_i = \sum_j Q_j \cdot DC_{ji} + (F_i + NM_i + BA_i + MO_i) \cdot B_i \qquad (4)$$

where *Pi* is the production of prey group *i*, *Qj* is the consumption of predator *j*, *DCji* is the diet composition contribution of *i* to *j*'s diet, *Fi* is the instantaneous rate of fishing mortality, *NMi* is the net migration rate of prey group *i*, *BAi* is the biomass accumulation rate for *i*, *M0i* is the other mortality rate for *i* (non-predation, non-fishery), and *Bi* is the biomass of *i*.

In addition, a diet matrix was constructed based on diet information from stomach content analysis from nekton collected in the model area (A. Adamack, unpublished data), supplemented by information available in the literature. To achieve mass balance, the diet matrix was adjusted to attain a plausible solution of the flow of biomass through the foodweb. The available diet information usually did not provide exact proportions of each diet item, which made the diet matrix the most suitable component to adjust in order to achieve mass-balance. For example, when previous studies indicated that a specific species was the dominant prey species for a predator, the exact proportion of this prey item was adjusted during the mass-balancing procedure while still maintaining its status as dominant prey item. The diet matrix is provides as supplemental material 1.

During the mass balancing procedure in Ecopath, the model calculates Ecoptrophic Efficiency (EE) of each group, which represents the amount of biomass of that group used in the system (Christensen et al., 2008). A mass-balanced solution of the model is presented in Table 1.

Spatial components

A model area of 44,890 km^2 was chosen, which encompasses the Louisiana coastal zone and the annually recurring hypoxic zone. This area was represented in Ecospace with 5 km^2 grid cells, and is the model area of our Ecospace model, which we have called the NGOMEX (Northern Gulf of Mexico) ecosystem model (Fig. 2).

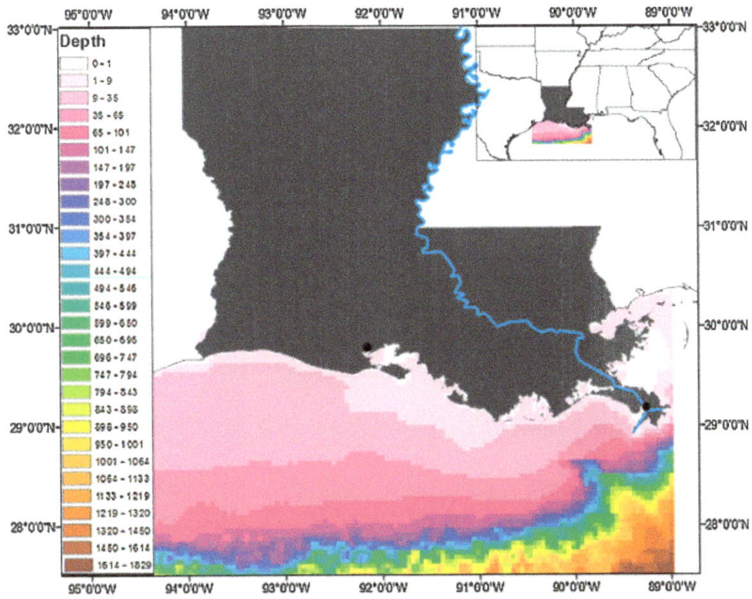

Figure 2: Model area of the NGOMEX (Northern Gulf of Mexico) Ecosystem model. Louisiana (USA) is indicated in gray, and the Mississippi River in blue. The coloration in the northern Gulf of Mexico indicates the bathymetry (source of bathymetry data: The Fish and Wildlife Research Institute).

For the spatial and temporal model simulations, dissolved oxygen and Chl *a* output from 1990 to 2004 of a coupled physical–biological hypoxia model (Fennel et al., 2011) was used as forcing functions. Chl *a* levels in Fennel et al. (2011) are affected by the nutrients entering the coastal zone from the Missisippi River and other freshwater sources. This output was averaged by month and matched to the Ecospace grid map so that one value of bottom dissolved oxygen and one of Chl *a* could be read into Ecospace per month per grid cell during a model simulation. In the few occasions where the model area of Fennel et al. (2011) did not overlap with our model, DO and Chl *a* output was extrapolated from nearby cells. This was done for the estuaries, while the focus area for our modeling effort had 100% overlap. Example DO output from Fennel et al. (2011) that is used as a spatial–temporal forcing function is shown in Fig. 3. A plug-in to Ecospace was used to read in this spatial–temporal forcing function (see Section 2.5 for more details). Dissolved oxygen affected the groups in the model as stipulated by the response curves, while Chl *a* was used as a driver of phytoplankton biomass, assuming a linear relationship.

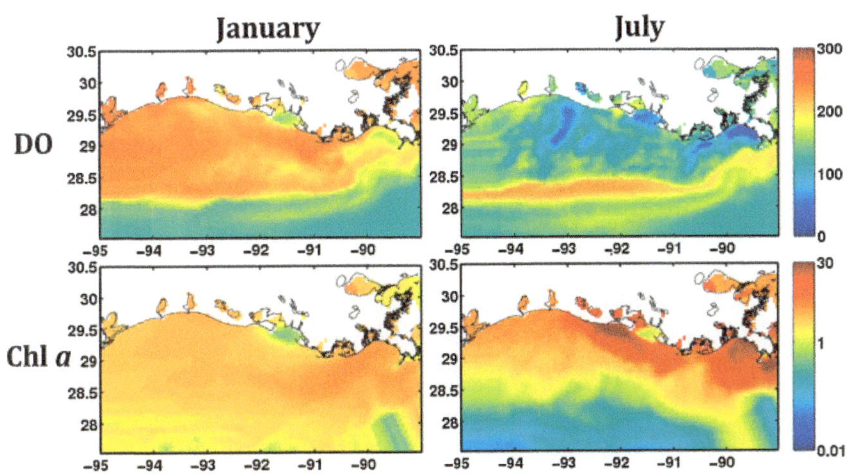

Figure 3: Example output of dissolved oxygen in mmol m^{-3} (top) and Chl a in mg m^{-3} (bottom) from Fennel et al. (2011) in a month without hypoxia (January) and a month with hypoxia (July). Monthly "maps" of this output are used as spatial–temporal forcing functions in the NGOMEX ecosystem model. Output was extrapolated in the estuaries as shown in the figure.

Two non-dynamic habitat features were included in the spatial model, depth and salinity area. Depth was based on the bathymetry of the model area; depth ranges were included to ensure (adult) offshore species would not enter shallow estuarine areas if they are not known to do that. While salinity is not modeled dynamically in this model, a 'marine', 'estuarine' and 'freshwater' zone is described loosely based on existing salinity gradients in the model area. While the focus of this model is on the marine coastal zone, these habitat features prevented species to escape coastal hypoxia by fleeing to areas that are too shallow or too fresh for them to enter in real life. A conceptual model of the NGOMEX ecosystem model is presented in Fig. 4.

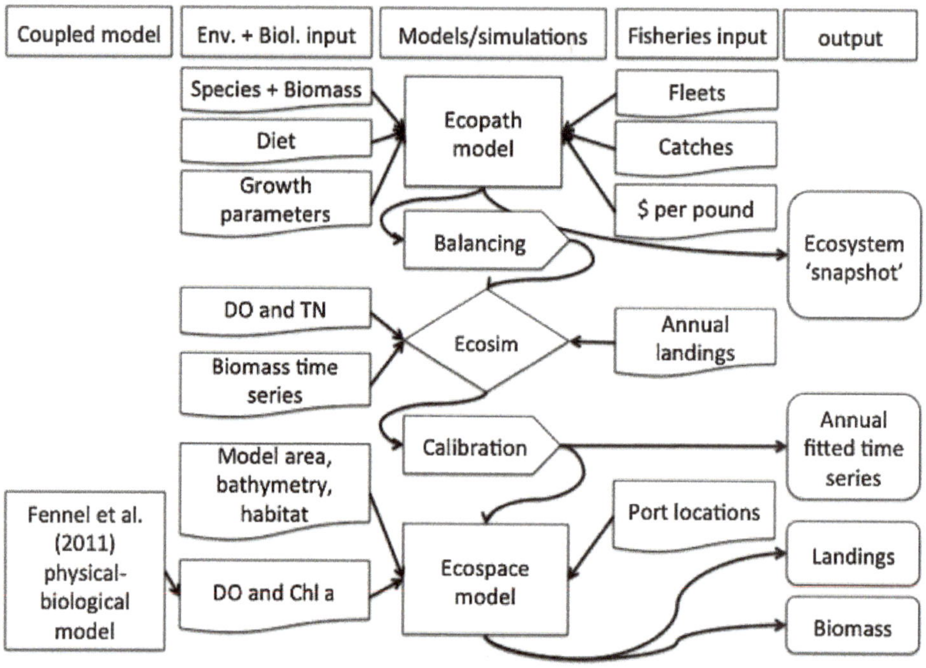

Figure 4: Conceptual diagram of the NGOMEX ecosystem model. DO = dissolved oxygen, TN = total nitrogen.

Model calibration

Temporal dynamic simulations were performed in Ecosim, the time-dynamic module of EwE, to calibrate the model. DO and NO*x* were included in the calibration runs as environmental forcing functions based on data described in Section 2.1. The level of dissolved oxygen affects the effective search rate of species in the model as described by the response curves in the same manner as salinity affected species in de Mutsert et al. (2012). The model was calibrated against biomass time series and landings data as described in Section 2.1. During calibration, the model was iteratively fitted to landings and biomass time series data by making vulnerability exchange rate adjustments until the smallest sum of squares (SS) was found using the fit-to-time-series feature in EwE (Christensen et al., 2008). Following the Foraging Arena Theory described in Walters and Martell (2004), each group is present in the model in a vulnerable (to predation) and invulnerable state. The vulnerability exchange rate determines how quickly the mass of a group can switch between those states, where high numbers (around 100) indicate Lotka–Volterra predator–prey interactions (all prey is vulnerable to predation because of the high exchange rate between the vulnerable and invulnerable portion), and low numbers (around 2) indicate a significant portion of the group is unavailable to predation. We used the fit-to-time series procedure to determine the vulnerability exchange rates that resulted in the best fit of model predictions to biomass and landings data. The metric used to determine model fit was the following:

$$SS = \sum_{i}^{nts} \left(\sum_{i}^{nobs_i} w_i \log \left(\frac{o_{it}}{p_{it}} \right)^2 \right) \tag{5}$$

where *SS* is sum of squares, *nts* is the number of time series loaded, *nobsi* the number of observations in time series *i* , w_i is the weight of the time series *i* (all time series weighted equal in our model), *oit* is the observed value in time series *i* at time step *t*and *pit* is the Ecosim predicted value for variable *i* at time step *t*.

Including DO and nutrient loading (in the form of NO*x*) as environmental forcing functions in Ecosim improved the fit of the model to time series, and decreased the total SS for all fits. Fig. 5 shows fits to time series (with SS) of a selection of species that are highly abundant in the area and/or have economic or ecological significance. The vulnerability exchange rates that were altered during this calibration procedure were carried over to Ecospace.

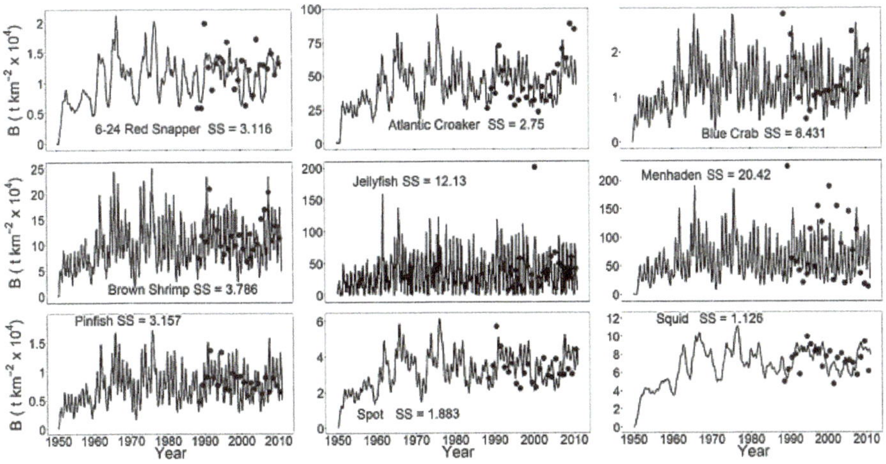

Figure 5: Model fits to observed biomass of selected groups/species in the model. The SS of the fit is indicated in each panel.

Model simulations

After calibration, spatial simulations were performed in Ecospace, the spatial–temporal module of EwE. In the new habitat foraging capacity model of Ecospace, dispersal rates of groups into a cell are affected by the cell suitability/capacity (Christensen et al., 2014). If the neighboring cell has a lower capacity then the dispersal rate to the cell will be proportional to the capacity difference. For example, if the capacity of a cell is 0.5 for a specific group, the maximum movement rate into this (in this model set to 300 km/yr for all groups) was adjusted by this proportion. The capacity of a cell was based on DO and habitat (depth and salinity area as described in Section 2.3). Fleets are dispersed by a gravitational model based on profitability per cell. Profitability per cell is based on the biomass of the target group(s) of a fleet, the price per pound of each target group in 2010, and the distance from port (fuel cost). Two ports with the highest landings in Louisiana were included in the model, Empire-Venice and Intracoastal City (www.oceanomics.org;Fig. 6).

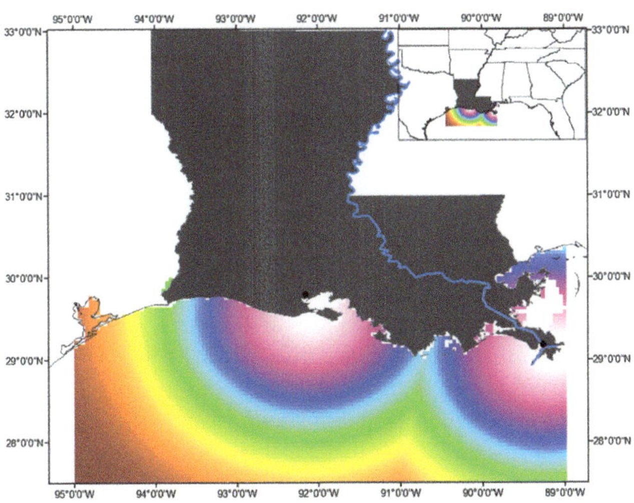

Figure 6: Location of ports in the NGOMEX ecosystem model, representing Intracoastal City on the left, and Empire-Venice on the right (black dots). The coloration indicates distance from port, which is included in the calculation of fisheries revenue.

To loosely link the physical–biological hypoxia model from Fennel et al. (2011) to Ecospace, a plug-in was added to the EwE source code. The plug-in reads in a DO and Chl *a* value per grid cell per time step (5 km^{-2} month^{-1}). This provides for spatial and temporal variation in the effective search rate and primary production. The DO values are fed into the environmental response functions defined in Ecosim. The values returned by the environmental response functions act as a forcing multiplier on the rate of effective search. This facility, provided by the plugin, works in the same manner as an Ecosim forcing function that has been applied to search rate (Christensen et al., 2008). The Chl *a*data is used to update the Ecospace Relative PP spatial layer, which allows for spatial shifts in primary production over time. The Ecospace Relative PP layer is a multiplier that is used to scale the primary production relative to the base productivity of the Ecopath model. During initialization the values in the Relative PP layer are normalized to scale the spatially averaged Ecospace productivity to the Ecopath base productivity rate (Christensen et al., 2008). The values read by the plug-in can shift from this baseline value to increase or decrease the spatially averaged productivity over time.

Scenarios simulated were 'no forcing', which simulated a coastal environment without nutrient fueling from the Mississippi River (or any other source of added nutrients) but also no formation of a hypoxic zone; 'enrichment only', which simulated nutrient loading effects on primary productivity, but where hypoxia had no effect on any organism; and 'enrichment + hypoxia', which included primary productivity forcing, and effects of DO (and thus hypoxia for part of the year) on fish biomass. Each scenario was run from 1950 to 2010; results presented reflect the output from simulation year 2010. While sixty groups were simulated, results are presented of a select group of species that are of economic or ecological interest.

RESULTS

Biomass and landings output of the scenarios 'no forcing', 'enrichment only', and 'enrichment + hypoxia' was compared. The scenario 'enrichment + hypoxia' simulates the real world scenario of Chl *a* concentration fueled by nutrient loading, and seasonal hypoxia in the coastal zone. The scenarios were run from 1950 to 2010, and output is presented as relative change, which is the change in biomass or landings of each group from the initial biomass or landings. The initial biomass and landings were the same for each scenario, so the scenario outcomes can be compared to each other. When looking at total landings and biomass, results indicate that the seasonal presence of hypoxia reduces both landings and biomass as compared to the 'enrichment only' scenario (Fig. 7). However, both 'enrichment only' and 'enrichment + hypoxia' had much higher increases from initial biomass and landings than the 'no forcing' scenario; the latter even showed a small decrease. The difference between 'enrichment only' and 'enrichment + hypoxia' is comparatively so small that these simulations suggest that the decrease in secondary production due to hypoxia (in indirect effect of nutrient loading) is trivial in comparison to the increase in secondary production due to the bottom up effect of nutrient loading. Overall, there was a 33% increase in total landings in the 'enrichment + hypoxia' scenario as compared to the 'no forcing' scenario, and a 13% increase in total biomass. Removing hypoxia only increased that amount by an extra 5% and 0.6% respectively.

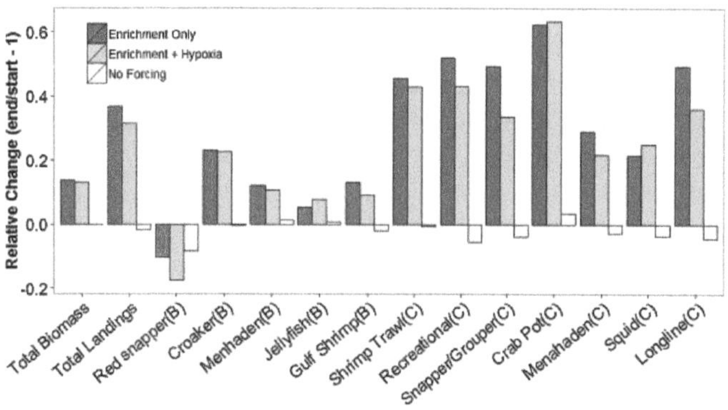

Figure 7: Total landings and total biomass results of three scenarios (no forcing, enrichment only, and enrichment + hypoxia) that ran from 1950 to 2010. The relative change from the same initial conditions is presented of total biomass and total landings, species-specific biomass of selected species (B), and catch from all fleets (C).

While total landings and biomass show the concurring trend of a small decrease in the 'no forcing' scenario, and large increases in 'enrichment only' and 'enrichment + hypoxia', individual groups vary in their response (Fig. 7). The biomass of common species in Louisiana; Gulf menhaden (*Brevoortia patronus*), Atlantic croaker (*Micropogonias undulatus*), and shrimp (brown shrimp – *Farfantepenaeus aztecus*, white shrimp – *Litopenaeus setiferus*, and pink shrimp – *Farfantepenaeus duorarum*) showed a response similar to what was seen in total biomass. Red snapper (*Lutjanus campechanus*) biomass however, decreased in all three scenarios, and decreased most in the 'enrichment + hypoxia' scenario (17.6%), followed by 'enrichment only' (10.4%) and 'no forcing' (8.3%). An opposite effect was seen in jellyfish, which displayed increases in all three scenarios, and the highest increase in the scenario with hypoxia (7.8%). Changes in landings do follow this pattern for almost all fleets, except for crab and squid fisheries, which see a small increase in landings when hypoxia is added as compared to enrichment alone.

DISCUSSION AND CONCLUSION

Our simulations suggest that reductions in landings and biomass due to hypoxia are an order of magnitude lower than increases seen due to the nutrient enrichment (which is the main cause of hypoxia). Some fisheries in the model even experience an increase in landings in the scenario that includes hypoxia, namely blue crab and squid landings. The crab pots are not set in areas affected by hypoxia, which could explain this pattern, while the increase in squid landings is likely an indirect effect, since squid had a slightly higher tolerance for low oxygen as most of its predators, and slightly higher biomass in the scenario with hypoxia as a result. In general, current simulations do not suggest that natural resource managers should take the hypoxic zone into account in fisheries management plans (e.g. by restricting effort during hypoxic events), as the occurrence of seasonal hypoxia in combination with fishing does not lead to unsustainable biomass reductions.

This study emphasizes the importance of the positive bottom-up effect of nutrient enrichment on secondary productivity (Nixon and Buckley, 2002). Some notable species that follow the pattern of large increases in biomass as a result of nutrient enrichment, and only a slight reduction in biomass as a result of hypoxia, include Atlantic croaker, which is the most abundant species in this area and knows to have a high tolerance for hypoxia (Bell and Eggleston, 2005), Gulf menhaden, which is the largest fishery in Louisiana, and gulf shrimp (brown, white, and pink shrimp), which is the fishery with the highest revenue in Louisiana.

Still, from these results cannot be inferred that nutrient load reduction is not an important restoration measure, and that it would necessarily reduce secondary productivity. Our scenario of 'no forcing' is not a real-world scenario, and no nutrient reduction plan would conceivably remove all nutrients from the freshwater sources flowing into the Gulf of Mexico. Therefore, the corresponding low secondary production seen in the 'no forcing' scenario would never be attained. In addition, the relationship between nutrient loading, primary productivity, and hypoxia is non-linear and complex (Fennel et al., 2011); a reduction in nutrient load would not necessarily reduce bottom up fueling of the foodweb and hypoxia to the same extent. Momentarily disregarding the 'no forcing' scenario, a consistent small decrease in biomass from the nutrient enrichment scenario to the nutrient enrichment scenario with summer hypoxia can be seen. This small reduction could be ecologically significant for some species.

One species that seems affected by nutrient loading as well as hypoxia in our simulations is red snapper. An increase in mortality due to higher shrimp landings – and thereby higher bycatch of juvenile red snapper – in the scenarios that include nutrient enrichment is a likely cause of a decrease in red snapper biomass in those scenarios. The model reflects the impact shrimp trawling has on red snapper, which has been reported in studies related to red snapper stock status (Cowan, 2011). The additional decrease in biomass when hypoxia is present does indicate a negative effect of hypoxia on red snapper. Weaker recruitment of red snapper in years of severe hypoxia has been observed in a previous study (Switzer et al., 2015).

Another interesting result is the increase in jellyfish biomass in the scenario with hypoxia. Jellyfish, often regarded as nuisance species, likely find refuge from predation in hypoxic areas due to their high tolerance of low oxygen conditions. Increases in jellyfish in response to hypoxia in coastal ecosystems has been predicted or observed in other studies (Breitburg et al., 2003, D'Elia et al., 2003 and Miller and Graham, 2012), and could exacerbate hypoxia effects on zooplankton by adding increased predation pressure.

This study concurs with some previous publications that hypoxia typically does not reduce overall fisheries landings or biomass, but that hypoxia should still be addressed in restoration plans (Breitburg et al., 2009). The use of novel spatial–temporal forcing functions in Ecospace allows for more realistic simulations of effects on fish and fisheries of environmental drivers that vary in space and time. Ecosystem models with this capability have only recently been described (Steenbeek et al., 2013 and Christensen et al., 2014), but are expected to increase in numbers rapidly. Their usefulness in developing restoration and/or natural resource management strategies, especially when linked to physical/chemical models seems evident, and has already been recognized (de Mutsert et al., 2014a and de Mutsert et al., 2014b). The model presented in this paper would be useful in restoration planning, and development of management strategies to reduce hypoxia without unacceptable losses to fisheries productivity. Models such as the physical–biological model of Fennel et al. (2011) could be used to simulate effects of nutrient load reductions on hypoxia and primary productivity in the coastal zone. The NGOMEX Ecospace model could then use those results to simulate effects of nutrient reductions on fish and fisheries in a scenario analysis. These loosely coupled models could thereby be used as a tool in nutrient reduction analyses to inform management decisions.

ACKNOWLEDGEMENT

KdM would like to thank Katja Fennel for providing model output, and Arnaud Laurent for reformatting the output to fit the NGOMEX ecosystem model grid, and extrapolating output into cells of the model area that were not included in the Fennel et al. (2011)model. KdM would also like to thank Carl Walters for providing input during model development. This research was funded by NOAA's Center for Sponsored Coastal Ocean Research (CSCOR) under grant nr. NA09NOS4780233.

REFERENCES

1. Altieri, A.H., 2008. Dead zones enhance key fisheries species by providing predation refuge. Ecology 89, 2808–2818, http://dx.doi.org/10.1890/07-0994.1.
2. Bell, G.W., Eggleston, D.B., 2005. Species-specific avoidance responses by blue crabs and fish to chronic and episodic hypoxia. Mar. Biol. 146, 761–770, http://dx.doi. org/10.1007/s00227-004-1483-7.
3. Breitburg, D.L., Adamack, A., Rose, K.A., Kolesar, S.E., Decker, B., Purcell, J.E., Keister, J.E., Cowan, J.H., 2003. The pattern and influence of low dissolved oxygen in the Patuxent River, a seasonally hypoxic estuary. Estuaries 26, 280–297.
4. Breitburg, D.L., Hondorp, D.W., Davias, L.A., Diaz, R.J., 2009. Hypoxia, nitrogen, and fisheries: integrating effects across local and global landscapes. Annu. Rev. Mar. Sci. 1, 329–349, http://dx.doi.org/10.1146/annurev. marine.010908.163754.
5. Chesney, E.J., Baltz, D.M., Thomas, R.G., 2000. Louisiana estuarine and coastal fisheries and habitats: perspectives from a fish's eye view. Ecol. Appl. 10, 350–366.
6. Christensen,V.,Coll,M., Steenbeek,J.,Buszowski,J.,Chagaris, D.D.,Walters,C.J., 2014. Representing variable habitat quality in a spatial food web model. Ecosystems 17, 1397–1412.
7. Christensen, V.,Walters, C.J., Pauly, D., Forrest, R., 2008. Ecopath with Ecosim version 6 user guide. Lenfest Ocean Futures Project 2008. Fisheries Centre, University of British Columbia, Vancouver, Canada, 235 pp.
8. Cowan Jr., J.H., 2011. Red Snapper in the Gulf of Mexico and U.S. South Atlantic: data, doubt, and debate. Fisheries 36 (7), 319–331.
9. Craig, J., 2012. Aggregation on the edge: effects of hypoxia avoidance on the spatial distributionof brownshrimpanddemersalfishes inthe Northern Gulf ofMexico. Mar. Ecol. Prog. Ser. 445, 75–95, http://dx.doi.org/10.3354/meps09437.
10. D'Elia, C.F., Boynton, W.R., Sanders, J.G., 2003. A watershed perspective on nutrient enrichment, science, and policy in the Patuxent River, Maryland: 1960–2000. Estuaries 26, 171–185.
11. de Mutsert, K., Cowan Jr., J.H., Walters, C.J., 2012. Using Ecopath with Ecosim to explore nekton community response to freshwater diversion into a Louisiana estuary. Mar. Coast. Fish. 4, 104–116.
12. de Mutsert, K., Lewis, K.A., Steenbeek, J., Buszowski, J., Milroy, S., 2014a. 2017 Coastal Master Plan: Ecosystem Outcomes, Community Model Description (4.5).Version I. Coastal Protection and Restoration Authority, Baton Rouge, LA, 117 pp.
13. de Mutsert, K., Lewis, K.A., Steenbeek, J., Buszowski, J., Milroy, S., 2014b. Delta Management Fish and Shellfish Ecosystem Model. Version I. Coastal Protection and Restoration Authority, Baton Rouge, LA, 98 pp.
14. Eby, L.A., Crowder, L.B., McClellan, C.M., Peterson, C.H., Powers, M.J., 2005. Habitat degradation from intermittent hypoxia: impacts on demersal fishes. Mar. Ecol. Prog. Ser. 291, 249–262, http://dx.doi.org/10.3354/meps291249.
15. Essington, T., Paulsen, C., 2010. Quantifying hypoxia impacts on an estuarine demersal community using a hierarchical ensemble approach. Ecosystems 13, 1035–1048, http://dx.doi.org/10.1007/s10021-010-9372-z.
16. Fennel, K., Hetland, R., Feng, Y., DiMarco, S., 2011. A coupled physical–biological model of the northern Gulf of Mexico shelf: model description, validation and analysis of phytoplankton variability. Biogeosci. Dis. 8, 1881–1899, http://dx. doi.org/10.5194/bg-8-1881-2011.
17. Goodman, L., Campbell, J., 2007. Lethal levels of hypoxia for gulf coast estuarine animals. Mar. Biol. 152, 37–42, http://dx.doi.org/10.1007/s00227-007-0685-1.
18. Gunter, G., 1963. The fertile fisheries crescent. J. Miss. Acad. Sci. 9, 286–290.
19. Libralato, S., Solidoro, C., 2009. Bridging biogeochemical and food web models for an End-to-End representation of marine ecosystem dynamics: the Venice lagoon case study. Ecol. Model. 220, 2960–2971, http://dx.doi.org/10.1016/j.ecolmodel. 2009.08.017.

20. Miller, M.-E.C., Graham, W.M., 2012. Environmental evidence that seasonal hypoxia enhances survival andsuccess ofjellyfishpolyps inthenorthern Gulf ofMexico.J. Exp. Mar. Biol. Ecol. 432–433, 113–120, http://dx.doi.org/10.1016/j.jembe.2012. 07.015.

21. Nixon, S.W., Buckley, B.A., 2002. "A strikingly rich zone" – nutrient enrichment and secondary production in coastal marine ecosystems. Estuaries 25, 782–796.

22. Pierson, J.J., Roman, M.R., Kimmel, D.G., Boicourt, W.C., Zhang, X., 2009. Quantifying changes in the vertical distribution of mesozooplankton in response to hypoxic bottom waters. J. Exp. Mar. Biol. Ecol. 381, S74–S79, http://dx.doi.org/10.1016/ j.jembe.2009.07.013.

23. Rabalais, N.N., Turner, R.E., 2006. Oxygen depletion in the Gulf of Mexico adjacent to the Mississippi River. In: Neretin, L.N. (Ed.), Past and Present Water Column Anoxia. Springer, Netherlands, pp. 225–245.

24. Rabalais, N.N., Turner, R.E., 2001. Coastal Hypoxia: Consequences for Living Resources and Ecosystems. American Geophysical Union, Washington, DC.

25. Robert, J.J., Brandt, S.B., Fanslow, D., Ludsin, S.A., Pothoven, S.A., Scavia, D., Hook, T.O., 2011. Effects of hypoxia on consumption growth and RNA:DNA ratio of young yellow perch. Trans. Am. Fish. Soc. 140 (6), 1574–1586.

26. Rose, K.A., 2000. Why are quantitative relationships between environmental quality and fish populations so elusive? Ecol. Appl. 10, 367–385.

27. Rose, K.A., Adamack, A.T., Murphy, C.A., Sable, S.E., Kolesar, S.E., Craig, J.K., Breitburg, D.L., Thomas, P., Brouwer, M.H., Cerco, C.F., Diamond, S., 2009. Does hypoxia have population-level effects on coastal fish? Musings from the virtual world. J. Exp. Mar. Biol. Ecol. 381, S188–S203, http://dx.doi.org/10.1016/j.jembe.2009.07.022.

28. Steenbeek, J., Coll, M., Gurney, L., Melin, F., Hoepffner, N., Buszowski, J., Christensen, V., 2013. Bridging the gap between ecosystem modeling tools and geographic information systems: driving a food web model with external spatial–temporal data. Ecol. Model. 263, 139–151.

29. Switzer, T.S., Chesney, E.J., Baltz, D.M., 2015. Habitat use by juvenile red snapper in the northern Gulf of Mexico: ontogeny, seasonality, and the effects of hypoxia. Trans. Am. Fish. Soc. 144 (2), 300–314, http://dx.doi.org/10.1080/00028487. 2014.991447.

30. Turner, R.E., Rabalais, N.N., Justic, D., 2006. Predicting summer hypoxia in the northern Gulf of Mexico: riverine N, P, and Si loading. Mar. Pollut. Bull. 52, 139–148, http://dx.doi.org/10.1016/j.marpolbul.2005.08.012.

31. Walters, C., Martell, S., 2004. Fisheries Ecology and Management. Princeton University Press, 399 pp.

32. Walters, C., Martell, S.J.D., Christensen, V., Mahmoudi, B., 2008. An Ecosim model for exploring Gulf of Mexico ecosystem management options: implications of including multistanza life history models for policy predictions. Bull. Mar. Sci. 83, 251–271.

CHAPTER 15

MARINE DEPENDENT LIVELIHOODS AND RESILIENCE TO ENVIRONMENTAL CHANGE: A CASE STUDY OF ANGUILLA

J. Forster, I.R. Lake, A.R. Watkinson, J.A. Gill

Keywords: Marine livelihoods, Resilience, Vulnerability, Environmental change, Marine resource-use, Caribbean

INTRODUCTION

Coastal communities throughout the developing world are recognised as being particularly vulnerable to environmental change [1], [2] and [3]. Many are dependent upon already depleted natural resources and degraded coastal systems, and may consequently be highly susceptible to changes in the condition of the natural resources upon which they rely [4] and [5]. In tropical regions, coastal ecosystems and communities may also be at risk from the impacts of natural phenomena such as earthquakes, tsunamis and hurricanes, with serious implications for human security and livelihood opportunities [6] and [7]. In addition, there is growing evidence of global climate change effects on coastal resource-dependent communities; for example through impacts on fish stocks and fisheries [3] and [8], increases in the frequency and severity of coral bleaching events [9] and [10] and threats to coastal resources from sea-level rise [11] and [12]. Consequently, there is considerable concern regarding the repercussions of global and local environmental change on coastal resource-dependent communities and industries [5] and [13].

Natural resource-dependency describes the direct association between the livelihoods of individuals, sectors or communities, and a natural resource and its local economy [14]. Close links between social and ecological systems, of which coastal resource-dependent users and industries are a prime example, can have major implications for managing and adapting to environmental change [4] and [15]. The ability of social-ecological systems to adapt to environmental and climatic change has gained prevalence, notably through the concept of 'resilience' for example, see [16], [17], [18] and [19]. The resilience of social-ecological systems is identified by their ability to cope with external stresses and disturbances resulting from social, political, or environmental change [20].

The resilience of many coastal communities largely depends on the flexibility of individual resource-users, or the 'social resilience' available to deal with and adapt to change or variability [14] and [21], as well as the ability of communities to act collectively i.e. their ‹social capital› see [16]. The social resilience of individuals can be influenced by a series of key components, including individual perception of risk associated with change, ability to plan, learn and reorganise, and social, economic and environmental dependencies such as the level of attachment to specific occupations and places, employability, family characteristics and financial status [22]. For example, it has been suggested that individuals with few family responsibilities, more financial security, and weak attachment to a resource-dependent occupation, may be more able or willing to change occupation, hence increasing their resilience and reducing their vulnerability [21]. The existence of diverse livelihood systems has also been identified as an important component that can enhance individual and community adaptability to disturbance and change [13]. Livelihood diversification can reduce the threat of livelihood failure in resource-dependent systems, by spreading risk across more than one source of income, and it may help to overcome variations associated with resource seasonality, market failures and uncertainties [23].

Several studies have investigated the vulnerability of coastal and marine resource-dependent communities and nations to climatic change [3], [4] and [24]. However, until recently, the implications of climate variability on the lives and livelihoods of marine resource-users at local scales have been less well explored [13] and [25]. Investigations of individual

Citation: J. Forster, I.R. Lake, A.R. Watkinson, J.A. Gill, Marine dependent livelihoods and resilience to environmental change: A case study of Anguilla, Marine Policy, Volume 45, March 2014, Pages 204-212, ISSN 0308-597X, http://dx.doi.org/10.1016/j.marpol.2013.10.017.

perceptions of environmental change have commonly used a livelihoods approach see [13] and [23]. This approach focuses on local-scale assets (land, stock, savings etc.), capabilities and activities of resource-dependent people, and assesses how different livelihood strategies can affect the ability of people or groups to withstand disturbance or change [23].

Here a livelihoods approach is used to assess the resilience of marine and coastal resource-users to environmental change on the Caribbean island of Anguilla, a country highly dependent on marine and coastal resources, with no other significant economic industries [26] and [27]. This study focuses on the effects of hurricanes to examine the resilience of communities to environmental change, as the islands of the Caribbean are particularly at risk from these extreme events [28] and [29]. The impacts from North Atlantic hurricane activity are expected to increase in the Caribbean region in response to changing global climate conditions [2] and [30], although specific changes in hurricane risk for the Caribbean are not yet fully understood e.g. see [31] and [32]. Nevertheless, hurricanes have considerable impacts on Caribbean islands and the increasing prevalence of these extreme events is a major concern for the region [28], [33] and [34].

The aim of this study is to explore the social-resilience of marine resource-dependent livelihoods on the Caribbean island of Anguilla to environmental stressors by (1) identifying the characteristics of marine and coastal resource-dependent users and livelihoods, (2) assessing the impacts of previous hurricane events on these resource-dependent livelihoods, and (3) investigating resource-user perceptions of future environmental change on the resource and livelihood security.

METHODS

The study location

The study was undertaken in Anguilla, a small island in the Lesser Antilles chain in the Caribbean Sea (Fig. 1). Like many islands in the Caribbean, the island of Anguilla depends heavily on its marine and coastal resources for fisheries and tourism [34] and [35]. Fishing in Anguilla is largely artisanal, and there are approximately 300 outboard-powered open-top fishing vessels, most of which are between 5 and 10 m in length. The majority of fishers operate close to shore, but due to low inshore catch rates, many vessels have expanded their range to within approximately 65 km radius of the island [27]. The inshore coral reef fishery principally targets reef fish (e.g. groupers (Serranidae), parrotfish (Scaridae)), and lobsters (spiny lobster (*Panulirus argus*) and spotted spiny lobster (*P. guttatus*), known locally as crayfish [35]).

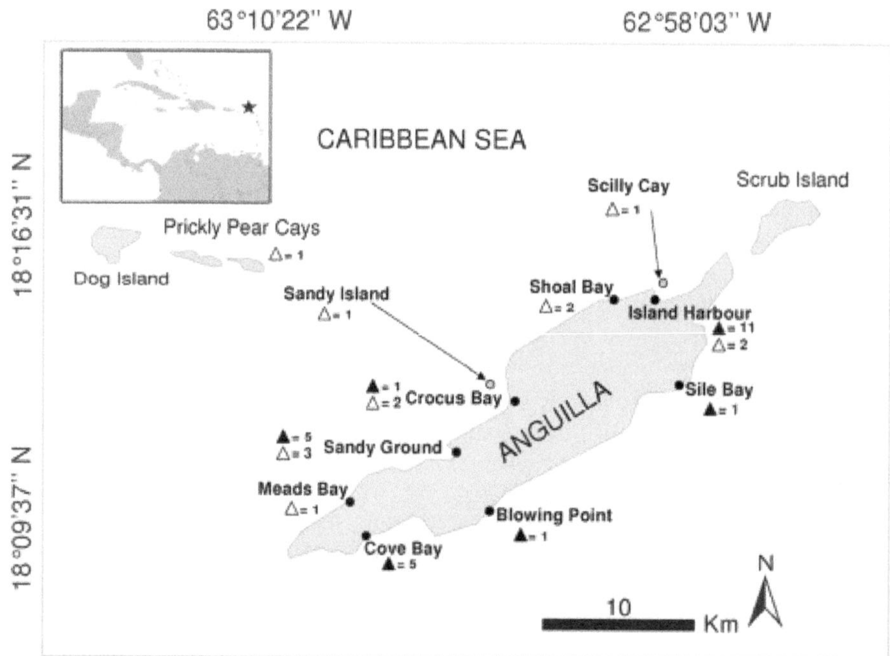

Figure 1: Anguilla and its associated islands and cays. The location and names of the study sites are indicated, with the numbers of respondents (fishers (black triangle) and tourist operators (white triangle)) interviewed at each site. The inset shows the location of Anguilla within the Caribbean region.

Tourism dominates Anguilla's economy and generates over 70% of the island's gross domestic product (GDP) and employment opportunities [36]. The tourist season in Anguilla is highly seasonal; the official tourist season runs from December to April, with peak tourist arrivals during December, March and April [34]. Since the late 1970s, Anguilla has developed into an upmarket beach tourist destination, and is renowned for its pristine white sand beaches [26]. The rapidly expanding luxury tourism industry on the island has also created a growing demand for seafood products such as lobster, in addition to the high demand for reef fish among the local population [35]. Evidently, the fishing industry has become an important contributor to the island's economy, currently employing an estimated 5% of the population (c. 400 individuals); although the vast majority fish part-time [27].

Interviews with marine resource-users

Interviews were conducted between February and April 2008, with 24 fishers from the six main harbours and 13 marine tourist operators (see Fig. 1). The fishers that were interviewed all relied on fishing coral reefs for all or part of their income, targeting the inshore reef fishery for reef fish and/or shellfish using fish or lobster traps (hereafter called traps) and hand-lines. The 13 marine-based tourist operators included all of the dive shops (n=3), glass-bottom boat (n=2), beach hire (n=1) and boat charter companies (n=4) as well as the three inshore tourist destinations with on-site restaurants on the island (Fig. 1).

Respondents were interviewed on the basis of recommendations from key informants (senior representatives from the Anguilla Department for Fisheries and Marine Resources (DFMR), and experienced local fishers), and through snowball sampling (whereby respondents recommended further potential interviewees [37]). Interviews with fishers and tourist operators consisted of a series of (a) structured closed questions to generate quantitative data on demographic variables and (b) open-ended, semi-structured questions to provide qualitative information on fishing practices and livelihood strategies, perceptions of marine ecosystem health and environmental change. Information on the impacts of previous hurricanes in Anguilla was gathered specifically as hurricanes are a particularly prevalent environmental stressor in the Caribbean. Interviews were tape-recorded and transcribed verbatim.

Data analyses

Responses to open-ended questions were manually coded and analysed using an 'open coding' method [38], in which similarities and differences in responses to questions are assessed. Conceptually similar responses or opinions were grouped together into 'categories' which were defined using a common theme. This method ensures that the response themes directly reflect the issues that emerged from the interviews. Triangulation was used to validate the results of interviews by cross-checking information with other interviewees and key informants from the wider fishing and tourist community, and government officials from the DFMR. Interviewee responses were also cross-validated with personal observations at the harbour and during fishing trips. Collectively, these practices affirmed the accuracy of the interview data [37].

Spearman rank correlations were used to explore associations between specific measures of fishing effort (number of traps, weight of catch and fuel expenditure) for individual fishers. Results are given for all 24 fishers where possible, but not all fishers provided all relevant data. Seasonal variation in tourist demand was quantified for the tourist operators, with each tourist operator providing an estimate of tourist demand for each month of the year, in $US or numbers of visitors. For individual respondents, tourism demand was standardised relative to the mean of all 12 months to give a relative monthly demand. This was then averaged across all 13 tourist operators.

RESULTS

Fishers

Demographic characteristics

All of the 24 fishers interviewed were male Anguillian nationals, with all but one having lived in Anguilla for their entire life. The majority of respondents had left education after secondary school (67%, n=14/21), with three completing high school and one holding a graduate qualification. Most of the respondents were married (71%, n=15/21) and of these the majority (93%) had children. With respect to these education and family status indicators, the respondents are typical of the male working population for the island[39] and [40]. In total, 81% (n=17/21) of respondents stated that they were responsible for dependents (children or family members).

The average age of the fishers was 46 years (±16 SD), with ages ranging between 19 and 70+ years. Most of the fishers were categorised in the 45–54 (n=8) and 55–64 year groups (n=4), with three fishers aged 65+ years. By comparison to the employed male population in Anguilla, these fishers are on average older, with 75% >35 years and 42% >50 years (the national census shows that 55% of working males on Anguilla are >35 years and 17% are >50 years [41]. Only six respondents were younger than 35 years. The majority of fishers started their fishing career in their late teens or straight after secondary school (mean age±SD, 18±6 years). Most respondents were from fishing families, following a hereditary occupation as demonstrated by 92% (n=22) with grandfathers or fathers that fished before them.

Fishing strategies

The majority of respondents (83%, n=20) considered fishing to be their main occupation and source of income, although half subsidised their fishing with alternative employment, including construction work and private boat charters. Fishers were relatively similar in terms of their fishing strategies; 20 respondents (83%) targeted both fish and lobster (two also target crayfish). Fewer (n=4) respondents targeted single species, with two fishers targeting only fish and two targeting only lobster. In addition, 14 fishers stated they also use hand- or long-lines to target species such as red snapper (*Lutjanus campechanus*). The number of traps used, however, differed substantially among fishers. For the months that the survey took place, total traps per fisher ranged from 20 to 380 (mean±SD, 82±75), with the number of fish traps ranging from 13 to 120 (48±59 traps), and lobster traps ranging from 8 to 300 (59±65 traps). Average daily catch for all gear types (i.e. combining fish and lobster traps and hand-lining) was 72 kg/day (SD±37 kg/day). Daily catch for only fish traps was 60 kg/day (± 35 kg/day) and for lobster traps was 53 kg/day (± 36 kg/day). Catch weight increased significantly with number of traps set by each fisher (total combined catch r_s=0.66, p<0.01, n=22; fish r_s=0.64, p<0.01, n=18; lobster r_s=0.86, p <0.001, n=15). Interviews revealed that, due to species-specific survival rates, fishers checked their lobster traps once a week whereas fish traps were checked every 2–3 days.

Fishing costs, income and assets

Anguillian fishers accumulate many occupation-specific assets. For example, interviews revealed that the cost of typical fishing boats, excluding the outboard-motor(s) was *c.*$US 25,000. All of the fishers stated that they fished using their own boat, and they all built their own traps. Respondents estimated that the cost of fish or lobster traps was between $US 135 and 225 per trap (excluding labour costs). Given the average total number of traps (82±75 SD), these fishers own between $US

11,020 (±10,125 SD) and $US 18,450 (±16, 875 SD) worth of traps.

After the initial costs incurred by building traps and boat acquisition, other running costs (e.g. bait, wages, general maintenance) were considered by respondents to be negligible compared to the cost of fuel. Weekly fuel expenditure ranged from $US 120 to 750 (mean±SD, 382±173), with the range reflecting the variation in the number of days respondents fished (between 1 and 6 days/wk, mean±SD, 3±1.4 days/wk). Weekly fuel expenditure was significantly positively associated with fishing days/ week (r_s=0.72, p<0.001, n=24).

The standard market price of catches varied according to species. During the time of surveying, lobster market price ($US 18.5 per kg) was higher than for reef fish ($US 11 per kg), reflecting a demand for lobster by the luxury tourism industry, compared with the local demand for reef fish. All fishers (n=24) commented that they could always sell their fish or lobster at any time of the year. The profitability of lobster varied according to season, with the price reducing by approximately half from the peak tourist season (November to April) to the off-peak tourist months. For this reason, and also because egg-bearing lobsters are present during the off-peak summer months, many fishers tended to switch to targeting only reef fish between May and November.

Using the above market prices, the estimated values of weekly catch per fisher at the time of sampling ranged from $US 600 to 3750 for reef fish (mean±SD, 2116±1023 $US) and from $US 298 to 2253 for lobster ($US 1106±584). Fishers average weekly takings after fuel costs, ranged from $US 450 to 3150 for fish ($US 1671±730) and from $US 210 to 1753 for lobster ($US 836±458), highlighting the profitability of fishing in Anguilla.

Environmental change effects on fishers

Previous hurricane impacts on fishing livelihoods

The most recent hurricanes that severely impacted Anguilla are hurricanes Luis in 1995 and Lenny in 1999. Hurricane Lenny caused significant flooding and damage to land-based infrastructure, but less impact at sea or on the fishing community. Consequently, when recounting impacts suffered from hurricanes, respondents predominantly focused their responses to the effects of hurricane Luis (Table 1). The accuracy of these recollections may be enhanced by both the age of these fishers and that many were fishing during hurricane Luis, in addition to the general significance of hurricane Luis for the whole island.

Table 1: Fisher response categories to the open question "If you were fishing during previous hurricanes did they affect your fishing?" All responses relate to the effects of hurricane Luis (1995) on Anguillian fishers and fishing. The number and percentage of respondents that mentioned each response is reported

Fisher response categories	Number of responses	Percentage of responses
Lost fish and/or lobster traps	18	75
The hurricane changed fishing grounds causing fish to move away	7	29
The hurricane destroyed fishing grounds and coral reefs	6	25
Hotels shut or had no electricity for two months, they could not freeze fish, and there was no demand for fish or lobster	3	12.5
The fishing industry took two years to fully recover	2	8
It took approximately two months to build traps and go fishing	2	8
The hurricane caused damage to, or the loss of, fishing boats	2	8
Lost income (thousands of dollars)	1	4
Hotels on the island took a year to return to pre-Luis state	1	4
Stopped fishing (three years)	1	4

The majority of respondents (75%) lost gear (fish and/or lobster traps) as a consequence of hurricane Luis, with losses per fisher ranging from 13 to 250 (mean±SD, 86±67) traps. The combination of lost gear and the impact of the hurricane on hotels meant that fishers were unable to fish for at least two months (Table 1), although one fisher stated he did not return to fishing for approximately three years. Respondents stated that the Anguilla government provided some financial assistance to the fishing community by giving each fisher three traps to re-start fishing, and offering subsidies on wire mesh and buoys to help fishers rebuild traps.

In addition to the substantial financial impacts accrued, six respondents stated that the fishing grounds had been altered by the hurricane. Another six respondents mentioned that the fishing grounds had been completely destroyed. All respondents continued to fish after the devastation of this hurricane, even though some took several years to return to fishing. It would appear that, despite the destruction of the hurricane, fishing remained a viable occupation, and the profitability of fishing in Anguilla will likely have influenced the decision of these fishers to continue fishing.

The personal and cultural ties that fishers have with their occupation, their 'fisher ethic', may provide an additional explanation for why fishers continued to fish after hurricane

Luis. When asked why they fished, 63% (n=15/24) of respondents stated their motive was because of an ingrained cultural or personal desire to fish. By comparison, fewer respondents (33%, n=8/24) mentioned the financial motivation. Examples of respondent response categories and selected quotes illustrating 'fisher ethic' are shown in Table 2.

Table 2: Fisher responses to the open question "Why do you fish?" Fifteen fishers stated their motive was related to the desire to fish rather than purely financial reward. These respondents had a strong 'fisher ethic'. The number of respondents that mentioned each 'fisher ethic' category is included, in addition to examples of respondent responses to contextualise each category

Response category	Number of respondents	Selected respondent responses
Love of fishing	7	"I just love it. I grew up around guys that do fishing, my father also"
		"The sea is where I should be…we love it, if we do not fish we don't feel right"
		"Fishing is great. Fishing is beautiful. I love fishing. And not only fishing, I love the fact of being on the sea. I could go on the ocean, sit down fishing, not catching anything and I feel good"
		"I fish because I love it…[and] I can make enough money"
Family tradition	6	"I grew up around the sea"
		"I kind of feel its [fishing] in my blood. I was drawn to fishing from very young. Right out of school I just wanted to do fishing for the rest of my life"
		"I took up fishing because I am from this area, born in Island Harbour"
		"I have been fishing ever since I have been born…grew up with my father fishing, I just continued as a little boy doing the same thing"
Own boss	3	"When you fish you don't have no boss"
		"Do not want to work for anyone else"
Only interested in fishing for a living	2	" Would not do anything else"
Relaxation	1	"For me it is most important livelihood because it pays and it is relaxation"
Expertise	1	"It is something I like and I'm good at it"

Longer term implications of hurricane Luis

The impact of hurricane Luis was manifest in seasonal changes in the fishing practices on Anguilla. Before hurricane Luis, fishers did not bring their traps inshore during the hurricane season; instead preferring to risk losing or damaging fishing gear and continuing to fish throughout the hurricane season. As a direct response to the losses sustained by the fishing community from this hurricane, fishers have developed more risk-averse fishing practices (Table 3). For example, the majority of fishers (n=16) now bring some or all of their traps inland or inshore at the start of the hurricane season, although specific strategies vary among individuals. A few fishers (n=2) also adjust their traps by adding different buoys or rope to increase trap robustness to storm impact. Only three respondents stated that they have not changed their fishing practices, and continue to leave their traps in the fishing grounds regardless of hurricane risk.

Table 3: Fishers' response strategies relating to the June-November hurricane season. The number and percentage of respondents that mentioned each response is reported

Fisher response categories	Number of responses	Percentage of responses
Bring all traps inland at the start of the hurricane season (avoids risk of hurricanes and the off-peak tourist season means there is limited demand)	8	33
Switch traps from lobster to reef fish (less money but can fish all year round)	5	21

Stop fishing for lobster April/May because of lobster spawning season. Take lobster traps out of the water but do not switch them to reef traps.	3	12.5
Leave the traps out regardless of hurricane forecast, too much effort to bring them in	3	12.5
Bring in traps closer/leave them in the inshore reef which is more sheltered from storms	3	12.5
Continue fishing for reef fish as normal, but fit traps with smaller buoys and longer rope (less drag/damage if there is a hurricane)	2	8
Stop fishing during the hurricane season, rely on alternative income	2	8
Bring some traps in and leave some out	2	8
Switch from inshore reef fishing to offshore hand-line fishing	1	4
Only bring in traps if a hurricane is predicted to hit	1	4
During hurricane season fish less but prepare for these months by saving more and spending less	1	4

Fishers' perceptions of future environmental change

Considering the impacts of previous hurricanes on the fishing community in Anguilla, fishers were understandably concerned about future hurricane impacts. When asked how they would feel if hurricane risk increased, 12 respondents stated that they would be very concerned. Five of these respondents stated that they would be concerned about the impact specifically on their fishing, and may consider another occupation. Perceptions relating to broader implications of environmental change elicited fewer responses (n=7), including specific concerns about coral bleaching (n=2), sea-level rise (n=1) and increases in sea temperature affecting fish movement (n=1).

With regards to environmental changes in the fishery, of greatest concern to these respondents were the changes in fish abundance. The majority of respondents considered that at present there are fewer fish (n=16) and smaller fish (n=6) than there were 20 years ago, particularly reef fish species such as groupers and parrot fish. Responses related the changes in fish abundance to an increase in the number of fishers (n=7) and irresponsible fishing practices (n=3), such as 'ghost fishing' from abandoned traps. One respondent perceived the problem of overfishing to be caused by the number of traps in use, rather than the number of fishers; while another stated that modern fishing gear has increased the effectiveness of fishing in Anguilla. As a result, a small number of respondents wanted to see changes to current fishery regulations, through implementing seasonal bans (n=5), no-take areas (n=1) and a ban on spear-fishing (n=1).

Marine-based tourist operators

Demographic characteristics

There was greater demographic variation among the 13 tourist operators in comparison to the fisher group. While most (n=10) of the tourist operators were male, there were also three women running marine-dependent tourist businesses. The majority of respondents in this group were Anguillian nationals (n=11) but two were European. Under half of the respondents were married (38%, n=5), and slightly more had children (46%, n=6). More respondents in this group had achieved a higher level of education, with three having been to university. The most common age category for these respondents was the 35–44 year group (n=7). Two respondents were younger than this modal age category, while four were older (n=1 in the 45–54 year and n=3 in the 55–64 year group).

Livelihood strategies of tourist operators

The majority of tourist operators owned their business (n=10), while the remaining three businesses were family-run enterprises. Respondents had worked in these businesses for between 1.5 and 27 years (mean±SD, 8±7 years). Almost all respondents (n=10) stated that their tourism business was their only or main source of employment. Five respondents had additional sources of income e.g. fishing, commercial diving, restaurant work (although this was predominantly during the off-peak tourist season, see below). There was general agreement among respondents (n=10) that there were other viable employment opportunities in Anguilla if necessary.

Seasonal variation in tourist demand

During the tourist season (December to April), the average number of days respondents work for their businesses is 5.5 days/wk (± 1 SD). For the off-season months when businesses are still open (May to July), the average number of days respondents work decreases slightly (4±3 days/wk). Tourist demand in Anguilla is strongly seasonal, with a sharp decline during the off-season summer months (August to November). Fig. 2highlights the seasonality in demand, determined by the tourism revenue or tourist numbers of each of the 13 tourist operators.

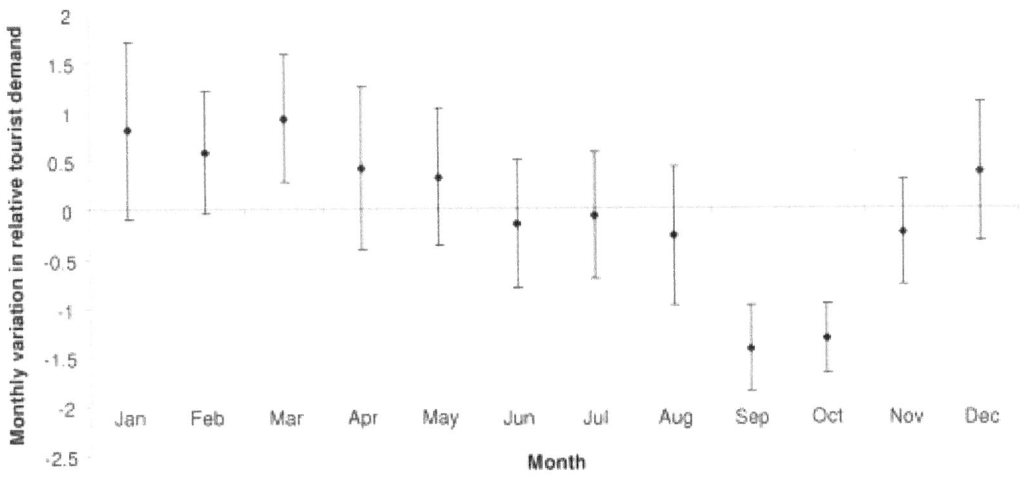

Figure 2: Monthly relative tourist demand for all (n=13) marine-based tourist operators in Anguilla. Bars show±1 SD.

The seasonality of the tourist industry affects many of the tourist operators' livelihoods (Table 4). Interviews revealed that, due to the combination of low tourist numbers and increased hurricane risk, most businesses close between August and October (n=9, 69%). Of the businesses that close, four respondents stated they take alternative seasonal employment during the off-peak season, or are able to rely on financial contributions from other family members. The majority of respondents (n=11, 85%) considered that the hurricane season (June to November) affects tourism in Anguilla to some degree. Two respondents stated that the hurricane season did not affect tourism, with one adding that the seasonality in tourism is principally driven by the lack of demand from American tourists who stay in the USA during their warmer summer months.

Table 4: Tourist operator responses to the open question "Do you think that the hurricane season affects tourism, and if so how? The number and percentage of respondents that mentioned each response is reported

Tourist operator response categories	Number of responses	Percentage of responses
Tourists are scared of hurricanes and avoid coming on holiday in the hurricane season	5	38
Hurricanes damage our beaches and tourism infrastructure	4	31
If a hurricane is forecast then tourists do not want to come here, or they cannot travel	4	31
After a hurricane, tourist arrivals decline	3	23
Hurricane season does not affect tourist demand, the summer tourist market changes with Americans staying in the USA.	2	15
If a hurricane impacts Anguilla and tourists are here, they cannot go to the beaches and so they will not return here again on holiday	1	8
The weather generally is not very good in the summer, it is hotter with more rainfall	1	8

Environmental change effects on tourist operators

Previous hurricane impacts on marine tourism livelihoods

Many of the tourist operators on the island were also severely affected by hurricane Luis in 1995 (Table 5). Five of the respondents' businesses suffered financial losses, both from direct damages and loss of earnings. As these marine tourist businesses vary substantially in terms of their infrastructure and assets, so too did their financial losses. For example, the direct impacts of the hurricane on Anguilla's surrounding islands and cays caused severe losses to the businesses located on Scilly Cay (estimated at $US 1 million in damages and $US 500,000 in lost earnings) and on Prickly Pear Cay (estimated as $US 20,000 in direct damages, plus loss of earnings) by their respective owners. Major losses were also sustained by the owners of Sandy Island, a small off-shore Cay, which was completely washed away by the hurricane; it took a decade to rebuild the Sandy Island restaurant and resort. Several other tourist businesses (n=5) suffered from the decline in numbers of tourists visiting Anguilla in the months and years following the hurricane. None of the respondents indicated that the damages caused their businesses to close permanently and, despite sustaining financial losses, these businesses have since been able to rebuild. Only two respondents indicated that the negative impact of the hurricane on their business meant they needed to rely on alternative income sources (e.g. carpentry and restaurant work) (Table 5). Several respondents noted that as a result of the severe impacts of hurricane Luis on Anguilla, it is now common-place for hotels on the island to close during the hurricane season (n=3).

Table 5: Tourist operator responses to the open question "Did hurricane Luis (1995) affect your business?" All responses are included, and the number and percentage of respondents that mentioned each is reported

Tourist operator response categories	Number of responses	Percentage of responses
The business sustained financial losses from direct damages and loss of earnings	5	38
There was a decline in the number of tourists (between six months and two years), causing loss of earnings	5	38
Damage to the beaches and coral reefs	5	38
The business was unaffected by the hurricane	3	23
Not working in this current business at that time	3	23
No electricity for two months	2	15
Relied on alternative sources of income	2	15
All of the hotels closed	1	8

Tourist operators' perceptions of environmental change

Many respondents (n=8) stated they would be concerned if hurricane risk increased, because of the implications of the hurricane season on tourism and the impacts sustained from hurricane Luis. Only two respondents said that they were not worried about hurricane risk. Like the fishers, perceptions regarding climate change elicited relatively few responses (n=5) from the tourist operators (Table 6). The climate change related threats that were of concern included increasing water temperature and coral bleaching (n=2), changing weather and tide patterns (n=2) and the increasing risk of hurricanes (n=1). When the tourist operators were asked specifically for their perceptions on the condition of the coral reef ecosystems, eight respondents stated that they had witnessed negative changes in the state of the reefs during their lifetime (i.e. physical damage to reefs (n=5), reduction in coral cover (n=3), and loss of colour or bleaching (n=2)).

Table 6: Tourist operator response categories to the open question "What do you feel are the causes of change to the coral reefs?" All responses are included, and the number and percentage of respondents that mentioned each is reported

Tourist operator response categories	Number of responses	Percentage of responses
Hurricane and storm damage	10	77
Overfishing/irresponsible fishing and no fisheries enforcement causing fewer and smaller fish and lobsters	8	62
Damage to coral from divers, snorkellers and spear-fishing	5	38
Warmer sea temperatures causing coral bleaching	3	23
More algae on the reef	1	8
Anchor damage to the reef	1	8
Pollution (e.g. suntan lotion and diesel from boats)	1	8

Hurricane and storm damage was mentioned by most respondents (n=10) as the primary cause of coral reef decline in Anguilla. The second most commonly mentioned stressor was fishing (n=8), and respondents spoke of the combination of too many fishers, irresponsible fishing practices and a lack of enforcement leading to major declines in fish and shellfish abundance, with knock-on implications for the coral reef. Increased prevalence of coral bleaching was a concern of some tourist operators (n=3). Additional changes to the coral reefs were also mentioned by individual respondents, including the growing prevalence of algae, damage caused to reefs by boat anchors and marine-based pollution. The majority of respondents (n=11, 85%) stated that coral reef condition affects their business, because unhealthy coral reefs mean there are fewer fish, and their client-base wishes to see fish and coral. Several respondents also referred to tourist demand for seafood, and that coral reef condition affects this aspect of the tourism market.

DISCUSSION

Many Caribbean islands are heavily dependent on tourism and fisheries for livelihood opportunities. Marine-dependent livelihoods on Anguilla are particularly susceptible to impacts on the marine environment from hurricanes, and the degradation of marine resources from over-exploitation and coral bleaching [29] and [42]. Previous hurricane events on Anguilla, in particular hurricane Luis in 1995 which represents the most significant environmental disaster in living memory, have demonstrated the vulnerability of these livelihoods to a variety of impacts, including the loss of fishing gear and damage to business infrastructure, reduced catch rates, and a decreased demand for seafood. Therefore, expected increases in hurricane risk due to changing global climate conditions which will cause further degradation of the marine environment, are likely to have major consequences for marine-resource livelihoods. The extent to which fishers and marine tourist operators responded to the impacts brought by hurricane Luis on Anguilla may have implications for their potential resilience to future changes in the marine environment.

Environmental threats to marine-dependent livelihoods

Hurricanes represent the most severe environmental threat affecting marine resource-users in Anguilla, causing both short- and long-term impacts. Immediate effects from hurricane Luis in 1995 included damage to fishing gear and boats, reducing the ability of fishers' to catch fish, and damages to business infrastructure and the decline in tourist arrivals causing major financial losses for tourist operators. In addition, the market-demand for seafood from hotels and restaurants was also significantly reduced, resulting in fishers being unable to sell what little catch they had. Both groups of marine-resource users are also vulnerable to the longer-term environmental impacts of hurricane events, in particular the destruction of coral reefs and fishing grounds, and associated changes in fish abundance.

Chronic environmental problems caused by the over-exploitation of marine resources and coral bleaching episodes are also an issue for both fishers and tourist operators. For example, the current depletion of the inshore reef in Anguilla may mean that more fishers are forced to start exploiting offshore fishing grounds, while other fishers may choose to leave the fishery altogether in the future. There may also be market-demand implications; if fish and shellfish become scarcer and/or if reliance on imports increases, then prices may increase on the island. Tourist operators that depend directly on the coral reefs (dive businesses, charter boat companies) are also expected to suffer from further coral reef decline. However, by comparison to the economic and environmental impacts sustained after a hurricane, issues of over-exploitation and coral bleaching may have smaller and more incremental effects on these marine-dependent livelihoods.

This study has shown that fishers and tourist operators were able to respond to the severe 1995 hurricane, through behavioural and livelihood adaptations, such as changes in fishing strategies, or reliance on alternative sources of income. However, if hurricanes become more frequent or severe, e.g. see [2] and [32] the effects on these marine resource-users may be critical. The coral reefs and fishing grounds surrounding Anguilla, as is common for the Caribbean region, are already in a highly degraded condition because of sustained over-fishing and hurricane damage [28] and [43]. Consequently, the extent to which the island's coral reefs and fishing grounds would be able to sustain another major hurricane is unclear. Indeed, several respondents in this study commented that, due to the present degradation of the coral reefs in Anguilla, they did not believe the reefs could withstand another extreme event like hurricane Luis. These resource-users thus consider the ecological resilience [44] and [45] of this marine system to be already heavily compromised.

Marine-dependent livelihoods and social resilience to environmental change

Despite variation among fishers in terms of personal characteristics and fishing-related assets and expenditures, their livelihood strategies and responses to hurricane Luis were largely similar. Indeed, the legacy of hurricane Luis has been manifest in a suite of direct responses by this sector (Table 3), and provides evidence of marine resource-users adapting livelihood strategies to withstand environmental uncertainty. The vast majority of respondents use mixed fishing strategies (fish and lobster traps, hand-lines) and many switch target species or fishing practices according to seasonal variations in prey abundance and hurricane risk. In addition, while most respondents considered fishing to be their principal occupation, approximately half subsidised their fishing with alternative employment. These features are all expected to contribute to fisher's social resilience to environmental variability or change.

In addition, the profitability of fishing in Anguilla, with some fishers earning many thousands of dollars each month, suggests that this is not the 'occupation of the last resort', and that it does not fit the typical characterisation of small-scale artisanal fishers as 'the poorest of the poor' [23]. The income that Anguillan fishers can make, together with the substantial asset-base that they can accumulate and the flexibility shown by their changes in behaviour post-hurricane Luis, may collectively enhance their intrinsic social resilience, by enabling them to buffer some of the consequences of change or variation in resource productivity [22]. The strong social cohesion within some of these respondents' fishing families and communities may also buffer individuals against uncertainty or fluctuations in the resource [46] and [47].

The fishers also share features that potentially may restrict their capacity to develop resilience. Family status and education can be important measures of how reliant resource-users are on a resource and therefore how resilient they might be to change [22]. For example, the majority of fishers in this study have families and children, which may mean that these individuals are less able to experiment with alternative employment options, as family responsibilities mean they need to retain employment stability. Consequently, these respondents may be less flexible to future changing conditions affecting their occupation [21]. Likewise, many of these respondents had left the education system early, worked in the fishery for most of their lives and may therefore have relatively few transferable skills. It is recognised that these factors may reduce the flexibility of individuals to move away from resource-dependent livelihoods such as fishing [21] and [48]. However, many fishers in this study stated that they are able to secure alternative employment and there was no indication that they were unable to support their families during the hardship brought by hurricane Luis, or during subsequent hurricane seasons. Importantly, all of these respondents were able to return to, or reconstruct, their marine-dependent livelihood after hurricane Luis.

The factor that may present the greatest limitation to adapting to change is, however, 'fisher ethic'; the expression of an entrenched attachment by fishers to their primary occupation. There was a strong desire among respondents to return to fishing after the events of hurricane Luis, even though these fishers sustained substantial losses in gear, the fishing grounds were damaged and the market-demand had declined substantially. Fishers in this study and others have shown that there is more to the occupation of fishing than financial incentive alone [49] and [50]. This connection to their occupation has been attributed to the psycho-cultural characteristics of people who fish (e.g. being adventurous, courageous, active, independent), and notably because fishing is more than an occupational preference, it is at the core of the self-identity of a fisher [49]. Fishers who are strongly attached to their resource-dependent livelihood are therefore potentially less resilient to change or uncertainty in the resource [21].

By comparison, the tourist operators may be more resilient to change and uncertainty than the fishers because many already have more diversified livelihood strategies, in addition to relatively high levels of education and greater transferable skills from working in other sectors. Fewer of these respondents have family responsibilities, and the vast majority of respondents also stated that there were other possible employment opportunities on the island if necessary; although again this may itself depend on the state of the marine environment in the future. The combination of these factors may make individuals more flexible and dynamic in their livelihood strategies and future planning [21] and [23]. The recovery of the tourist operators following the devastating events of hurricane Luis, in some cases losing many years of income, and rebuilding their entire business infrastructure, suggests that these tourist operators have the financial buffer to withstand stress. Finally, while all of these tourist operators have strong personal or family ties to their businesses, by comparison to the fishers, there did not appear to be the same degree of social or cultural dependency on their occupation. Collectively, these attributes imply that the marine tourist operators may have potentially more social resilience to environmental change.

However, in general there was little variation between the fishers and tourist operators with regards to their livelihood strategies, their strong dependence on the marine environment, and their susceptibility to environmental impacts from hurricanes and coral reef degradation. Of particular importance was the dependence by all of these respondents on the tourism industry. For example, even though many of the fishers and tourist operators stated they had the means to generate income aside from their primary occupation, the vast majority of their alternative occupations were also tourism-dependent.

This dependence on the tourism industry may have the most significant implications for the vulnerability of these marine resource-users to environmental change. As has been shown, tourists visit Anguilla primarily for the beaches and not for the coral reefs [34]; which might indicate some resilience by the island's tourism industry (and tourism operators) to cope with changes in coral reef health. The implications of hurricanes on tourism-dependent livelihoods may, however, be more substantial. For example, as the seasonality in tourism demand on Anguilla (Fig. 2) may be driven by the risk of hurricanes and favourable summer conditions in the home countries of the tourists that visit the island (mainly USA nationals), tourism-dependent livelihoods are potentially vulnerable if future environmental change negatively affects tourism demand. For instance, if hurricane risk in Anguilla increases (or is perceived to increase), tourists may choose not to holiday on the island [34]. On the other hand, global warming may also result in altered climate conditions in the countries of the tourists that currently visit Anguilla e.g. USA, Europe; [51], which could also affect future travel patterns and demand (and is clearly unrelated to hurricane activity). Consequently, the strong dependence by all of the marine resource-users in Anguilla on the tourism industry may ultimately undermine their capacity to develop social resilience to future environmental change.

CONCLUSIONS

Fishers and tourist operators in Anguilla are highly dependent on marine and coastal resources. The capacity of these marine-dependent livelihoods to use resources is significantly affected by hurricane impacts and marine resource degradation. Marine-dependent livelihoods in Anguilla have been able to respond and rebuild their livelihoods after past impacts from hurricanes through adaptations such as changes in fishing strategies and livelihood diversification, which suggests a capacity for resilience in the face of environmental stress. However, their ability to cope with future stresses will clearly depend on the extent of the environmental changes. The high-end tourism on Anguilla boosts the demand for expensive seafood and provides fishers with stable and often high levels of income. However, the threat of environmental change on marine-dependent livelihoods is common throughout the Caribbean. Indeed, Caribbean-wide changes in the marine environment show that issues of marine degradation are widespread throughout the region [43] and [52], and are expected to worsen with climate change [2] and [53]. Urgent attention is required to provide sustainable and resilient futures for the many thousands of marine-dependent livelihoods throughout the Caribbean threatened because of already depleted marine resources and future environmental changes.

ACKNOWLEDGEMENTS

Thank you to all of the individuals who gave up their time to participate in this study, and to the staff at Anguilla DFMR who provided invaluable local information and logistical support. Thanks also to Katie Newton who assisted with data collection. Johanna Forster was supported by a joint studentship from the Economic and Social Research Council and the Natural Environment Research Council (UK).

REFERENCES

1. Hassan R, Schole R, Ash N. Coastal systems. Millennium ecosystem assessment (MEA), Ecosystem and human well-being, Current state and trends assessment. Washington, D.C.: World Resources Institute; 2005 chapter19.
2. Mimura N, Nurse L, McLean RF, Agard J, Briguglio L, Lefale P, et al. Small Islands. In: Parry ML, Canziani OF, Palutikof JP, van der Linden PJ, Hanson CE, editors. Climate Change 2007: Impacts, Adaptation and Vulnerability Cambridge. Cambridge University Press; 2007. p. 687–716.
3. Allison EH, Perry AL, Badjeck M-C, Adger WN, Brown K, Conway D, et al. Vulnerability of national economies to the impacts of climate change on fisheries. Fish and Fisheries 2009. http://dx.doi.org/10.1111/j.467-2979.008.00310x.
4. Thomas DSG, Twyman C. Equity and justice in climate change adaptation amongst natural-resource-dependent societies. Global Environ ChangeHuman Policy Dimensions 2005;15:115–24.

5. Marshall NA. Understanding social resilience to climate variability in primary enterprises and industries. Global Environ Change-Human Policy Dimensions 2010;20:36–43.
6. Adger WN. Social vulnerability to climate change and extremes in coastal vietnam. World Dev 1999;27:249–69.
7. Pomeroy RS, Ratner BD, Hall SJ, Pimoljinda J. Coping with disaster: rehabilitating coastal livelihoods and communities. Mar Policy 2006;30:786–93.
8. Dulvy NK, Allison E. A place at the table? Nature 2009;3:68–70.
9. Wilkinson C., Linden O., Cesar H., Hodgson G., Rubens J., Strong A.E. Ecological and socioeconomic impacts of 1998 coral mortality in the Indian Ocean: an ENSO impact and a warning of future change? Ambio. 1999;28:188–196.
10. McWilliams JP, Cote IM, Gill JA, Sutherland WJ, Watkinson AR. Accelerating impacts of temperature-induced coral bleaching in the Caribbean. Ecology 2005;86:2055–60.
11. Mimura N, Nunn PD. Trends of beach erosion and shoreline protection in rural Fiji. J Coastal Res 1998;14:37–46.
12. Fish MR. Coastal development and sea-level rise: impacts on sandy beaches and sea turtles. University of East Anglia; 2006 (Ph.D. thesis).
13. Badjeck MC, Allison EH, Halls AS, Dulvy NK. Impacts of climate variability and change on fishery-based livelihoods. Mar Policy 2010;34:375–83.
14. Adger WN. Social and ecological resilience: are they related? Prog Human Geogr 2000;24:347–64.
15. Adger WN, Hughes TP, Folke C, Carpenter SR, Rockström J. Social-ecological resilience to coastal disasters. Science 2005;309:1036–9.
16. Adger WN. Social capital, collective action, and adaptation to climate change. Econ Geogr 2003;79:387–404.
17. Fraser EDG, Mabee W, Slaymaker O. Mutual vulnerability, mutual dependence – The reflexive relation between human society and the environment. Global Environ Change-Human Policy Dimensions 2003;13:137–44.
18. Hughes TP, Bellwood DR, Folke C, Steneck RS, Wilson J. New paradigms for supporting the resilience of marine ecosystems. Trends Ecol Evol 2005;20: 380–386.
19. Folke C. Resilience: the emergence of a perspective for social-ecological systems analyses. Global Environ Change-Human Policy Dimensions 2006;16: 253–267.
20. Gallopin GC. Linkages between vulnerability, resilience, and adaptive capacity. Global Environ Change-Human Policy Dimensions 2006;16:293–303.
21. Marshall NA, Fenton DM, Marshall PA, Suffon SG. How resource dependency can influence social resilience within a primary resource industry. Rural Sociol 2007;72:359–90.
22. Marshall NA, Marshall PA. Conceptualizing and operationalizing social resilience within commercial fisheries in northern Australia. Ecol Soc 2007:12.
23. Allison EH, Ellis F. The livelihoods approach and management of small-scale fisheries. Mar Policy 2001;25:377–88.
24. McClanahan TR, Cinner JE, Maina J, Graham NAJ, Daw TM, Stead SM, et al. Conservation action in a changing climate. Conserv Lett 2008;1:53–9.
25. Schwarz AM, Bene C, Bennett G, Boso D, Hilly Z, Paul C, et al. Vulnerability and resilience of remote rural communities to shocks and global changes: Empirical analysis from Solomon Islands. Global Environ Change-Human Policy Dimensions 2011;21:1128–40.
26. Carty B, Petty C. Anguilla: Tranquil Isle of the Caribbean. London, UK: Macmillan Eductaion Ltd.; 2000; 138.
27. Mukhida F, Gumbs JC. Linking the social with the natural: a socio-economic review of Shoal Bay and Island Harbour, Anguilla. Rep Anguilla Natl Trust Anguilla Dep Fish Mar Res 2008:52.
28. ECLAT. Anguilla: An assessment of the economic and other damages caused by Hurricane Lenny November 1999. United Nations, Economic Commission for Latin America and the Caribbean, Caribbean Development and Cooperation Committee. 2000; 30pp.
29. Wilkinson C, Souter D. Status of Caribbean coral reefs after bleaching and hurricanes in 2005. Global Coral Reef Monitoring Network Reef and Rainforest Research Centre, Townsville 2008:152.
30. Goodess CM. How is the frequency, location and severity of extreme events likely to change up to 2060? Environ Sci Policy 2013;27:S4–14.
31. Trenberth K. Uncertainty in hurricanes and global warming. Science 2005;308:1753–4.
32. Webster P.J., Holland G.J., Curry J.A., Chang H.R. Changes in tropical cyclone number, duration, and intensity in a warming environment. Science. 2005;309:1844-1846.
33. Becken S, Hay JE. Tourism and climate change: risks and opportunities. U.K.: Channel View Publications; 2007; 329.
34. Forster J, Schuhmann PW, Lake IR, Watkinson AR, Gill JA. The influence of hurricane risk on tourist destination choice in the Caribbean. Clim Change 2012;114:745–68.
35. Abernethy KE, Allison EH, Molloy PP, Cote IM. Why do fishers fish where they fish? Using the ideal free distribution to understand the behaviour of artisanal reef fishers Canadian J Fish Aquat Sci 2007;64:1595–604.
36. WTTC. The Caribbean: the impact of travel and tourism on jobs and the economy. World Travel and Tourism Council, London, UK. 84p. ⟨http://www. wttc.org/eng/Download_Centre/#⟩accessed 8.11.09.. 2004.
37. Bunce L., Townsley P., Pomeroy R., Pollnac R.B. Socioeconomic manual for coral reef management. Aust Inst Mar Sci, Townsville. 2000.
38. Bryman A. Social Research Methods. 2nd ed. Oxford: Oxford University Press; 2004.

39. Government of Anguilla. Demography, culture, migration, crime, marital status and fertility of the resident population of Anguilla according to the May 2001 census. Statistics Department, Ministry of Finance, Government of Anguilla. URL: ⟨http://gov.ai/statistics/⟩ accessed19.04.10.. 2001.
40. Government of Anguilla. Education and training of the resident population of Anguilla according to the May 2001 census. Statistics Department, Ministry of Finance, Government of Anguilla. ⟨http://gov.ai/statistics/⟩ accessed 19.04.10.. 2001.
41. Government of Anguilla. Economic activity of the resident population of Anguilla according to the May 2001 census. Statistics Department, Ministry of Finance, Government of Anguilla. ⟨http://gov.ai/statistics/⟩ acccessed 19.04.10.. 2001.
42. Spalding MDA. Guide to the coral reefs of the Caribbean. Berkeley, USA: University of California Press; 2004.
43. Paddack MJ, Reynolds JD, Aguilar C, Appeldoorn RS, Beets J, Burkett EW, et al. Recent region-wide declines in caribbean reef fish abundance. Curr Biol 2009;19:590–5.
44. Bellwood DR, Hughes TP, Hoey AS. Sleeping functional group drives coral-reef recovery. Curr Biol 2006;16:2434–9.
45. Hughes TP, Rodrigues MJ, Bellwood DR, Ceccarelli D, Hoegh-Guldberg O, McCook L, et al. Phase shifts, herbivory, and the resilience of coral reefs to climate change. Curr Biol 2007;17:360–5.
46. Hicks CC, McClanahan TR, Cinner JE, Hills JM. Trade-offs in values assigned to ecological goods and services associated with different coral reef management strategies. Ecol Soc 2009:14.
47. Ramirez-Sanchez S, Pinkerton E. The impact of resource scarcity on bonding and bridging social capital: the case of fishers' information-sharing networks in Loreto, BCS, Mexico. Ecol Soc 2009:14.
48. Allison HE, Hobbs RJ. Resilience, adaptive capacity, and the "Lock-in trap" of the western Australian agricultural region. Ecol Soc 2004:9.
49. Pollnac RB, Poggie JJ. Job satisfaction in the fishery in two southeast Alaskan towns. Human Organ 2006;65(3):329–45.
50. Pollnac RB, Poggie JJ. Happiness, well-being, and psychocultural adaptation to the stresses associated with marine fishing. Human Ecol Rev 2008;15: 194–200.
51. IPCC. Climate Change 2007: The Physical Science Basis. In: Solomon S, M. Manning DQin, Chen Z, Marquis M, Averyt KB, Tignor M, Miller HL, editors. Contribution of Working Group 1 to the Forth Assessment Report of the Intergovernmental Panel on Climate Change. UK and New York, USA: Cambridge University Press, Cambridge; 2007. p. 996.
52. Gardner TA, Cote IM, Gill JA, Grant A, Watkinson AR. Long-term region-wide declines in Caribbean corals. Science 2003;301:958–60.
53. Hoegh-Guldberg O. Coral reefs in a century of rapid environmental change. Symbiosis 2004;37:1–31.

CHAPTER 16

RECREATIONAL FISHER COMPLIANCE IN TEMPERATE MARINE CONSERVATION AREAS

D. Lancaster, P. Dearden, N.C. Ban

Keywords: Rockfish Conservation Area (RCA), Marine Protected Area (MPA), Compliance, Recreational fishing, Sport fishers, Bycatch

INTRODUCTION

The last half-century has seen global declines in marine resources, with large-bodied and long-lived species experiencing the most noticeable decreases (Pauly et al., 1998, Molfese et al., 2014). Many countries have started using fisheries closures, harvest refugia, and marine protected areas (MPAs) (hereafter jointly referred to as marine conservation areas) in an effort to protect and recover depleted marine resources. Although such conservation areas vary in size, distribution, and protection levels, their growing prevalence highlights increasing concern and actions around marine resource conservation (Allison et al., 1998, Marinesque et al., 2012).

Spatial management is an important tool in marine conservation, but lack of compliance, even from comparatively small recreational fisher populations (Post et al., 2002), may reduce effectiveness significantly (Little et al., 2005, Arias, 2015). For example, a simulation of line fishing infringement in the Great Barrier Reef Marine Park showed that even a few fishers intensively fishing in an area made marine reserves ineffective at conserving fish species biomass (Little et al., 2005). With continued fishing activity in conservation areas, fish populations often fail to recover (Kritzer, 2004). While many commercial fisheries have on-board observers, recreational fisheries are difficult to monitor. Recreational fishing makes up 12% of annual global marine fish catches (Cooke and Cowx, 2004 and Granek et al., 2008) and has been implicated in fish stock collapses (Post et al., 2002 and Lewin et al., 2006). Recreational fisher noncompliance could have a significant impact on the success of management areas, especially in popular, and typically overfished coastal areas. However, few studies have attempted to quantify recreational fisher's compliance and its drivers (Bergseth et al., 2013).

Several factors may influence fisher compliance (Pollnac et al., 2010 and Bergseth et al., 2013). Consultation, education, and communication enhance positive perceptions of marine conservation areas and help to foster high fisher compliance (Cinner et al., 2005, Read et al., 2011 and Pita et al., 2013). For example, in a review of 127 international MPAs, local perceptions and understanding of marine reserves was the main factor leading to high compliance (Pollnac et al., 2010). Furthermore, in the south of France, the involvement of fishing guilds in MPA planning and communication between fishers and scientists led to high levels of social acceptance and compliance (Leleu et al., 2012). Compliance in the south of France was found to be dependent on positive perceptions of marine conservation areas (Leleu et al., 2012). Additionally, an Australian study found that a thorough knowledge of marine conservation area existence, regulations, and the possible consequences of noncompliance was essential for high compliance (Read et al., 2011).

Compliance research has focused mainly on commercial and artisanal fishers in tropical MPAs (Gribble and Robertson, 1998, McClanahan et al., 2005, Cinner et al., 2005, Pollnac et al., 2010, Leleu et al., 2012 and Bergseth et al., 2013). Very few studies target recreational fisher noncompliance and its drivers (Schill and Kline, 1995, Read et al., 2011, Smallwood and Beckley, 2012, Arias and Sutton, 2013), and recreational compliance studies in temperate environments are rare (Blank and Gavin, 2009 and Read et al., 2015). Our study fills an important knowledge gap by assessing recreational fisher compliance and its drivers in non-tropical marine conservation areas.

Citation: D. Lancaster, P. Dearden, N.C. Ban, Recreational fisher compliance in temperate marine conservation areas, Canada, Global Ecology and Conservation, Volume 4, July 2015, Pages 645-657, ISSN 2351-9894, http://dx.doi.org/10.1016/j.gecco.2015.11.004.

This study uses Rockfish Conservation Areas (RCAs) in British Columbia (BC), Canada as a case study of drivers of compliance in marine conservation areas. RCAs are a network of Canadian fisheries closures that were implemented rapidly with minimal public outreach after the fact (Haggarty, 2014). Our case study objectives were to (1) assess levels of recreational fisher (hereafter referred to as fisher) RCA knowledge and compliance, (2) explore factors influencing fisher RCA knowledge and compliance, (3) quantitatively assess levels of fisher rockfish bycatch and release rates, (4) elicit fisher perceptions of RCAs, and (5) obtain fishers suggestions for improving rockfish conservation.

Case study description: rockfish conservation areas in BC

In BC many marine fishes–especially those susceptible to overfishing such as rockfish–have declined substantially over the past fifty years (Love et al., 2002, Post et al., 2002, Cooke and Cowx, 2004, Granek et al., 2008 and Hutchings et al., 2012). Rockfish (*Sebastes*) are a long-lived (inshore species live from 50 to 120 years), benthic genus that experience severe barotrauma when captured at depth and brought to the surface (Lotterhos et al., 2013). Once a fish has been captured, it rarely survives if released unless redescended to depth ((Hannah and Matteson, 2007), Parker et al., 2006, Jarvis and Lowe, 2007). The ~37 species of rockfish found along the coast of BC play an important role in both predator and prey relationships within a variety of food webs (Love et al., 2002). Rockfish are primarily non-migratory, with most individuals occupying a single reef. Their low mobility means that they should respond well to spatial protection (Love et al., 2002 and Haggarty, 2014).

Between 2003 and 2007, a network of 164 Rockfish Conservation Areas (RCAs) was created to protect rockfish populations as part of the Rockfish Conservation Strategy (Yamanaka and Logan, 2010). The Rockfish Conservation Strategy addressed concerns of fishers and NGOs that inshore rockfish populations had been greatly depleted by the development of a successful commercial rockfish fishery in the 1980s (Marliave and Challenger, 2009 and Yamanaka and Logan, 2010). The RCAs were designed to protect five inshore rockfish species: Yelloweye (*Sebastes ruberrimus*), Quillback (*Sebastes malinger*), Tiger (*Sebastes nigrocinctus*), Copper (*Sebastes caurinus*), and China (*Sebastes nebulosus*) rockfish (Yamanaka and Logan, 2010). The Committee on the Status of Endangered Wildlife in Canada (COSEWIC) currently lists Quillback rockfish as threatened and Yelloweye rockfish as a species of concern (COSEWIC, 2009a and COSEWIC, 2009b).

Rockfish populations have seen the most dramatic declines in the Salish Sea (Strait of Georgia, Puget Sound, and Strait of Juan de Fuca), with Yelloweye rockfish at 12% of their 1918 biomass (DFO, 2012). The Salish Sea also experiences the highest levels of recreational fishing pressure. Recreational rockfish catch accounts for 89% of the total annual rockfish catch within the Strait of Georgia (Data from Fisheries and Oceans Canada in Haggarty 2014). As such, two thirds of RCAs were placed in this region to address special concerns for rockfish. RCAs restrict many recreational and commercial fishing activities within their boundaries (Table 1), with recreational fishing limited to invertebrate trapping and hand picking, and smelt gillnetting (all other recreational fishing is restricted). The Rockfish Conservation Strategy also reduced total allowable commercial rockfish catches in all inside waters (all waters between Vancouver Island and the mainland) by 75%, and recreational fisher bag limits were reduced from five to one rockfish per day in the Strait of Georgia (Yamanaka and Logan, 2010).

Table 1: Permitted and prohibited commercial and recreational fishing activities within RCAs. Aboriginal fisheries are not included in the table as their fishing activities are unrestricted within RCAs due to their constitutional right to harvest (Haggarty, 2014 and DFO, 2014)

Commercial		Recreational	
Permitted fisheries	Prohibited fisheries	Permitted fisheries	Prohibited fisheries
• Hand picking and diving for invertebrates • Prawn and crab trapping • Smelt by gillnet • Scallop trawling • Salmon by seine or gillnet • Herring by seine, gillnet, and spawn-on-kelp • Sardine by gillnet, seine, and trap • Krill by mid-water trawl • Opal squid by seine • Groundfish by mid-water trawl	• Groundfish bottom trawl • Groundfish hook and line for halibut, inside rockfish, outside rockfish, lingcod, dogfish • Sablefish by trap • Salmon Trolling • Opal squid by hook and line or ring net • Shrimp by trawl	• Hand picking of invertebrates • Prawn and crab trapping • Smelt by gillnet	• Groundfish by hook and line • Salmon by hook and line (e.g. Trolling, Jigging, Mooching) • Spearfishing

Despite these conservation measures, preliminary studies on the effectiveness of RCAs at rebuilding inshore rockfish populations have not consistently shown a significant difference inside and outside RCAs (Challenger and Marliave, 2009, Cloutier, 2010, Chalifour, 2012 and Haggarty, 2014). The introduction of rockfish total allowable catch reductions and the integration of the commercial groundfish fishery that occurred simultaneously with RCA creation could be confounding these results. The commercial Groundfish fishery jointly manages all Groundfish stocks (e.g. halibut, rockfish, sablefish) and sets catch restrictions. Thus, these recent dramatic reductions on rockfish total allowable catch quotas (≥50% reductions) mean rockfish stocks could be rebuilding both inside and outside RCAs. However, low levels of recreational compliance to RCA regulations could also be significantly impacting the ability of RCAs to effectively rebuild rockfish populations (Haggarty et al. in review). Commercial rockfish total allowable catch reductions limit the ability of commercial fishers to catch rockfish even as bycatch, and on-board and dockside observer programs, in tandem with GPS tracking on commercial fishing vessels, means commercial fishing compliance is extensively monitored and believed to be nearly 100% (Haggarty, 2014). Although commercial fishers could be negatively impacting rockfish populations through habitat degradation or rockfish prey reduction, commercial noncompliance is a lesser concern in RCAs. However, very little is known about recreational fisher compliance (Haggarty, 2014). Research by Haggarty et al. (in review), using Fisheries and Oceans Canada (DFO) aerial fly over data, suggested that suspected low levels of recreational fisher compliance could be significantly impacting the effectiveness of RCAs in the Salish Sea.

METHODS

We carried out structured interviews with recreational fishers in the southern Canadian Salish Sea (Southern Gulf Islands and Victoria area). This region was selected for its high density of RCAs, popularity amongst fishers, its accessibility, and concerns about high, localized rockfish declines. Surveys were conducted in 16 locations at marinas, docks and boat launches during the peak recreational fishing season (July and August 2014) (Fig. 1). We used clustered convenience sampling to conduct structured 5-min interviews with saltwater fishers. Individuals who were likely fishers–based on boat type, fishing gear, or presence at popular fishing sites–were approached by one of two researchers and asked to participate in a 5-min voluntary survey on Rockfish Conservation Areas.

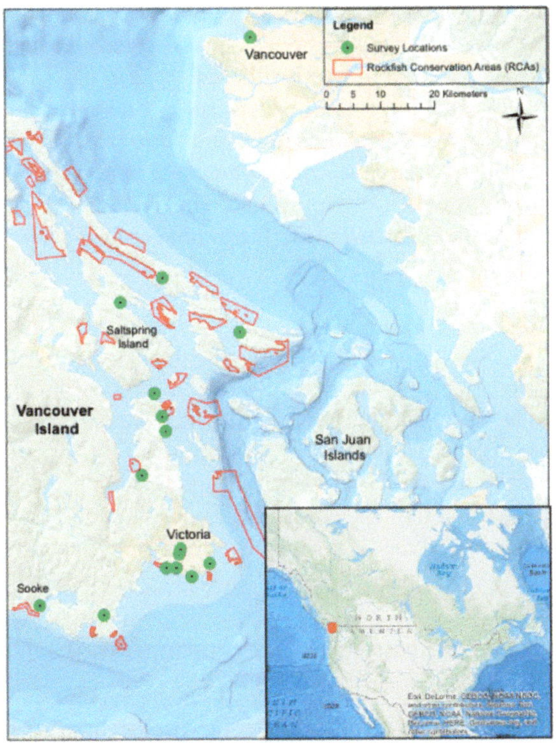

Figure 1: Recreational fisher survey locations at marinas and boat launches. Red square on locator map marks study region. (For interpretation of the references to colour in this figure legend, the reader is referred to the web version of this article.)

Quantifying RCA knowledge and compliance

We assessed levels of RCA knowledge and compliance through structured (closed and open) questions and the randomized response technique (RRT). Questions about RCA knowledge asked, for example, if fishers had knowledge of RCA existence before beginning the survey and if they felt confident of RCA boundary locations. We explained all permitted fishing activities to participants and then used charts of RCA locations as a visual aid when asking respondents if they had accidentally fished using prohibited techniques in an RCA in the past 2 years. We asked respondents to locate their typical fishing area(s) on the charts and then asked if they had ever accidentally fished using prohibited techniques in nearby RCAs. We reminded respondents that surveys were completely anonymous and that the researchers were exclusively associated with the University of Victoria.

We used two techniques to assess levels of intentional noncompliance. First, we used a non-sensitive direct question technique (Arias and Sutton, 2013). We asked the question, "Do you personally know anyone who has intentionally fished, using prohibited techniques, in a Rockfish Conservation Area in the past two years?". This technique is considered non-sensitive since it does not ask respondents if they have personally illegally fished and it does not ask them to identify who they know that has illegally fished. However, by tallying the number of people who know someone who illegally fishes in RCAs, intentional noncompliance can be estimated indirectly. For example, given a large and varied sample, if ten respondents personally know someone who has intentionally illegally fished in an RCA this method assumes that there are ten people who intentionally illegally fish in RCAs. However, since this technique does not ask respondents to specify the number of people they know illegally fish in RCAs, this method could underestimates levels of intentional noncompliance. Second, we used the Randomized Response Technique (RRT) (Horovitz and Greenburg 1976, Fox and Tracey 1986, Arias and Sutton, 2013) to ask respondents if they had personally intentionally fished, using prohibited techniques, in an RCA in the past 2 years. RRT is effective at obtaining more honest answers from respondents by assuring them response anonymity (i.e. the interviewer does not know if participants are telling the truth). We used a 20-sided die inside an opaque cup. Out of sight of the interviewer, respondents were instructed to shake the cup and look at the number they rolled. If they rolled a 1 they were instructed to answer the question with "yes". If they rolled a 2 they were instructed to answered "no". If they rolled any other number from 3 to 20, they were instructed to answer the sensitive question honestly: "Have you ever intentionally fished, using prohibited techniques, in a Rockfish Conservation Area in the past 2 years?". The noncompliance rate can then be estimated statistically using the known, "assumed" probabilities for predetermined responses (Krumpal, 2012 and Arias and Sutton, 2013). We call these "assumed" probabilities because they depend upon participants following the RRT instructions without error. For our study, the assumed probability of an honest answer was $P1=0.9$, the assumed probability of a predetermined "yes" answer was $P2=0.05$, and the assumed probability of a predetermined "no" answer was $P3=0.05$. The expected value θ of observing a "yes" answer can be calculated as $\theta=P2+P1\pi$, where π is the–in this case unknown–proportion of survey respondents who agree with the sensitive question (i.e. respondents who answer the sensitive question with a yes). Since π is unknown, we can use the observed yes responses as an estimate of θ to calculate $\hat{\pi}$ (*estimated* proportion of respondents who answer the sensitive question honestly with yes):

$$\hat{\pi} = \frac{\hat{\theta} - P2}{P1}$$

where $\hat{\theta}$ is the observed number of yes responses (including both predetermined and honest sensitive question yes responses). The variance can be calculated using:

$$\text{Var}(\hat{\pi}) = \frac{\hat{\theta}(1 - \hat{\theta})}{nP1^2}.$$

We tallied participant responses to three knowledge-based survey questions: (1) knowledge of RCA boundaries; (2) knowledge of RCA fishing restrictions; (3) knowledge of rockfish bag limits. Responses to these questions were first converted to binomial variables. For example, only fishers who stated they were confident of RCA boundaries were given a knowledge point. Fishers who thought they knew RCA boundaries or were unsure of boundaries were not given a knowledge point. Fishers who answered all three questions 100% correctly received a score of 3. Fishers who answered 2 of 3 questions 100% correctly received a score of 2, etc. We combined participant responses to these questions to create a gradient of fisher RCA and rockfish regulation knowledge.

Factors contributing to RCA knowledge and compliance

We used the following dependent variables–included as questions in the survey–as measures of RCA and rockfish fishing knowledge and compliance: (1) knowledge of RCA boundaries, (2) knowledge of rockfish bag limits in the Strait of Georgia, (3) previous knowledge of RCA existence, (4) admission of accidental illegal fishing in RCAs, and (5) personally knowing someone who had intentionally illegally fished in an RCA. We hypothesized that the following independent variables would lead to higher knowledge of RCAs and/or higher compliance: (1) greater number of years fishing, (2) greater number of days fishing (past 2 years), (3) higher percentage of time fishing rockfish (past 2 years), (4) obtaining fishing information directly from the fisheries management agency (Fisheries and Oceans Canada, DFO), (5) residence nearby (Vancouver Island or in British Columbia), (6) gender, and (7) previous knowledge of RCAs.

We used several analyses to assess factors that influence RCA knowledge and compliance. First, we used generalized linear models (GLMs) in R (R Core Team, 2013), using the package MASS (Venables and Ripley, 2002) and the function glm with binomial errors and logit-link (Crawley, 2007), to identify significant contributors to RCA compliance and knowledge. We removed seven partially completed surveys from our GLM dataframe (n=318) to facilitate accurate comparison. We used a subtractive method and the Akaike Information Criterion (AIC) to create models with the lowest AIC score for the greatest number of significant predictor variables. We used the standard alpha level 0.05 (Fisher, 1934). We calculated the deviance explained for each GLM with significant predictor variables using the equation:

$$de = \frac{(nd - rd)}{nd}$$

where *de* is deviance explained, *nd* is null deviance, and *rd* is residual deviance.

We used deviance explained, effect size, and p-value to determine how well our models predicted RCA knowledge and behaviour.

Second, we used two methods to assess whether fisher characteristics align with high or low RCA knowledge and compliance. We used multi-dimensional scaling (MDS) in R to determine if RCA knowledge was clustered with other fisher characteristics. We used the package vegan (Oksanen et al., 2015) and the function vegdist to create a dissimilarity matrix, which we then used to calculate the MDS using the function metaMDS. We also ran a hierarchical cluster model in R, using the package clustsig (Whitaker and Christman, 2014) and the function simprof, to determine if the characteristics: years fishing, days fishing, percent of time fishing rockfish, number of accidentally caught rockfish, place where fishing regulations are obtained, gender, age, First Nations status, and place of residence were clustered into meaningful groups.

Rockfish bycatch and release rates

We assessed levels of recreational fisher rockfish bycatch and release rates with survey questions. We extrapolated rockfish bycatch rates for the recreational fishing population in BC based on the number of annual, 5 day, 3 day, and 1 day fishing licenses sold in 2014 (DFO, 2014). Our study region is one of the more intensive recreational fishing regions in BC (Haggarty et al. in review), but many of our survey participants fished in multiple regions throughout BC. Inshore rockfish populations are also more depleted in the Salish Sea (Haggarty, 2014), thus, although there are fewer recreational fishers in other regions, chances of catching rockfish in these less depleted areas is likely higher. We assumed that the number of rockfish caught as bycatch per year was representative of other regions. We accounted for limitations on fishing time among 5-day, 3-day, and 1-day fishing licenses with the following formula:

$$RC = \left(\left(\frac{ARC}{365\ \text{days}}\right) DF\right) NL$$

where *RC* is accidentally caught rockfish per year per fishing license type (e.g. 1 day license), *ARC* is average number of accidentally caught rockfish per fisher per year, *DF* is number of days fishing annually as determined by license type (e.g. 1 day), and *NL* is number of specific fishing licenses sold annually (e.g. 1 day licenses). We then summed the *RC* totals for each license type to calculate the total annual recreational rockfish bycatch in BC.

Fisher perceptions and recommendations for RCAs

We elicited fisher perceptions of RCAs and obtained suggestions for improvements through six open-ended short-answer survey questions. We coded categories of answers to highlight important themes and patterns among responses. We then calculated the proportion of participants mentioning each category, providing an overview of RCA perceptions and suggestions for improvement. We also selected direct quotes from survey respondents to highlight a variety of different viewpoints.

RESULTS

Quantifying RCA knowledge and compliance

We found that 25.5% of fishers had never heard of RCAs before taking our survey, and 59% of fishers were not confident of RCA boundaries in the places they typically fish. Thirty eight percent of fishers incorrectly believed that salmon fishing is permitted in RCAs and 23% of fishers believed that halibut fishing is permitted in RCAs. Knowledge of permitted fishing activities within RCAs is very low with less than one percent (n=3) of fishers knowing all permitted activities. We found that 67% of fishers did not know rockfish bag limits (1 rockfish per day) in the Strait of Georgia. Our composite variable (Fig. 2) showed that 44% of fishers did not correctly answer any RCA regulation questions and less than one percent of fishers (n=2) answered all three knowledge questions correctly.

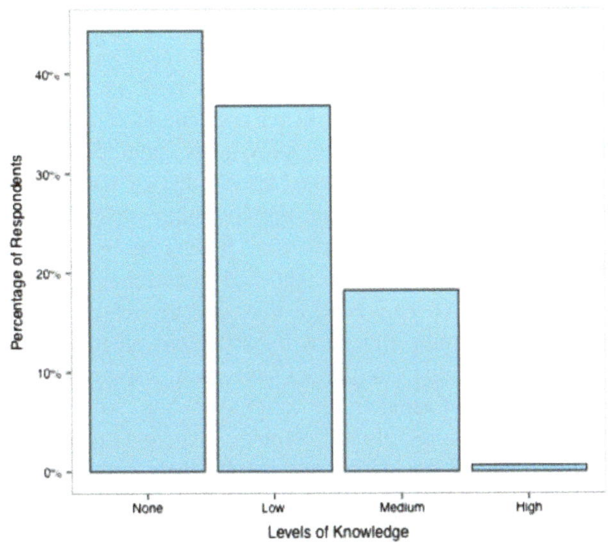

Figure 2: Knowledge of RCA regulations based on fisher responses to three knowledge based questions.

Using maps of RCAs as a reference, 16% of fishers admitted to accidentally illegally fishing within an RCA in the last two years. Seven percent of fishers answered yes to the non-sensitive direct question, "Do you personally know anyone who has intentionally fished, using prohibited techniques, in a Rockfish Conservation Area in the past two years?". We combined the percent of accidental noncompliers and intentional noncompliers (estimated from the percent of respondents who know someone who has intentionally illegally fished in an RCA) for an overall, estimated (accidental and intentional) noncompliance rate of 23% within RCAs. We did not include the results of our RRT compliance question in this estimate.

Of 318 fishers who completed the RRT section, only eight answered the question "Have you ever intentionally fished, using prohibited techniques, in a Rockfish Conservation Area in the past two years" with a yes, including predetermined responses. Using the RRT probability equation to account for predetermined "yes" and "no" responses, the RRT question showed no intentional noncompliance ($\hat{\pi} = -0.028$) in RCAs. The variance was <0.001. This negative intentional noncompliance rate is the result of the probability calculation which assumes that if a 20 sided die is rolled 318 times, it will land on each number ~16 times. We had only eight yes responses, which is lower than the necessary 16 yes responses required to arrive at a positive number. However, three of the eight yes responses we received were confirmed intentional fishing based on unprompted admission of intentional illegal fishing activity from survey participants.

Factors contributing to RCA knowledge and compliance

We used GLMs to determine what factors may contribute to better RCA and rockfish regulation knowledge, as well as higher RCA compliance. We created a GLM for each of our measures of knowledge and compliance (Table 2).

Table 2: Overview of final GLMs for each dependent variable

RCA knowledge/compliance GLMs (dependent variable)	Final model (independent variables)	P-values	Effect size	Deviance explained
Knowledge of RCA boundaries	Years fishing	<0.05	0.014	7%
	Place where fishing information is obtained	<0.01	−0.94	
	Gender	<0.05	−1.44	
	Residence (Canada outside BC)	<0.05	−1.849	
	Residence (USA)	<0.05	−1.727	
Knowledge of rockfish bag limits in the Strait of Georgia	Years fishing	<0.05	−0.016	11%
	Days fishing	<0.001	0.009	
	Previous knowledge of RCAs	<0.001	1.667	

Previous knowledge of RCA existence	Years fishing	<0.01	0.022	6%
	Place where fishing information is obtained	<0.001	−1.032	
Admission of accidental illegal fishing in RCAs	Days fishing	Not significant		9%
	Percent of time fishing rockfish	<0.001	0.021	
	Previous knowledge of RCAs	Not significant		
	Gender	<0.05	1.112	
Personally knowing someone who has intentionally illegally fished in an RCA	No significant independent variables			

Knowledge of RCA existence and RCA boundaries was significantly correlated with greater years fishing and obtaining information directly from Fisheries and Oceans Canada (DFO), the government agency that manages fisheries, either through their website or printed materials. Knowledge of rockfish bag limits was negatively significantly correlated with years fishing, with those who fished more years more frequently mistaking rockfish bag limits. However, in the same GLM, greater days fishing was significantly correlated with increased knowledge of rockfish bag limits. Female fishers and fishers who reside outside of British Columbia are significantly less likely to be confident of RCA boundaries. The deviance explained for each of these GLMs was low (Sechrest and Yeaton, 1982).

More time spent fishing rockfish was significantly correlated with accidental illegal fishing in RCAs, as was gender, with women admitting to accidental illegal fishing in RCAs more frequently than men. Women comprised less than 10% of fishers surveyed and 33% of female fishers admitted to accidental illegal fishing in RCAs, compared to 15% of male fishers. No factors were significantly correlated with personally knowing someone who intentionally illegally fished in an RCA.

Multi-dimensional scaling of the survey data did not reveal any clear clustering of fisher characteristics associated with our five measures of knowledge and compliance (knowledge of RCA boundaries, knowledge of rockfish bag limits in the Strait of Georgia, previous knowledge of RCA existence, admission of accidental fishing in RCAs, and personally knowing someone who had intentionally illegally fished in an RCA). However, a hierarchical cluster analysis identified three main groups of fishers based on the number of days spent fishing, amount of time spent fishing rockfish, and the number of accidentally caught rockfish. We identified these clusters as: Group (1) Novice fishers (Low fishing effort, knowledge, accidental rockfish bycatch, and accidental noncompliance); Group (2) Average fishers (Moderate fishing effort, knowledge, accidental rockfish bycatch, and accidental noncompliance); Group (3) Intensive fishers (Highest fishing effort, overall knowledge, accidental rockfish bycatch, and accidental noncompliance) (Table 3).

Table 3: Summary of main differences between hierarchical clustering fisher groups

Variable	Group 1 (novice fishers) n=20	Group 2 (average fishers) n=246	Group 3 (intensive fishers) n=52
Years fishing	2.5	31	28
Days fishing (*past* 2 *years*)	5	25	154
Percent of time fishing rockfish (*past* 2 *years*)	0.2%	13%	12%
Number of accidentally caught rockfish (*past* 2 *years*)	1	9	39
Accidental rockfish caught per day fishing (rockfish caught/days fishing)	0.2	0.36	0.25
Had no previous knowledge of RCAs	60%	23%	23%
Was not confident of RCA boundaries	85%	56%	61%
Does not know fisher bag limits	70%	70%	48%
Gets fishing information directly from DFO	60%	84%	75%
Had fished in an RCA accidentally (*past*2 *years*)	5%	16%	21%

Composite knowledge variable (*zero knowledge to high knowledge*)	Zero = 60	Zero = 44%	Zero = 40%
	Low = 35%	Low = 39%	Low = 29%
	Moderate = 5%	Moderate = 17%	Moderate = 29%
	High = 0	High = 1%	High = 2%

Fisher rockfish bycatch and release rates

We found that 59% of fishers never deliberately target rockfish, with less than 11% of fishers spending between 50%–100% of their fishing time specifically targeting rockfish. In the past two years, fishers had accidentally caught, on average, 14 rockfish (7 rockfish per year per fisher). One individual accidentally caught approximately 240 rockfish in the past two years (Fig. 3).

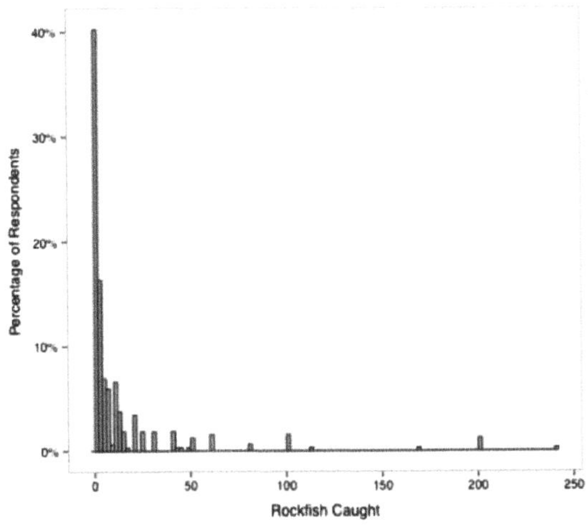

Figure 3: Histogram of accidentally caught rockfish in the past 2 years.

Sixty percent of fishers who accidentally caught rockfish released them 100% of the time. However, less than three percent of fishers redescended rockfish to their capture depth, a practice that has been associated with decreased rockfish mortality after accidental capture (Hannah and Matteson, 2007, Jarvis and Lowe, 2007). Most fishers had never heard of rockfish recompression and needed the question explained before responding.

In the 2013/2014 fishing license year, 307,157 recreational fishing licenses were sold, including annual, 5-day, 3-day, and 1-day resident and non-resident licenses (DFO, 2014). Extrapolating the fisher average of 7 accidentally caught rockfish per year, there were approximately 1,438,388 rockfish accidentally caught by fishers in Canadian Pacific waters in the 2013/2014 license year.

Recreational fisher perceptions of RCAs and recommendations for RCA improvement

Most fishers (77%) believe that there is a need for rockfish conservation in BC, and only four percent indicated that there is not a need for conservation (Table 4). Many fishers suggested rockfish conservation could be improved through increased knowledge and awareness through signage, education campaigns, and general advertising as well as permanent or temporary rockfish fishing moratoriums. Fishers cited ignorance and poaching as the most common reasons why fishers would fish in RCAs. Here poaching is defined as intentional illegal fishing as opposed to accidental illegal fishing. Rockfish are an important fish in many Asian cultures, and specific reference to Asian communities poaching rockfish was mentioned by 6% of respondents.

Table 4: Full results of open ended, short answer question coding analysis. Respondents often offered more than one answer to each question, and thus responses do not add up to 100%

Fisher responses	Response frequency (%)
Question 1: Do you think there is a need for rockfish conservation in British Columbia?	
Yes	77%
Yes—Overfished/depleted stocks	45%
Yes—Biologically sensitive fish	16%
Yes—Need for all conservation	10%
Yes—Because of this survey	5%
Yes—Poaching	4%
Yes—Asian depletion of stocks	4%
No	4%
No—populations are stable	3%
Probably	6%
No idea/not sure	*10%*
Not sure—no data on rockfish	*1%*
Not sure—never heard of rockfish	*>1%*
Other	*13%*
Question 2: What do you think are the best ways to improve rockfish conservation in general?	

Advertising of RCAs	23%
Education (rockfish bio and significance/overfishing concerns)	19%
Monitoring	17%
Moratorium (mention permanent or temporary full closures of rockfish fishing)	17%
RCAs (as currently implemented—also includes mention of protected areas in general)	16%
Dock Signs	10%
Target Asian overfishing	6%
Fines/penalties	5%
More RCAs	4%
Lower catch limit	4%
Information with fishing license (info and quizzes)	4%
Commercial fishery is the problem	3%
Marker Buoys	1%
Don't know/not sure	*7%*
Other	*22%*
Question 3: Why do you think some recreational fishers fish in Rockfish conservation areas?	
Ignorance	55%
Poaching/don't care about conservation	36%
Better fishing in RCAs	16%
Asian populations targeting them	6%
Targeting other fish in RCAs (i.e. Salmon)	4%
Aboriginal fishing in RCAs	1%
Don't know/not sure	*10%*
Other	*2%*
Question 4: What do you think is the best way to stop recreational fishers from intentionally fishing in Rockfish conservation areas?	
Monitoring	46%
Fines	33%
Confiscations	17%
Advertising/signage	12%
Education	9%
Citizen reporting/patrolling	6%
Intentional fishing is not occurring	3%
Clearer regulations	2%
Jail	1%
Don't know	*4%*
Other	*6%*
Question 5: What do you think is the best way to stop recreational fishers from accidentally fishing in Rockfish conservation areas?	
Advertising/info with licence	32%
Signs on docks/marinas	31%
Education	22%
Monitoring	18%
Marker Bouys	12%
Clearer website/regulations	12%
Fines/penalties	7%
Warnings	7%
App/Navionics overlay	4%
Don't know	*4%*

Other	*7%*
Question 6: Do you think the Rockfish conservation Areas are currently an effective conservation tool?	
Yes	40%
Partially	13%
Good start	5%
What else can be done	3%
Hopefully	3%
Mostly	2%
Better than nothing	2%
No	8%
No, nobody knows about them	3%
Not sure	*26%*

Fishers suggested more monitoring and frequent fines were necessary to reduce intentional noncompliance in RCAs. Suggested reasonable fines ranged from \$50 to \$20,000 (mean = \$1700). One respondent stated, "We need bigger fines and we need to take their boats. A few big cases like that will get people talking". Six percent of respondents also suggested citizen reporting and patrolling as a solution to intentional noncompliance. One respondent stated, "We should encourage self-reporting, DFO just doesn't have the budget, we need community monitoring".

Fishers suggested accidental illegal RCA fishing could be reduced with signage and advertising in addition to more monitoring. Twelve percent of fishers thought DFO should improve and simplify their website and make fishing regulations clearer and simpler. One respondent stated, "The DFO website is bad. The information is bad too. It's too hard to understand. We need postings on marinas, an area-specific regulations booklet, an App with RCA notifications". Most fishers (54%) believed that the RCAs are currently either a fully or partially effective conservation tool.

Many fishers were surprised that RCAs restrict nearly all fishing since the name seems to imply that only rockfish fishing is prohibited. Many salmon and halibut fishers had heard of RCAs while looking at DFO fishing regulations and assumed, based on the name, that RCA regulations would not apply to them.

DISCUSSION

Noncompliance can compromise the ability of marine conservation areas to protect and rebuild target species, and thus areas with noncompliance–intentional or unintentional–are unlikely to protect and recover target species (Bergseth et al., 2013). We found that knowledge of RCAs and RCA regulations by recreational fishers in BC is low, likely contributing to relatively high levels of accidental fishing within RCAs. With a combined intentional and accidental noncompliance rate of 23%, RCAs are unlikely to fulfill their mandate to protect inshore rockfish stocks (Yamanaka and Logan, 2010), especially in regions with high levels of recreational fishing pressure (Haggarty, 2014).

The relatively few studies that have investigated factors contributing to compliance found that education and positive perceptions of marine conservation areas are essential for high compliance (Alder, 1996, Gribble and Robertson, 1998, Cinner et al., 2005, Pollnac et al., 2010, Read et al., 2011, Leleu et al., 2012, Smallwood and Beckley, 2012, Arias and Sutton, 2013, Bergseth et al., 2013, Pita et al., 2013). Support for conservation actions are only useful when fishers know the rules and locations of marine conservation areas (Read et al., 2011). Our data showed that low levels of RCA knowledge are present across all recreational fishers (occasional to expert), and accidental noncompliance levels do not vary greatly between groups. Knowledge about RCAs was uniformly low, and thus potential drivers of noncompliance did not explain much of the variation. Although knowledge increased slightly with more days spent fishing, accidental fishing also increased, indicating that outreach and education about RCA are lacking overall. A similar study in the Great Barrier Reef Marine Park (GBRMP) determined that the noncompliance rate within no-take zones in the park was eight percent (Arias and Sutton, 2013). In the GBRMP, managers have gone to great lengths to educate fishers and other stakeholders over the past four decades with television advertisements, radio, signage, and public seminars (Alder, 1996). There may always be a small percentage of the population that will not respond to education and awareness campaigns so enforcement is still important (Alder, 1996). However, given that the majority of fishers we surveyed (77%) already believe rockfish conservation is necessary, a thorough outreach campaign, similar to that in the GBRMP, could significantly reduce accidental noncompliance. Additionally, education and outreach campaigns are generally less costly than enforcement and monitoring (Alder, 1996).

Reducing noncompliance within conservation areas is an urgent priority. Our research showed that bycatch of rockfish by recreational fishers was high and knowledge of re-descenders negligible. Given the low levels of awareness of RCAs, many of the rockfish caught as bycatch annually (~1,400,000) may be caught in RCAs. The mortality of rockfish with barotrauma released on the surface is assumed to be 100% by DFO (DFO, 2012). Although rockfish survival rates after being re-descended vary greatly depending on species, survival rates of 97% were found for Black rockfish (Parker et al., 2006). Many rockfish with severe barotrauma, typically assumed to be dead by fishers, were alive and able to swim away without major behavioural impairment after being re-descended (Hannah and Matteson, 2007). Given the high bycatch rate, and low knowledge of rockfish re-descenders, an education program should include information about RCA locations and regulations, rockfish biology, their susceptibility to barotrauma, and how to re-descend rockfish after capture. Simple and clear naming of RCAs could be an easy and important first step towards higher levels of voluntary compliance, as many fishers were confused about permitted fishing activities.

While our study is one of the few to quantitatively assess recreational fisher compliance, it has several limitations. First we focused on only one region. Thus our findings may not be representative of all recreational fishing demographics. However, we provide a snapshot of recreational fishing knowledge and compliance in a relatively new and little-studied network of marine conservation areas. Further research could examine recreational fishing patterns and RCA knowledge in other regions and other fishing groups (e.g., indigenous fishers). Second, our study focused on individual recreational fishers, not guided fishing tours. In other regions of BC, recreational fishers often use fishing guides, and thus a study of knowledge and compliance of guides should be considered. If guides are not knowledgeable and compliant, they could be an important focus of education and enforcement efforts. Conversely, if they are knowledgeable and compliant, they could act as RCA educators. Third, the discrepancy between the noncompliance rate determined by the RRT question and the non-sensitive, direct noncompliance question, may suggest that survey respondents answered "no" to the RRT question even when the device instructed them to say "yes". This may indicate respondents were either not comfortable answering the RRT question honestly, or did not fully understand how the technique would protect their confidentiality. Arias and Sutton (2013) found comparable intentional noncompliance rates between both techniques. The difference in results across techniques in our study could be the result of random chance or the low degree of confidentiality offered by our RRT design. However, this result may also suggest that people who illegally fish in RCAs are unlikely to participate in a voluntary survey about RCAs. Further RCA research should also quantify illegal fishing effort to determine the overall impact of this high recreational noncompliance on rockfish stocks.

Our study suggests that effective education and outreach is an essential first step in creating marine conservation areas with high compliance. In the case of RCAs, outreach was brief and minimal and regulation information is hard to access or impractical for use on the water (e.g. no hardcopy manuals available) (Haggarty, 2014). We suggest that marine managers designing education materials carefully consider the complex social dynamics and psychological aspects of human behaviour to avoid repercussions from poorly designed outreach material (Arias, 2015). Marine management plans, including outreach initiatives, should integrate local representatives (e.g. sport fish advisors) and interdisciplinary experts (e.g. ecologists, sociologists, psychologists) in decision making processes (Arias, 2015).

CONCLUSION

This study emphasizes the need for education of recreational fishers about regulations in, and locations of, marine conservation areas (Read et al., 2011). Knowledge about RCAs was uniformly low, and thus potential drivers of noncompliance did not explain much of the variation, although fishing experience, source of information about regulations, gender and nearby residence contributed to knowledge about RCAs. The high recreational noncompliance rate in RCAs–primarily accidental illegal fishing–is likely compromising the ability of these marine conservation areas to protect inshore rockfish stocks in BC. The ecological usefulness of marine conservation areas hinges upon users knowing about, and understanding, protected area rules and regulations. Even a perfectly designed network of conservation areas cannot be effective if users are not informed through education, outreach, and enforcement. Additionally, managers should take care to make protected area names and regulations as clear and easy to understand as possible. Although recreational fishing populations take a smaller percentage of the total global fish catch than commercial fishers (Cooke and Cowx, 2004 and Granek et al., 2008), they can heavily impact near-shore fish stocks and, as such, should not be ignored (Post et al., 2002 and Cooke and Cowx, 2004). Efforts to educate and involve this fishing sector in marine conservation efforts should be a management priority. Compliance research, especially on recreational fisher populations in temperate regions, has received little research attention. We recommend researchers continue to explore compliance and its drivers in order to maximize both the social and ecological effectiveness of marine conservation areas.

ACKNOWLEDGMENTS

Funding was provided by SSHRC (766-2013-0161), the Sara Spencer Foundation, the University of Victoria School of Environmental Studies, and Port Metro Vancouver. Thank you to BC Parks, SIMRES, and the Galiano Conservancy Association for fieldwork assistance and support. Thank you to field research assistant Kaia Bryce, to Keith Erikson and Lily Burke for GIS assistance, and to Nancy Shackelford and Francis Stewart for statistical analysis assistance.

REFERENCES

1. Alder, J., 1996. Costs and effectiveness of education and enforcement, Cairns Section of the Great Barrier Reef Marine Park. Environ. Manag. 20, 541–551. http://dx.doi.org/10.1007/BF01474654.
2. Allison, G.W., Lubchenco, J., Carr, M.H., 1998. Marine reserves are necessary but not sufficient for marine conservation. Ecol. Appl. 8 (1), S79–S92.
3. Arias, A., 2015. Understanding and managing compliance in the nature conservation context. J. Environ. Manag. 153, 134–143. http://dx.doi.org/10.1016/j.jenvman.2015.02.013.
4. Arias, A., Sutton, S.G., 2013. Understanding recreational Fisher compliance withno-take zones in the Great Barrier Reef Marine Park. Ecol. Soc. 18, http://dx.doi.org/10.5751/ES-05872-180418.
5. Bergseth, B.J., Russ, G.R., Cinner, J.E., 2013. Measuring and monitoring compliance in no-take marine reserves. Fish Fish. 16, 240–258. http://dx.doi.org/10.1111/faf.12051.
6. Blank, S.G., Gavin, M.C., 2009. The randomized response technique as a tool for estimating non-compliance rates in fisheries: a case study of illegal red abalone (Haliotis rufescens) fishing in Northern

California. Environ. Conserv. 36, 112–119. http://dx.doi.org/10.1017/S037689290999004X.

7. Chalifour, L., 2012. GCA Final Report—Freedom to Swim: Research component for rockfish recovery project 11-8552. Galiano Conservancy Association.

8. Challenger, W., Marliave, J., 2009. Monitoring and evaluating rockfish conservation areas in British Columbia. Can. J. Fish. Aquat. Sci. 66 (6), 995–1006.

9. Cinner, J.E., Marnane, M.J., McClanahan, T.R., 2005. Conservation and community benefits from traditional coral reef management at Ahus Island, Papua New Guinea. Conserv. Biol. 19, 1714–1723. http://dx.doi.org/10.1111/j.1523-1739.2005.00209.x-i1.

10. Cloutier, R.N., 2010. Direct and indirect effects of marine protection: rockfish conservation areas as a case study (Master's Thesis), Simon Fraser University: Department of Biology.

11. Cooke, S.J., Cowx, I.G., 2004. The role of recreational fishing in global fish crises. BioScience 54, 857–859. http://dx.doi.org/10.1641/00063568(2004)054[0857:TRORFI]2.0.CO2. COSEWIC, 2009a.

12. COSEWIC assessment and status report on the Yelloweye rockfish, Sebastes ruberrimus: Pacific Ocean inside waters population, Pacific Ocean outside waters population in Canada. Committee on the Status of Endangered Wildlife in Canada, Ottawa.

13. COSEWIC, 2009b. COSEWIC assessment and status report on the Quillback Rockfish Sebastes maliger in Canada. Committee on the Status of Endangered Wildlife in Canada.

14. Crawley, M., 2007. The R Book. John Wiley and Sons Ltd., http://www.kharms.biology.lsu.edu/CrawleyMJ_TheRBook.pdf.

15. DFO, 2012. Stock assessment for the inside population of yelloweye rockfish (sebastes ruberrimus) in Britsh Columbia, Canada for 2010. Fisheries and Oceans Canada. http://www.dfo-mpo.gc.ca/Library/346394.pdf.

16. DFO, 2014. Statistics for Tidal Water Sport Fishing Licences—1999 to Current. Fisheries and Oceans Canada. http://www.pac.dfo-mpo.gc.ca/fmgp/rec/licence-permis/index-eng.htm#Stats.

17. Fisher, A.R., 1934. Statistical Methods for Research Workers, fifth ed. Oliver and Boyd, Edinburgh.

18. Granek, E.F., Madin, E.M.P., Brown, M.A., Figueira, W., Cameron, D.S., Hogan, Z., Kristianson, G., de Villiers, P., Williams, J.E., Post, J., Zahn, S., Arlinghaus, R., 2008. Engaging recreational fishers in management and conservation: Global case studies. Conserv. Biol. 22 (5), 1125–1134.

19. Gribble, N., Robertson, J.W.A., 1998. Fishing effort in the far northern section cross shelf closure area of the Great Barrier Reef Marine Park: the effectiveness of area-closures. J. Environ. Manag. 52, 53–67. http://dx.doi.org/10.1006/jema.1997.0160.

20. Haggarty, D., 2014. Rockfish conservation areas in BC - Our current state of knowledge. David Suzuki Foundation. http://www.davidsuzuki.org/publications/RockfishConservationAreas-OurCurrentStateofKnowledge-Mar2014.pdf.

21. Hannah, R.W., Matteson, K.M., 2007. Behavior of nine species of Pacific Rockfish after hook-and-line capture, recompression, and release. Trans. Am. Fisheries Society 136, 24–33. http://dx.doi.org/10.1577/T06-022.1.

22. Hutchings, J.A., Cote, I.M., Dodson, J.J., Fleming, I.A., Jennings, S., Mantua, N.J., Peterman, R.M., Riddell, B.E., Weaver, A.J., VanderZwaag, D.L., 2012. Sustaining anada's marine biodiversity: Responding to the challenges posed by climate change, fisheries, and aquaculture. The Royal Society of Canada Expert Panel. http://www.rsc.ca/sites/default/files/pdf/RSCMarineBiodiversity2012_ENFINAL.pdf.

23. Jarvis, E.T., Lowe, C.G., 2007. The effects of barotrauma on the catch-and-release survival of southern California nearshore and shelf rockfish (Scorpaenidae, Sebastes spp.). Can. J. Fish. Aquat. Sci. 65, 1286–1296. http://dx.doi.org/10.1139/F08-071.

24. Kritzer, J.P., 2004. Effects of noncompliance on the success of alternative designs of marine protected area networks for conservation and fisheries management. Conserv. Biol. 18 (4), 1021–1031.

25. Krumpal, I., 2012. Estimating the prevalence of xenophobia and anti-Semitism in Germany: A comparison of randomized response and direct questioning. Soc. Sci. Res. 41, 1387–1403. http://dx.doi.org/10.1016/j.ssresearch.2012.05.015.

26. Leleu, K., Alban, F., Pelletier, D., Charbonnel, E., Letourneur, Y., Boudouresque, C.F., 2012. Fishers' perceptions as indicators of the performance of Marine Protected Areas (MPAs). Mar. Policy 36, 414–422. http://dx.doi.org/10.1016/j.marpol.2011.06.002.

27. Lewin, W.C., Arlinghaus, R., Mehner, T., 2006. Documented and potential biological impacts of recreational fishing: Insights for management and conservation. Rev. Fish. Sci. 14, 305–367. http://dx.doi.org/10.1080/10641260600886455.

28. Little, L.R., Smith, A.D.M., Mcdonald, A.D., Punt, A.E., Mapstone, B.D., Pantus, F., Davies, C.R., 2005. Effects of size and fragmentation of marine reserves and Fisher infringement on the catch and biomass of coral trout, Plectropomus leopardus, on the Great Barrier Reef, Australia. Fish. Manag. Ecol. 12, 177–188. http://dx.doi.org/10.1111/j.1365-2400.2005.00440.x.

29. Lotterhos, K.E., Dick, S.J., Haggarty, D.R., 2013. Evaluation of rockfish conservation area networks in the United States and Canada relative to the dispersal distance for black rockfish (Sebastes melanops). Evol. Appl. 7 (2), 238–259.

30. Love, M.S., Yoklavich, M., Thorsteinson, L.K., 2002. The Rockfishes of the Northeast Pacific. Univ of California Press. Marinesque, S., Kaplan, D.M., Rodwell, L.D., 2012. Global implementation of marine protected areas: Is the developing world being left behind? Mar. Policy 36, 727–737. http://dx.doi.org/10.1016/j.marpol.2011.10.010.

31. Marliave, J., Challenger, W., 2009. Monitoring and evaluating rockfish conservation areas in British

Columbia. Can. J. Fish. Aquat. Sci. 66, 995–1006. http://dx.doi.org/10.1139/F09-056.

32. McClanahan, T., Davies, J., Maina, J., 2005. Factors influencing resource users and managers' perceptions towards marine protected area management in Kenya. Environ. Conserv. 42–49. http://dx.doi.org/10.1017/S0376892904001791.

33. Molfese, C., Beare, D., Hall-Spencer, J.M., 2014. Overfishing and the replacement of demersal finfish by shellfish: An example from the english channel. PLoS One 9, e101506.

34. Oksanen, J., Blanchet, F.G., Kindt, R., Legendre, P., Minchin, P.R., O'Hara, R.B., Simpson, G.L., Solymos, P., Stevens, M.H.H., Wagner, H., 2015. Vegan Community Ecology Package. R package version 2.2-1. http://CRAN.R-project.org/package=vegan.

35. Parker, S.J., McElderry, H.I., Rankin, P.S., Hannah, R.W., 2006. Buoyancy regulation and barotrauma in two species of nearshore rockfish. Trans. Am. Fish. Soc. 135 (5), 1213–1223.

36. Pauly, D., Christensen, V., Dalsgaard, J., Froese, R., Torres, F., 1998. Fishing down marine food webs. Science 279 (5352), 860–863.

37. Pita, C., Theodossiou, I., Pierce, G.J., 2013. The perceptions of Scottish inshore fishers about marine protected areas. Mar. Policy 37, 254–263. http://dx.doi.org/10.1016/j.marpol.2012.05.007. Social and cultural impacts of marine fisheries.

38. Pollnac, R., Christie, P., Cinner, J.E., Dalton, T., Daw, T.M., Forrester, G.E., Graham, N.A.J., McClanahan, T.R., Gaines, S.D., 2010. Marine reserves as linked social–ecological systems. Proc. Natl. Acad. Sci. USA 107 (43), 18262–18265.

39. Post, J.R., Sullivan, M., Cox, S., Lester, N.P., Walters, C.J., Parkinson, E.A., Paul, A.J., Jackson, L., Shuter, B.J., 2002. Canada's recreational fisheries: The invisible collapse? Fisheries 27, 6–17. http://dx.doi.org/10.1577/1548-8446(2002)0272.0.CO2.

40. R Core Team, 2013. R: A language and environment for statistical computing. R Foundation for Statistical Computing, Vienna, Austria. http://www.Rproject.org/.

41. Read, A.D., West, R.J., Haste, M., Jordan, A., 2011. Optimizing voluntary compliance in marine protected areas: A comparison of recreational Fisher and enforcement officer perspectives using multi-criteria analysis. J. Environ. Manag. 92, 2558–2567. http://dx.doi.org/10.1016/j.jenvman.2011.05.022.

42. Read, A.D., West, R.J., Kelaher, B.P., 2015. Using compliance data to improve marine protected area management. Mar. Policy 60, 119–127. http://dx.doi.org/10.1016/j.marpol.2015.06.008.

43. Schill, D.I., Kline, P.A., 1995. Use of random response to estimate angler noncompliance with fishing regulations. N. Am. J. Fisheries Manag. 15 (4), 721–731. http://dx.doi.org/10.1577/1548-8675(1995)0152.3.CO;2.

44. Sechrest, L., Yeaton, W.H., 1982. Magnitudes of experimental effects in social science research. Eval. Rev. 6, 579–600. http://dx.doi.org/10.1177/0193841X8200600501.

45. Smallwood, C.B., Beckley, L.E., 2012. Spatial distribution and zoning compliance of recreational fishing in Ningaloo Marine Park, North-Western Australia. Fisheries Res. 125–126. 40–50.

46. Venables, W.N., Ripley, B.D., 2002. Modern Applied Statistics with S, fourth ed.. Springer, New York, ISBN: 0-387-95457-0, http://www.stats.ox.ac.uk/pub/ MASS4.

47. Whitaker, D., Christman, M., 2014. clustsig: Significant Cluster Analysis. R package version 1.1. http://CRAN.R-project.org/package=clustsig.

48. Yamanaka, K.L., Logan, G., 2010. Developing British Columbia's inshore rockfish conservation strategy. Mar. Coast. Fish. 2 (1), 28–46.

CHAPTER 17

HYBRID MODEL OF GENETIC ALGORITHM AND SUPPORT VECTOR MACHINE FOR PREDICTION OF CHLOROPHYLL A IN RESERVOIRS

Jieqiong Su, Xuan Wang, Shouyan Zhao, Bin Chen, Chunhui Li and Zhifeng Yang

Keywords: Concentration prediction, chlorophyll a, support vector machine, genetic algorithm, Miyun Reservoir

INTRODUCTION

Water conflicts are key issues for sustainable water resources management. Under the dual effects of climate change and human activities, many water bodies are polluted to varying degrees, further exacerbating water conflicts [1,2]. Ecosystem studies such as water enhancement, water quality risk assessment, and early warnings have drawn much attention across the world [3]. As important engineering measures are developed to guarantee water supply, irrigation, electricity, and other functions, reservoirs can help solve these issues through the redistribution of runoff in both time and space; therefore, they are widely used throughout the world. Although water demands of each production department (e.g., industrial department, agricultural department, and so on) correspond to different water quality requirements, water quality always needs to be up to its appropriate standard in different water usage [4]. Accordingly, it is important to forecast water quality accurately, which could provide a scientific decision basis for reservoir management.

Chlorophyll *a* is an important component of algae organisms, and its concentration in water bodies is closely related to the type and the quantity of algae [5]. Therefore, as an important symbol of phytoplankton stock, concentration of chlorophyll *a* can reflect water nutritional status, making chlorophyll *a* one of the indicators in controlling the eutrophication of lakes and reservoirs. The minimum threshold concentration of chlorophyll *a* for eutrophic lakes identified by the Organization for Economic Cooperation and Development (OECD) is 0.008 mg/L. Consequently, there is a need to control the concentration of chlorophyll *a* in water to prevent potential eutrophication. For this reason, accurate prediction of chlorophyll *a* is a worldwide concern.

The generation mechanism of chlorophyll *a* in water is accordingly complex, which is closely related to ecological, environmental, and societal activities. Therefore, the elements involved in the prediction for chlorophyll *a* in water are complex accordingly. In the existing literature, prediction models for chlorophyll *a* mainly included two categories: statistical regression models [6] and mechanism models [7]. Statistical regression models were established with the applications of statistical correlation analysis theory and methods. This means that the sample size had a major influence on prediction accuracy. Moreover, these models usually applied a linear relationship to simplify complex problems, leading to unsatisfactory prediction results under the situation when the limiting factors of chlorophyll *a* changed. Mechanism models mainly included the nutrient model, phytoplankton model, and ecological dynamic models such as CE-QUAL-ICM, WASP, CAEDYM, AQUATOX, and ECOPATH [8,9,10]. Based on the principle of hydrodynamics and ecosystem dynamics, these models comprehensively considered the interaction mechanisms among indicators of water resources system and ecosystem, and then predicted the future development of the system accurately. However, these models also had a high demand for data quantity, which was inconvenient for model

Citation: Jieqiong Su, Xuan Wang, Shouyan Zhao, Bin Chen, Chunhui Li and Zhifeng Yang, Hybrid Model of Genetic Algorithm and Support Vector Machine for Prediction of Chlorophyll a in Reservoirs, doi:10.3390/w7041610.

calibration and verification, leading to a decline in reliability and applicability. Furthermore, due to uncertainty factors such as the concentration of phosphorus input from non-point source pollution, the prediction of chlorophyll *a* based on deterministic differential equations was not reasonable [11]. For this reason, the uncertainty of input parameters and the nonlinearity of the system required further consideration when constructing models.

To improve the accuracy and efficiency of nonlinear system simulations, intelligent algorithms have been applied in recent years [12]. Widely used intelligent algorithms include the artificial neural network (ANN), the genetic algorithm (GA), the particle swarm algorithm, the wavelet theory, and the projection pursuit algorithm, *etc.*; these intelligent algorithms overcome the uncertainty to a certain extent with high simulation precision. In recent years, the support vector machine (SVM) algorithm, which is a new type of machine learning tool based on statistical learning theory, has drawn more attention [13]. This intelligent algorithm can solve nonlinear system problems and has reasonable generalization ability when using small samples, ameliorating the weaknesses of the above intelligent algorithms, e.g., large sample size requirements and being susceptible to underfitting and overfitting the data for the ANN. The SVM has demonstrated promise for applied studies of water environments, especially for the prediction of hydrologic factors, such as wave height [14], inflow [15], and water levels [16]. Previous studies of chlorophyll *a* based on the SVM algorithm often focused on the retrieval of chlorophyll *a* in water [17], although very few results of chlorophyll *a* prediction have been reported [18]. Furthermore, chlorophyll *a* is affected by many factors, and irrelevant and redundant information is often hidden in the time series of high dimensional feature vectors, leading to structurally complex models and a decrease of analysis precision and application efficiency of the SVM model when using conventional modeling processes [19]. To simplify the model structure and avoid the interference of redundant information in chlorophyll *a* forecasts, it is desirable to obtain more accurate and reliable prediction results by using SVM models with the most relevant influence factors as input vectors and simple structures. Feature selection is an important approach for getting structurally simplified model by removing those redundant parameters. Cho *et al.* [20] used principal component analysis (PCA) to extract variables for the prediction model of chlorophyll *a*. Compared to conventional parameter extraction approaches such as PCA, the GA has a distinct advantage in fast random search. Thus, the SVM model can be expected to get satisfactory prediction results through combing the GA to extract feature variables and simplify the model structure in such complex water bodies as reservoirs, whose water quality variations likely result from a combination of multiple factors. However, the GA-SVM hybrid model needs to be further developed for nonlinear water resources system.

This study developed a hybrid model of GA and SVM algorithms to predict chlorophyll *a* in the Miyun Reservoir of northern China. Based on the feature selection with the GA, we extracted appropriate input vectors, so that the redundant information was effectively eliminated with the simplified model structure. This model could analyze water quality and its change trend with reliable results, and was of great practical significance in preventing water pollution accidents.

STUDY AREA AND DATA DESCRIPTION

The Miyun Reservoir is located in the Miyun County of Beijing City. Built in 1960, it is the largest reservoir and is a unique surface source of drinking water in Beijing City (Figure 1). In addition to functioning as a water supply, the Miyun Reservoir also provides irrigation, flood control, power generation, aquaculture, tourism, and other comprehensive benefits. The Miyun Reservoir's surface area is 183.6 km^2, its maximum depth is 153.93 m, and the maximum volume of reservoir storage is 3.349×10^{10} m^3. Monitoring data shows that in recent years, the total phosphorus concentration in the reservoir fluctuated between 0.010 and 0.025 mg/L, which means the nutrition status of the water is at a mesotrophication to oligotrophication level. The total nitrogen concentration ranges between 0.62 and 1.43 mg/L, indicating that the nutrition status is at a mild or moderate eutrophication level. Planktonic algae have rich diversities, and the dominant population in various periods is different in the Miyun Reservoir. As for cyanobacterium, from 2001 to 2003 it was the dominant algae from September to October [21]; from 2008 to 2010, it was the dominant algae from June to September [22,23]. Considering the current water quality situation, we should take effective measures to alleviate adverse influences resulting from climate change and human activities on the reservoir.

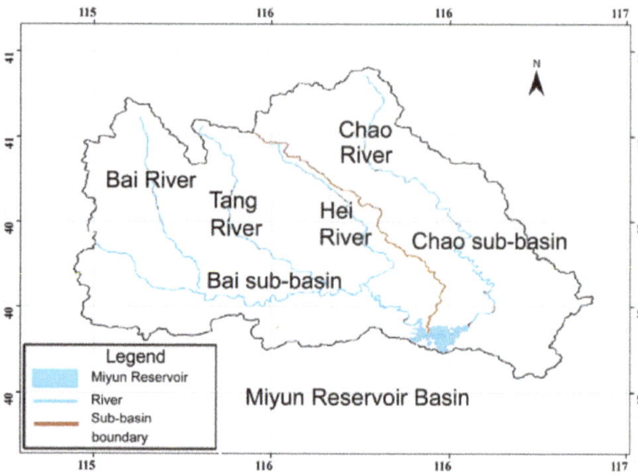

Figure 1: The Miyun Reservoir in northern China.

In this paper, the Baihe Key Dam in the Miyun Reservoir is taken as the research area. Data for the model establishment and calibration are from the monitored data, including water quality indicators (*i.e.*, chlorophyll *a* in water, total nitrogen, total phosphorus, permanganate index, and dissolved oxygen), hydrological indicators (*i.e.*, water temperature, pH, transparency, flow, reservoir storage, inflow, outflow, and water level), and meteorological indicators (*i.e.*, precipitation and temperature). The monthly data from 2000 to 2010 were obtained from the Miyun Reservoir Management Office. Because the Miyun Reservoir is frozen for the period from December to March, the prediction of chlorophyll *a* focused on the period from April to November in each year, and other indicators in the SVM model corresponded to these months. The average monthly variations of parts of these indicators are shown in Figure 2.

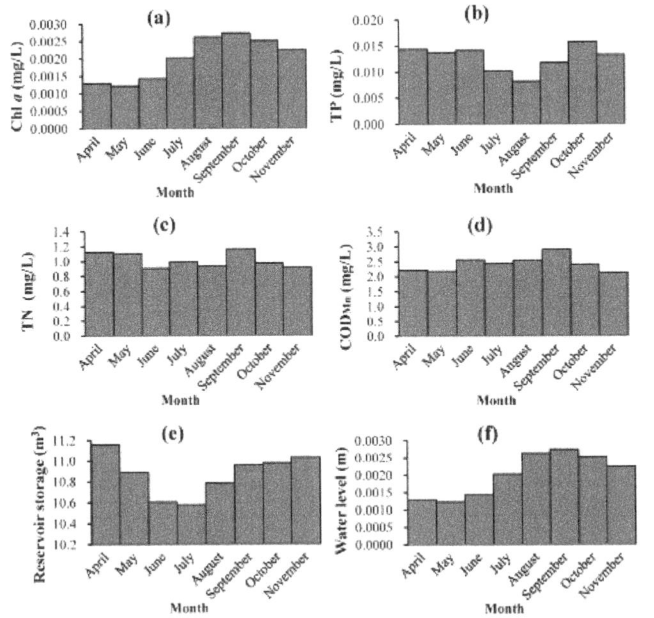

Figure 2: Water environmental situation in Miyun Reservoir: (**a**) Concentration of chlorophyll *a*; (**b**) TP concentration; (**c**) TN concentration; (**d**) COD_{Mn} concentration; (**e**) reservoir storage; and (**f**) water level.

METHODS

Originally proposed in 1985 by Cortes and Vapnik, the SVM algorithm was widely used to solve highly nonlinear classification and regression problems with good generalizability [24]. The SVM algorithm can be easily applied to other machine learning problems such as function fitting. It is based on the VC dimension theory and the structure risk minimum principle of the statistical learning theory. By seeking the best compromise among the complexities in the model with a limited sample (*i.e.*, learning accuracy of particular training samples) and learning ability (*i.e.*, learning ability to identify random sample), the SVM algorithm can achieve the best generalization ability. There are many meteorological and hydrological parameters that influence chlorophyll *a*. To avoid blindness in selecting the input vector during the process of chlorophyll *a* prediction, this study firstly took feature selection to determine the best input vectors of prediction model with the GA, and then constructed the SVM model with a simplified model structure to achieve the purpose of improving prediction accuracy.

The Flow Chart for Developing a Simplified Structural GA-SVM Hybrid Model

The penalty factor and kernel function parameters of the SVM model may directly influence simulation results. This study developed a SVM model by using the GA to optimize the input parameters in the SVM model and extract feature parameters with the aim of simplifying the model structure. The flow chart of the simplified structural SVM model based on the GA is shown in Figure 3. After data pre-treatment, the input and output vectors were determined, and the sample set was divided into a training data set and a testing data set. The GA was applied to optimizing the parameters of the SVM model and extracting input vectors. The SVM model was trained and calibrated with optimal parameters, then used to predict chlorophyll *a* in the water. This study applies the LIBSVM software package developed by Lin Chih-Jen *et al.*, of Taiwan University to run the program on the MATLAB platform [25].

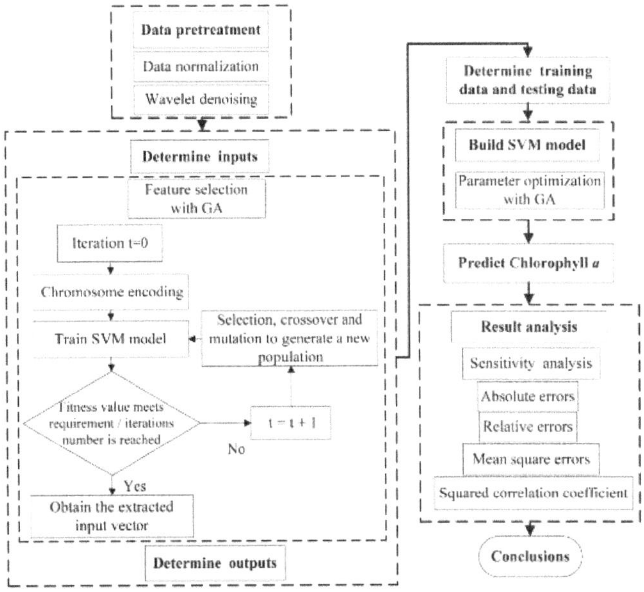

Figure 3: The flow chart for developing simplified structural GA-SVM hybrid model for chlorophyll *a* prediction.

Construction of Chlorophyll a Prediction Model Based on the SVM Algorithm

This paper applied the principle of the SVM algorithm to establish a prediction model for chlorophyll *a* in the Miyun Reservoir. The basic principle of the SVM algorithm was to first select a nonlinear mapping algorithm as a kernel function, through which the input vectors were mapped into a high dimensional feature space, and in this space simpler linear regressions can replace complex nonlinear regressions of the original input space [26]. Then an optimal decision function was produced in the feature space to realize the nonlinear decision function of the original input space, and finally the linear learning method can be applied to solve the classification and regression problems in the input space. This process can be expressed as:

$$y = f(x, w) = w \cdot \emptyset(x) + b, \tag{1}$$

where y is the output, $y \in R$; x is the input vector, $x \in R^n$; w is the matrix of the regression weight vector; $\emptyset$ is a non-linear function by which x is mapped into a high dimensional feature space; b is a bias; and b and w can be obtained with Equation (3). In the mapping process, a kernel function $k(*,*)$ can be constructed by $k(x,x') = (\emptyset(x) \cdot \emptyset(x'))$. Therefore, we only need to replace the x or x_i of the original space with $\emptyset(x)$ or $\emptyset(xi)$, while it is not necessary to know the explicit expression of nonlinear mapping $\emptyset$. In this study, we selected radial basis function (RBF) as the kernel function:

$$K(x, x_i) = exp(-\gamma \|x - x_i\|^2), \tag{2}$$

where x_i is the input vector, $x \in R^n$; and γ is the parameter of the RBF kernel function.

In Equation (1), the concentration of chlorophyll *a* in reservoir water was selected as y in the SVM model, whereas other water quality factors, hydrological factors, and meteorological factors were selected as x in the SVM model. In this way, the concentration of chlorophyll *a* was predicted based on the other factors. To solve Equation (1), the following regularized risk function (*i.e.*, Equation (3)) was used. These constraints ensured the regression errors of the samples being within the area that was delineated by the error tolerance and the slack variables. Equation (3) can be solved with the Lagrange technique.

Minimize

$$\frac{1}{2}\parallel w \parallel^2 + C\sum_{n=1}^{N}(\xi_n + \xi_n^*)$$

subject to

$$\begin{cases} y_i - W^T\emptyset(X_i) - b \leq \varepsilon + \xi_i \\ W^T\emptyset(X_i) + b - y_i \leq \varepsilon + \xi_i^* \\ \xi_i,\ \xi_i^* \geq 0 \end{cases} i = 1,2,\ldots,N, \tag{3}$$

where C is a penalty parameter that determines the penalty degree for the sample classification errors in the optimization problem; N is the number of the samples of $\{x_n, y_n\}_{n=1}^{N}$; ξ_n and ξ_n^* are slack variables that penalize training errors by the loss function over the error tolerance (ε); ξ_n represents the upper training errors subject to ε; ξ_n^* represents the lower training errors subject to ε; ξ_n and ξ_n^* can be calculated with Equation (3); and was normally set to 0.001.

Overall, the chlorophyll *a* simulation with the SVM model depended on the ability to learn the nonlinear causality between the historical data of the concentration of chlorophyll *a* and its influencing factors (*i.e.*, other water quality factors, hydrological factors, and meteorological factors). The modeling process for the SVM model was introduced below.

Step 1. Determine chlorophyll *a* as the output value of the prediction model, with the other indicators as input values:

$$\rho_{Chla} = F(S, W_I, W_O, L, TP, TN, COD_{Mn}, DO, T_W, pH, SD, T_A, P), \tag{4}$$

where S is reservoir storage, W_I is inflow, W_0 is outflow, L is water level, TP is the concentration of total phosphorus in water, TN is total nitrogen in water, COD_{Mn} is permanganate index in water, DO is the dissolved oxygen concentration in water, TW is water temperature, pH is hydrogenion concentration of water, SD is water transparency, T_A is temperature, and P is precipitation.

Step 2. Before establishing the SVM model, to extract useful information in the original data and determine the most reasonable and relevant input vectors of the prediction model, this study applied data normalization, wavelet denoising, and feature selection for the data pre-treatment in the MATLAB software. To test the prediction ability of the SVM model, the sample set was divided into separate training and testing sets. Data between 2000 and 2004 were used as the training set, and those between 2005 and 2010 as the testing set. Thus, the testing set data were independent and not used to train the model. In the parameter optimization, the initial conditions of the GA were set: the biggest evolution generation was 100, the largest population was 20, the gap of genetic algorithm was 0.9, and the k-fold cross-validation number was 5.

Step 3. To determine the effect of each input indicator to the prediction model, we carried out sensitivity analysis of the chlorophyll *a* prediction model. The analysis method was to change a particular input variable (increase or decrease by 10%) while the other input variables remained fixed and then applied the established SVM model to re-predict; the variable of the sensitivity model was obtained by calculating the relative changes in chlorophyll *a* with the output value.

Step 4. To eliminate the irrelevant and redundant information hidden in the time series of high dimensional feature vectors, and reveal the more representative features that influenced the concentration of chlorophyll *a* in the Miyun Reservoir, we applied feature selection to the input vectors of the SVM model for chlorophyll *a* simulation by using the GA. Subsequently, we established the hybrid SVM model using the extracted feature vectors and improved prediction accuracy, generalization ability, and efficiency.

Feature Selection and Parameter Optimization Based on Genetic Algorithm Optimization

The feature means each attribute of the data set. Too many features will increase the complexity of the work, while the accuracy of data mining may not be improved. To pick out the most representative and effective feature vectors of the chlorophyll *a* prediction model, we used the GA for feature selection. In addition, considering that the SVM model did not provide a method for selecting the parameter in the RBF kernel function and the penalty parameter (C), we used the GA for optimizing these two parameters in the SVM model. The principle of the GA is based on a specific operation for the structure of objectives, according to a predefined criteria function, to improve the new population by comparing it with the original one. In the process of generation, proper coding was used and the operator was applied to imitate the path of natural selection. Reproduction, crossover, and mutation were taken to operators in the current population [27]. Procedures of feature selection and parameter optimization with applications of the GA were as follows.

Step 1. Chromosome encoding. In the selection and optimization process, the iterations were set to zero. Chromosomes were encoded with binary coding. Each operator was composed of N codes, where N is the number of characteristic vectors or SVM parameters that need to be optimized. When a number in the operator was 1, it represented the characteristic vector and the parameter was selected; otherwise it was not selected, and the initial population was generated randomly.

Step 2. Evaluation of the fitness function value. Determine the square of the root mean square error in the training phase as the objective function for the fitness value. Then, calculate the fitness function value of the current generation. Choose a certain adaptation level, retaining the individuals whose fitness function value is greater than the adaptation level; these individuals compose the next generation.

Step 3. Selection, crossover and mutation of operators. Apply genetic operation of selection, crossover, and mutation to individuals in the group; the next generation was produced after the genetic optimization.

Step 4. Termination judgment. If the iteration number was greater than the set value, or the accuracy of the fitness function value reached the expected value, then terminate the iteration [28]. The extracted features and the optimal model parameters were then determined.

Model Calibration

In order to analyze the performance of the model, four indicators—The absolute error (AE), relative error (RE), root mean square error (RMSE), and square of correlation coefficient (R^2)—Were selected to evaluate the fit and prediction effect of the model. AE represented the deviation between monitoring and prediction values, and RE was the ratio of AE and monitoring values, reflecting the objective accuracy of measurement results. RMSE reflected the performance of the prediction model, *i.e.*, generally, the smaller the RMSE the better the performance. R^2 represented the degree of linear relevance among the variables, *i.e.*, the closer R^2 was to 1 the higher the relevance. The expressions of these four indicators were as follows:

$$AE = |y_i - \hat{y}_i| \quad (5)$$

$$RE = \frac{AE}{y_i} \times 100\% \quad (6)$$

$$RMSE = \sqrt{\frac{1}{n}\sum_{i=1}^{n}(y_i - \hat{y}_i)^2} \quad (7)$$

$$R^2 = 1 - \frac{\sum_{i=1}^{n}(y_i - \hat{y}_i)^2}{\sum_{i=1}^{n}(y_i - \bar{y})^2} \quad (8)$$

where y_i is the real value of the data set, $\hat{y}_i$ is prediction value, $\bar{y}$ is the average of the original data, and n is the amount of data for the testing set.

RESULTS AND DISCUSSION

Wavelet Denoising

The results of applying the wavelet denoising method to the original data of chlorophyll *a* are shown in Figure 4. The upper and lower figures represented the time series data before and after the denoising, respectively. It was clear to see that after wavelet decomposition and reconstruction the low frequency characteristics of the original data were preserved, while eliminating the high frequency data. After this process, the abrupt change points of the time series data were smoothed, and the main information regarding the concentration of chlorophyll *a* was well preserved.

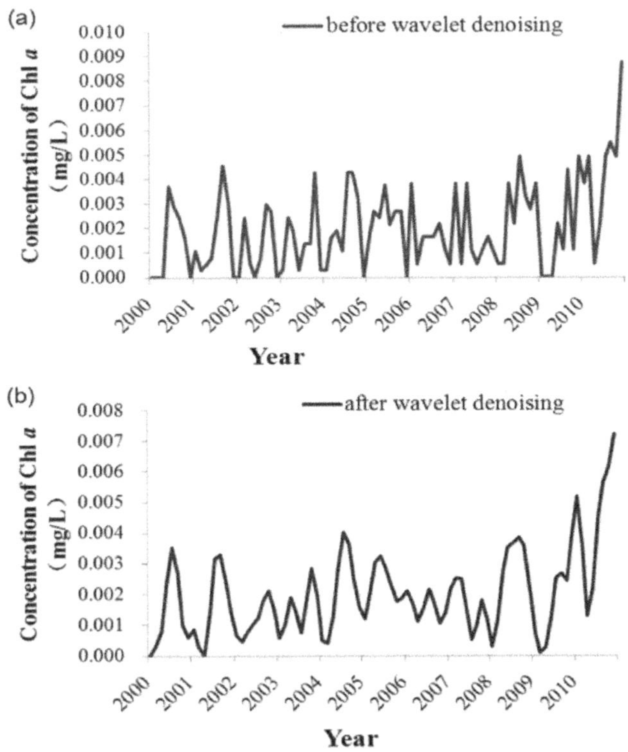

Figure 4: Comparisons of before and after the wavelet denoising.

The noise of the original data about chlorophyll a was mainly caused by the errors in sampling and deviations in experiments due to unsuitable experimental conditions and improper operation caused by human errors. The wavelet denoising was realized through multi-scale decomposition of sequence data and reconstruction. The original signal was decomposed into a series of low frequency and high frequency components by using the wavelet decomposition, and the noise of chlorophyll a's raw data was concentrated in the high frequency components. The high frequency components were processed with threshold, and the low frequency components were reconstructed to obtain the smooth data series. Because the low frequency components could preserve the details of the original data, excessive deviations can be avoided in data applications. Therefore, it was reasonable and concise to use the time series data of chlorophyll a after the wavelet denoising as input variables of the SVM model, without the loss of important information.

Results of Sensitivity Analysis and Feature Selection

The use of the GA aimed to automate and enhance SVM designing process. The results of parameter optimization indicated that the optimal penalty factor C was 1.0737, and the optimal parameter γ in RBF kernel function was 1.0005. The final SVM model was established based on these results. Figure 5 shows sensitivity analysis results for input vectors of the SVM model. It can be seen that when the values of eight parameters (i.e., rainfall, water level, inflow, pH, water temperature, permanganate index, total nitrogen, and total phosphorus) increased by 10%, the sensitivity degree was greater than zero, which meant the prediction of chlorophyll a showed a positive correlation. Alternatively, when the values of the other five parameters (i.e., temperature, outflow, reservoir storage, dissolved oxygen, and transparency) increased by 10%, the sensitivity degree was less than zero, which meant the prediction of chlorophyll a showed a negative correlation. These correlations coincided with the mechanism of action for hydrology and water quality. Compared with other feature vectors, the prediction model for chlorophyll a in the Miyun Reservoir water had greater sensitivity to dissolved oxygen, transparency, permanganate index, pH, temperature, total nitrogen, and total phosphorus. It should be noted that the variations of transparency and dissolved oxygen were the results of the change in chlorophyll a.

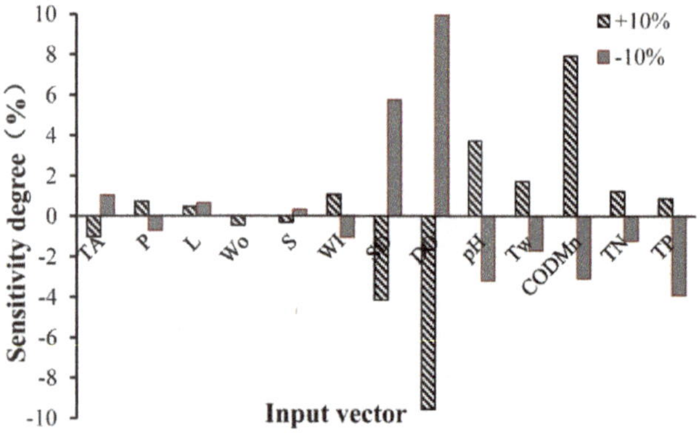

Figure 5: Sensitivity analysis for input vectors of the SVM model.

Four parameters, *i.e.*, total phosphorus, total nitrogen, permanganate index, and reservoir storage, were extracted through feature selection. The results of the feature selection were consistent with those of the sensitivity analysis. Compared with other studies, although the influence factors of chlorophyll *a* were different for various research objects, two factors including the concentration of TN and TP were always main factors. For example, Canfield [29] applied statistical analysis to pick out the total phosphorus concentration and the total nitrogen concentration as the explanatory variables in their developed prediction model of chlorophyll *a*.

With these four parameters as the input vectors, the simplified SVM model was constructed. Figure 6a shows the results of the model training process. It can be seen that the simulation values are perfectly consistent with the monitored values, with the exception of a bias in the extreme points. Accordingly, the RMSE was only 0.00017, and the R^2 was 97.33%. After the SVM prediction model was trained, chlorophyll *a* in the Miyun Reservoir was predicted for the period between 2005 and 2010. In Figure 6b, we can see that from 2005 to 2009, the simulation effect was passable. However, in 2010 the simulation effect was not satisfactory. Calculations showed that the RMSE of the testing set was 0.000641 and the R^2 was 81.97%. From 2005 to 2009, the RMSE was 0.0004 and the R^2 was 85.96%; however, in 2010, the RMSE was 0.0013 and the $R^2$was only 79.00%. This was primarily related to the fluctuations and periodicity of the monitored data. For the training data set, from 2000 to 2004, the concentration of chlorophyll *a* generally showed a peak in the middle of the year, but there was no obvious periodic trend for the concentration of chlorophyll *a* in the testing data set. In addition, in 2010 the concentration of chlorophyll *a* in the Miyun Reservoir was relatively higher compared with previous years. During April and for the period from August to November, the concentration of chlorophyll *a* was anomalously high, exceeding 0.004 mg/L, whereas in the training data set, the concentration of chlorophyll *a* had never achieved this level. Therefore, the SVM model was sensitive to the data. To explore which indicators were most relevant to chlorophyll *a*, sensitivity analysis for each input vector of the model was conducted.

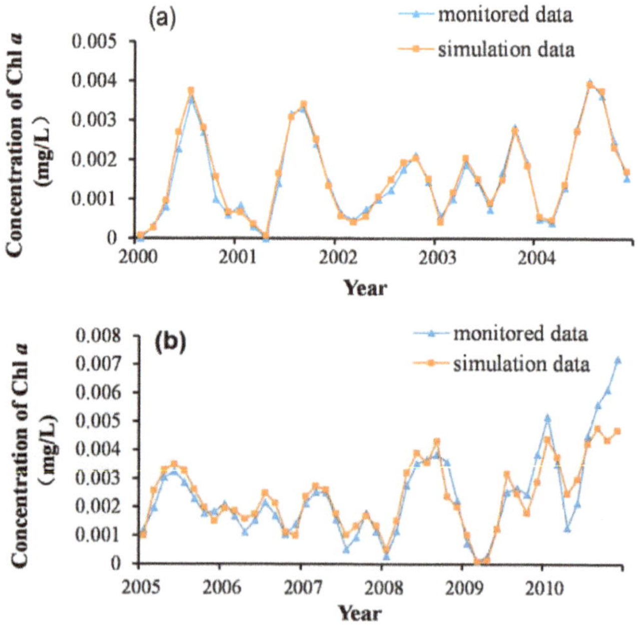

Figure 6: Training and prediction results of the SVM model: **(a)** training results and **(b)** prediction results.

Relative Errors of the SVM Model

Figure 7 shows the relative errors in each testing year of the SVM prediction model. Overall, the relative error between 2005 and 2009 was smaller than that in 2010. This indicated that the model had improved prediction accuracy during the early part of the testing data. This result was related to the variations in the testing data. The first five years of the testing data were consistent with the training data with regard to the amplitude, cycle, and peak values. However, data in the last year did not exhibit regularity, as evidenced by the notable fluctuations in each month. Given the obvious fluctuation in the overall trend, the average relative error of the prediction in 2010 was the largest.

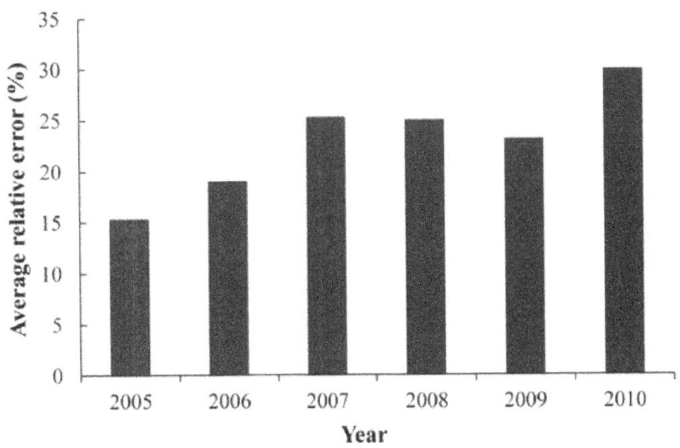

Figure 7: Average relative errors for each year.

To analyze the errors of chlorophyll *a* prediction in different months, the monthly average relative error of the SVM model was compared between April and November (Figure 8). Prediction results showed that the average relative errors from April to November were 29.10%, 15.90%, 35.84%, 14.64%, 27.53%, 19.41%, 19.15%, and 22.24%, respectively. It can be seen that in these six years, the biggest relative error occurred in June, followed by April, August, November, September, October, May, and July. Differences among these months may be due to the precipitation during summer. Inflow to the reservoir increased with rainfall, augmenting the frequency of soil and water erosion and leading to more nutrients being deposited into the reservoir. It required a significant amount of time for these nutrients to be used by the microorganisms in the water. Considering the cumulative effect, in autumn the concentration of chlorophyll *a* fluctuated markedly. Therefore, due to the synergy of rainfall and inflow, as well as the cumulative effect of the nutrients, the prediction error of chlorophyll *a* in summer and autumn was larger.

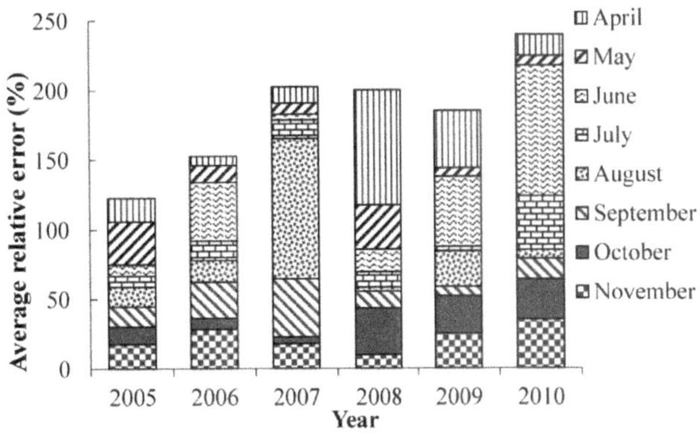

Figure 8: Average relative errors for each month.

Comparisons of Model with Feature Selection and Model without Feature Selection

To determine the effect of feature selection on the model, we also established a prediction model without feature selection. Table 1 shows the comparisons of model structure and prediction results between these two models. With feature selection, the input vectors were simplified from 13 to 4, and the extracted features were consistent with the sensitivity analysis. For the prediction results, in the training process, the mean AE, mean RE and R^2 of the model with feature selection were slightly larger, and the RMSE was slightly smaller; in the testing process, the mean AE, mean RE, and RMSE of the model with feature selection were smaller, and the R^2 was significantly higher than that of model without feature selection.

Table 1: Comparisons of model with feature selection and model without feature selection

Description		Model with Feature Selection	Model without Feature Selection
Number of input vectors		4	13
Input vectors		TP, TN, COD_{Mn}, S	TP, TN, COD_{Mn}, S, W_1, W_0, L, DO, T_W, pH, SD, T_A, P
Training process	Mean AE	0.00014244	0.00013824
	Mean RE	12.35%	10.64%
	RMSE	0.00017	0.00018
	R^2	97.33%	97.19%
Testing process	Mean AE	0.00045199	0.00057325
	Mean RE	22.98%	26.16%
	RMSE	0. 000641	0.000836
	R^2	81.97%	69.36%

Notes: S is reservoir storage; W_I is inflow; W_O is outflow; L is water level; TP is the concentration of total phosphorus in water; TN is total nitrogen in water; COD_{Mn} is permanganate index in water; DO is the dissolved oxygen concentration in water; T_W is water temperature; pH is hydrogenion concentration of water; SD is water transparency; T_A is the temperature; and P is precipitation.

It can be seen that feature selection picked out the representative and effective feature vectors from the original features that were more related with the chlorophyll *a*, so the dimensions of feature space were reduced. When the redundant or irrelevant information was deleted and the data set was simplified, the model was more concise and understandable [30]. Though the simulation accuracy in the training process was similar or even smaller than that prior to feature selection, relevant input vectors and reasonable model structure led to better prediction results in the testing process. As a whole, the SVM model with feature selection showed better performance both in model structure and prediction effect, and this indicated the model with feature selection had great potential in prediction ability, which had close relation with the internal structure of the model. In brief, feature selection with the GA in this study played important roles in three specific aspects. Firstly, it determined the feature vectors that were most relevant to chlorophyll *a* concentration. Their information was preserved accordingly in the simplified model, leading to the accuracy improvement with the decreased amount of calculation. It was a rather feasible way to improve calculation efficiency, especially for large-scale computing with multiple parameters.

Secondly, it reduced the dimensions of the input vectors to avoid dimension disaster (*i.e.*, with the increase of the input vectors' dimensions, the complexity of the calculation would greatly increase), while revealing the representative factors that influenced the chlorophyll *a* in the Miyun Reservoir. Thirdly, it was easy to combine with other algorithms (e.g., SVM) to improve the generalization ability of the SVM prediction model, reflecting good convergence and robustness.

Besides feature selection, the proposed SVM model showed good performance mainly due to the proper initial settings of parameters, which would affect the computational complexity and convergence rate directly. Although the GA used in the prediction model effectively avoided falling into a local optimal solution and producing a low convergence speed, the initial values of the GA's parameters were determined through the trial method in this study. Recent research mainly used two advanced approaches to optimize the initial settings for the GA's parameters: one approach was to optimize the initial population's characteristics and quantity by combining other approaches, such as the heuristic algorithm [31]; another approach was to improve the crossover and mutation rates with adaptive GA [32], such as clustering-based adaptive GA [33]. In future research, the determination of reasonable initial values of the GA's parameters would combine with these approaches.

CONCLUSIONS

A GA-based SVM model for predicting the monthly concentration of chlorophyll *a* of the Miyun Reservoir was constructed. We firstly carried out a sensitivity analysis of the prediction model, and identified that the concentration of chlorophyll *a* had great sensitivity to seven input indicators. With the GA being used for the removal of redundant features and the feature selection of input vectors, the four most relevant influence factors of chlorophyll *a* (*i.e.*, total phosphorus, total nitrogen, permanganate index, and reservoir storage) were screened as new input vectors, which were consistent with the results of the sensitivity analysis. With these new input vectors, the prediction model had simpler structure and better prediction accuracy than the model without the feature selection. Due to the stronger correlation of the input vector structure, the simplified GA-SVM model showed improved calibration and prediction ability. This proved that the SVM prediction model was sensitive to the structure of the input variables. In brief, this study proposed an intelligent algorithm for predicting the concentration of chlorophyll *a* of the reservoir water, which provided an effective tool for the management of reservoirs, especially for an early warning of eutrophication. Besides, this model could solve practical problems with different nutritional load conditions, and its applications can be extended to other reservoirs. In future research, the interaction mechanism of influence factors should be further considered to optimize the parameters used in the developed hybrid model of GA-SVM algorithm to get more reliable results for the prediction of chlorophyll *a*, and empirical models will be explored to get better application performance in chlorophyll *a* prediction.

ACKNOWLEDGMENTS

This study was supported by the Fund for Innovative Research Group of National Natural Science Foundation of China (No. 51421065), the National Science and Technology Support Program (No.2011BAC12B02) and the key project of National Natural Science Foundation of China (No. 51439001). We would like to extend special thanks to the editors and anonymous reviewers for all their detailed comments and valuable suggestions in greatly improving the quality of this paper.

AUTHOR CONTRIBUTIONS

Xuan Wang determined the focus of this work; Jieqiong Su performed the numerical simulations and drafted the initial manuscript under the guidance of Xuan Wang; Shouyan Zhao provided the data of the Miyun Reservoir; Bin Chen and Chunhui Li reviewed the work and helped bring it to its final form; and Zhifeng Yang was supporting this work as an expert in model building and fault detection.

REFERENCES

1. Scholz, M. Sustainable water systems. *Water* 2013, *5*, 239–242.
2. Cai, Y.P.; Huang, G.H.; Tan, Q.; Yang, Z.F. An integrated approach for climate-change impact analysis and adaptation planning under multi-level uncertainties. Part I: Methodology. *Renew. Sustain. Energy Rev.* 2011, *15*, 2779–2790.
3. Cai, Y.P.; Huang, G.H.; Yang, Z.F.; Tan, Q. Identification of optimal strategies for energy management systems planning under multiple uncertainties. *Appl. Energy* 2009, *86*, 480–495.
4. Tan, Q.; Huang, G.H.; Cai, Y.P. Radial interval chance-constrained programming for agricultural non-point source water pollution control under uncertainty. *Agric. Water Manag.* 2011, *98*, 1595–1606.
5. Mulia, I.E.; Tay, H.; Roopsekhar, K.; Tkalich, P. Hybrid ANN-GA model for predicting turbidity and chlorophyll-*a*concentrations. *J. Hydro-Environ. Res.* 2013, *7*, 279–299.
6. Liu, Y.; Guo, H.; Yang, P. Exploring the influence of lake water chemistry on chlorophyll *a*: A multivariate statistical model analysis. *Ecol. Model.* 2010, *221*, 681–688.
7. Cerco, C.F.; Noel, M.R. Twenty-one-year simulation of Chesapeake Bay water quality using the CE-QUAL-ICM eutrophication model. *J. Am. Water. Resour. Assoc.* 2013, *49*, 1119–1133.
8. Blancher, E.C. Modeling nutrients and multiple algal groups using AQUATOX: Watershed management implications for the Braden River Reservoir, Bradenton Florida. *Proc. Water Environ. Feder.* 2010, *10*, 6393–6410.
9. Rinke, K.; Yeates, P.; Rothhaupt, K.O. A simulation study of the feedback of phytoplankton on thermal structure via light extinction. *Freshw. Biol.* 2010, *55*, 1674–1693.
10. Seo, D.G.; Ahn, J.H. Prediction of chlorophyll-a changes due to weir constructions in the Nakdong River using EFDC-WASP modelling. *Environ. Eng. Res.* 2012, *17*, 95–102.

11. Chen, Q.; Han, R.; Ye, F.; Li, W. Spatio-temporal ecological models. *Ecol. Inform.* 2011, *6*, 37–43.

12. Gandomi, A.H.; Yun, G.J.; Yang, X.S.; Talatahari, S. Chaos-enhanced accelerated particle swarm optimization. *Commun. Noulinear Sci. Numer. Simul.* 2013, *18*, 327–340.

13. Maity, R.; Bhagwat, P.P.; Bhatnagar, A. Potential of support vector regression for prediction of monthly streamflow using endogenous property. *Hydrol. Process.* 2010, *24*, 917–923.

14. Malekmohamadi, I.; Bazargan-Lari, M.R.; Kerachian, R.; Nikoo, M.R.; Fallahnia, M. Evaluating the efficacy of SVMs, BNs, ANNs and ANFIS in wave height prediction. *Ocean. Eng.* 2011, *38*, 487–497.

15. Karamouz, M.; Ahmadi, A.; Moridi, A. Probabilistic Reservoir Operation Using Bayesian Stochastic Model and Support Vector Machine. *Adv. Water Resour.* 2009, *32*, 1588–1600.

16. Çimen, M.; Kisi, O. Comparison of Two Different Data-driven Techniques in Modelling Lake Level Fluctuations in Turkey. *J. Hydrol.* 2009, *378*, 253–262.

17. Zhang, Y.C.; Qian, X.; Qian, Y.; Liu, J.P.; Kong, F.X. Application of SVM on Chl-*a* concentration retrievals in Taihu Lake. *China Environ. Sci.* 2009, *29*, 78–83. (In Chinese).

18. Xiang, X.Q.; Tao, J.H. Eutrophication Model of Bohai Bay Based on GA-SVM. *J. Tianjin Univ.* 2011, *44*, 215–220. (In Chinese).

19. Liu, C.; Tang, D. Spatial and temporal variations in algal blooms in the coastal waters of the western South China Sea. *J. Hydro-Environ. Res.* 2012, *6*, 239–247.

20. Cho, K.H.; Kang, J.H.; Ki, S.J.; Park, Y.; Kim, J.H. Determination of the optimal parameters in regression models for the prediction of chlorophyll-*a*: A case study of the Yeongsan Reservoir, Korea. *Sci. Total Environ.* 2009, *407*, 2536–2545.

21. Wang, L.; Yang, M.; Guo, Z.H.; Zhang, Y.; Jiang, Y.; Fan, K.P. Study on water quality transformation in Miyun Reservoir. *China Water Wastewater* 2006, *22*, 45–48. (In Chinese).

22. Jia, D.M.; Wang, J.S.; Xue, X.J.; Qi, Z.Y. Research on phytoplankton characteristics of Miyun Reservoir. *Beijing Water*2013, *1*, 12–15. (In Chinese).

23. Pan, K.M.; Wang, J.M. Control and management of eutrophication of the Miyun reservoir. *Beijing Water* 2010, *6*, 25–27. (In Chinese).

24. Cortes, C.; Vapnik, V. Support-vector networks. *Mach. Learn.* 1995, *20*, 273–297.

25. Chang, C.C.; Lin, C.J. LIBSVM: A library for support vector machines. *ACM Trans. Intell. Syst. Technol.* 2011, *2*, 1–39.

26. Maus, A.; Sprott, C. Neural network method for determining embedding dimension of a time series. *Commun. Nonlinear Sci. Numer. Simul.* 2011, *16*, 3294–3302.

27. Pourbasheer, E.; Riahi, S.; Ganjali, M.R.; Norouzi, P. Application of genetic algorithm-support vector machine (GA-SVM) for prediction of BK-channels activity. *Eur. J. Med. Chem.* 2009, *44*, 5023–5028.

28. Fernandez, M.; Caballero, J.; Fernandez, L.; Sarai, A. Genetic algorithm optimization in drug design QSAR: Bayesian-regularized genetic neural networks (BRGNN) and genetic algorithm-optimized support vectors machines (GA-SVM). *Mol. Divers.* 2011, *15*, 269–289.

29. Canfield, D.E., Jr. Prediction of chlorophyll *a* concentrations in Florida lakes: The importance of phosphorus and nitrogen. *J. Am. Water Resour. Assoc.* 1983, *19*, 255–262.

30. Noori, R.; Karbassi, A.R.; Moghaddamnia, A.; Han, D.; Zokaei-Ashtiani, M.H.; Farokhnia, A.; Gousheh, M.G. Assessment of input variables determination on the SVM model performance using PCA, Gamma test, and forward selection techniques for monthly stream flow prediction. *J. Hydrol.* 2001, *401*, 177–189.

31. Besalatpour, A.A.; Ayoubi, S.; Hajabbasi, M.A. Feature selection using parallel genetic algorithm for the prediction of geometric mean diameter of soil aggregates by machine learning methods. *Arid Land Res. Manag.* 2014, *28*, 383–394.

32. Zandieh, M.; Karimi, N. An adaptive multi-population genetic algorithm to solve the multi-objective group scheduling problem in hybrid flexible flowshop with sequence-dependent setup times. *J. Intell. Manuf.* 2011, *22*, 979–989.

33. Halder, U.; Das, S.; Maity, D. A cluster-based differential evolution algorithm with external archive for optimization in dynamic environments. *IEEE Trans. Cybern.* 2013, *43*, 881–897.

CHAPTER 18

VEGETATION MANAGEMENT WITH FIRE MODIFIES PEATLAND SOIL THERMAL REGIME

Lee E. Brown, Sheila M. Palmer, Kerrylyn Johnston, Joseph Holden

Keywords: Burning, Heather, Moorland, Peat, Soil, Temperature

INTRODUCTION

Temperature plays an important role in soil ecosystem biogeochemistry by directly moderating rates of mineral weathering and soil water solution reactions (Brady and Weil, 2013), and indirectly by influencing the decomposition of organic matter (Davidson and Janssens, 2006 and Grosse et al., 2011) and uptake of nutrients by soil dwelling flora and fauna (Allison et al., 2010, Conant et al., 2011 and Melillo et al., 2002). Variations in soil thermal regime have been linked to changes in the abundance and diversity of soil microfauna (Allison and Treseder, 2011 and Darby et al., 2011), seed germination and vegetation growth/production (Glinski and Lipiec, 1990), and nutrient uptake by plants (Dong et al., 2001 and Pregitzer and King, 2005). Soil surface temperature can also influence latent heat fluxes and thus soil moisture (Kettridge et al., 2012), and is crucial for geomorphological processes such as freeze-thaw weathering (Holden, 2007). Soil temperature therefore plays a major role in global biogeochemical cycles, and so a clear understanding of the processes influencing soil thermal regime is a key requirement for ecosystem scientists and land managers to manage terrestrial environments effectively.

Fires are a common occurrence throughout the world, both naturally and for management purposes, in landscapes dominated by forest, grassland (e.g. prairie, moorland) and shrubs (e.g. chaparral, moorland) (Anderson, 1989 and Yibarbuk et al., 2001), and in wetland and peatland environments in the UK (Ramchunder et al., 2013) and boreal North America, northern Europe and Siberia, and South-East Asia (Aksamit and Irving, 1984 and Turetsky et al., 2011). Vegetation removal with fire can significantly alter the energy balance of the land surface by exposing underlying soils to more incoming solar radiation, altering long-wave radiation emission, and increasing exposure to wind thus changing latent heat fluxes. There are also likely to be changes in evapotranspiration which will influence soil moisture and thus specific heat capacity (Kettridge et al., 2012). Observational and modelling studies on soil temperature in some northern hemisphere peatlands have showed increases in temperature for the near-surface parts of the soil profile following wildfire (Harden et al., 2006 and Kettridge et al., 2012). However, wildfires may exert more serious damage to peat soils than prescribed vegetation burning because they typically burn at far higher temperatures and often ignite the underlying peat in addition to the surface vegetation. Prescribed burns are utilised by land managers to control vegetation succession but no studies to date have considered their effects on soil temperature in the years after fire. Vegetation burning on UK peatlands has been associated with increased dissolved organic carbon (DOC) release to rivers in several studies with implications for potable water treatment (Holden et al., 2012), but the processes that underpin changes in soil C cycling and DOC release to rivers following fires are still poorly understood.

Peatlands cover around 4.4 million km^2 of the Earth's surface but they store a disproportionate amount of soil carbon (C); an estimated 500 (±100) Gt has accumulated since the Last

Citation: Lee E. Brown, Sheila M. Palmer, Kerrylyn Johnston, Joseph Holden, Vegetation management with fire modifies peatland soil thermal regime, Journal of Environmental Management, Volume 154, 1 May 2015, Pages 166-176, ISSN 0301-4797, http://dx.doi.org/10.1016/j.jenvman.2015.02.037.

Glacial which is equivalent to > 2/3 of the atmospheric store (Yu, 2012). These systems act as a long term C sink at a rate >5 Gt C per century (equivalent to 11 g C m^{-2} y^{-1}; Yu, 2012). In peatland soils, temperature is associated strongly with the carbon balance, and in particular, the production and oxidation of CH_4 which responds exponentially to increased temperature (Yvon-Durocher et al., 2014), the production of CO_2 (Lafleur et al., 2005 and Moore et al., 1998) and the production and release of dissolved organic carbon (DOC) in pore waters (Freeman et al., 2001). Management changes to peatlands that lead to alterations in soil temperature can therefore be expected to have significant effects on the C balance of these systems.

Around 15% of the UK is covered with peat, with 87% of this being blanket peat (Baird et al., 2009). An estimated 3150 km^2 (18%) of UK peatlands have been subjected to prescribed burning (Worrall et al., 2010), although the use of this technique is regionally variable (Hester and Sydes, 1992 and Yallop et al., 2006). Here, vegetation burning is undertaken mainly to improve production of red grouse (Lagopus lagopus scotica) populations but also to benefit grazing livestock and deer (Ramchunder et al., 2013) and occasionally for forest regeneration purposes (Hancock et al., 2005), and typically occurs in 10- to 20-year rotations. There have been several recent studies of prescribed burning effects on peatland vegetation, soil hydrology and soil solution chemistry (e.g. Clay et al., 2009 and Worrall et al., 2007), and the evidence suggests that this management practice is altering fluvial C loads in upland peatland streams (Holden et al., 2012). To date, though, there has not been any consideration of how managed vegetation removal with fire and exposure of the underlying peat modifies the thermal regime of the soil mass in peatlands.

This study aimed to assess the role of prescribed vegetation burning on peat soil thermal regime at a series of intensively monitored locations in northern England. Specifically, thermal regimes were monitored across 12 burned peatland soil plots over an 18-month period to (i) quantify thermal dynamics between burned plots of different ages, and (ii) develop statistical models to determine the magnitude of thermal change caused by vegetation management. Soil thermal regimes were monitored in spatially independent soil plots spanning a period of <2 to >15 years since prescribed vegetation burning. Three hypotheses were tested: (H_1) upper soil-profile mean and maximum temperatures would be elevated in recently burned plots compared with areas of mature vegetation, but the effect would decrease as time since burning increased and vegetation cover regenerated; (H_2) upper soil-profile minimum temperatures would be lowest in recently burned plots compared with mature vegetation, which retain their 'insulating' canopy cover, and this combined with H_1 would mean a wider temperature range under recent burn plots; (H_3) enhanced soil surface maxima and minima in burned plots would be transmitted down into the soil at recently burned sites, but strong thermal attenuation would be evident with depth at all monitoring points.

METHODS

Study area and field data collection

The study was undertaken in the southern Pennine hills of northern England between 28 March 2010 and 24 October 2011. The main study site was located in the Bull Clough catchment (53°28′24.8″N; 1°42′46.2″W) on Midhope Moor ~16 km SW of Barnsley, South Yorkshire. Much of the catchment is managed using prescribed burning of vegetation patches (typically ranging from 2500 to 5000 m^2) on a 20- to 25-year rotation, and it is considered to be a relatively typical peatland managed specifically for grouse. Soil plots in four age classes were studied: vegetation patches where burning took place <2 years prior to the beginning of the study (B2; recent burn), patches with burning 3–4 years prior to the study (B4; vegetation early growth phase), patches with burning 5–7 years prior to the study (B7; later growth phase) and patches that were burnt 15–25 years prior to the experiment (B15+; mature heather). Patch age was confirmed by the gamekeeper who knew when burning had taken place across the catchment. However, it was not possible to narrow down the age range of the patches that were burnt at some point between 15 and 25 years prior to sampling, although it is thought that most were last burnt around 20 years before our study.

For each age class (B2, B4, B7, B15+), three plots of approximately 400 m^2 were chosen with respect to the topographic index, ln(tanß/α), where ß is the slope and α is drainage length per unit contour width (Beven and Kirkby, 1979). The use of the topographic index allowed us to control for any possible slope position effects, which is important because Holden (2009) showed that slope position can control the proportion of flow through macropores in several soil types, and thus soil moisture movement which would serve to moderate thermal variability. Each burn age class had one plot in a low topographic index setting, one in a mid-topographic index setting and the other in a high topographic index setting. Effectively, this was equivalent to a top-, mid- and foot-slope position. The same topographic index values were determined for plots chosen for each treatment so that slope position effects were controlled and were equal between treatments. Therefore data were collected at twelve spatially independent soil plots in total (4 age classes × 3 slope positions).

Soil temperature was measured in each plot at four depths: surface layer (0–1 cm), 5 cm, 20 cm and 50 cm. Traditional meteorological station measurements that obtain soil temperature do so at the surface, 5, 10, 20, 50 and 100 cm. However, it was not possible to instrument at all of these depths due to cost constraints and so 10 cm and 100 cm were not included. The depths we sampled conform well to typical plant structure and rooting depths in peatlands for mosses, grasses/sedges and dwarf shrubs. Soil temperature was measured at the end of each 15 min time interval using Gemini PB-5001 thermistors interfaced with Gemini Tinytag TGP-4520 dataloggers. The manufacturer›s published accuracy is within ±0.3 °C across the temperature range experienced in this study. For surface measurements, probes were placed horizontally within the top 1 cm of the peat-litter complex and checked for position every three weeks. For sub-surface measurements, probes were inserted into the previously undisturbed soil using a wooden dowel (which was then subsequently retrieved) so that the probe tip containing the thermistor element was located at the desired measurement depth. Air temperature was measured using the same models of Gemini probe and logger as described above, but housed in a radiation shield mounted 1 m above ground surface. Air temperature data were collected at the same intervals and using the same logging procedure as for soil temperatures. For comparative purposes, soil temperature data were also collected from soil plots with mature Calluna cover on a peatland without prescribed burning at Oakner Clough, Close Moss (53°36′11.1″N; 1°58′03.4″W) near Marsden ~15 km north west of Bull Clough. Parts of the Oakner catchment were damaged in a wildfire on 9th April 2011, thus, when comparing between Bull Clough and Oakner Clough records, only those data collected pre-wildfire were used in this study to avoid any confounding effects. All dataloggers were synchronised prior to deployment in the field, and thereafter every three weeks following data download.

All soil plots at both Bull Clough and Oakner Clough had blanket peat depth of >1 m and were subject to very light sheep grazing at <0.5 sheep ha^{-1} (with no sheep November–February). Vegetation surveys conducted after the study period showed the site was dominated by Calluna, with widespread Vaccinium myrtillus, a high but patchy cover of Rubus chamaemorus, but very little Sphagnum cover and a sparse cover of Eriophorum angustifolium and Eriophorum vaginatum. Water tables were monitored automatically for each plot using Trafag DL/N 70 pressure transducers positioned in a dipwell located in the centre of each plot and recording at 15-min intervals. Shallower water tables were, in general, found in the B15 + plots.

Data analysis

Raw data were summarised by calculating descriptive statistics from 15-min temperature datasets. Thereafter, all analyses used daily mean, maximum and minimum statistics calculated from 15-min resolution time series. Gaps in some temperature records (days 255–298, 2010 for B15 + plots) were applied to all other time-series to ensure comparable results. Scatter plots of air versus soil temperature data were created to illustrate the nature of thermal dynamics at all plots and depths. All graphical and statistical analyses were implemented in R 2.14 (R Development Core Team, 2014).

Regression model construction and evaluation

To assess whether vegetation removal with fire acted as a significant modifier of soil thermal regime, a generalised regression approach was developed to predict soil temperature time-series at burned plots (Bull Clough) from those measured at the unburned site (Oakner Clough). The approach is similar to that adopted elsewhere to model river water temperature dynamics following disturbance (Dickson et al., 2012 and Gomi et al., 2006) and allows direct statistical comparison of time-series data where data are not temporally independent. Regressions were established to predict daily mean and maximum soil temperature under mature heather cover (B15+) at Bull Clough from soil temperature records collected at Oakner Clough. This approach was preferred over models utilising air temperature because it allowed the use of linear rather than logistic regressions. Datasets were split into equal length odd and even day datasets, with odd day temperatures used to build models and even day data used subsequently to evaluate model performance prior to application across fire impacted patches (see below). Initial exploratory ordinary least squares (OLS) regression, autocorrelation analyses and Durbin–Watson statistics highlighted significant residual autocorrelation for all time-series, and so generalised least squares (GLS) regression in the nlme package (Pinheiro et al., 2006) was used subsequently.

Models were developed as:

$$T_{soil[Bull]} = \alpha + \beta_1 T_{soil[Oak]} + \beta_2 \sin(2\pi j/T) + \beta_3 \cos(2\pi j/T) + \varepsilon \quad (1)$$

where $T_{soil[Bull]}$ = soil temperature at either 0 cm, 5 cm, 20 cm or 50 cm measured under B15 + plots at Bull Clough, α = regression intercept, ß = regression coefficients, $T_{soil[Oak]}$ = soil temperature measured at Oakner Clough for the same depth, j = calendar day of year, T = number of days in year (i.e. 365), and ε = error term. For some plots/depths, the sine and cosine terms were statistically insignificant ($p > 0.05$) and so these were omitted from analyses for those plots/depths. Error terms were modelled predominantly as either first- or second-order autoregressive processes based on a priori examination of autocorrelation and partial autocorrelation functions.

A measure of random disturbance ($\hat{u}_t$) was calculated as:

$$\hat{u}_t = (y_t - \hat{y}_t) - \rho_1(y_{t-1} - \hat{y}_{t-1}) - \rho_i(y_{t-i} - \hat{y}_{t-i}) \quad (2)$$

where y = observed soil temperature for a given depth, $\hat{y}$ = predicted soil temperature for a given depth on day t, and ρ = the lag autocorrelation coefficient(s) from the GLS regression.

Regression models developed using odd day data were evaluated in two ways; first, confidence intervals (95%) of odd day model estimates were calculated as 1.96(σ $\hat{u}_t$) and compared visually with the distribution of $\hat{u}_t$ data. Second, soil temperatures under B15 + at Bull Clough for even day data were predicted following an odd-even day approach similar to Hannah et al. (2006). The production of $\hat{u}_t$ explicitly accounted for autocorrelation through the incorporation of lag coefficients derived from the GLS regressions. Finally, two-sample Kolmogorov–Smirnov (K–S) tests were used (Gomi et al., 2006) to assess if $\hat{u}_t$ had a similar distribution (i.e. $p > 0.05$) for both odd and even days.

Model application to recently burned soil plots

Odd day models derived for each slope position/depth were used thereafter to predict soil temperatures at equivalent slope positions and depths in rotationally burned heather patches B2, B4 and B7. Models were used to predict even day data for burned plots, and measures of random disturbance were calculated. K–S tests were used to compare distributions of $\hat{u}_t$ between B15+ and either B2, B4 or B7; if $\hat{u}_t$ had a similar distribution for these comparisons (i.e. $p > 0.05$) there would be no detectable effect of burning on soil thermal regime. For comparisons that produced statistically significant K–S test results, non-parametric effect sizes were calculated with Cliff's δ using the orddom package (Rogman, 2013). δ values range from −1 to 1, with a value of −1 denoting all observations were lower, and +1 all observations higher, than at B15+. ANOVA with Tukey HSD comparison was then used to determine whether there were differences in average estimates of daily mean and daily maximum $\hat{u}_t$ at a given depth for the three plots (i.e. disregarding slope position) in each burn age category.

RESULTS

Soil temperatures

Mean, maximum and variability (St. Dev) of temperature decreased with depth into the soil at all 12 plots, while minima were greater as depth increased (Table 1). With respect to fire effects, a general pattern of higher mean and maximum temperatures was found in B2 and B4 plots for all depths compared to B15 + plots (Table 1). The greatest differences were most evident at the surface, with mean temperatures being greater by 0.5–0.9 °C under the most recent burn plots compared with under mature heather. Maximum temperature differences were extremely pronounced for B2 compared to B15 + plots, being >20 °C warmer in the top-slope plot, >12 °C warmer mid-slope and >33 °C warmer at the bottom-slope plot. In addition, minimum soil surface temperatures were consistently lowest in B2 plots by between 1.9 °C at the bottom-slope plot and 4.3 °C at the top-slope plot compared to B15 + plots. Fire effects were also notable at 5 cm depth, although the magnitude of difference was far lower than at the surface, with mean temperature increases of 0.3 °C in B2 compared to B15+, and maximum temperature differences ranging from 1.7 °C mid-slope to 7.4 °C in bottom-slope plots (Table 1).

Table 1: Descriptive statistics derived from 15-min datasets for all combinations of slope position and soil depth. For each measurement location, data are presented in °C as Mean; Maximum; Minimum and St. Dev for the entire record

Slope-position/depth	B15+	B7+	B4	B2
Top-slope				
Surface	8.1; 29.0; −2.8; 5.5	8.4; 37.7; −2.4; 5.5	8.6; 46.2; −5.0; 6.5	8.9; 49.1; −7.1; 7.3
5 cm	7.9; 18.2; −0.1; 4.5	8.0; 17.3; −0.1; 4.2	8.2; 22.4; −2.0; 4.8	8.2; 20.6; −0.9; 4.9
20 cm	7.8; 13.8; 1.0; 3.7	7.8; 12.8; 1.1; 3.5	8.1; 14.0; 0.4; 3.9	8.0; 15.4; 0.3; 4.2
50 cm	7.3; 11.4; 2.4; 2.9	7.5; 11.2; 2.3; 2.7	7.9; 11.7; 2.1; 3.0	7.6; 12.4; 1.4; 3.3
Mid-slope				
Surface	8.2; 32.2; −2.0; 5.4	8.0; 26.3; −1.1; 5.1	9.0; 49.5; −4.3; 7.3	8.7; 44.6; −5.2; 6.5
5 cm	8.0; 18.5; −0.3; 4.5	7.9; 17.3; −0.1; 4.3	8.1; 21.6; −1.9; 4.9	8.3; 20.2; −0.7; 4.8
20 cm	7.7; 13.6; 0.9; 3.8	7.6; 13.8; 0.7; 3.6	8.4; 15.5; 0.7; 4.3	7.7; 14.6; 0.6; 4.1
50 cm	7.3; 11.4; 2.5; 2.9	7.4; 11.6; 1.8; 3.1	7.7; 11.7; 2.5; 3.0	7.3; 11.8; 2.1; 3.2
Bottom-slope				
Surface	8.2; 29.0; −2.0; 5.4	8.5; 44.6; −2.9; 6.0	8.3; 39.0; −3.8; 6.0	9.3; 52.8; −3.9; 7.5
5 cm	7.8; 15.4; 0.1; 4.1	8.3; 21.3; −0.5; 4.5	8.3; 17.7; −0.2; 4.4	8.1; 22.8; −0.5; 4.6
20 cm	7.7; 13.5; 1.1; 3.7	7.9; 13.1; 1.1; 3.6	8.1; 14.1; 0.8; 3.9	8.1; 13.8; 1.1; 3.8
50 cm	7.2; 11.0; 2.6; 2.8	7.4; 11.1; 2.3; 2.9	7.7; 11.5; 2.5; 3.0	7.7; 11.8; 2.5; 3.0

Scatter plots showing relationships between air and soil temperature showed clearly that increasing soil depth led to a strong reduction in soil temperature response both for daily mean and daily maximum temperatures (Fig. 1 and Fig. 2). For surface and 5 cm depth temperatures, soil temperature response was clearly damped when air temperature dropped below 0 °C. For each depth/slope combination, relationships between air and soil temperature were quite similar across burn ages with the main differences evident at air temperature extremes, with B2 and B4 plots displaying elevated soil temperatures at high air temperature, particularly for surface top-slope and foot-slope plots (Fig. 1). These effects were also seen in maximum soil surface temperatures, albeit much more pronounced (Fig. 2).

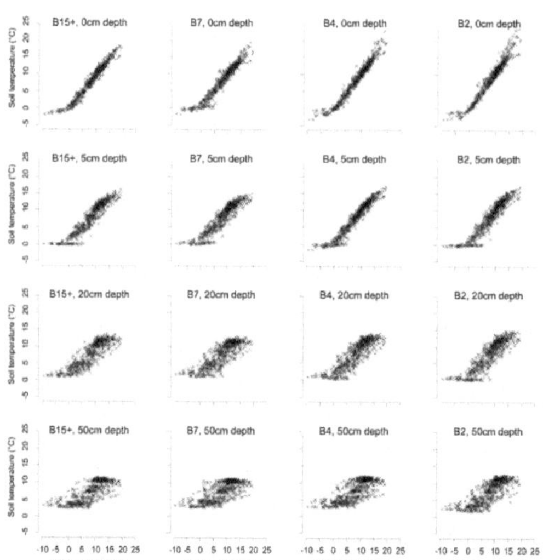

Figure 1: Relationships between mean daily air and soil temperatures at four depths in top-slope plots.

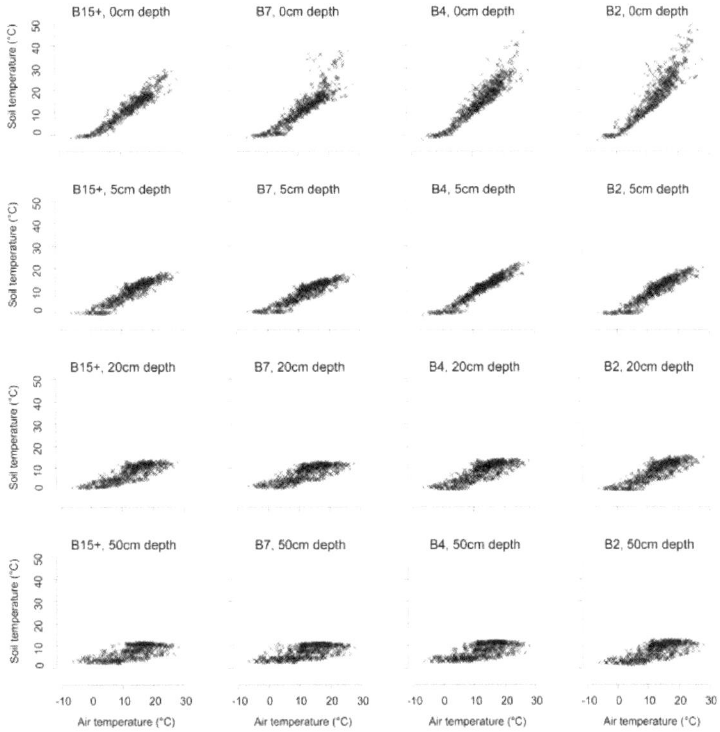

Figure 2: Relationships between maximum daily air and soil temperatures at four depths in top-slope plots.

Model evaluation

Mean daily temperature models showed strong synchronicity between soil temperatures at Bull Clough B15 + plots and those at Oakner Clough, with β_1 coefficients >0.90 and in several cases equal to 1.0 for top and mid-slope positions. Maximum temperature models produced similar β_1 coefficients for 5 cm, 20 cm and 50 cm depths, although surface temperatures were more variable and β_1 coefficients were between 0.55 and 0.59, although in all cases the relationships were nevertheless statistically significant. Mean $\hat{u}_t$ estimates for both mean and maximum daily temperature models were all close to zero, with standard deviations of <0.73 °C for all slope positions and depths, with the exception of maximum soil surface temperatures which were more varied (Table 2). Every model developed using odd day data produced no significantly different $\hat{u}_t$ distributions when applied to even day data, thus it was possible subsequently to use the odd day models to determine the magnitude of thermal regime change in B2, B4 and B7 plots relative to B15+.

Table 2: Mean (±1 St.Dev) $\hat{u}_t$ estimates for mean and maximum daily temperature model validations, with significance results from K–S tests

Slope-position/depth	Daily mean			Daily maximum		
	Odd	Even	p	Odd	Even	p
Top-slope						
Surface	0.0007 ± 0.66	0.002 ± 0.72	0.99	−0.005 ± 1.91	0.01 ± 1.93	0.59
5 cm	0.007 ± 0.44	0.01 ± 0.44	0.69	0.006 ± 0.63	0.03 ± 0.64	0.92
20 cm	0.0002 ± 0.32	0.0005 ± 0.32	0.92	0.0002 ± 0.32	−0.007 ± 0.31	0.92
50 cm	0.001 ± 0.08	0.0008 ± 0.08	0.51	0.001 ± 0.09	0.001 ± 0.10	0.97
Mid-slope						
Surface	−0.005 ± 0.69	0.01 ± 0.69	0.78	−0.005 ± 1.97	0.002 ± 1.82	0.92
5 cm	0.003 ± 0.38	0.02 ± 0.39	0.99	0.0001 ± 0.56	0.04 ± 0.59	0.35
20 cm	−0.003 ± 0.23	−0.008 ± 0.23	0.51	−0.004 ± 0.23	−0.01 ± 0.26	0.97
50 cm	−0.002 ± 0.05	−0.004 ± 0.05	0.92	−0.003 ± 0.05	−0.005 ± 0.05	0.97
Foot-slope						
Surface	−0.003 ± 0.70	0.005 ± 0.73	0.51	−0.007 ± 2.02	0.02 ± 1.86	0.50
5 cm	0.002 ± 0.27	0.002 ± 0.26	0.69	0.0002 ± 0.33	0.001 ± 0.37	0.92
20 cm	−0.003 ± 0.20	−0.007 ± 0.20	0.78	−0.005 ± 0.20	−0.01 ± 0.23	0.99
50 cm	−0.001 ± 0.05	−0.002 ± 0.05	0.86	−0.002 ± 0.05	−0.004 ± 0.05	1.00

Model application

Disturbances for mean daily temperature data were significant at all depths and in all slope positions in B2 and B4 burn plots with the exception of B4 foot-slope surface (Table 3). Only top-slope surface measurements in B7 plots were significantly different to those in B15 + plots. Surface effect-size estimates were between 0.12 and 0.27. The magnitude of both the mean and standard deviation of significant disturbances was greatest at the surface in all B2 plots and top/mid-slope plots for B4, and decreased at greater depths (Table 3, Fig. 3). Boxplots illustrated that the effect of burning was most pronounced at top and foot-slope plots, with clear upward shifts in $\hat{u}_t$ median, inter-quartile ranges and outliers from B15 + through to B2. Maximum $\hat{u}_t$ estimates of 5.2 °C and 6.2 °C were observed at the surface for top-slope and foot-slope plots, respectively (Fig. 3).

Table 3: Mean (±1 St. Dev) $\hat{u}_t$ estimates for odd day mean daily temperature predictions, with significance results from K–S tests for each age plot relative to B15+ [* = p < 0.05; **p < 0.01; ***p < 0.001]. Values in square parentheses are Cliff's δ estimates of effect size

Slope-position/depth	B15+	B7+	B4	B2
Top-slope				
Surface	0.0007 ± 0.66	0.21 ± 0.70** [0.20]	0.36 ± 0.97** [0.23]	0.54 ± 1.32*** [0.27]
5 cm	0.007 ± 0.44	0.02 ± 0.43	0.18 ± 0.89*** [0.12]	0.14 ± 0.69* [0.12]
20 cm	0.0002 ± 0.32	−0.07 ± 0.27* [-0.12]	0.02 ± 0.43** [0.12]	−0.02 ± 0.67** [-0.03]
50 cm	0.001 ± 0.08	0.05 ± 0.41* [0.11]	0.11 ± 0.19*** [0.57]	0.03 ± 0.43*** [-0.01]
Mid-slope				
Surface	−0.005 ± 0.69	−0.008 ± 0.74	0.25 ± 1.11** [0.12]	0.23 ± 0.75*** [0.20]
5 cm	0.003 ± 0.38	0.003 ± 0.38	−0.07 ± 0.73** [-0.07]	0.17 ± 0.66*** [0.18]
20 cm	−0.003 ± 0.23	−0.06 ± 0.38** [-0.15]	0.06 ± 0.34*** [0.20]	0.08 ± 0.44*** [0.20]
50 cm	−0.002 ± 0.05	−0.01 ± 0.21*** [-0.15]	0.01 ± 0.07*** [0.23]	0.006 ± 0.10*** [0.18]
Foot-slope				
Surface	−0.003 ± 0.70	0.12 ± 0.92	0.01 ± 0.91	0.53 ± 1.47*** [0.23]
5 cm	0.002 ± 0.27	0.03 ± 0.54*** [0.03]	0.02 ± 0.86*** [0.07]	0.18 ± 0.76*** [0.12]
20 cm	−0.003 ± 0.20	0.003 ± 0.41* [-0.04]	0.09 ± 0.40*** [0.23]	0.10 ± 0.49*** [0.21]
50 cm	−0.001 ± 0.05	0.0005 ± 0.23*** [0.09]	0.02 ± 0.09*** [0.28]	0.02 ± 0.12*** [0.24]

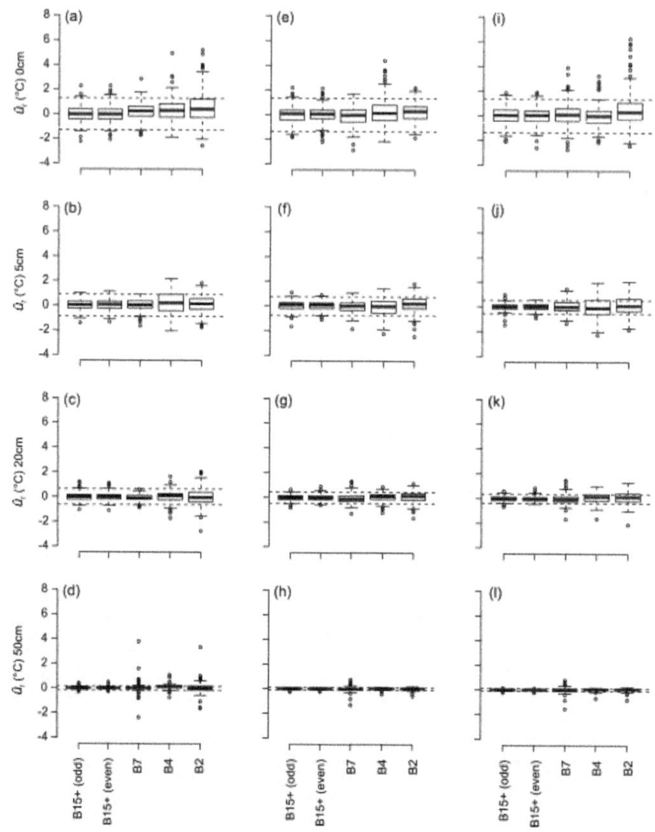

Figure 3: Boxplots for mean daily soil temperature $\hat{u}_t$ estimates at: (a–d) top-slope, (e–h) mid-slope, (i–l) foot-slope plots. Broken horizontal lines denote the 95% confidence intervals of predictions from B15 + odd day models.

Similar to daily mean temperature data, disturbances for maximum daily temperature data were significant in all B2 and B4 burn plots with the exception of B4 foot-slope surface (Table 4). Both top-slope and foot-slope surface maximum $\hat{u}_t$ measurements in B7 plots were significantly different to those in B15 + plots. Surface effect size estimates for B2 and B4 maxima were generally higher than for mean daily data, ranging between 0.26 and 0.42. The magnitude of both the mean and standard deviation of significant disturbances was greatest at the surface in all plots and generally decreased with increasing depth (Table 4, Fig. 4). Boxplots illustrated strong effects of time since burning on maximum $\hat{u}_t$ values at all slope positions, and clear upward shifts in $\hat{u}_t$ median, inter-quartile ranges and outliers from B15 + through to B2. Maximum $\hat{u}_t$ estimates reached 15.6 °C (top-slope), 9.8 °C (mid-slope) and 19.6 °C (foot-slope; Fig. 4). The range of temperature maxima also increased notably in recently burned plots, with minima being much lower in B2+ plots (−3.9 to −7.1 °C) compared to B15+ (−2.0 to −2.8 °C). Significant and relatively high temperature maxima effects were also recorded at 5 cm depth with up to 4.0 °C (top-slope), 3.1 °C (mid-slope) and 8.5 °C (foot-slope) difference. ANOVA showed significant differences in the magnitude of average daily mean $\hat{u}_t$ between B2 plots and B15 + at the surface, and between B2 and both B7 and B15 + at 5 cm depth (Table 5). For daily maximum $\hat{u}_t$ the only significant differences were observed between B2 and B15 + plots at the surface.

Table 4: Mean (±1 St. Dev) $\hat{u}_t$ estimates for odd day maximum daily temperature predictions, with significance results from K–S tests for each age plot relative to B15+ [^{x} = p > 0.05; * = p < 0.05; **p < 0.01; ***p < 0.001]. Values in square parentheses are Cliff›s δ estimates of effect size

Slope-position/depth	B15+	B7+	B4	B2
Top-slope				
Surface	−0.005 ± 1.91	0.63 ± 2.37* [0.15]	1.88 ± 3.65*** [0.35]	2.77 ± 4.49*** [0.42]
5 cm	0.006 ± 0.63	0.009 ± 0.64	1.21 ± 1.67*** [0.47]	0.48 ± 1.13*** [0.27]
20 cm	0.0002 ± 0.32	−0.08 ± 0.27** [-0.18]	0.03 ± 0.44* [0.10]	0.06 ± 0.68*** [0.07]
50 cm	0.001 ± 0.09	0.05 ± 0.46	0.12 ± 0.18*** [0.58]	0.03 ± 0.33*** [0.02]
Mid-slope				
Surface	−0.005 ± 1.97	−0.21 ± 1.99	1.46 ± 3.50*** [0.26]	1.04 ± 2.44*** [0.26]
5 cm	0.0001 ± 0.56	0.006 ± 0.66	0.48 ± 1.17*** [0.27]	0.47 ± 0.99*** [0.34]
20 cm	−0.004 ± 0.23	−0.04 ± 0.38** [0.11]	0.07 ± 0.33*** [0.21]	0.10 ± 0.43*** [0.24]
50 cm	−0.003 ± 0.05	−0.01 ± 0.21*** [0.14]	0.01 ± 0.07*** [0.23]	0.006 ± 0.10*** [0.18]
Foot-slope				
Surface	−0.007 ± 2.02	0.89 ± 3.46* [0.12]	0.30 ± 2.65	2.27 ± 4.66*** [0.31]
5 cm	0.0002 ± 0.33	0.80 ± 1.40*** [0.33]	0.39 ± 0.74*** [0.35]	0.37 ± 1.48*** [0.15]
20 cm	−0.005 ± 0.20	0.08 ± 0.23** [0.22]	0.14 ± 0.35*** [0.34]	0.16 ± 0.34*** [0.42]
50 cm	−0.002 ± 0.05	0.0008 ± 0.07	0.02 ± 0.09*** [0.31]	0.03 ± 0.12*** [0.34]

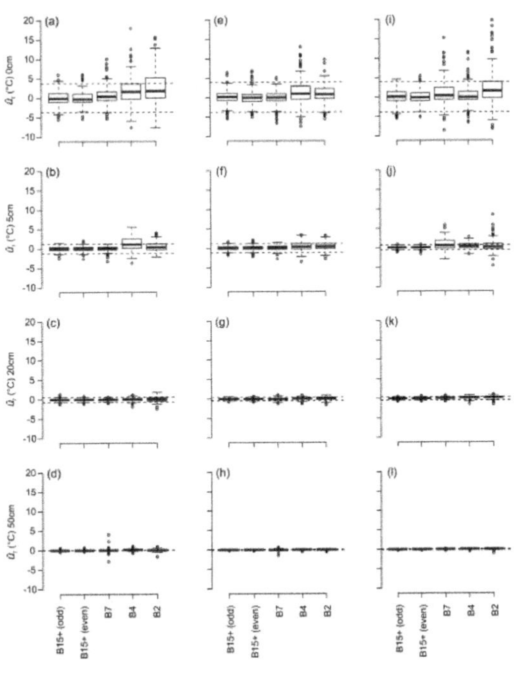

Figure 4: Boxplots for maximum daily soil temperature $\hat{u}_t$ estimates at: (a–d) top-slope, (e–h) mid-slope, (i–l) foot-slope plots. Broken horizontal lines denote the 95% confidence intervals of predictions from B15 + odd day models.

Table 5: Adjusted p-values from the ANOVA of average daily mean and maximum $\widehat{u}_t$. Significant differences are highlighted in bold font

	Daily mean				Daily maximum			
	B2	B4	B7	B15	B2	B4	B7	B15
Surface								
B2	–	0.26	0.08	0.02	–	0.44	0.09	0.02
B4	–	–	0.81	0.31	–	–	0.62	0.20
B7	–	–	–	0.76	–	–	–	0.72
B15	–	–	–	–	–	–	–	–
5 cm								
B2	–	0.23	**0.01**	**0.008**	–	0.77	0.92	0.40
B4	–	–	0.24	0.14	–	–	0.43	0.11
B7	–	–	–	0.97	–	–	–	0.74
B15	–	–	–	–	–	–	–	–
20 cm								
B2	–	0.84	0.85	0.14	–	0.84	0.62	0.05
B4	–	–	1.00	0.42	–	–	0.97	0.15
B7	–	–	–	0.40	–	–	–	0.27
B15	–	–	–	–	–	–	–	–
50 cm								
B2	–	0.70	1.00	0.87	–	0.75	1.00	0.88
B4	–	–	0.73	0.31	–	–	0.71	0.37
B7	–	–	–	0.34	–	–	–	0.91
B15	–	–	–	–	–	–	–	–

DISCUSSION

Virtually all biogeochemical processes occurring in soils are temperature dependent (Pregitzer and King, 2005). However, near-surface processes within the peat can be extremely important in peatlands for both C processing (Cole et al., 2002) and hydrological fluxes (Holden and Burt, 2003). Therefore, our findings that prescribed peatland vegetation burning leads to significant increases in mean and maximum near-surface soil temperatures (H_1) as well as lower minima and thus wider thermal variability (H_2), provide a clear indication that the use of fire to remove patches of vegetation is likely to be having unintended consequences for C processing and release. Our study spanned two summers but only one winter, therefore there are more observations of high versus low temperatures in our dataset. Nevertheless, the extreme warming seen in recently burned plots relative to those with mature vegetation may be one of the contributory factors associated with enhanced DOC release from peatlands subject to prescribed burning (Holden et al., 2012); this is because DOC generated in the upper few centimetres of the peat often dominates the stream water DOC signal (Clark et al., 2008).

Upper soil profile temperatures were elevated in recently burned patches compared with patches of mature vegetation, thus H_1 was accepted. The effects were most likely due to the exposure of soils to incoming solar radiation and reduced albedo due to the dark colour of exposed peat (Post et al., 2000). The exposure is partly related to the removal of vegetation by burning, but it is also due to the lack of protective litter cover in recently burnt plots compared to B15 + plots. The protective canopy and litter of plots that have not been burnt for many years would also insulate against heat loss at night and on cold days, reducing the temperature range compared with recently burnt peat (thus supporting H_2). While transpiration would be expected to be much lower in recently burnt patches, evaporation might be greater if water tables are sufficiently close to the surface and so these effects might, in some circumstances, cancel each other out and not greatly impact soil temperatures (Thompson, 2012).

The exact nature of any evapotranspirative cooling effects and related feedbacks requires further study in these peatland systems, but evidence from other northern peatlands suggest evapotranspiration (ET) may depend on antecedent soil moisture conditions (Bridgham et al., 1999) and also on whether sites are Sphagnum-rich or not (Thompson and Waddington, 2013). At our sites, which had an almost complete lack of Sphagnum cover, recently burnt plots had lower near-surface hydraulic conductivity, smaller proportions of bypassing macropore flow and higher bulk density compared to B15 + or unburnt plots (Holden et al., 2014) demonstrating related physical and hydrological impacts of fire management. Slowed through flow of water and a reduction in bypassing flow would mean that recently burnt plots would have less moderation of their temperature because water moving through the peat would be warmed or cooled more by local plot conditions than water moving more quickly through the plot from upslope.

It is not clear why the soil temperatures of mid-slope plots were less affected by burning than the other plots, but one explanation may be related to water residence times. Soil water flowing through mid-slope plots may remain on the plot for a shorter period of time, and have less chance to be influenced by plot surface conditions, than water on the flatter top or foot slope areas. This effect may be enhanced by differences in hydraulic conductivity such as those found by Lewis et al. (2012) who suggested that near-surface hydraulic conductivity may decrease towards the centre of a peatland. Topographic exposure to wind may also contribute to temperature differences across the peatland system (Kettridge et al., 2012) such that more exposed top-slope plots may undergo enhanced evaporative cooling which may counteract (slightly) the warming effects on burnt top slopes on hot days when compared to mid-slope or foot-slope plots.

Overall there were very large surface temperature disturbances in B2 plots based on daily data. Even at 5 cm depth, mean and maximum daily temperature disturbances were of the order of 4–6 °C in B2 plots, supporting H_3 that surface thermal disturbances would be transmitted into the soil, whilst at B15+ and B7 plots temperature disturbances were less pronounced at 5 cm. Hence, the effect of surface vegetation modification by fire on peat temperatures is strong and appears to persist in plots that were burned several years previously. In line with our predictions for H_1 and H_2, however, there is strong 'space for time' evidence that the effects decrease over time since burning, as temperature disturbances from model predictions were typically lower for B7 plots compared with B2 and B4 plots at equivalent slope positions. This 'recovery' has also been seen in soil hydraulic conductivity and macropore flow analyses across the Bull Clough plots (Holden et al., 2014). The area of bare peat was generally greatest on B2 plots with less than 4% on all B4 plots to 0% on B7 plots, thus the loss of the burn effect on soil temperatures appears to coincide with canopy closure which would alter surface albedo (Post et al., 2000). However, there were some small areas of bare peat (<4%) on one of B15 + plots where vegetation became more degenerated (as at the Oakner control plot) but at the same time there was a large increase in litter cover across two of the B15 + plots to a similar level to that at the Oakner plot, compared to all other plots. Litter cooling effects on surface peat during hot days has been observed during peat restoration projects (Price et al., 1998) and these effects, plus the insulating effects on cold days, have long been known from studies of soils in other environments (MacKinney, 1929).

The wider range of temperatures observed in recently burned plots compared to older plots is likely to contribute towards peat decomposition and erosion. At higher temperatures, microbial breakdown processes that enhance the production and supply of dissolved organic compounds to stream water could proceed more quickly (Holden et al., 2012), though so could the respiration of this organic matter (Yvon-Durocher et al., 2012). The balance between DOC production and respiration as a function of temperature needs to be studied in more detail in these systems to determine whether the warming effect is concomitant with net DOC increases or decreases (or no change) in soils, especially as some studies have found significant interactive effects of temperature and moisture availability on soil metabolism (O'Donnell et al., 2009). Alternatively, at the lower temperatures there could be enhanced generation of more particulates by freeze-thaw mechanisms. These particulates could subsequently be eroded and transported by wind and water to nearby streams (Imeson, 1971), where some recent studies have noted higher quantities of fine sediment in river bed sediments compared with streams in catchments that have no burning at all (Brown et al., 2013 and Ramchunder et al., 2013). Thus, both dissolved and particulate C losses need to be considered when linking altered soil thermal regime to net changes in C.

The temperature maxima recorded in this study exceeded 50 °C in some plots, which was similar to peak temperatures that have been recorded in Canadian peatlands impacted by wildfire (Kettridge et al., 2012). Significant reductions (>50%) in mycorrhiza biomass are thought to occur only when vegetation burning results in soil temperatures >50–60 °C (Allison and Treseder, 2011 and Certini, 2005). However, while temperatures were not recorded concurrently with any vegetation burning, our results illustrate that temperatures above 50 °C occur during warm periods long after fires have passed. In a study of burned vegetation plots elsewhere in northern England, Ward et al. (2012)reported a significant reduction (83%) in fungal phospholipid fatty acids which may have been due to the loss of ericoid mycorrhiza. While Ward et al. (2012) did not report data on the thermal behaviour of their study sites, the soil temperatures recorded in our plots under similar burn regimes suggest that alterations to thermal regimes are likely to be a key driver of altered plant-soil function until re-vegetation with heather occurs.

The statistical modelling approach adopted as part of this study shows clearly the magnitude of effect that vegetation removal has on soil thermal regime. However, its main limitation is that it does not provide us with a mechanistic basis for understanding the thermodynamic processes that are driving such alterations (Kettridge et al., 2012). Such approaches are necessary to quantify, for example, the magnitude of any changes in incoming solar radiation reaching the peat surface under different canopy structures (Williams and Quniton, 2013), long-wave radiation emission, and the interactions between soil moisture, temperature, microtopography and air movement in driving latent heat fluxes (Bridgham et al., 1999). Future studies should therefore aim to monitor numerous meteorological and hydrological parameters at the soil surface across different aged plots.

The significant alteration of soil thermal regime following vegetation removal by burning indicates a potential need to rethink the ways in which vegetation management could be optimised to provide the desired open canopy for growth of fresh shoots, but still retain shading of the soil surface to minimise warming. One possibility that could be considered is to switch to cutting of vegetation (Glaves et al., 2013 and Worrall et al., 2013) and leave a thick brash cover on the soil surface. However, while this may moderate any thermal effects it could prevent heather shoot regrowth and thus make foraging harder for young grouse, or lead to enhanced DOC generation from the decomposition of the cut vegetation litter. An alternative, that maintains aspects of burning but could still possibly mitigate against significant soil thermal regime alteration, is the burning of narrow strips that run east-west rather than burning larger open blocks. Thus, mature vegetation would provide some shading to the adjacent cleared strip, and it could also be a useful technique for reducing losses of game birds to aerial predators (The Heather Trust, 2013). Other considerations would be avoiding vegetation removal on south-facing slopes to minimise warming effects on exposed soils. Before any major changes in management approaches are considered though, it would be prudent to study in more detail the process links between energy inputs/losses, soil thermal regime and biogeochemical cycling in peatlands.

CONCLUSIONS

This study has demonstrated that prescribed vegetation burning in the UK uplands leads to significant changes in soil thermal regime. The greatest effects were observed at the soil surface, and in most plots were transmitted into the soil to depths of up to 20 cm. Thermal regime changes were most notable in soil plots that had been burned within 2 years of the study, but showed a clear tendency to recover in plots as vegetation regrows. Our findings that prescribed peatland vegetation burning alters soil thermal regime should provide an impetus for further research to understand the consequences of thermal regime change for carbon processing and release, and hydrological processes, in peatlands.

ACKNOWLEDGEMENTS

The research was conducted as part of the EMBER (Effects of Moorland Burning on the Ecohydrology of River basins) project funded by the UK Natural Environment Research Council (NE/G00224X/1). Yorkshire Water provided additional project funding. The authors thank the landowners and gamekeepers and The National Trust for granting permission to access the field sites. Natural England are thanked for commissioning the vegetation surveys and Stuart Hedley of Wild Plants and Vegetation for carrying them out. Catherine Wearing and

Martin Gilpin provided much appreciated assistance with the collection of field data.

REFERENCES

1. Aksamit, S.E., Irving, F.D., 1984. Prescribed burning for lowland black spruce regeneration in northern Minnesota. Can. J. For. Res. 14, 107e113.
2. Allison, S.D., Treseder, K.K., 2011. Climate change feedbacks to microbial decomposition in boreal soils. Fungal Ecol. 4, 362e374.
3. Allison, S.D., Wallenstein, M.D., Bradford, M.A., 2010. Soil-carbon response to warming dependent on microbial physiology. Nat. Geosci. 3, 336e340.
4. Anderson, R.C., 1989. The historic role of fire in the north American grassland. In: Collins, S.L., Wallace, L.L. (Eds.), Fire in North American Tallgrass Prairies. University of Oklahoma Press, Oklahoma.
5. Baird, A.J., Holden, J., Chapman, P.J., 2009. Literature Review of Emissions of Methane from Peatlands, Final Report to Defra. Project code: SP0574.
6. Beven, K.J., Kirkby, M.J., 1979. A physically based, variable contributing area model of basin hydrology. Hydrol. Sci. Bull. 24, 43e69.
7. Brady, N.C., Weil, R.R., 2013. The Nature and Properties of Soils. Pearson.
8. Bridgham, S.D., Pastor, J., Updegraff, K., Malterer, T.J., Johnson, K., Harth, C., Chen, J., 1999. Ecosystem control over temperature and energy flux in northern peatlands. Ecol. Appl. 9, 1345e1358.
9. Brown, L.E., Johnston, K., Palmer, S.M., Aspray, K.L., Holden, J., 2013. River ecosystem response to prescribed vegetation burning on blanket peatland. PLoS One 8, e81023.
10. Certini, G., 2005. Effects of fire on properties of forest soils: a review. Oecologia 143, 1e10.
11. Clark, J.M., Lane, S.N., Chapman, P.J., Adamson, J.K., 2008. Link between DOC in near surface peat and stream water in an upland catchment. Sci. Total Environ. 404, 308e315.
12. Clay, G.D., Worrall, F., Fraser, E.D.G., 2009. Effects of managed burning upon dissolved organic carbon (DOC) in soil water and runoff water following a managed burn of a UK blanket bog. J. Hydrology 367, 41e51.
13. Cole, L., Bardgett, R.D., Ineson, P., Adamson, J.K., 2002. Relationships between enchytraeid worms (Oligochaeta), climate change, and the release of dissolved organic carbon from blanket peat in northern England. Soil Biol. Biochem. 34, 599e607.
14. Conant, R.T., Ryan, M.G., Ågren, G.I., Birge, H.E., Davidson, E.A., Eliasson, P.E., Evans, S.E., Frey, S.D., Giardina, C.P., Hopkins, F.M., Hyvonen, R., € Kirschbaum, M.U.F., Lavallee, J.M., Leifeld, J., Parton, W.J., Steinweg, J.M., Wallenstein, M.D., Wetterstedt, J.A.M., Bradford, M.A., 2011. . Temperature and soil organic matter decomposition rates e synthesis of current knowledge and a way forward. Glob. Change Biol. 17, 3392e3404.
15. Darby, B.J., Neher, D.A., Housman, D.C., Belnap, J., 2011. Few apparent short-term effects of elevated soil temperature and increased frequency of summer precipitation on the abundance and taxonomic diversity of desert soil micro- and meso-fauna. Soil Biol. Biochem. 43, 1474e1481.
16. Davidson, E.A., Janssens, I.A., 2006. Temperature sensitivity of soil carbon decomposition and feedbacks to climate change. Nature 440, 165e173.
17. Dickson, N.E., Carrivick, J.L., Brown, L.E., 2012. Flow regulation alters alpine river thermal regimes. J. Hydrology 464e465, 505e516.
18. Dong, S., Scagel, C.F., Cheng, L., Fuchigami, L.H., 2001. Soil temperature and plant growth stage influence nitrogen uptake and amino acid concentration of apple during early spring growth. Tree Physiol. 21, 541e547.
19. Freeman, C., Evans, C.D., Monteith, D.T., Reynolds, B., Fenner, N., 2001. Export of organic carbon from peat soils. Nature 412, 785.
20. Glaves, D., Morecroft, M., Fitzgibbon, C., Owen, M., Phillips, S., Leppitt, P., 2013. The Effects of Managed Burning on Upland Peatland Biodiversity, Carbon and Water. Natural England Evidence Review NEER004, Peterborough, UK. Glinski, J., Lipiec, J., 1990. Soil Physical Conditions and Plant Growth. CRC Press, Boca Raton, Florida.
21. Gomi, T., Moore, R.D., Dhakal, A.S., 2006. Headwater stream temperature response to clear-cut harvesting with different riparian treatments, coastal British Columbia, Canada. Water Resour. Res. 42, W08437.
22. Grosse, G., Harden, J., Turetsky, M., McGuire, A.D., Camill, P., Tarnocai, C., Frolking, S., Schuur, E.A.G., Jorgenson, T., Marchenko, S., Romanovsky, V., Wickland, K.P., French, N., Waldrop, M., Bourgeau-Chavez, L., Striegl, R.G., 2011. Vulnerability of high-latitude soil organic carbon in North America to disturbance. J. Geophys. Res. Biogeosciences 116, G00K06.
23. Hancock, M., Egan, S., Summers, R., Cowie, N., Amphlett, A., Rao, S., Hamilton, A., 2005. The effect of experimental prescribed fire on the establishment of Scots pine Pinus sylvestris seedlings on heather Calluna vulgaris moorland. For. Ecol. Manag. 212, 199e213.
24. Hannah, D.M., Bower, D., McGregor, G., 2006. Associations between Western European air-masses and river flow regimes. IAHS Publ. 308, 344e349.
25. Harden, J.W., Manies, K.L., Turetsky, M.R., Neff, J.C., 2006. Effects of wildfire and permafrost on soil organic matter and soil climate in interior Alaska. Glob. Change Biol. 12, 2391e2403.
26. Hester, A.J., Sydes, C., 1992. Changes in burning of Scottish heather moorland since the 1940s from aerial photographs. Biol. Conserv. 60, 25e30.
27. Holden, J., 2007. A plea for more careful presentation of near-surface temperature data in geomorphology. Earth Surf. Process. Landforms 32, 1433e1436.
28. Holden, J., 2009. Flow through macropores of different size classes in blanket peat. J. Hydrology 364, 342e348.

29. Holden, J., Burt, T.P., 2003. Runoff production in blanket peat covered catchments. Water Resour. Res. 39, 1191. DOI: 1110.1029/2003WR002067.

30. Holden, J., Chapman, P.J., Palmer, S.M., Kay, P., Grayson, R., 2012. The impacts of prescribed moorland burning on water colour and dissolved organic carbon: a critical synthesis. J. Environ. Manag. 101, 92e103.

31. Holden, J., Wearing, C., Palmer, S., Jackson, B., Johnston, K., Brown, L.E., 2014. Fire decreases near-surface hydraulic conductivity and macropore flow in blanket peat. Hydrol. Process. 28, 2868e2876.

32. Imeson, A.C., 1971. Heather burning and soil Erosion on the North Yorkshire Moors. J. Appl. Ecol. 8, 537e542.

33. Kettridge, N., Thompson, D.K., Waddington, J.M., 2012. Impact of wildfire on the thermal behavior of northern peatlands: observations and model simulations. J. Geophys. Res. 117, G02014.

34. Lafleur, P.M., Moore, T.R., Roulet, N.T., Frolking, S., 2005. Ecosystem respiration in a Cool Temperate bog depends on peat temperature but not water table. Ecosystems 8, 619e629

35. Lewis, C., Albertson, J., X, X., Kiely, G., 2012. Spatial variability of hydraulic conductivity and bulk density along a blanket peatland hillslope. Hydrol. Process. 26, 1527e1537.

36. MacKinney, A.L., 1929. Effects of forest litter on soil temperature and soil freezing in autumn and winter. Ecology 10, 312e321.

37. Melillo, J.M., Steudler, P.A., Aber, J.D., Newkirk, K., Lux, H., Bowles, F.P., Catricala, C., Magill, A., Ahrens, T., Morrisseau, S., 2002. Soil warming and carbon-Cycle feedbacks to the climate system. Science 298, 2173e2176.

38. Moore, T.R., Roulet, N.T., Waddington, J.M., 1998. Uncertainty in predicting the effect of climatic change on the carbon cycling of Canadian peatlands. Clim. Change 40, 229e245.

39. O›Donnell, J., Turetsky, M., Harden, J., Manies, K., Pruett, L., Shetler, G., Neff, J., 2009. Interactive effects of fire, soil climate, and Moss on CO2 fluxes in black Spruce ecosystems of Interior Alaska. Ecosystems 12, 57e72.

40. Pinheiro, J., Bates, D., DebRoy, S., Sarkar, D., 2006. Package Nimes. Version 3.1e73. Linear and Nonlinear Mixed Effects Models. http://www.r-project.org/.

41. Post, D.F., Fimbres, A., Matthias, A.D., Sano, E.E., Accioly, L., Batchily, A.K., Ferreira, L.G., 2000. Predicting soil albedo from soil color and spectral reflectance data. Soil Sci. Soc. Am. J. 64, 1027e1034.

42. Pregitzer, K.S., King, J.S., 2005. Effects of soil temperature on nutrient uptake. In: Bassirirad, H. (Ed.), Nutrient Acquisition by Plants: an Ecological Perspective, Ecological Studies, vol. 181, pp. 277e310.

43. Price, J., Rochefort, L., Quinty, F., 1998. Energy and moisture considerations on cutover peatlands: surface microtopography, mulch cover and Sphagnum regeneration. Ecol. Eng. 10, 293e312.

44. R Development Core Team, 2014. R: a Language and Environment for Statistical Computing. R Foundation for Statistical Computing, Vienna, Austria.

45. Ramchunder, S.J., Brown, L.E., Holden, J., 2013. Rotational vegetation burning effects on peatland stream ecosystems. J. Appl. Ecol. 50, 636e648.

46. Rogman, J.J., 2013. Ordinal Dominance Statistics (Orddom): an R Project for Statistical Computing Package to Compute Ordinal, Nonparametric Alternatives to Mean Comparison (Version 3.1). http://cran.r-project.org/.

47. The Heather Trust, 2013. Annual Report 2013. The Heather Trust, Dumfries, Scotland.

48. Thompson, D.K., 2012. Wildfire Impacts on Peatland Ecohydrology. PhD. McMaster University.

49. Thompson, D.K., Waddington, J.M., 2013. Wildfire effects on vadose zone hydrology in forested boreal peatland microforms. J. Hydrology 486, 48e56.

50. Turetsky, M.R., Kane, E.S., Harden, J.W., Ottmar, R.D., Manies, K.L., Hoy, E., Kasischke, E.S., 2011. Recent acceleration of biomass burning and carbon losses in Alaskan forests and peatlands. Nat. Geosci. 4, 27e31.

51. Ward, S., Ostle, N.J., Oakley, S., Quirk, H., Stott, A., Henrys, P.A., Scott, W.A., Bardgett, R.D., 2012. Fire accelerates assimilation and transfer of photosynthetic carbon from plants to soil microbes in a northern peatland. Ecosystems 15, 1245e1257.

52. Williams, T.J., Quniton, W.L., 2013. Modelling incoming radiation on a linear disturbance and its impact on the ground thermal regime in discontinuous permafrost. Hydrol. Process. 27, 1854e1865.

53. Worrall, F., Armstrong, A., Adamson, J.K., 2007. The effects of burning and sheepgrazing on water table depth and soil water quality in a upland peat. J. Hydrology 339, 1e14.

54. Worrall, F., Clay, G.D., Marrs, R., Reed, M.S., 2010. Impacts of Burning Management on Peatlands Scientific Review. IUCN UK Peatland Programme.

55. Worrall, F., Rowson, J., Dixon, S., 2013. Effects of managed burning in comparison to vegetation cutting on dissolved organic carbon concentrations in peat soils. Hydrol. Process. 27, 3994e4003.

56. Yallop, A.R., Thacker, J.I., Thomas, G., Stephens, M., Clutterbuck, B., Brewer, T., Sannier, C.A.D., 2006. The extent and intensity of management burning in the English uplands. J. Appl. Ecol. 43, 1138e1148.

57. Yibarbuk, D., Whitehead, P.J., Russel-Smith, K., Jackson, D., Godjuwa, C., Fisher, A., Cooke, P., Choquenot, D., Bowman, D.M.J.S., 2001. Fire ecology and aboriginal land management in central Arnhem Land, Northern Australia: a tradition of ecosystem management. J. Biogeogr. 28, 325e343.

58. Yu, Z.C., 2012. Northern peatland carbon stocks and dynamics: a review. Biogeosciences 9, 4071e4085.

59. Yvon-Durocher, G., Allen, A.P., Bastviken, D., Conrad, R., Gudasz, C., St-Pierre, A., Thanh-Duc, N., del Giorgio, P.A., 2014. Methane fluxes show consistent

temperature dependence across microbial to ecosystem scales. Nature 507, 488e491.

60. Yvon-Durocher, G., Caffrey, J.M., Cescatti, A., Dossena, M., del Giorgio, P., Gasol, J.M., Montoya, J.M., Pumpanen, J., Staehr, P.A., Trimmer, M., Woodward, G., Allen, A.P., 2012. Reconciling the temperature dependence of respiration across timescales and ecosystem types. Nature 487, 472e476.

CHAPTER 19

ADAPTATION PLANNING SUPPORT TOOLBOX: MEASURABLE PERFORMANCE INFORMATION BASED TOOLS FOR CO-CREATION OF RESILIENT, ECOSYSTEM-BASED URBAN PLANS WITH URBAN DESIGNERS, DECISION-MAKERS AND STAKEHOLDERS

Frans H.M. van de Ven, Robbert P.H. Snep, Stijn Koole, Reinder Brolsma, Rutger van der Brugge, Joop Spijker, Toine Vergroesen

Keywords: Urban climate adaptation, Collaborative planning, Green infrastructure, Ecosystem-based adaptation, Planning support system, Performance indicators

INTRODUCTION

Adaptation of urban areas

The need for adaptation of urban areas to changing climatic conditions is widely recognized (Deltaprogramma, 2015, IPCC, 2007, IPCC, 2012 and PROVIA, 2013). Flooding, drought, heat stress and related problems with water quality, water supply and land subsidence, aggravated by the UHI effect, are increasing hazards threatening the liveability of our urban areas as well as our social and economic urban systems (Albers et al., 2015, Jha et al., 2012, Rovers et al., 2014, World Bank, 2010 and Zevenbergen et al., 2010). Risks are further increased by on-going urbanization (Nichols et al., 2007 and UN DESA, 2014) and by intensification of urban land use; the invested capital and the asset value of buildings, infrastructure and industrial facilities has increased drastically over the past decades (Kind, 2013). Although the need for adapting our urban environments is clear, in practice adaption is difficult. Opportunities for adaptation are often limited to new development projects, to large infrastructural renovation and renewal projects or to initiatives from individual residents (Van der Brugge and De Graaf, 2010).

Adaptation requires the construction of structural or "hard" adaptation measures (Hallegatte, 2009 and Pelling, 2011). Such measures are physical or technological interventions, constructed facilities that require space and therefore are subject of spatial planning and design (Taylor and Wong, 2002). This article will focus on the right design of structural adaptation measures, as embedded in a planning process that leads to a decision on a spatial adaptation plan.

The pallet of adaptation measures has extended dramatically over the past decades. Earlier, Sustainable Urban Drainage Systems (SUDS) (CIRIA, 1998 and Svenske Vatten- och Aflopsverksföreningen, 1983) and Water Sensitive Urban Design(WSUD) for urban drainage (Brown et al., 2008 and Engineers Australia, 2006), nowadays also known as green or blue-green infrastructure, were

Citation: Frans H.M. van de Ven, Robbert P.H. Snep, Stijn Koole, Reinder Brolsma, Rutger van der Brugge, Joop Spijker, Toine Vergroesen, Adaptation Planning Support Toolbox: Measurable performance information based tools for co-creation of resilient, ecosystem-based urban plans with urban designers, decision-makers and stakeholders, Environmental Science & Policy, Available online 12 July 2016, ISSN 1462-9011, http://dx.doi.org/10.1016/j.envsci.2016.06.010.

introduced. Maksimovic et al. (2014)recently argue that a new concept of Multiple-Use Water Services (MUS) is emerging. MUS solutions enhance the synergy of urban water (blue) infrastructure with green assets and ecosystem services, are economically viable and climate (environmental) adaptive.

Ecosystem-based Adaptation (EbA) is at the heart of this MUS development. EbA- measures integrate the use of biodiversity and ecosystem services into an overall strategy for helping people adapt to climate change (Munroe et al., 2012). In addition to flood control, drought mitigation and heat stress reduction they provide e.g. aesthetic quality, recreational and restorative capacity and health benefits (Opdam et al., 2009, Van den Berg et al., 2007 and Van den Berg et al., 2015). This article shows how planning 'blue-green' EbA measures is used to advance climate resiliency, while maximizing their co-benefits.

Adaptation planning

Urban planning exists of a series of more or less consecutive phases starting from system analysis and program development (initiative phase), via conceptual, preliminary and final design (design phase) up to implementation (Fig. 1). The process ends with a final decision on an adaptation or (re)development plan. Although shown as a straightforward, stepwise process in theory, the process in practice often reiterates to an earlier stage to investigate alternative adaptation pathways.

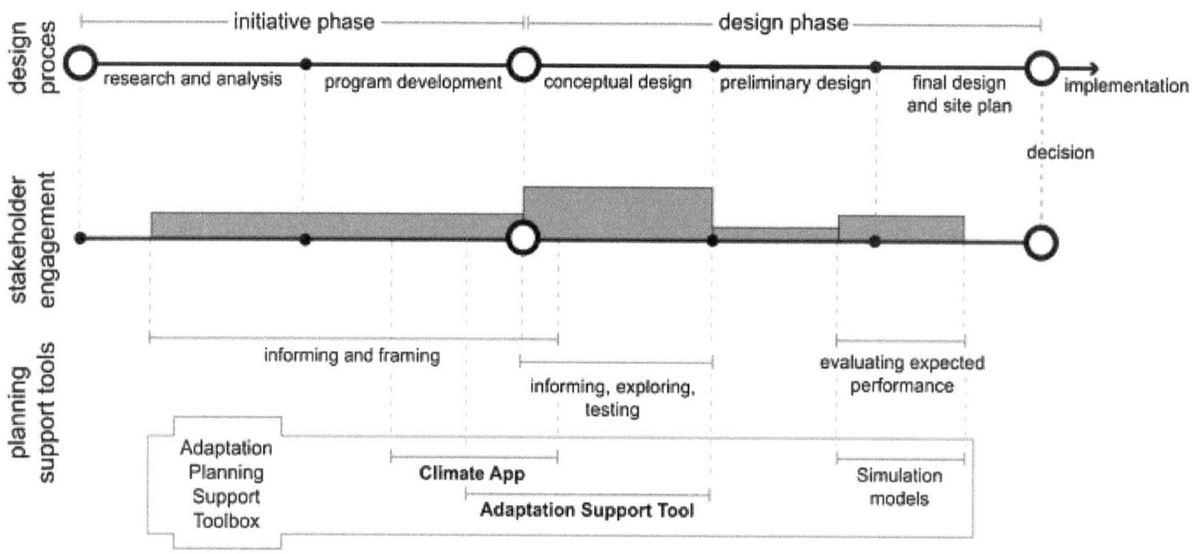

Figure 1: Adaptation planning process, stakeholder engagement and planning support tools. Both tools (**bold**) in the Adaptation Planning Support Toolbox will be discussed in this article.

Many guidelines on climate resilient urban planning provide procedures for hazard, exposure and vulnerability analysis and an overview of potential solutions and/or best practices (Challenge for Sustainability; Climate-ADAPT; Deltaprogramma N&H, 2014; EPA; Great Lakes and St. Lawrence Cities Initiative; PROVIA, 2013). They however lack guidance where it comes to the selection of appropriate packages of adaptation measures during the initiative and design phases (Voskamp and Van de Ven, 2015). For these phases tools seems unavailable to support stakeholders to make hard choices which adaptation measures are attractive and effective for the project area (Bours et al., 2014 and PROVIA, 2013); this while complex simulation models to evaluate the expected hydraulic and hydrological performance of the final plan are readily available (Lerer et al., 2015)

In the initiative phase, urban planners are often in the lead of the process. Eliasson (2000) showed that climatology so far has a low impact on the planning process; urban planners' use of climatic information is unsystematic as the urban climatologists fail to provide them with good arguments, suitable methods and tools. This underlines the need for a planning support system that bridges the gap between urban planners and engineers; she makes a plea for a "communicative approach" to the planning process.

Adaptation support tools for collaborative planning

Involvement of local stakeholders, land & water engineers, experts from other disciplines and decision-makers is considered essential in particular in planning reconstruction of existing urban areas. Each of them not only has different interests, agendas and roles in the process. They differ in their sense of urgency of the problem, their approach to the problem, their language and knowledge level, and their rationality regarding potential solutions (Van Stigt et al., 2015). Design workshops during the initiative phase are meant to get to know each other, to share each other's knowledge and understanding of the problems and to collectively identify interesting adaptation solutions.

Question is how to support the planners, stakeholders and decision-makers in this analysis – dialogue – design-engineering process with knowledge and information, in order to get a converging learning process that leads to a final positive decision on an adaptation plan? Such planning support tools should raise awareness, present the broad range of adaptation options, let participants explore the impact of different design choices on the climate resiliency of their project area (Pelzer et al., 2013) and maximize the co-benefits of adaptation measures.

Goal of our study was to develop a toolbox that supports the incorporation of climate adaptation in the actual planning and design practice in cities. This Adaptation Planning Support Toolbox was developed to provide urban planners, landscape architects, civil engineers and local stakeholders and decision makers with quantified, evidence-based information on the climate resilience of their ideas in early phases of the planning process and to facilitate decision-making during conceptual design workshops. In design workshops the toolbox should supports them in how to share their knowledge and discuss alternative measures, including location, size, costs and (co) benefits.

TOOLBOX TO SUPPORT ADAPTATION PLANNING

An integrated 'dialogue – design – engineering' planning process

The Adaptation Planning Support Toolbox was developed to effectively support the collaborative planning process in the phases of program development and/or conceptual planning. Two actual tools were developed to support the 'dialogue – design – engineering' planning process (Fig. 1). The Climate Adaptation App (climateApp) informs participants about more than 120 potential adaptation measures and produces a long list of relevant measures. The Adaptation Support Tool (AST) guides stakeholders in the next step, the conceptual design. Resulting conceptual plans are input for urban planners and designers, to make detailed preliminary designs.

The climateApp and the AST are both web-based software tools running on touch enabled hardware. This because a touch table facilitates 'reasoning together', is community supportive, empirically based, experimentally oriented and information and knowledge disseminating (Geertman, 2006).

Climate adaptation app

The Climate Adaptation App was developed to start the design workshop with overview and pre-ranking of potential measures for all participants (www.climateapp.org or Appstore/ Playstore). From different publications (Pötz and Bleuzé, 2012, Van de Ven et al., 2009 and Vergroesen et al., 2013) a list of over 120 structural adaptation measures was composed. The app provides information on each measure and ranks measures for potential applicability based on local circumstances and preferences by toggling the different filters (Fig. 2).

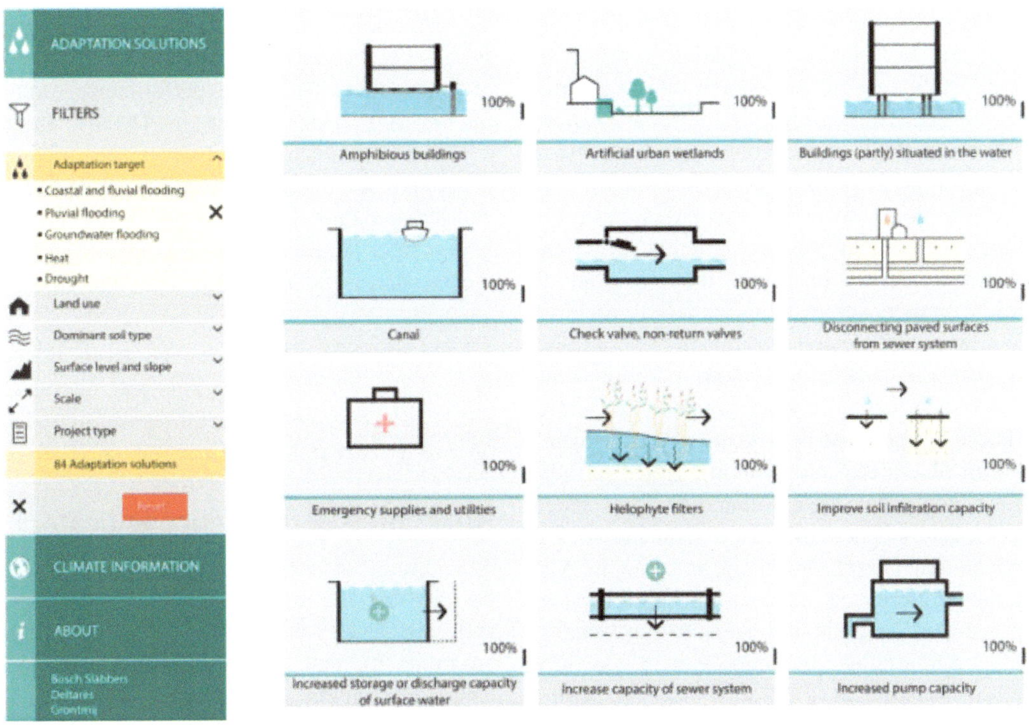

Figure 2: Screen of the Climate Adaptation App (www.climateapp.org). Adaptation measures are ranked by toggling the filters. More information on a measure is obtained by clicking the tile.

Design workshop participants go through the list and discuss applicability and attractiveness of potential measures to create a long list for their project area.

Adaptation support tool

The Adaptation Support Tool (AST) is a touch-table based platform that design workshop participants may use to select adaptation interventions, situate them in their project area and immediately see an estimate of their effectiveness and costs (Fig. 3). The AST consists of a left panel for input, a middle panel for design (map of project area) and a right panel as an "AST dashboard" for output.

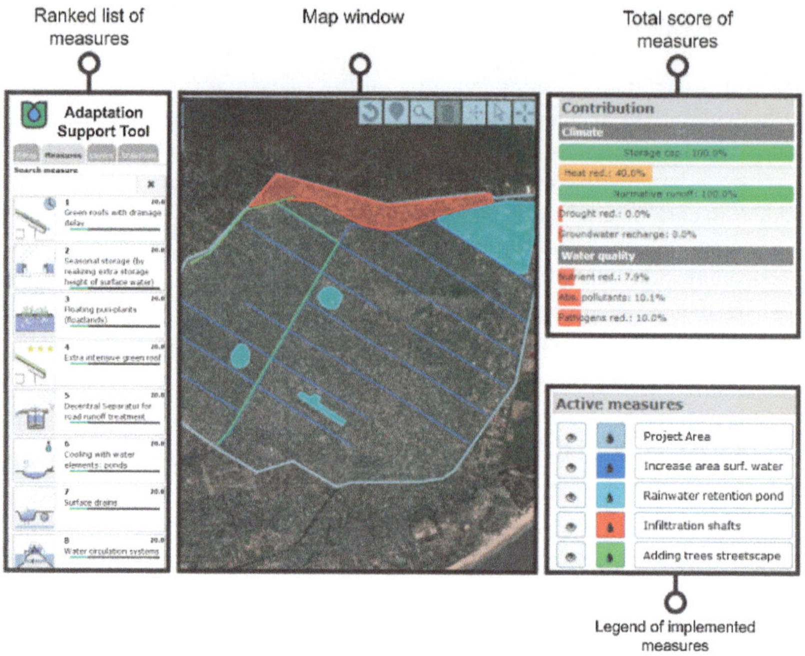

Figure 3: Screen components of the Adaptation Support Tool. Left on the touch screen is the ranked list of 62 adaptation measures. Selected measures are planned in the project area (middle). At the right side the AST dashboard, showing the resilient performance of the total package of measures and of each active measure. Shown is the application in of the AST in Beira, Mozambique.

The current AST version includes a long list 62 blue, green and grey adaptation measures for reduction of pluvial flooding, drought and heat stress, a selection assistant for ranking their applicability and an assessment tool to estimate the effectiveness of applied measures. The left panel shows a ranked list of adaptation interventions. The long list of measures has been composed from multiple inventories found in literature. The selection of measures was based on criteria that differ for blue-green and for grey adaptation interventions. As many blue-green interventions were included that the authors and project partners are aware of from both literature and practice. We however selected traditional/grey solutions in such a way that a comparison between traditional and blue green solutions can be made when planning alternative solutions and because traditional interventions can enhance the effectiveness of blue-green interventions. Based on local common practice additional interventions can be added. Ranking of the measures is determined on characteristics of the area and adaptation targets (Voskamp and Van de Ven, 2015). These targets differentiate between threshold capacity for damage prevention and coping capacity for damage minimization in case of a failing protection system (De Graaf et al., 2007).

In the central panel different map layers can be shown. Default a Google Earth and OpenStreetMap layer are provided, with layers like surface elevation, land ownership, flood depth, heat stress maps or future land use as additional. Design workshop participants can now select a measure from the list left and draw it in the project area on a map layer, on the location where they think that it would provide added value. For example, the user can apply a green roof on a large flat roof, install permeable pavement on sidewalks and artificial wetlands near the outlet of a tributary drain. Next, the tool requests the water storage depth of the measure and the additional contributing inflow area.

On the basis of this input, the AST estimates a number of performance indicators, e.g. storage capacity, normative runoff, heat stress reduction, water quality effects, costs and additional benefits. These performance estimates are shown on the right panel. Under the Details tab (not shown) the contribution of each proposed measure to the adaptation targets is given in combination with the estimated costs for realization and maintenance. Users can also switch to the Overview tab of the right panel, as shown in Fig. 3, to get a summary of the measures and their total effectiveness in relation to the adaptation assignment.

Results of a session can be saved as snapshots and re-opened at a later moment. This way alternative plans can be created and compared. The tool is web-based and can run both on a webserver and standalone.

Adaptation performance indicators

The current selection of performance indicators was based on the demand of participants of the design workshops and the role of water as the key to a climate resilient urban environment. The quantified performance indicators include estimated changes of physical characteristics that are relevant for damage reduction, resilience, public health and feasibility.

- Prevention of flooding due to extreme rainfall requires effective storage (retention) of water as well as peak flow reduction. Created storage volume is shown, as this has to comply with the target volume that our water managers set to reduce pluvial flood risk. The normative runoff frequency allows for estimation of flood risk reduction in terms of a reduction in frequency of a certain peak flow. Estimates of these flood prevention indicators are based on the result of simulation of the effect of a specific adaptation measure, using long time series of rainfall and evaporation – 30 years or more –, a climate change scenario, a multi-reservoir rainfall-runoff water balance simulation model, a theoretical design of the intervention and extreme value analysis to quantify changes in effective storage capacity and peak flow reduction. Parameters characterising the hydrological performance of the specific adaptation measures were taken from experimental results reported in the international scientific literature.

- Drought control requires groundwater recharge information and inter-seasonal storage of water, in particular in areas prone to land subsidence or a lack of replenishment due to soil sealing. On the other hand, in case of very shallow groundwater tables high recharge rates would lead to the need for subsurface drains. Estimated groundwater recharge also results from output of the multi-reservoir simulation model and a theoretical design of the intervention. Average annual recharge change is calculated as a performance indicator.
- Heat stress reduction is achieved by provision of shade and evaporative cooling from vegetation and water surfaces; though, to that end vegetation has to have enough water available, which is related to groundwater recharge. Heat stress reduction is based on the reported observed cooling effect of blue-green infrastructure in Dutch urban areas and scaling based on the dimensions of the measure.
- The quality of the water is essential for the functions and services it can provide. To evaluate potential functionality water quality improvement of the blue, green and grey adaptation measures is expressed by three indicators: nutrient reduction, absorbed pollutants reduction and pathogen reduction. These water quality performance indicators are determined as a pollution reduction factors based on recorded effectiveness of treatment processes in a facility and scaling based on the dimensions of the measure. Nature based treatment processes included in the pollution reduction factor include natural degradation, settling and soil filtration. For intensive green roofs fertilization was included as a negative pollution reduction factor for nutrients.
- Average costs of construction and costs of management and maintenance are estimated for each adaptation measure based on unit prices on the Netherland's market.

The purpose of the AST is to provide estimates on the effectiveness and costs of adaptation interventions in the early planning phase of urban (re)development projects, in order to meet adaptation targets. Such targets can be met by different packages of measures. No framework or guidelines are provided for the selection specific adaptation measures; the AST allows for any strategy to reduce its vulnerability (De Graaf et al., 2007). The actual effectiveness and costs will depend on the implementation which is determined by exact local physical conditions, and specific wishes and ambitions of the stakeholders.

AST APPLICATIONS

In the period 2014–2015 the Adaptation Planning Support Toolbox has been used in adaptation processes in different cities (Table 1). Being both AST developers and participant, we learned valuable lessons concerning the optimal use of the Toolbox for local adaptation process. Two examples are briefly addressed here.

Table 1: Overview of applications of the Adaptation Planning Support Toolbox

Area	Project area type and size	Type of project/ spatial planning process	Climate challenges	Phase of planning process	Participants	Experiences, Lessons learned
Chota- Beira (Mozambique)	788ha, district+detail	Redevelopment & development	Flooding	Program formulation	Municipality, citizens, politicians, local university, NGOs	Stakeholder sessions including non-professionals demand a combination of high tech (AST) and low tech process tools
Decoy Brook,London (United Kingdom)	292ha district+detail	Research	Flooding	Conceptual design	Environment Agency, Borough repres., university	Make plans for both total area and-detailed design for specific hot spots
Oaxaca (Mexico)	195ha district	development &redevelopment	Flooding, drought	Program formulation	State and municipal authorities, citizens, university, NGOs,	AST is a very handy tool as a catalogue of possibilities in district zones with different characteristics.
Utrecht (Netherlands)	49ha neighbourhood	Redevelopment	Flooding, heat stress	Conceptual design	Municipality, real estate owners, architect; urban water and green experts;	Participants go for urban quality rather thanfor cost reduction
Dordrecht (Netherlands)	65ha neighbourhood	Student Climate Resilient Urban Design workshop	Flooding, heat stress	Conceptual design	Students,	AST toolbox is effective training tool

Tilburg (Netherlands)	46ha, neighbourhood	Research	Flooding, heat stress	Program formulation	Municipality; urban water experts	The AST can also be applied as a quick-scan method to assess if e.g. green roofs have an added value for specific urban areas.

Beira, Mozambique

The city of Beira (Mozambique) frequently floods by heavy rain, having serious health and economic impacts for the 0.5 million residents. Blue-green adaptation measures may increase water retention capacity and will improve the liveability. Discussing adaptation strategies with local Beira stakeholders in a workshop setting has been done based on the following steps (Picketts et al., 2012):

Building capacity

Municipal civil servants, representatives of the Chota neighbourhood (pilot area), and local university staff (UCM) were briefed by the authors (acting as facilitators) on climate adaptation and the key role the workshop participants have in adaptation planning as experts with important local knowledge.

Identifying local impacts and vulnerabilities

Climate information was distributed before and during the workshop. It included information on hydrology in urbanized delta regions, flooding maps of Beira based on 3D aerial information, historical climate information and future predictions. The maps and explanation provided a good overview of the impacts and vulnerabilities of Chota and surroundings, including underlying mechanisms. For most workshop participants especially the hydrodynamic information was new, enabling them to better identify the causes of flooding and the ways flooding can be prevented.

Determining priorities and outlining implementation

The workshop participants defined short and long term targets to prevent frequent and large-scale flooding of their residential areas in the future. The facilitators calculated the overall retention capacity to achieve these goals. The facilitators then explained about the AST: the goal, the lay out of the AST tools, the range of measures and underlying data. Based on local knowledge the participants selected a number of measures that fit local physical conditions and culture: surface water bodies (channels, small lakes, lagoon), multifunctional green (public green fields that can be inundated temporary). Measures demanding high-level construction and maintenance (e.g. green roofs, technical installations) were rejected, not fitting the local possibilities in water management. Locations within the Chota area where measures could be implemented were identified (Fig. 3). For each location and accompanying measures the AST calculated water retention capacity and other parameters, based on local meteorological data. By doing so, it became clear for the participants that additional retention nearby Chota was needed, resulting in a proposal for a lagoon development adjacent to Chota. Through field visits the workshop participants together with the municipal board verified whether implementation of the measures (including lagoon) was indeed possible. Most of the recommended interventions were accepted by the municipal council; in one occasion however a land development claim became the topic of discussion, because this development would decrease retention opportunities for the larger area. The mayor of Beira expressed his intention to reject that claim. The total set of measures was further elaborated on a map and – together with the other information – presented in a report (Kalsbeek, 2015). See this report for more details and background information on this case. The Chota adaptation plan as composed by the workshop participants and their facilitators was also welcomed at an international financing meeting in September 2015; it now serves as the outline for detailed design of drainage improvement works.

Utrecht, the Netherlands

In the redevelopment of the Utrecht City Centre – West, there is a need for a more climate resilient, attractive and pleasant accommodation area. Using the AST, stakeholders sketched three alternative plans, selecting different adaptation measures they deemed applicable and effective. Two of these alternatives can be seen in Fig. 4. To make the area more attractive and to reduce the heat stress emphasis was put on greening the area, both at street level and by creating green roofs and urban agriculture on the roofs of the large exhibition halls in the area. Stormwater retention capacity was also created by installing rain tanks, a water square and application of porous pavements. The design workshop participants managed to meet the climate adaptation targets they had set in advance, while creating substantial co-benefits for themselves, for future residents and for the numerous visitors of this area (Van de Ven et al., 2016a and Van de Ven et al., 2016b).

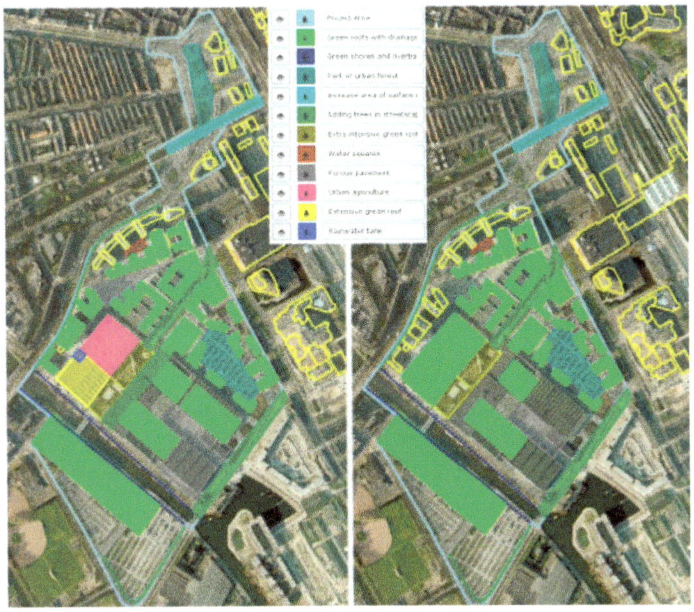

Figure 4: Example of AST application: Two of the alternative conceptual adaptation plans for the Utrecht City Centre—West, each with its own set of adaptation measures and, consequently, a different contribution to adaptation targets and co-benefits. The Green alternative (right) proved less effective than the Plus alternative (left) (Van de Ven et al., 2016a and Van de Ven et al., 2016b). (For interpretation of the references to colour in this figure legend, the reader is referred to the web version of this article.)

Building capacity

Representatives of different municipal offices (including urban planning, health, water management, urban green and project development) and representatives of the private stakeholders participated in the Climate KIC Smart Sustainable District project on the sustainable and climate resilient redevelopment of the Utrecht Centre West area and two design workshops. These parties learned about the vulnerability of the area for flooding, drought and heat stress and about the many potential solutions that can be used to strengthen resilience, meanwhile delivering substantial ecosystem, economic and social services.

Identifying local impacts and vulnerabilities

National climate change scenarios are available for the Netherlands (KNMI, 2015). Flood hazard maps and heat stress maps showed significant climate risks in the project area. Drought however turned out to be less of an issue. An attempt to map all critical and vulnerable objects, networks and population groups for a risk assessment turned out to be complicated. Information is scattered over very many desks. Impacts and vulnerable spots were recognized by the participants of the design workshop.

Adaptation targets for stormwater retention, peak flow reduction and heat stress reduction were quantified on the basis of these climate and land use projections. These targets, though negotiable, are used to evaluate performance of the packages of adaptation measures.

Determining priorities and outlining implementation

The workshop participants first used the climateApp to get an overview of potentially applicable solutions; most of them were not familiar with the large variety of potential adaptation measures They discovered and learned about other solutions. After that first step they started discussing the applicability and attractiveness of implementing specific adaptation measures on specific sites in the project area. Two alternative plans emerged from this discussion: a blue-green alternative and a high density urban alternative. Both alternatives did not completely meet adaptation targets. That is why a third alternative was produced, the Plus alternative. This alternative combines measures from both the Green and the Urban alternative and meets the adaptation targets on storage/retention capacity and peak flow reduction. Heat stress reduction targets are met at all places where people stay, walk or bike. A first analysis was made of the ecosystem services, the economic and social benefits of the proposed alternative adaptation plans as well as a qualitative analysis of who benefits from implementing the Plus alternative and in which way.

The blue-green adaptation plans are now being merged with the mobility adaptation plan and the energy transition plan for the project area to produce comprehensive redevelopment plan alternatives. These alternatives will be used in 2016 (a) to evaluate if adaptation targets are still being met, (b) as input for public engagement sessions and (c) as basis for a value case analysis. This value case analysis is meant to specify the benefits and the beneficiaries of the redevelopment plan in more detail and use this as a basis for a fair distribution of investment and maintenance costs. Results of this value case analysis are meant to support final decision making in 2017 by the City of Utrecht and private stakeholders and project developers on the urban and economic development of the Utrecht City Centre-West area.

DISCUSSION

Addressing adaptation in city planning and design

Local adaptation of our urban infrastructure, buildings and environment is required to minimize negative consequences of climate change. A wide variety of blue, green and grey infrastructural measures is available to strengthen resilience against flooding, drought and heat stress. Decisions are to be taken about adaptation targets and about where and how which adaptation measures are to be located. Such an adaptation plan is to be produced in a collaborative planning process of urban planners, engineers other experts, local stakeholders and political decision makers.

Overall, more and more cities recognize the need for adaptation at a policy-level, but lack the practical instruments to go from vulnerability assessments towards adaptation-inclusive urban planning – see e.g. [ND-GAIN, 2016] – and lack of support for adaptation investments. Moreover, adaptation is a relative new phenomenon, not considered by everyone as his/her responsibility (Nalau et al., 2015). Investors seem to focus on cost reduction rather than on long term benefits of implementing adaptation measures. The fact that most ecosystem-based adaptation measures not only reduce vulnerability of the urban environment to extreme weather events but also produce substantial economic, ecological and social benefits for the citizens is often overlooked, let alone maximized in spatial planning, partly due to the fact that these benefits are hard to quantify. This lack of quantitative information is partly overcome by implicit evaluations that take place while the participants in this collaborative planning process evaluate the performance information produced by the AST.

Role of tools in planning for climate resilience

Urban planning and design routine is not equipped yet to easily incorporate climate proofing. To gain public support, there is a need for stakeholder participation when addressing adaptation in city practice (Hurlbert and Gupta, 2015). In a collaborative planning workshop based setting local stakeholders are able to provide their implicit knowledge of the area and of the community's preferences (Picketts et al., 2012 and Van Stigt et al., 2015). Many stakeholders however are not aware of the large variety of adaptation options to choose from – the AST contains 62 –, each with their own pro's and con's. Planning and decision support tools for climate resilient urban design should therefor support knowledge sharing and collaborative exploration of alternative adaptation solutions in community-based meetings. To effectively support policy making, planning support tools should bridge the gap between the worlds of scientific expertise and self-organised adaptation in urban reality (Larsen et al., 2012 and Löschner et al., 2016) and that of the creative urban planner.

Pyke et al. (2007) conclude that the existing decision support systems are more effective when they balance the provision of information with concern for organizational and political processes.

Application experiences with the Adaptation Planning Support Toolbox

The Adaptation Planning Support Toolbox has effectively supported climate-proof planning in several cases on different continents. Participants of the design workshops expressed their satisfaction with the way the planning process was structured, with the ranked overview of potential blue, green and grey adaptation measures and with the estimates of the effectiveness and the costs of proposed measures; this information supported a learning process and informed decision making. Concerns on organizational or political issues around details of the plan were discussed among participants at the design table. As such there seemed no need to include such issues explicitly in the tool.

The toolbox builds on the results of vulnerability assessments and on the willingness to adapt, as e.g. analysed with the Uniform Adaptation Assessment (Chenchen, 2015) or the Climate Stress Test (Deltaprogramma N&H, 2014). Flood hazard maps, heat stress maps and water balance calculations provide valuable information on where to concentrate adaptation efforts. In practice it turned out to be hard to formulate adaptation targets for drought and heat stress. The AST was in such cases used to explore the feasibility of a certain impact reduction.

The use of the AST in design workshops requires skilled facilitation. The dialogue that takes place around the design table benefits from an independent facilitator. Moreover, the use of the AST proved to be complex for participants that are not familiar with design workshops and/or with the wide range of potential adaptation measures. In practice, the facilitator or another professional that is trained in the use of the tool assists the application. The Climate Adaptation App is available as a stand-alone tool, because this tool can be used for individual learning by professionals and non-professionals around the globe.

And although decisions on the application of adaptation measures suffer from deep uncertainties on expected climate change and exposure, we have seen in practice that many adaptation measures are selected because of the expected co-benefits of the blue, green and grey measures for the liveability and economic functioning of the urban environment; climate resilience was dealt with as a valuable co-benefit rather than a primary target – as long as adaptation targets were met.

As concluded by Pelzer et al. (2013), the use of a touch table during the design workshops proved to be effective in supporting the planning process. The use of the touch table supports learning processes and stimulates thinking beyond the own professional roles. Moreover, the performance indicators shown on the touch table forced participants to be explicit about their proposed interventions and the expected effectiveness. The struggle they reported of the urban planners with the application of the touch table is interesting. Designer's working practice, to which intuitive sketching and visualization are central, is disrupted by the use of the touch table. This was solved by having regular maps, transparent and drawing pens next to the touch table, so that they could sketch their ideas when they felt the need for it. According to Pelzer et al. (2013)designers also felt the integral approach as a barrier to their creativity. This could not be confirmed in our workshop, potentially because the objective of our workshop was more specific than the objective of their workshop – create a more climate resilient and attractive urban area versus planning a more sustainable new urban area.

Usability and reliability

Performance indicators produced by the AST and used to select and plan adaptation measures are based on evidence-based key figures on the characteristics, performance and costs of each adaptation measure retrieved from international literature

(De Jong et al., 2014, De Jong et al., 2015 and Geisler and Barjenbruch, 2015; Kosteninformatie.nl, 2015; Vergroesen et al., 2013). They are also based on conceptual modelling of the measure's performance using local climate and land use conditions. Although the accuracy is limited we argue that this information is reliable enough to compare different measures and different alternatives and to find a common preference with all participants. Arguments to decide on a specific choice are exchanged, while keeping an eye on their contribution to the adaptation targets and on their cost-effectiveness. Conceptual designs are so far made without quantified information on performance of proposed adaptation measures; the availability of a more or less reliable performance and cost estimation is a valuable contribution to informed decisions on the selection and design of adaptation measures.

The Toolbox is used for planning problems at building to district scale; use at larger scale level is questionable because the tools do not consider interconnections and flow capacities between adaptation measures. Estimated performance at larger scale could consequently be misleading.

Moreover, the AST shows only performance indicators regarding climate resilience in relation to the water system and estimated costs; other benefits and co-benefits of the measures – e.g. landscape quality, added economic and social value – are not quantified, but in practice play an important role in the dialogue and selection decisions of the workshop participants. Quantified information would give the benefits a more equal treatment in the selection and decision making process as compared to the costs. Research to find out which information on co-benefits session participants would like to see on the AST dashboard is on-going.

Measures against heat stress tend to have local effects. In order to evaluate heat stress control measures we would like to visualise the local cooling effects of planned blue green measures in a map instead of presenting a general decrease in average areal temperature as a figure on the dashboard. We planned to realise this functionality in 2016.

Another relevant question is who should participate in the design workshops. Participation of urban planners, landscape architects, water managers, civil engineers, local stakeholders and other experts is evident. But how about participation of city council members and commercial developers? The fact that city council members participated in the design workshop in Beira turned out very effective for further decision making. In other cases political decision makers were not invited by the host of the design workshop; further study is required to evaluate the impact of their participation.

The toolbox was used both in the Netherlands and abroad. For the applications in Beira and London the key figures for calculating the performance indicators of each measure had to be calculated with the local climate and local land use data. So far this has been done manually and has required substantial effort. For easier applications abroad this process could be automated. Cost figures remained unchanged so far; if local unit cost figures are available these can be brought in the tool without much effort. Moreover a stronger coupling (export-import function) of the AST with hydraulic and hydrological simulation models for plan evaluation would be convenient.

CONCLUSIONS

There is a gap in the tools available to support resilient, climate-proof urban planning. Tools and procedures are available for climate vulnerability assessment and for evaluating the performance of final designs with the help of simulation models. But tools that have the ability to support implementing adaptation in the actual urban planning and design practice, i.e. to support defining the program of demands, setting adaptation targets, for selecting adaptation measures from a wide variety of blue, green and grey adaptation measures and for informed co-creation of a conceptual design, seem to be missing.

To close this gap and support the planning of a climate resilient urban environment we developed and tested an Adaptation Planning Support Toolbox. The toolbox contains a Climate Adaptation App (climateApp) and the Adaptation Support Tool (AST). From our applications so far we conclude that this Toolbox meets the demands of local policymakers, planners, designers and practitioners to provide evidence-based support for their collaborative analysis –dialogue-design-engineering (= planning) sessions. Participants appreciated the AST because its overview and pre-ranking of a wide range of potential adaptation measures, the possibility to create different adaptation design options (scenarios) for their own project area, and to explore the contribution of these options on adaptation targets and co-benefits. Discussions on the design table were focussed on the opportunities and the benefits of specific interventions, rather than on the costs. The combination of informing, exploring and testing at the same time, and doing this in a collaborative dialogue with relevant stakeholders, is considered as of added value to current adaptation planning practice.

Essential is that urban planners, landscape designers, water managers, urban green managers have to learn how to combine their working practice in such a collaborative planning and design process. This transition requires courage and perseverance from all parties, and will lead to further development of the toolbox or similar tools. With more and more cities worldwide that will make the step from climate policymaking to an actual adaptation-inclusive urban (re) development practice we foresee a growing demand of tools like the climateApp and the AST to ensure that adaptation will be seriously adopted by the local actors while maximizing the social and economic co-benefits of the adaptation measures.

ACKNOWLEDGEMENTS

The Adaptation Support Tool was developed by Deltares, Alterra Wageningen University & Research and Bosch Slabbers landscape architects as part of the EU-EIT Climate-KIC project Blue Green Dream, funded by the European Union and the three organisations. The application in Utrecht was funded partly by the EU-EIT Climate-KIC project Smart Sustainable District. The Climate Adaptation App was developed by Deltares, Grontmij, Witteveen + Bos, Bosch Slabbers and KNMI, in a project funded by the Netherlands' Climate Changes Spatial Planning Knowledge Programme.

REFERENCES

1. Albers, R.A.W., Bosch, P.R., Blocken, B., Van den Dobbelsteen, A.A.J.F., Van Hove, L.W. A., Spit, T.J.M.,

Van de Ven, F.H.M., Van Hooff, T., Rovers, V., 2015. Overview of challenges and achievements in the climate adaptation of cities and in the Climate Proof Cities program. Build. Environ. 83 (January), 1–10.

2. Bours, D., McGinn, C., Pringle, P., 2014. Monitoring & Evaluation for Climate Change Adaptation and Resilience: A Synthesis of Tools, Frameworks and Approaches, 2nd edition SEA Change CoP, Phnom Penh and UKCIP, Oxford. http://www. seachangecop.org/sites/default/files/documents/2014%2005%20 15%20SEA% 20Change%20UKCIP%20Synthesis%20 Report%202nd%20edition.pdf.

3. Brown, R., Keath, N., Wong, T., 2008. Transitioning to water sensitive cities: historical, current and future transition states. Proc. 11th International Conference on Urban Drainage, Edinburgh, Scotland.

4. CIRIA, 1998. CDM regulations: practical guidance for planning supervisors. CIRIA Report 173. CIRIA, London (Connection with CIRIA research project 555).

5. Chenchen (2015) The Urban Adaptation Assessment (UAA) Framework: Three Levels of Analysis. https://www.climate-eval.org/blog/urban-adaptationassessment-uaa-framework-three-levels-analysis (accessed 22.01.16)

6. De Graaf, R.E., Van de Giesen, N.C., Van de Ven, F.H.M., 2007. Alternative water management options to reduce vulnerability for climate change in the Netherlands. Nat. Haz. doi:http://dx.doi.org/10.1007/s11069-007-9184-4.

7. De Jong, J.J., Van Raffe, J.K., Griffioen, A.J., De Boer, T.A., 2014. Normenboek. Natuur, Bos en Landschap 2014 (in Dutch). Alterra Wageningen UR, pp. 135.

8. De Jong, J.J., Smidt, R.A., Van Raffe, J.K., Spijker, J.H., 2015. Databank Gemeentelijk Groenbeheer. Rapportage boekjaar 2014 (in Dutch). Alterra, Wageningen UR, pp. 80.

9. Deltaprogramma N&H, 2014. Handreiking voor de uitvoering van een Stresstest Klimaatbestendigheid (Guideline for Stress Testing Climate Resilience of Urban Areas). Ministerie van Infrastructuur en Milieu, De Haag http://www. ruimtelijkeadaptatie.nl/en/vulnerability (accessed 22.01.16.).

10. Deltaprogramma, 2015. Deltaprogramma 2016; Werk aan de delta; en nu begint het pas echt/Delta Programme 2016. Working on the Delta; Now It Really Starts. Ministerie van Infrastructuur en Milieu, Den Haag.

11. Eliasson, I., 2000. The use of climate knowledge in urban planning. Landscape Urban Plann 48, 31–44. http://dx.doi.org/10.1016/S0169-2046(00)00034-7.

12. Engineers Australia, National Committee for Water Engineering (2006), Australian Runoff Quality: A Guide to Water Sensitive Urban Design. Engineers Australia, Barton, ACT 2600 Australia [online]. http://www.eabooks.com.au/epages/ eabooks.sf/en_AU/?ObjectID=40244481 (accessed 25.01.16).

13. Geertman, S., 2006. Potentials for planning support: a planning-conceptual approach. Environ. Plan. B: Plan. Design 33, 863–880. doi:http://dx.doi.org/ 10.1068/b31129.

14. Geisler, D., Barjenbruch, M. (2015) (personal communication) Literature review on key characteristics of blue-green adaptation measures, made for the Climate KIC project Blue Green Dream. TU Berlin.

15. Hallegatte, S., 2009. Strategies to adapt to an uncertain climate change. Global Environ. Change 19, 240–247. doi:http://dx.doi.org/10.1016/j. gloenvcha.2008.12.003.

16. Hurlbert, M., Gupta, J., 2015. The split ladder of participation: a diagnostic, strategic, and evaluation tool to assess when participation is necessary. Environ. Sci. Policy 50 (June), 100–113. doi:http://dx.doi.org/10.1016/j.envsci.2015.01.011.

17. IPCC, 2007. Climate change 2007: impacts, adaptation and vulnerability. In: Parry, M.L., Canziani, O.F., Palutikof, J.P., Van der Linden, P.J., Hanson, C.E. (Eds.), Contribution of Working Group II to the Fourth Assessment Report of the Intergovernmental Panel on Climate Change, 2007. Cambridge University Press, Cambridge, UK.

18. IPCC, 2012. Managing the risks of extreme events and disasters to advance climate change adaptation. In: Field, C.B., Barros, V., Stocker, T.F., Qin, D., Dokken, D.J., Ebi, K.L., Mastrandrea, M.D., Mach, K.J., Plattner, G.-K., Allen, S.K., Tignor, M., Midgley, P.M. (Eds.), A Special Report of Working Groups I and II of the Intergovernmental Panel on Climate Change. Cambridge University Press Cambridge, UK, New York, NY, USA, pp. 582.

19. Jha, A.K., Bloch, R., Lamond, J., 2012. Cities and Flooding; A Guide to Integrated Urban Flood Risk Management for the 21 st Century. The World Bank doi:http://dx.doi. org/10.1596/978-0-8213-8866-2.

20. KNMI, 2015. KNMI Climate Scenarios for the Netherlands 2014 Revised Edition 2015). Royal Netherlands Meteorological Institute. http://www.climatescenarios.nl/images/Brochure_KNMI14_EN_2015.pdf.

21. GreenInfra 4 Beira. Deltares Report 185. In: Kalsbeek, P. (Ed.), Deltares, Delft. Kind, J.M., 2013. Economically efficient flood protection standards for the Netherlands. J. Flood Risk Manage. 7 (June (2)), 103–117. doi:http://dx.doi.org/ 10.1111/jfr3.12026.

22. Löschner, L., Nordbeck, R., Scherhaufer, P., Seher, W., 2016. Scientist-stakeholder workshops: a collaborative approach for integrating science and decisionmaking in Austrian flood-prone municipalities. Environ. Sci. Policy 55 (January (2)), 345–352. http://dx.doi.org/10.1016/j.envsci.2015.08.003.

23. Larsen, R.K., ÅG, Swartling, Powell, N., May, B., Plummer, R., Simonsson, L., Osbeck, M., 2012. A framework for facilitating dialogue between policy planners and local climate change adaptation professionals: cases from Sweden, Canada and Indonesia. Environ. Sci. Policy 23 (November), 12–23. doi:http://dx.doi.org/ 10.1016/j.envsci.2012.06.014.

24. Lerer, S.M., Arnbjerg-Nielsen, K., Mikkelsen, P.S., 2015. A mapping of tools for informing water sensitive urban design planning decisions—questions, aspects

and context sensitivity. Water 7, 993–1012;. doi:http:// dx.doi.org/10.3390/ w7030993.

25. Maksimovic, C., Kurian, M., Ardakanian, R., 2014. Rethinking infrastructure design In: rethinking infrastructure design for multi-Use water services. In: Maksimovic, C., Kurian, M., Ardakanian, R. (Eds.), SpringerBriefs in Environmental Science. Springer doi:http://dx.doi.org/10.1007/978-3-319- 06275-4_1 ISBN: 978-3-319-06274-7 (Print) 978-3-319-06275-4 (Online).

26. Munroe, R., Roe, D., Doswald, N., 2012. Review of the evidence base for ecosystembased approaches for adaptation to climate change. Environ. Evid. 1, 13. doi: http://dx.doi.org/10.1186/2047-2382-1-13.

27. Nalau, J., Preston, B.L., Maloney, M.C., 2015. Is adaptation a local responsibility? Environ. Sci. Policy 48, 89–98. doi:http://dx.doi.org/10.1016/j. envsci.2014.12.011.

28. Nichols, R.J., Wong, P.P., Burkett, V.R., Codignotto, J.O., Hay, J.E., McClaen, R.F., Ragoonaden, C.D., Woodroffe, C.D., 2007. Coastal systems and low-lying areas. In: Parry, M.L., Canziani, O.F., Palutikof, J.P., Van der Linden, P.J., Hanson, C.E. (Eds.), Climate Change 2007: Impacts, Adaptation and Vulnerability. Contribution of Working Group II to the 4th Assessment Report of the Intergovernmental Panel on Climate Change, 2007. Cambridge University Press, Cambridge, UK, pp. 315–356.

29. Opdam, P.F.M., Elands, B., Grashof-Bokdam, C.J., Haan, H.J., Hagens, J.E., Van der Knaap, W.G.M., Snep, R.P.H., Steingrover, E.G., 2009. Groen-blauwe netwerken in duurzame gebiedsontwikkeling (Green-blue networks in sustainable area development). Habiforum/WageningenUR, Gouda ISBN 978-94-90287-03-0. Pötz, H., Bleuzé, P., 2012. Urban green-blue grids for sustainable and dynamic cities. Coop for Life. Delft ISBN 978-90-818804-0-4.

30. PROVIA, 2013. PROVIA guidance on assessing vulnerability, impacts and adaptation to climate change. Consultation Document. United Nations Environment Programme, Nairobi, Kenya, pp. 198 ISBN: 978-92-807-3335-8.

31. Pelling, M., 2011. Adaptation to Climate Change, from Resilience to Transformation. Routhledge, London and New York ISBN 978 0 415 477505.

32. Pelzer, P., Arciniegas, G., Geertman, S., De Kroes, J., 2013. Using MapTable1 to Learn about sustainable urban development. In: Geertman, S., et al. (Ed.), Planning Support Systems for Sustainable Urban Development, Lecture Notes in Geoinformation and Cartography. Springer-Verlag, Berlin Heidelberg doi:http:// dx.doi. org/10.1007/978-3-642-37533-0_10.

33. Picketts, I.M., Werner, A.T., Murdock, T.Q., Curry, J., Déry, S.J., Dyer, D., 2012. Planning for climate change adaptation: lessons learned from a community-based workshop. Environ. Sci. Policy 17 (March), 82–93. doi:http://dx.doi.org/10.1016/ j.envsci.2011.12.011.

34. Pyke, C.R., Bierwagen, B.G., Furlow, J., Gamble, J., Johnson, T., Julius, S., West, J., 2007. A decision inventory approach for improving decision support for climate change impact assessment and adaptation. Environ. Sci. Policy 10 (7–8), 610– 621. doi:http:// dx.doi.org/10.1016/j.envsci.2007.05.001.

35. Climate proof cities; final report 2010–2014. In: Rovers, V., Bosch, P., Albers, R. (Eds.), KfC Report Nr: 129/2014 Knowledge for Climate. Utrecht.

36. Svenske Vatten- och Aflopsverksföreningen, (1983). Lokalt omhändertagande av Dagvatten—LOD; Anvisninggar och kommentarer. Publication VAV, P 46, June 1983 Stockholm.

37. Taylor, A.C., Wong, T.H.F., 2002. Non-structural stormwater quality best management practices-an overview of their use, value, cost and evaluation. Technical Report 02/11. Cooperative Research Centre for Catchment Hydrology, Melbourne.

38. UN DESA, 2014. World Urbanization Prospects: The 2014 Revision, Highlights (ST/ ESA/SER.A/352). United Nations, Department of Economic and Social Affairs, Population Division.

39. Van Stigt, R., Driessen, P.P.J., Spit, T.J.M., 2015. A user perspective on the gap between science and decision-making. Local administrators' views on expert knowledge in urban planning. Environ. Sci. Policy 47, 167–176. doi:http://dx.doi.org/ 10.1016/j.envsci.2014.12.002.

40. Van de Ven F.H.M., Luyendijk E., De Gunst M., Tromp E., Schilt M., Krol L., Gersonius B., Vlaming C., Valkenburg L., Peeters R. (2009). Waterrobuust Bouwen (Water resilient urban design; in Dutch). Artikelnr 604.08, SBR, Rotterdam ISBN 978 90 5367 496 3

41. Van de Ven, F.H.M., Bosch, P., Brolsma, R., Keijzer, E., Kok, S., Van der Meulen, S., Schasfoort, F., Ten Velden, C., Vergroesen, T., 2016a. Green, comfortable, attractive and climate resilient Utrecht Centre-West area. Deltares Report 1220357-001, TNO Report 2016 R10158. Deltares, TNO, Utrecht.

42. Van de Ven, F.H.M., Bosch, P., Brolsma, R., Keijzer, E., Kok, S., Van der Meulen, S., Schasfoort, F., Ten Velden, C., Vergroesen, T., 2016b. Green, comfortable, attractive and climate resilient Utrecht Centre-West area. Deltares Report 1220357-000-BGS-004, TNO Report 2016 R10158. Deltares, TNO, Utrecht. Van den Berg, A.E., Hartig, T., Staats, H., 2007. Preference for nature in urbanized societies: stress, restoration, and the pursuit of sustainability. J. Soc. Issues 63 (1), 79–96. doi:http:// dx.doi.org/10.1111/j.1540-4560.2007.00497.x.

43. Van den Berg, M., Wendel-Vos, W., Van Poppel, M., Kemper, H., Van Mechelen, W., Maas, J., 2015. Health benefits of green spaces in the living environment: a systematic review of epidemiological studies. Urb. For. Urb. Green. 14 (4), 806– 816. doi:http://dx.doi. org/10.1016/j.ufug.2015.07.008.

44. Van der Brugge, R., De Graaf, R., 2010. Linking water policy innovation and urban renewal: the case of Rotterdam, The Netherlands. Water Policy 12 (3), 381–400. doi:http://dx.doi.org/10.2166/wp.2010.037.

45. Vergroesen, T., Brolsma, R., Tollenaar, D., 2013. Verwerking van extreme neerslag in stedelijk gebied. Deltares Rep. 12 1202270–009.

46. Voskamp, I.M., Van de Ven, F.H.M., 2015. Planning support system for climate adaptation: composing effective sets of blue-green measures to reduce urban vulnerability to extreme weather events. Build. Environ. 83, 159–167. doi:http:// dx.doi.org/10.1016/j.buildenv.2014.07.018.

47. World Bank, 2010. Climate Risk and Adaptation in Asian Coastal Megacities: a Synthesis Report. International Bank for Reconstruction and Development/The World Bank, Washington DC.

48. Zevenbergen, C., Cashman, A., Evelpidou, N., Pasche, E., Garvin, S., Ashley, R., 2010. Urban Flood Management. CRC Press ISBN 9780415559447.

CHAPTER 20

IMPACT OF CONTRASTED MAIZE ROOT TRAITS AT FLOWERING ON WATER STRESS TOLERANCE

Daniel Leitner, Félicien Meunier, Gernot Bodner, Mathieu Javaux, Andrea Schnepf

INTRODUCTION

Crop production must double by 2050 to keep pace with global population growth (Passioura, 1977 and OECD/FAO, 2012). Extension of cultivated land and further management intensification is limited by growing environmental concerns and increasing costs of production factors. New ways to food security therefore require a more efficient resource use (Rockström et al., 2007 and Raza et al., 2012), particularly in water-limited environments that make up 45% of global land (Safriel et al., 2005). The improvement of cropping systems by targeting plant-soil interactions is a largely unexploited field of agricultural management (Sposito, 2013). In this context the root system is the key plant organ for more efficient use of available resources (Lynch, 2007), particularly under stress conditions (Waines and Ehdaie, 2007). Crops have increasing competitive advantage from optimized root systems in resource limited environments due to improved water and nutrient uptake efficiency (Lynch, 2007, Hinsinger et al., 2011 and Smith and de Smet, 2012).

Historic advance in breeding has largely optimized the aboveground stature of crop plants as well as their radiation use (Araus et al., 2008). It is recognized that superior root water uptake can be a crucial trait explaining superior yield of drought resistant cultivars, e.g. contributing to (flag) leave duration, optimum pollination and grain filling (e.g. Borrell et al., 2001). Within this framework, Blum (2009) points to water uptake maximization as a focus for breeding because of its general compatibility with high yield. Taking into account the existing variability in root traits in different crops (e.g. chickpea, Kashiwagi et al., 2005; rice, Kato et al., 2006; durum wheat, Nakhforoosh et al., 2014), it can be expected that improvement of root water uptake is a feasible agronomic option. Manschadi et al. (2008) demonstrated diversity of wheat seedling root architecture. Based on field evidence and simulations studies using APSIM, they suggested that selection for growth angle and number of seminal roots may help identify genotypes with a root system that is better adapted to drought conditions. Using a model to explain historic yield trends in US maize, Hammer et al. (2009) found that changes in root architecture associated with higher water uptake to be the main reason for better abiotic stress tolerance. Using APSIM-Wheat Lilley and Kirkegaard (2011) showed that in simulations across 109 years, 5 soils, 6 management options and 6 combinations of (virtual) wheat root modifications of root system traits could result in more yield gain in dryland cropping regions. The root traits that were modified were rate of downward root penetration, efficiency of water extraction from the subsoil, an APSIM parameter for the effects of root length density, soil diffusivity and root-soil contact on the rate of water extraction. For systematic crop improvement it is therefore imperative to understand which root architecture would best contribute to maximize water acquisition under the prevalent drought stress conditions. Thus, different authors have discussed the optimal root architecture traits needed to improve plant drought tolerance (Lobet et al., 2014). For rain-fed wheat production with high importance of stored subsoil moisture, Wasson et al. (2012) showed that maximum rooting depth and shifting of rooting density to deeper layers were most relevant root traits for yield. Vadez et al. (2008) on the other hand did not find a strong correlation between root length density distribution and water uptake for groundnut in a breeding programme for a tropical residual moisture environment. They suggested to consider root functionality via water uptake and to only subsequently characterize root morphological differences between genotypes. Some authors defined root traits of apparently broad relevance for better

Citation: Daniel Leitner, Félicien Meunier, Gernot Bodner, Mathieu Javaux, Andrea Schnepf, Impact of contrasted maize root traits at flowering on water stress tolerance – A simulation study, Field Crops Research, Volume 165, 15 August 2014, Pages 125-137, ISSN 0378-4290, http://dx.doi.org/10.1016/j.fcr.2014.05.009.

crop growth under drought stress, such as steep, deep and metabolically cheap roots (Lynch, 2013) or root systems with fine root dominance (Comas et al., 2013). Root plasticity itself has been proposed as a beneficial trait especially when there are trade-offs between different traits (Zhang et al., 2011 and Trachsel et al., 2013).

Beyond the architecture itself, different authors have also investigated the question of ideal axial and radial conductance in roots. In simulation studies, the ratio between axial and radial conductivity proofed useful for classification, since this number determines if root uptake is uniformly, or preferentially from topsoil (Doussan et al., 2006, Levin et al., 2007 and Draye et al., 2010). Richards and Passioura (1989) showed that reduced axial conductance of seminal roots allows water saving during vegetative growth and help alleviate stress during the generative stage of yield formation. Other distributions of root hydraulic properties that favour water uptake are greater axial and radial conductivities in deep roots in order to increase the uptake and transport capacity of water from deep soil layers (according to Wasson et al., 2012) and, decrease of axial conductance in order to save water for the end of the crop cycle (according to Comas et al., 2013). Draye et al. (2010) pointed out the importance of the root hydraulic architecture for root water uptake, i.e. the combination of the architecture and of the distribution of the root hydraulic properties between root segments. This implies that root contribution to drought adaptation will depend on how root hydraulic properties are connected and how these eventually affect critical water use trait (Vadez, 2014).

However, the above cited studies demonstrated that not always genotypes with apparently better rooting ability provided better water uptake and/or yield improvement. A precondition to obtain better water stress tolerance via the root system is a precise definition of the target trait(s) for a clearly defined environment. In this sense root systems do not differ from what (Tardieu (2012)) stated for abiotic stress tolerance in general – any trait can confer drought tolerance, just design the right drought scenario. In climates where crops depend on stored subsoil soil water, narrow growth angle and larger numbers of seminal roots are beneficial, in climates with regular in-season rainfall, less narrow growth angles with shallower root system provide better adaptation. Seasonal drought during (late) reproductive growth is most effectively mitigated by drought escape (early maturity; e.g. Thomson and Siddique, 1997 and Francia et al., 2011) as well as water saving during the vegetative stages to improve generative water availability (Mori and Inagaki, 2012). Richards and Passioura (1989) demonstrated that reducing seminal root axial conductions conserve deep water resources for grain filling. However, the precondition for successful water saving is an effective later uptake to avoid unutilized deep water after harvest. Kirkegaard et al. (2007) and Lilley and Kirkegaard (2007) quantified the value of available soil water in the subsoil for yield (i.e. additional yield per millimetre subsoil water use) and showed that its efficiency depends on management strategies that defer water use from early in the season to late in the season so that the crop uses subsoil water during grain filling. Strategies for optimizing agricultural water management first of all depend on the site specific hydrology. Yield levels in rain-fed cropping system from Mediterranean and semi-arid tropical climates largely rely on stored soil water from off-season rainfalls (e.g. French and Schultz, 1984). In these cropping environments part of the later (generative) development stages coincide with the onset of the dry season. Thus, stress sensitive processes such as flowering and grain filling are strongly related to the use of previously stored subsoil water (Kirkegaard et al., 2007). In most crops water shortage during the reproductive stage, particularly when flowering, is most detrimental for yield (Farooq et al., 2009).

Maize is a crop growing in a wide range of target environments from temperate to tropical climates, but Heisey and Edmeades (1999) estimated that one-quarter of the global maize area is affected by drought in any given year. Tollenaar and Lee (2002)demonstrated that main yield gains in maize did not come from higher yield potential or heterosis per se, but mainly from better stress tolerance. It has been shown that Maize is most susceptible to stress at flowering. Shaw (1977) showed that stress in the period from about 7 days before to 15 days after anthesis reduces maize grain yield two to three times more than at other growth stages. Indeed drought stress during this period increases the period between male and female flowering, widening the anthesis-silking interval (ASI) (Campos et al., 2004). Yield under stress has been showed to be strongly dependent to kernel number per plant, barrenness and ASI in tropical Maize (Bolanos and Edmeades, 1996). Our assumption in this simulation study is that stress occurring during this period will have a major impact on the final grain yield.

Facing this complex root trait–target environment interaction and the inherent adaptive plasticity of root systems, Draye et al. (2010) suggested to develop in-silico experiments to test how root traits may increase drought stress tolerance. To be able to investigate the impact of root architecture on water stress tolerance, models with root growth, soil and root water flow and water uptake modules are thus required. The most sophisticated approach is to describe water movement in soil and roots simultaneously, considering the 3-dimensional root architecture, heterogeneous water distribution in the rooting zone and water potential gradients between soil and roots (Javaux et al., 2011). Root water uptake is then driven by the water potential gradients between root surface and soil.

The objective of this study is to use a mathematical model to quantitatively compare how contrasting root systems of Zea mays affect root water uptake pattern and soil water dynamics. We focus on the flowering period by comparing how three types of maize root architecture (called phenotypes) with different traits will suffer from stress under contrasted water conditions in a silt loam: (a) when most of the water is located at the top of the profile (SWtop) and (b) when the water content is increased in depth (SWbot). A reference phenotype (P1) is compared with two contrasted root system architectures having the same volume (proxy of the amount of invested carbon): a second type (P2) with steeper main roots and a third one (P3) based on P2 with longer lateral roots.

MATERIAL AND METHODS

In the following we outline the mathematical model in detail, including all parameters and units. For the numerical implementation of the model we refer to the corresponding software packages that we have used, R-SWMS (Javaux et al., 2008) and RootBox (Leitner et al., 2010a).

Root architecture modelling

For each root, we consider three zones: the basal zone l_b at the base of the root, the branching zone, and the apical zone l_a at the end of the root (Fig. 1). After the basal and apical zones have developed their lengths stay constant during root growth, while the branching zone develops. New lateral roots are only created within the branching zone with the inter-spaces l_n.

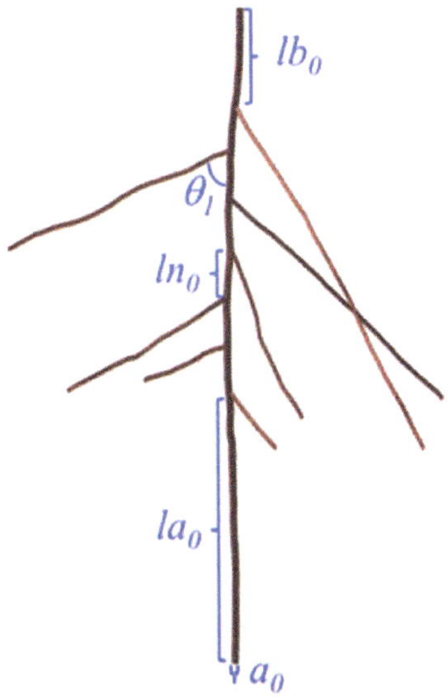

Figure 1: Root growth parameters of the RootBox model. Each root order or root type is described by the length of the apical zone l_a, basal zone l_b, inter-root distance l_n, branching angle θ, root radius a, and (not visualized) the maximal number of branches nob, and root elongation rate r.

Following Pagès et al. (1989) the root length λ, is dependent of the root age τ by

$$\lambda(\tau) = k\left(1 - \exp(-\frac{r}{k}\tau)\right), \tag{1}$$

where k is the maximal root length, and r is the initial growth speed. The maximal root length is calculated for each individual root by

$$k = l_b + l_a + l_n \cdot (nob - 1), \tag{2}$$

where nob is the maximal number of lateral branches. New lateral roots are initiated with an insertion angle γ which describes the angle between the parent root and the initial direction of the new lateral root. Each root has a constant radius a.

Therefore, we describe each root type in a specific soil by seven parameters: l_a [L], l_b [L], l_n [L], nob [1], r [LT^{-1}], γ[1] and a [L]. To add randomness to this deterministic approach each parameter is given by its mean value and its standard deviation.

Additionally, we describe random changes in direction during root tip elongation by randomly choosing two angles (Fig. 2). First, we choose the angle α, which describes the angle between current and new root tip heading by a rotation around the local axis (h_y). The angle α is a random number and normally distributed with a mean of 0 and a standard deviation of σ per root length. Thus, the expected angular change in root tip heading per unit length is σ. Second, we choose the radial angle β uniformly distributed between 0 and 2π, which rotates the root tip heading around the local axis (h_x).

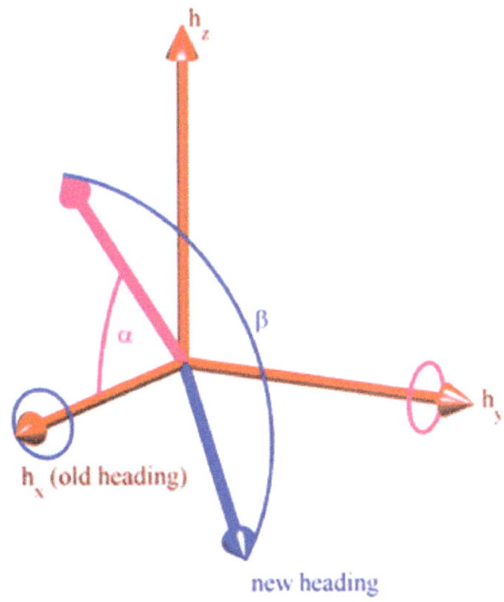

Figure 2: Change of root tip heading due to two random angles α and β. h_x, h_y and h_z are the axes of the local coordinate system of a root segment, where h_x points into the direction of the old heading.

To describe root growth towards a preferential direction (i.e. tropisms), we randomly calculate N pairs of angles (α, β) and then choose the most suitable pair regarding to an objective function. N is the number of trials to find optimal values for α and β, the larger N, the more likely the root will grow in the preferred direction. The objective function is dependent on the type of tropism: for gravitropism headings into the z-direction are favoured, for exotropism headings into the initial direction, and for hydrotropism headings towards higher water content. Therefore, tropism is modelled by three parameters for each root type: σ [L^{-1}], N [1], and type [1].

The main root axes of the maize plant can be implemented as crown or brace roots that start to grow after predefined times and predefined depths. The parameters for the initiation times are called $t_{0,1}$–$t_{0,}$M, and the predefined depths are given by the z-coordinates z_1–z_M, where M is the maximal number of main root axes. The root axes are initiated at a predefined angle θ_0.

We used the L-System implementation of RootBox (Leitner et al., 2010b) to simulate the root growth model in Matlab. The dynamic growth of root systems was simulated until root systems reached a given volume. Those root systems were then used as input for simulating soil and root water flow.

Root water uptake and soil water flow modelling

To describe water uptake by the root system we follow the approach of Javaux et al. (2008), who combined the model of water movement in soil of Somma et al. (1998) with the model describing water flow inside a root branching structure of Doussan et al. (1998). In plants, it is common to express water potential as energy per unit volume (yielding pressure units) while it is common to describe the status of water in soils as energy per unit weight (head units). For model coupling, we chose to consistently express water potential in both systems on a volume basis.

Modelling soil water flow

Water movement in the soil is calculated with the Richards equation:

$$\frac{\partial \theta}{\partial t} = \nabla \cdot \left(\frac{K}{\rho g} (\nabla \psi_m - \rho g e_z) \right) - S, \tag{3}$$

where $\theta:=\theta(t,x)$ is the volumetric water content [$L^3\ L^{-3}$], t is time [T], x is the spatial coordinate [L], $K:=K(\theta)$ is the hydraulic conductivity tensor [$L\ T^{-1}$], $\psi_m:=\psi_m(t,x)$ is the matric potential [$M\ L^{-1}\ T^{-2}$], ρ is the fluid density [$M\ L^{-3}$], g is the gravitational acceleration [$L\ T^{-2}$], e_z is the unit vector pointing in z -direction [1], and $S:=S(t,x,\psi)$is a volumetric sink term representing water depletion due to root water uptake per volume of soil [$L^3\ L^{-3}\ T^{-1}$]. At the top and the bottom of the domain we choose a no-flux boundary condition. At the sides, we assume periodic boundary conditions. As initial condition, we predefine a water content distribution $\theta_0(x):=\theta(0, x)$.

The solution for the Richards equation Eq. (3) is based on the SWMS_3D algorithm (Šimůnek et al., 1995). The finite mesh for soil is made of cubic voxels which are automatically subdivided into 6 tetrahedral elements. A Galerkin finite element approach is applied that uses tetrahedral elements for its spatial discretization. The time component is incorporated using an implicit backward finite difference method. A solver based on a conjugate gradient method is integrated and the solution $\psi_m(x, t)$ is obtained from a Picard iterative numerical scheme. Additional measures are taken to improve solution efficiency in transient problems, including automatic time step adjustment.

Eq. (3) together with the initial condition and the boundary conditions allow us to simulate soil water flow, if the volumetric sink term S is known. In the following, we describe how we calculate the sink term S, taking the soil water potential, as well as the water potential distribution inside the root system into account.

Water potential distribution within the xylem network

We describe the root water pressure $\psi_p(t,\zeta)$ in the xylem tubes, where ζ is the arc length of the centreline of a single root segment with $\zeta = 0$ at the base of each root segment. For each ζ, the corresponding position in Cartesian space is given by its parameterisation $x_{root}(\zeta)$. For a single root segment, we adapt the equation for the axial xylem water flow J_a(Javaux et al., 2008, Eq. (2b)):

$$J_a = -K_a \left(\frac{\partial \psi_p}{\partial \zeta} - \rho g \left| \frac{\partial x_{root}}{\partial \zeta} \right|_z \right), \tag{4}$$

where $J_a:=J_a(t,\zeta)$ describes the rate of axial xylem water flow [$L^3\ T^{-1}$], K_a is the axial root hydraulic conductance [$M^{-1}\ L^5\ T$], $\psi_p:=\psi_p(t,\zeta)$ is the pressure potential in the root xylem [$M\ L^{-1}\ T^{-2}$], whereζ is the distance from the root base [L], and $|\partial x_{root}/\partial\zeta|_z$ is the absolute value of the z -component of the unit tangent vector at position ζ[1]. The total water potential in the xylem (neglecting the osmotic potential) is $\psi_{total}:=\psi_p-\rho g$.

Following Landsberg and Fowkes (1978) the radial water flux, which enters radially into root segments is described by

$$q_r = 2\pi a k_r(\psi_m - \psi_p), \tag{5}$$

where $q_r:=q_r(t, \zeta)$ is the rate of water flow into the root per root length [$L^3\ L^{-1}\ T^{-1}$], $a:=a(\zeta)$ is the root radius [L], k_r is the root radial hydraulic conductivity [$M^{-1}\ L^2\ T$], and $\psi_m:=\psi_m(t,x_{root}(\zeta))$ is the matric potential at the soil root interface [$M\ L^{-1}\ T^{-2}$].

When no water storage is considered, the conservation of water within the root yields

$$\frac{\partial}{\partial \zeta} J_a = q_r. \tag{6}$$

This 1-dimensional partial differential equation describes the water potential in a single root section in dependence on the soil matric potential at the soil root interface.

To describe the flux distribution in a root system we stitch together multiple 1-dimensional root sections. At the nodes between the sections (i.e. the branching points) we apply Kirchhoffs law stating that the water volume flowing into the node is equal to the water volume flowing out of the node. The remaining boundaries are located at the root collar, and at the root tips. At the root collar we apply flux boundary condition under non-stressed conditions (Neumann boundary condition, $J_{a,c} = T_{pot}$, where T_{pot} is the potential transpiration), changing to a fixed water pressure potential when stress appears (Dirichlet boundary condition, $\psi_{total,}c = 1.5$ MPa). At the remaining boundaries we apply a no-flux boundary condition. It results in a system of $n_n = n_s + 1$ equations (n_n and n_s being respectively the number of root nodes and root segments), which writes in matrix notation (Doussan et al., 1998),

$$C\psi_{total} = Q, \tag{7}$$

where C (dimensions $n_p \times n_p$) is called the conductance matrix which contains the root hydraulics characteristics, Q (dimensions $n_p \times 1$) contains the terms related to soil and ψ_{total} (dimensions $n_p \times 1$) is the unknown xylem water pressure vector (Doussan et al., 1998). Due to the large size of the system, these matrices are stored under a sparse format to reduce memory consumption. A biconjugate gradient method is used to solve this asymmetric linear system (Press, 1996).

Implementation

The coupling between soil and root and the root water uptake was implemented by Javaux et al. (2008) and the software is called R-SWMS. The sink term S is calculated by summing up the radial uptake q_r of the root sections located within a soil voxel V [L^3].

$$S(t,V) = \frac{1}{|V|} \int_V \delta_2(\chi_{roots}(x)) q_r(t,x) dx, \tag{8}$$

where $|V|$ is the volume of V [L^3], $\chi_{roots}(x)$ is a characteristic function, which we define to be 0 if the Cartesian point x lies on the centreline of the root system, and else 1, $\delta_2$is the 2-dimensional Dirac delta function [L^{-2}], and $q_r(t, x)$ is the volumetric water flow per root length if x is on the root system, else 0 [$L^3\ L^{-1}\ T^{-1}$].

For coupling the water flow in both soil and root system, an implicit iterative scheme is used. First the root water flow equations given by Eq. (8) are solved given an initial estimate of the pressure head distribution in the soil. This generates a first estimate of the xylem potential ψ_{total} and water flow distribution (Eqs. (4) and (5)). A 3-dimensional sink term distribution can thus be calculated from Eq. (8), which allows the numerical solution of the Richards equation (Eq. (3)). We simultaneously solve these equations for root and soil pressure head using fix point iteration. Note that solving Eqs. (5) and (7) requires the assessment of the soil water potential in the vicinity of each root node ψ_m, defined as the average of the soil water potential of the 8 nodes of the cuboid around the root node weighted by the inverse distance between root and soil nodes.

Simulation scenarios

Root water uptake scenarios were run for a 7 day or a 10 day periods (for the high K_{rs} cases, see hereafter) under sinusoidal potential transpiration conditions (daily T_{pot} = 5.33 mm), corresponding to typical Mediterranean climatic conditions in June–July. Simulation scenarios were built from (i) three different root system architectures (P1, P2, and P3), (ii) two distinct values for root axial conductance and radial conductivities of the root segments (indicated by the subscripts R and H), and (iii) two hydrological situations (SWtop and SWbot). This allowed studying the interactions of root system properties with the pedo-climatic environment. Given the fact that the root architecture parameters are associated with random variation (cf. > 2.1), each scenario was run in five replicates of root architecture. Each of these replicates share the same parameter sets but differ in the precise location of each root segment. Thus a total number of 60 simulations were run (3 root system architectures × 5 replicates × 2 root segment hydraulic properties × 2 soil moisture conditions).

Root architecture

Three different types of root architectures were used and are referred to as phenotypes P1, P2 and P3 in the following text. We simulated root architectures of approximately 60 days old plants, corresponding to the flowering stage, however ensuring that root volumes were equal between phenotypes. For sake of simplicity, we modelled root systems only based on first (main roots) and secondary root orders (laterals). The first P1 is considered as a standard phenotype. The second one (P2) had steeper and deeper principal roots (as suggested by Lynch, 2013), while P3 was created based on the root system architecture of P2 but was changed to have another beneficial trait for drought resistance, i.e. fewer but longer lateral roots at depth (Wasson et al., 2012). Fig. 3a shows one realisation of the architectural model for each phenotype P1, P2 and P3.

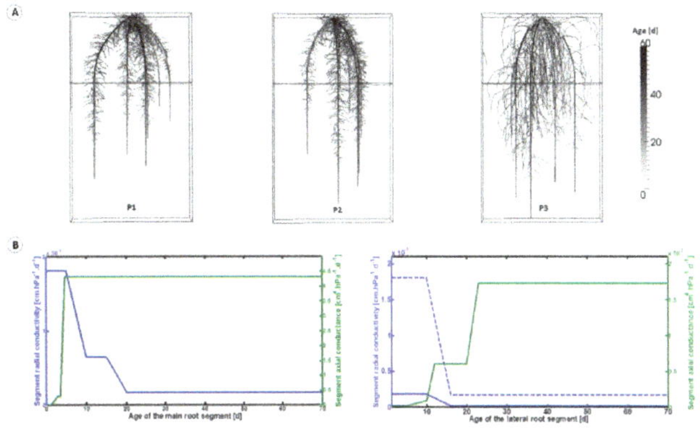

Figure 3: (A) Architecture and age distribution of the root segments for a realization of each phenotype (P1, P2 and P3, respectively) and the separation line for the initial water content. (B) Age dependent root radial conductivity and axial conductance of primary (left) and lateral roots (right). In subplot B, solid and dashed lines stand for reference and increased conductivities, respectively.

The 15 root systems (3 phenotypes × 5 replicates) were generated using the root growth model described in section 2.1 based on the parameters presented at Table 1. The parameterisation of P1 corresponded to the standard parameter set for maize according to Pagès et al. (1989). Two parameters of the main root axes were changed to form the parameter set describing P2: initial growth rate of the main axes was increased while their growth angle was decreased. The magnitude of these parameter changes was evaluated based on available data of the DROPS project (http://www.dropsproject.eu/), which screens maize genotypes for yield and root system traits. One hundred Maize genotypes were screened in aeroponics (de Dorlodot, 2005) and the extent of growth rate and insertion angle was estimated from image analysis (X. Draye, personal communication).

Table 1: Root architectural parameters for the phenotypes 1–3

Symbol (unit)	Meaning	Parameters (primary root)	Parameters (secondary root)
l_a(cm)	Length of apical zone	$18.1_{(P1)}, 24.2_{(P2)}, 21.3_{(P3)}$	$10_{(P1)}, 10_{(P2)}, 100_{(P3)}$
l_b(cm)	Length of basal zone	$0.5_{(P1)}, 0.5_{(P2)}, 2_{(P3)}$	0
l_n(cm)	Length of branch inter-space	$0.5_{(P1)}, 0.5_{(P2)}, 2_{(P3)}$	1.2
nob(−)	Maximal number of branches	$324_{(P1)}, 432_{(P2)}, 109_{(P3)}$	0
r(cm d^{-1})	Initial root elongation	$2.94_{(P1)}, 3.9_{(P2)}, 3.55_{(P3)}$	$0.75_{(P1)}, 0.75_{(P2)}, 1.5_{(P3)}$
γ(rad)	Insertion angle	$1.4_{(P1)}, 1.06_{(P2)}, 1.06_{(P3)}$	1.5
a(cm)	Root radius	0.13	0.05
σ(cm^{-1})	Standard deviation of the root tip heading	0.1	0.1
N(−)	Number of trials	2	2
Type(−)	Tropism type	geotropism	exotropism

P1: phenotype 1, P2: phenotype 2, P3: phenotype 3.

Table 2 summarizes the averaged root characteristics of the three phenotypes. As P2 has a higher initial growth rate of main axis, its root system is younger than that of P1 when it reaches the same volume. As P3 has longer laterals with the same root radius, the number of roots is smaller in P3 than in P2. In this way we simulate the same amount of carbon investment in different kinds of root architectures. Their effect on soil water dynamics and root water uptake during one week was simulated and in this time the root system was taken to be static.

Table 2: Averaged root statistics

Symbol (units)	Meaning	Mean values[a]
L_0(cm)	Total root length	$6468_{(P1)}$ $6426_{(P2)}$ $7002_{(P3)}$
L_{comp}(cm)	Total root length by compartment	$5104_{(P1)}$ $4606_{(P2)}$ $4348_{(P3)}$ (top[b])
		$1364_{(P1)}$ $1820_{(P2)}$ $2654_{(P3)}$ (bottom)
L_{ord}(cm)	Total root length by order	$618_{(P1)}$ $664_{(P2)}$ $759_{(P3)}$ (primary)
		$5850_{(P1)}$ $5762_{(P2)}$ $6243_{(P3)}$(secondary)
rVol(−)	Relative volume of the root system	$1_{(P1)}$ $1.02_{(P2)}$ $1.01_{(P3)}$
K_{rs}(cm^3 cm^{-1} d^{-1})	Equivalent conductance of the root system	$0.06_{(P1r)}$ $0.062_{(P2r)}$ $0.051_{(P3r)}$
		$0.17_{(P1h)}$ $0.166_{(P2h)}$ $0.091_{(P3h)}$
Age(d)	Age at the beginning of the simulation	$64_{(P1)}$ $60_{(P2)}$ $62_{(P3)}$

[a]References of the scenarios are given under parentheses: P1, P2 and P3 refer respectively to phenotype 1, phenotype 2 and phenotype 3. Subscript r and h refer respectively to reference and high radial conductivity of the laterals.

[b]Type of initial water content distribution: top; Soil Water Top (SWtop), bottom; Soil Water Bottom (SWbot).

When needed, the root length density was calculated by summing up the root length per 5 cm-deep soil compartments, considering the soil surface of 75 cm × 15 cm.

Root hydraulic properties

In order to investigate the role of the plant root hydraulic properties on root water uptake and stress onset, two cases were analysed for each phenotype: a reference scenario with hydraulic properties similar to Doussan et al. (1998) and another scenario with increased root radial conductivity. They will be referred to by the subscripts R and H: for instance, $P3_H$ refers to phenotype 3 with higher conductivity values.

The local conductivity and conductance of the root segment for the reference scenario are taken from Doussan et al. (1998) with modifications according to Couvreur et al. (2012). Fig. 3b shows the reference (R scenario) radial conductivity and axial conductance of main and lateral roots, respectively in function to segment age. The H-scenario is based on the same values except the radial conductivity of the laterals, which is increased by a factor 10 (dotted line in Fig. 3b, right subplot) as these segments appear to be the most resistive part to water transfer. This factor ten is in the range of variations observed in experimental studies (Draye et al., 2010).

By considering a root system as a network of resistances as in an electrical circuit, the plant root equivalent conductance (K_{rs}; cf. Table 2 for values of the three phenotypes) can be estimated using the Thevenin theorem (Couvreur et al., 2012). The actual transpiration rate [L^3 T^{-1}] of the plant can then be defined as (Javaux et al., 2013)

$$T_{act} = K_{rs}(\psi_L - \psi_{t,eq}), \tag{9}$$

where ψ_L is the water potential at the root collar and $\psi_{t,eq}$ is an equivalent total soil water potential, which is defined as the effective soil water potential at the soilroot-interface (Couvreur et al., 2012). The equivalent root soil water potential is the mean soil water potential of soil-root interfaces, weighted by the relative local extraction rate of the root segments in a homogeneous soil (see Eq. (17) in Couvreur et al., 2012). It is a physically sound way to define a root-sensed water potential in a heterogeneous soil. The higher $\psi_{t,eq}$ the less energy is required to extract water from the soil. For a given amount of extracted water (represented by the cumulative transpiration), a plant with a higher (less negative) $\psi_{t,eq}$ undergoes a less negative suction, and thus is in more favourable conditions of survival or for extracting soil water.

The evaluation of phenotype performance is done based on different indicators: (i) the equivalent soil water potential at the soil-root interface, (ii) the instantaneous stress fraction being defined as the ratio between water uptake (i.e. actual transpiration) and potential transpiration, and (iii) the cumulative water uptake as an indicator of the carbon acquired for the period.

Soil hydraulic properties

The soil was defined as a silt loam with soil hydraulic parameters from Carsel and Parrish (1988) (see Table 3). Its water retention curve, which relates the soil matric head $\left(h = \frac{\psi_m}{\rho g}\right)$ to the volumetric soil water content θ, and the functional form of the hydraulic conductivity tensor K(h) are shown in Fig. 4. They were modelled using the van Genuchten–Mualem functions (Mualem, 1976 and van Genuchten, 1980) using six parameters: saturated water content θ_s [L^3 L^{-3}], residual water content θ_r [L^3 L^{-3}], the saturated hydraulic conductivity K_{sat} [L T^{-1}] and the parameters α_{vg} [L^{-1}], n_{vg} [1], and λ_{vg}[1], which describe the curvature of the function.

Table 3: Parameter values for the Mualem–Van Genuchten model

Symbol (unit)	Meaning	Parameter values
K_{sat} (cm d^{-1})	Saturated hydraulic conductivity	10.8
θ_s (cm^3 cm^{-3})	Saturated water content	0.45
θ_r (cm^3 cm^{-3})	Residual water content	0.067
α_{vg} (cm^{-1})	Shape parameter	0.02
n_{vg} (–)	Shape parameter	1.41
λ_{vg} (–)	Shape parameter	0.5

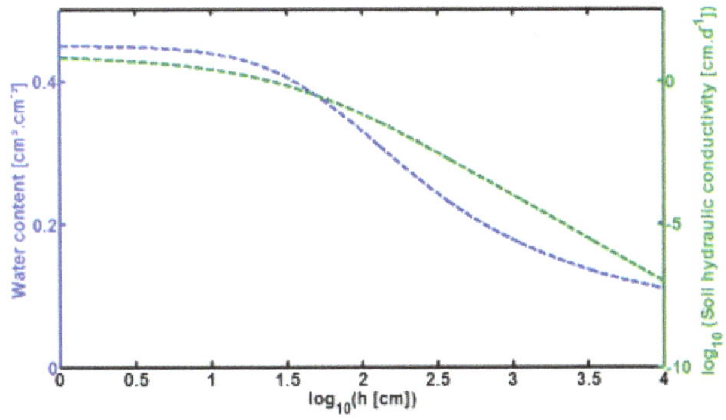

Figure 4: Water retention and hydraulic conductivity curves of the soil used for the simulation.

Hydrological regimes

To account for different climatic conditions in a very simple way, we consider two possible conditions for the initial water content $\theta_0(x)$. In the first case (called SWbot), most of the water is located at the bottom of the profile. This hydrological condition is representative for summer dry climates with high importance of subsoil water, e.g. a Mediterranean climate or an oceanic climate during a summer dry spell. The separation between the wet soil ($\psi_{m,bot} = -100$ hPa) and the dry one ($\psi_{m,top} = -3000$ hPa) was set at the first third of the soil depth (42 cm). In the second scenario (SWtop), the boundary is in the same location but the water content is higher close to the soil surface ($\psi_{m,top} = -100$ hPa and $\psi_{m,bot} = -3000$ hPa). This second case mimics a climatic situation with dominant role of regular rainfall input for plant water supply and low subsoil water reserves. Such conditions are found particularly for climates with dominant in-season rainfall during the vegetation period (e.g. semi-arid continental climates in the temperate zone), semi-arid tropical conditions before refilling of deep soil profile layers during the rainy season and for irrigation during a dry season.

RESULTS

We will first present the impact of different types of root traits (phenotypes) using the reference hydraulic properties, while in the second part we will compare the impact of increased root radial conductivity. For the sake of clarity, the three phenotypes will be associated with a unique colour code in the subsequent plots: P1 will be in red, P2, in green and P3 in blue.

Influence of root system architecture

Fig. 5 shows the distribution of the water content for the scenario SWbot after 6.5 days for one given replicate of each phenotype. We observe that the impact of the initial water distribution remains visible over the whole period, and that the high hydraulic pressure gradient between the two soil layers generates upward water flux. The effect of the root water uptake is also clearly visible, and we can distinguish through the water depletion pattern the main features of the root architectures of the three phenotypes. In the videos, the day–night dynamics is visible through the redistribution of the water content during night.

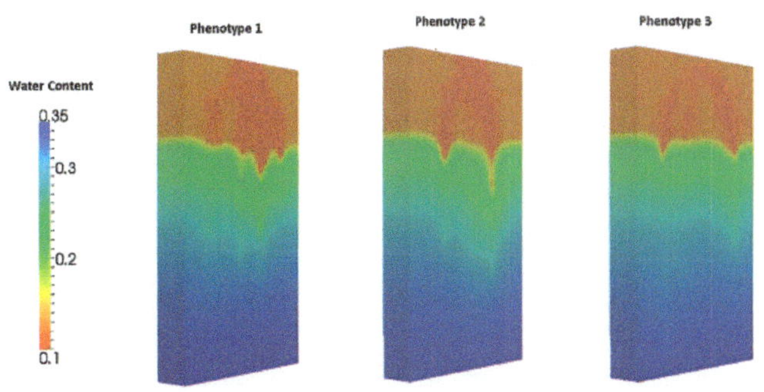

Figure 5: 3-dimensional visualisation of soil water content as influenced by reference hydraulic architecture of the three phenotypes after 6.5 days for the SWtop scenario.

To investigate the difference between phenotype behaviours better, we can look at the transpiration dynamics of the three phenotypes under the two opposite hydrological regimes. In Fig. 6, the averaged transpiration of the five replicates is shown for each phenotype and for the SWbot case (Fig. 6a) and for the SWtop case (Fig. 6b). As standard deviations between replicates are small (always smaller than 3% of the mean value), they are not shown on the plots. For all phenotypes and cases, a water stress (i.e. a decrease of the actual transpiration as compared to the potential demand) is observed after the beginning of the first day of simulation and its magnitude increases with time.

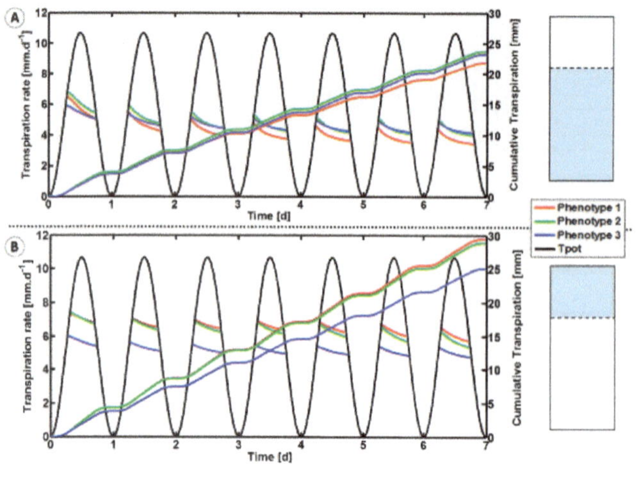

Figure 6: Average actual and cumulative transpiration of the three phenotypes in SWbot (A) and SWtop (B) scenarios. Red: phenotype 1; green: phenotype 2 and blue: phenotype 3. The potential transpiration is represented by the black solid line. (For interpretation of the references to colour in this figure legend, the reader is referred to the web version of this article.)

If we compare both scenarios, we can see that generally these three phenotypes transpire more when the upper layer is wetter: all the phenotypes transpire more than 25 mm after 7 days for SWtop conditions (Fig. 6b), while none of them could reach 25 mm for SWbot (Fig. 6a). When we compare the change of total transpiration between SWbot and SWtop, we observe differences between phenotypes. For P3, the change of cumulative transpiration is very small (2 mm), for P2 of about 5 mm and for P1 of about 8 mm, demonstrating a different sensitivity to climatic conditions.

In the SWbot scenario (Fig. 6a), the ranking between the three phenotype transpiration rates evolves with time. While P1 and P2 have the highest transpiration rate at the beginning of the period, transpiration of P3 is highest at the end. Here P1 with a surface near root distribution undergoes a more severe stress and transpiration falls below the other phenotypes at the second day. The phenotypes with deeper root system allocation behaved similar, with the phenotype P3 having a similar behaviour between days of the whole simulation period.

In the scenario with wetter top soil layer (SWtop, Fig. 6b), it is observed that the two phenotypes P1 and P2 behave similarly while the phenotype P3 suffers from a more drastic transpiration decrease.

Fig. 7 shows the change of the equivalent soil water potential $\psi_{t,eq}$ with increasing water extraction from soil between the beginning and the end of the period, expressing the global moisture deficit that builds up at the root–soil interface. Logically, we see that $\psi_{t,eq}$ has a decreasing trend with time and cumulated uptake. During the night however, $\psi_{t,eq}$ gets less negative, as the system relaxes: water potential gradients built up during day around the root segments decreases as the uptake is reduced and soil water redistribution occurs.

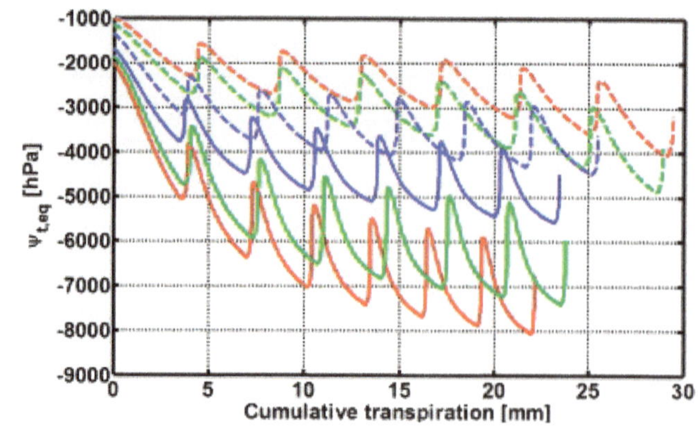

Figure 7: Equivalent total soil water pressure as a function of the cumulative transpiration for SWbot (solid lines) and SWtop (dashed lines). Red: P1, green: P2 and blue: P3. (For interpretation of the references to colour in this figure legend, the reader is referred to the web version of this article.)

In the SWbot case (continuous lines), all phenotypes experience a lower soil water potential (and a lower total transpiration) than in the SWtop scenario. However, in SWbot the phenotype P3 is most efficient, as it transpires almost the same amount of water as P2 and experiences higher (less negative) soil water potential. Thus, it could be expected that P3 would be able to sustain a higher transpiration compared to P2 over a longer period of stress as it can access water more easily at depth. In the case where the water is in the upper layer (dashed lines), P1 has always a higher (less negative) total equivalent water potential than P2 and P3 for a similar given amount of cumulated transpiration. As can be expected, the deep rooted phenotype P3 (continuous blue line) not only transpires less (i.e. experiences a more severe stress) but also undergoes a lower (more negative) total equivalent water potential. However, as already observed for the transpiration dynamics, P3 has a relatively similar behaviour between SWbot and SWtop cases.

Fig. 8 a and b show the averaged sink term profiles of the three phenotypes at two different times (t = 0.5 and 6.5 days) and under variable hydrological conditions. Fig. 8d shows the root length density profile of the three phenotypes: P1 has more roots at the top, P3 deeper rooting distribution and P2 an intermediate distribution. When the water is in the bottom layer (SWbot scenario, Fig. 8a), all plant phenotypes decrease their water uptake in the upper layer after 6.5 days while remaining stable in the bottom layer (Fig. 8a: dashed and continuous lines overlap in the bottom layer). Although the water content in the upper layer is low in this scenario, the three phenotypes are able to extract it at the beginning of the period, as they have a lot of roots there. The phenotype P3 has more ability to extract the soil water from deeper layer below 80 cm from the beginning of the period on. On the opposite, phenotype P1 extracts soil water less deep and suffers from stronger stress (see Fig. 7).

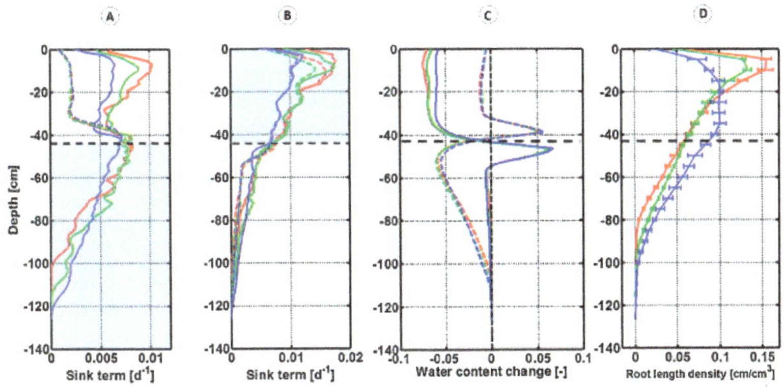

Figure 8: 1-D sink term profile under scenario SWbot (A) and SWtop (B) after 0.5 (solid lines) and 6.5 (dashed lines) days and corresponding water changes (C) and root length densities (D). Red: P1, green: P2 and blue: P3. (For interpretation of the references to colour in this figure legend, the reader is referred to the web version of this article.)

When water is located in the upper layer, P1 and P2 are able to take up more water from that horizon as compared to P3 (Fig. 8b). As P3 cannot compensate this lower uptake in the upper horizon by a much higher uptake in the drier bottom horizon, it suffers from earlier stress.

Fig. 8c shows the water changes over the whole simulation period for case SWbot (continuous lines) and SWtop (dashed lines). Although the sink terms are very different between phenotypes, this is not reflected in the soil water content changes. Indeed water content changes not only in function of the sink term S but also due to the water pressure distribution in the soil which generate water flow. This is particularly clear near the interface between the wet and dry zones, where water changes occur due to water and thus matric potential differences.

When we compare the sink to the root length density profiles (Fig. 8d), we hardly see any direct correspondences for any phenotypes. This is due to the fact that water is preferentially extracted where the water is available. In addition, the root length density profile only partially reflects the distribution of the root ability to extract water. The hydraulic architecture i.e. how the root ability to extract water is distributed into the soil, combined with the water availability control the uptake (Javaux et al., 2013).

To summarize, we observed that (1) P1 which has more roots in the upper layer is very efficient in the SWtop case, but has low ability to deal with a drier upper layer; (2) P3, which has more deeper roots outperforms for SWbot; (3) P2 transpires as much as P1 for SWtop and as much as P3 for SWtop. However, its equivalent soil water potential is always more negative than the two other phenotypes.

Influence of root hydraulic properties

It was shown in section 3.1 that root architecture determines the extraction pattern. In this section, we investigate how distinct root hydraulic properties influence water extraction and stress onset. By increasing the radial local conductance of the laterals (H case) and comparing the performance of the 3 genotypes as compared to their initial conductance values (R case), we can assess the impact of the hydraulic properties on root water uptake.

The root equivalent conductance K_{rs} (see Eq. (8)) is a key parameter which characterizes the hydraulic architecture, i.e. the impact of the root system architecture and of the distribution of the root segment hydraulic properties. As the local hydraulic properties depend to the root order (main or lateral) and to the age of the root segments (see Table 2and Fig. 4d), each phenotype has a different K_{rs}. Values of the mean plant equivalent conductance K_{rs} for the six cases are given in Table 2. Phenotypes P1 and P2 typically have a relatively similar K_{rs}, as they have a similar proportion of root segment age and order. We observe the following ranking: $K_{rs}(P3_R) < K_{rs}(P1_R) < K_{rs}(P2_R) < K_{rs}(P3_H) < K_{rs}(P2_H) < K_{rs}(P1_H)$. Multiplying the radial conductivity of the laterals by 10 increases the K_{rs} by a factor 2.8, 2.7 and 1.8 for P1, P2 and P3, respectively, as root segment age and order differ between phenotypes.

As compared to $P1_R$, $P2_R$ and $P3_R$, the extraction patterns are reinforced. Table 4 shows the amount of transpiration after 7 days for the two hydrological regimes, the three phenotypes and the two K_{rs}. Higher K_{rs} always increases the amount of extracted water. Under the SWtop scenario, we do not observe any stress (100% of the T_{pot} is extracted) for $P1_H$ and $P2_H$, while $P3_H$ suffers from a slight stress. For SWtop, the different root system architectures generate a actual transpiration difference of 12% between phenotypes. By increasing the conductance of the laterals, there is almost no impact of the root system architecture anymore. For SWBot, stress still occurs for all phenotypes with higher K_{rs}. However, differences between root architecture generate only 5–8% difference of actual transpiration while a difference in K_{rs} increases the actual transpiration by 20–30%.

Table 4: Cumulative transpiration [mm] and percentage of the total potential transpiration effectively transpired after seven days for each scenario

	P1		P2		P3	
	R	H	R	H	R	H
SWtop	29.5/79.1	37.3/100	28.9/77.5	37.3/100	25.2/67.5	36/96.5
SWbot	21.9/58.7	32.5/87.1	23.8/63.8	34.6/92.7	23.3/62.5	29.7/79.6

In the following, we focus on the scenario with predominant subsoil water dependence SWbot, and on the two root phenotypes P2 and P3 with adapted root distribution for such hydrological conditions.

Fig. 9 compares the transpiration dynamics of P2 and P3 with reference and increased root radial conductivities for a 10 days period. When the local radial conductivity of a plant root is increased, its capacity to extract water is improved (i.e. K_{rs} increases and roots pose less resistance to water uptake and movement inside the root system), and thus transpiration is higher for a given potential difference between soil and root (see Eq. (9)). We observe that the ranking of cumulative transpiration over 10 days follows the ranking of the K_{rs} explained here before.

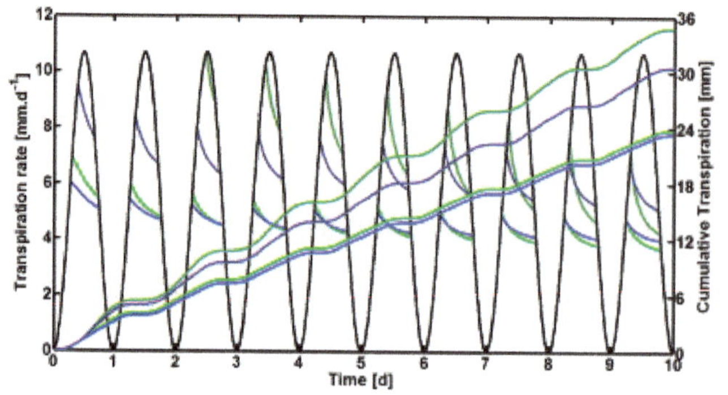

Figure 9: Actual and cumulative transpiration change for P2 with reference (light green, $P2_R$) and increased (dark green, $P2_H$) root radial conductivity and P3 with reference (light blue, $P3_R$) and increased (dark blue, $P3_H$) root radial conductivity for the SWbot scenario. The black solid line is the potential transpiration. (For interpretation of the references to colour in this figure legend, the reader is referred to the web version of this article.)

While the two phenotypes behave rather similar in the SWbot scenario when the reference hydraulic properties are used (denoted a $P2_R$ and $P3_R$), there is a more distinct differentiation when considering an increased K_{rs} (denoted as $P2_H$ and $P3_H$). Indeed, $P2_H$ is superior in transpiration compared to $P3_H$ over the 10 days of simulation. Still $P2_H$ also shows a stronger decrease in transpiration rate compared to $P3_H$, which would result in higher stress over prolonged dry periods. Indeed, after 10 days, $P3_H$ gets a higher daily transpiration rate than $P2_H$. In general, whatever its K_{rs}, P3 remains more stable over the simulation period, and always has a higher transpiration rate than P2 at the end of the simulation period.

Fig. 10 gives another perspective on the comparison between both phenotypes. It shows the instantaneous stress fraction (T_{act}/T_{pot}) as a function of the fraction of the water extracted from the soil (called hereafter soil water depletion D) defined as:

$$D(t) = 100\frac{\theta_{ini} - \theta_{(t)}}{\theta_{ini} - \theta_r} \tag{10}$$

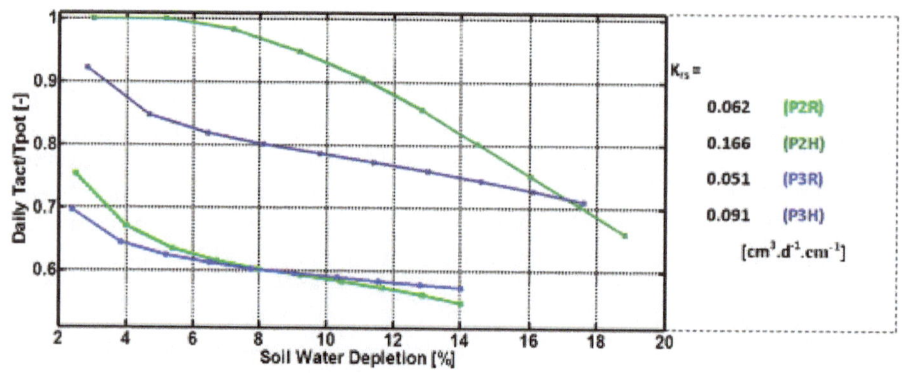

Figure 10: Reduction of the actual to potential transpiration ratio for a ten days long simulation as a function of the soil water depletion in the soil (in %). Blue lines are for P3 and green lines for P2. Dark colours stand for root systems with an increased radial conductivity and light colours for reference lateral conductivities, both are for the SWtop scenario. The initial classification of the root systems is determined by the value of the root equivalent conductance (recalled on the right). (For interpretation of the references to colour in this figure legend, the reader is referred to the web version of this article.)

with θ_{ini}, θ_r, $\theta(t)$, the initial, residual and at time t water content of the soil, respectively. The instantaneous stress fraction reflects the reduction in transpiration and thus, the stomatal closure at a given time t. A plant with higher stress fraction keeps its stomata open so that carbon acquisition is not decreased and thus suffers less from water shortage. In Fig. 10, we observe that the water depletion is higher for phenotypes with higher root conductance K_{rs}. We also see that for a given amount of extracted water, the high K_{rs} phenotypes ($P2_H$ and $P3_H$) experience less stress. It clearly illustrates that the water availability for plants is not only a question of soil type but also of root conductance. $P2_H$ for instance is able to extract the potential transpiration rate at the beginning of the period without reduction, due to its high K_{rs}. If we compare this with Fig. 9, we see that $P2_H$ is able to extract more water over first part of the simulation period and thus has a stronger trend to deplete available water. $P3_H$ on the contrary has a sustained water extraction over longer times. At the end of the periods P3 phenotypes have always higher cumulated water extracted with a higher water potential than P2 phenotypes, i.e. a better resistance to prolonged drought period.

This corresponds to two different strategies for the considered period: $P2_H$ takes as much water as possible at the beginning but reduces more strongly its transpiration stream, while $P3_H$ has a more conservative behaviour, keeping a constant transpiration rate with time. Note that there is no difference in the way how stress onset is represented or parameterised between scenarios. This difference is only due to a different root hydraulic architecture, i.e. $P2_H$ has a higher K_{rs} than $P3_H$ which allows more water uptake at the beginning but more stress at the end of the simulation time.

DISCUSSION

Agriculture has to achieve a more efficient use of water to cope with future needs of food production and competition between different fresh water users (Rockström et al., 2007). This can be achieved by reducing losses via soil management change (e.g. mulch coverage) and more efficient water acquisition by improved crops (e.g. Passioura, 1977). Root architecture and functioning are key crop traits in rain-fed cropping systems where yield depends on optimum use of site-specific water availability during the growing season.

Two main challenges to breeding for efficient root systems so far are: (i) selection of an appropriate root target trait, and (ii) ensuring that this trait effectively conveys improved crop water supply. Vadez et al. (2008) demonstrated that structural knowledge of root systems alone is not sufficient to understand different water uptake capacity among chickpea cultivars. Our study clearly sustains this empirical observation: although there is a relation between the distribution of water availability and root axes allocation in the soil profile, root functionality as expressed by root equivalent conductance appeared as key variable to infer superior water uptake of a given phenotype. A main finding of our study is that root length density distribution alone is not sufficient to describe root water uptake from soil. Root systems consist of dynamic populations of roots of different types and age, with age dependent root axial conductances and radial conductivities (Draye et al., 2010). As in Doussan et al. (2006) our model is based on combining root architectural, root system hydraulic and soil hydraulic properties including the possibility of compensation and redistribution. In our study, the difference in water acquisition induced by different architectural root traits (the 3 phenotypes) was much smaller than the increase induced by increasing radial hydraulic conductivity of lateral roots by a factor 10 (Table 4). This demonstrates that root architectural traits stricto sensu are not sufficient to explain water uptake from soil by plant roots, but root architecture and related age distribution of root segments is important for overall conductance. This is in line with the conclusions of Vadez (2014).

Within the dehydration avoidance type of drought resistance strategies, Levit (1980)distinguished between water savers and water spenders. Our study showed that root traits can confer both, water saving and water spending on a short period. The P3 root phenotype was a water saver. Concerning root system architecture (distribution) P3 had comparatively highest allocation of the sink to deeper layers and comparatively less sink strength in upper layers (Fig. 8). Whenever upper layers contributed essentially to water supply, P3 thus had lowest transpiration. However with increasing depletion P3 could sustain uptake from deeper parts of the profile over longer times (Fig. 9 and Fig. 10). The prolonged uptake from the deeper layer is related to the lower sink strength leading to a less intense depletion. The other phenotypes had a more water spending behaviour with quick initial depletion and stronger decrease of transpiration over time. For overall crop performance the importance of high short term water extraction vs. low sustained uptake depends on the drought scenario. If only intermittent dry spells have to be buffered, a water spending depletion root phenotype is more efficient as such a behaviour will keep stomata more open. In case of longer dry periods however a root phenotype conferring water saving will improve crop water supply. Here we do not consider potential competitive effects between plant water uptake and transpiration losses from upper layers. Generally water saving is only effective for stored soil moisture in deeper layers not prone to evaporation losses.

Traditionally, water uptake simulation mainly considered root distribution, with more recent architecture models giving a better insight into the 3D structure of root systems (Draye et al., 2010). More recently a focus was on integrating root functionality into architecture models. While some cropping system models like APSIM (Keating et al., 2003) used an empirical parameter for functionality, more mechanistic approaches strive to consider root hydraulic architecture as such (e.g. Sperry et al., 1998). Our results showed that differentiation between phenotypes P2 and P3 became more evident when considering root conductance (Table 4). Under a high conductance scenario, P2 showed a substantial increase in initial water uptake with values near the potential transpiration, followed by a quick decrease. Thus the water spending behaviour was accentuated. For P3 there was also an increase in the transpiration level, still the extent was less compared to P2 and this rooting phenotype maintained a more stable transpiration level even at higher root hydraulic conductance. The stronger differentiation between P2 and P3 in the high conductance scenario was due to the age dependence of this root functional trait. As P3 contained younger root segments in deeper layers with still lower axial conductance, the bulk increase in root conductance (K_{rs} cf. Table 2) is less important. This is in line with the findings on the role of root conductance for transpiration resumed by Comas et al. (2013) who stated that generally highly conductive large diameter xylem vessels are related to high levels of (unstressed) transpiration, while lower diameter vessels induce water saving.

The differentiation between P2 and P3 at the high conductance scenario also indicates that still there was significant uptake capacity from the upper soil compartment. The substantially higher initial transpiration of P2 was mainly a result of enhanced uptake for upper layer. As revealed by the sink distribution (Fig. 8), the higher root length density in the upper compartment resulted in substantial proportion of water extracted for the upper part of the profile particularly during the first days. The sink distribution reveals that with the advance of depletion over time, upper layer extraction was increasingly source limited, while towards lower layers sink strength was limited by the lower rooting density, i.e. it was more sink limited. The sink strength in the upper compartment was further increased in the high conductance scenario, leading to an efficient and quick exploitation of extractable water resources (not shown). In the lower layers sink strength was not only limited by root distribution but further decreased in proportion to the upper layer by the lower conductance of young segments. Following Fitter›s terminology (Fitter, 2002), root system architecture thus determines the overall exploration capacity of the soil profile via synchronization of water availability and depth distribution of root segments (source-sink synchronization), while root functionality (partially reflected in the root equivalent conductance) defines its capacity to exploit the available water, supporting the spending vs. saving behaviour.

In order to sustain yield under drought conditions, breeders need to blend together all knowledge on drought tolerance-related traits and their phenotypic expression (Cattivelli et al., 2008). It remains difficult to assess the overall effect of traits contributing to drought tolerance, because several traits can play roles, and these traits are likely to interact with one another and with the environment.

CONCLUSIONS

We simulated two contrasting hydraulic regimes to analyse the performance of root phenotypes with surface near root allocation (P1) vs. steeper root allocation (P2 and P3), where P2 and P3 correspond to deep and steep phenotypes with high lateral growth for P3. As expected, superiority of a given root architecture was dependent on the water regime. For the soil conditions used here, our simulations showed that a deep and steep phenotype is beneficial only in situations where crop water supply depends on subsoil moisture while surface-near allocation of roots is more preferential when crops can rely on high in-season rainfall. This corresponds to findings from vegetation ecology and ecohydrology (e.g. Schenk and Jackson, 2002 and Preti et al., 2010).

Although the root systems of the three virtual phenotypes considered in this study have the same volume, they behave quite differently in terms of water uptake from soil.

In general, the same amount of carbon spent can create different root system architectures with different hydraulic properties that will affect overall water uptake behaviour.

In particular, the root length density profile is not sufficient to characterize root system water uptake efficiency. In this simulation case study we kept the overall amount of carbon spent constant and compared the effect when the carbon was spent into more or less deep and steep root system architectures. However, this results in the fact that we did not model the cheap aspect of the ideotype of Lynch (2013), which could also affect the conclusions of this study regarding root system water uptake efficiency.

Regarding the hydrological conditions, phenotype P3 that had fewer and longer laterals was better in a situation that can be found in climates with a dry summer and high importance of subsoil water: it has a more conservative behaviour on a longer term. For plants with comparable rooting depth, the overall root system conductance K_{rs} is a key parameter to explain transpiration dynamics.

The plant root systems we created were generic, although the variability of parameters was chosen in correspondence with the genotypic variations of maize root systems analysed in the DROPS project for drought tolerant yielding plants. We demonstrated how root architecture and hydraulic properties can induce different behaviour, also as a function of the environment considered, and thus one should be careful while designing an ideotype based on root architectural traits alone. Beyond the root architecture itself, the age distribution and related root hydraulic properties need to be considered. These results could feed into cropping system models to see the effect of these processes on crop yield.

ACKNOWLEDGMENTS

This work is a contribution of the Transregio Collaborative Research Center 32, Patterns in Soil–Vegetation–Atmosphere Systems: Monitoring, Modelling and Data Assimilation, which is funded by the German Research Association, DFG. This work has also been funded in part by the project DROPS (DROught tolerant yielding PlantS: http://www.drops-project.eu), which received funding from the European Communitys Seventh Framework Programme under the grant agreement no. FP7-244374. F. M. is supported by the Fonds National de la Recherche Scientifique (FNRS) of Belgium as a research fellow. D. L. is recipient of an APART – fellowship of the Austrian Academy of Sciences at the Computational Science Center. We acknowledge the Austrian Science Fund (FWF) through grant no. V220–N13 (A.S.).

REFERENCES

1. Araus, J.L., Slafer, G.A., Royo, C., Serret, M.D., 2008. Breeding for yield potential and stress adaptation in cereals. Crit. Rev. Plant Sci. 27, 377–412.
2. Blum, A., 2009. Effective use of water (EUW) and not water-use efficiency (WUE) is the target of crop yield improvement under drought stress. Field Crops Res. 112, 119–123.
3. Bolanos, J., Edmeades, G.O., 1996. The importance of the anthesis-silking interval in breeding for drought tolerance in tropical maize. Field Crops Res. 48, 65–80.
4. Borrell, A., Hammer, G., Van Oosterom, E., 2001. Stay-green: a consequence of the balance between supply and demand for nitrogen during grain filling? Ann. Appl. Biol. 138, 91–95.

5. Campos, H., Cooper, M., Habben, J.E., Edmeades, G.O., Schussler, J.R., 2004. Improving drought tolerance in maize: a view from industry. Field Crops Res. 90, 19–34.
6. Carsel, R.F., Parrish, R.S., 1988. Developing joint probability distributions of soil water retention characteristics. Water Resour. Res. 24, 755–769.
7. Cattivelli, L., Rizza, F., Badeck, F.W., Mazzucotelli, E., Mastrangelo, A.M., Francia, E., Marè, C., Tondelli, A., Stanca, A.M., 2008. Drought tolerance improvement in crop plants: an integrated view from breeding to genomics. Field Crops Res.105, 1–14.
8. Comas, L.H.B., SR, Von Mark, V.C., Byrne, P.F., Dierig, D.A., 2013. Root traits contributing to plant productivity under drought. Front. Plant Sci. 4.
9. Couvreur, V., Vanderborght, J., Javaux, M., 2012. A simple three-dimensional macroscopic root water uptake model based on the hydraulic architecture approach. Hydrol. Earth Syst. Sci. 16, 2957–2971.
10. de Dorlodot, S., 2005. Scaling up quantitative phenotyping of root system architecture using a combination of aeroponics and image analysis. Aspects Appl. Biol. 73, 41–54.
11. Doussan, C., Pagès, L., Vercambre, G.,1998. Modelling of the hydraulic architecture of root systems: an integrated approach to water absorption – model description. Ann. Bot. 81, 213–223.
12. Doussan, C., Pierret, A., Garrigues, E., Pagès, L., 2006. Water uptake by plant roots: II – modelling of water transfer in the soil root-system with explicit account of flow within the root system – comparison with experiments. Plant Soil 283, 99–117.
13. Draye, X., Kim, Y., Lobet, G., Javaux, M., 2010. Model-assisted integration of physiological and environmental constraints affecting the dynamic and spatial patterns of root water uptake from soils. J. Exp. Bot. 61, 2145–2155.
14. Farooq, M., Wahid, A., Lee, D.J., Ito, O., Siddique, K.H.M., 2009. Advances in drought resistance of rice. Crit. Rev. Plant Sci. 28, 199–217.
15. Fitter, A.H., 2002. Characteristics and functions of root systems. In: Waisel, Y., Eshel, A., Kafkaf, U. (Eds.), Plant Roots: The Hidden Half. Marcel Dekker Inc., New York, NY, pp. 21–50.
16. Francia, E., Tondelli, A., Rizza, F., Badeck, F.W., Li Destri Nicosia, O., Akar, T., Grando, S., Al-Yassin, A., Benbelkacem, A., Thomas, W.T.B., van Eeuwijk, F., Romagosa, I., Stanca, A.M., Pecchioni, N., 2011. Determinants of barley grain yield in a wide range of Mediterranean environments. Field Crops Res. 120, 169–178.
17. French, R.J., Schultz, J.E., 1984. Water use efficiency of wheat in a Mediterraneantype environment. I. The relation between yield, water use and climate. Aust. J. Agric. Res. 35, 743–764.
18. Hammer, G.L., Dong, Z., McLean, G., Doherty, A., Messina, C., Schussler, J., Zinselmeier, C., Paszkiewicz, S., Cooper, M., 2009. Can changes in canopy and/or root system architecture explain historical maize yield trends in the U. S. corn belt? Crop Sci. 49, 299–312.
19. Heisey, P.W., Edmeades, G.O., 1999. Maize Production in Drought Stressed Environments: Technical Options and Research Resource Allocation. International Maize and Wheat Improvement Centre, Mexico D.F, pp. 1–36.
20. Hinsinger, P., Brauman, A., Devau, N., Gérard, F., Jourdan, C., Laclau, J.P., Le Cadre, E., Jaillard, B., Plassard, C., 2011. Acquisition of phosphorus and other poorly mobile nutrients by roots. Where do plant nutrition models fail? Plant Soil 348, 29–61.
21. Javaux, M., Couvreur, V., Vanderborght, J., Vereecken, H., 2013. Root water uptake: from three-dimensional biophysical processes to macroscopic modeling approaches. Vadose Zone J. 12. Javaux, M., Draye, X., Doussan, C., Vanderborght, J., Vereecken, H., 2011. Root water uptake: toward 3-D functional fpproaches. In: Glinski, J., Horabik, J., Lipiec, J. (Eds.), Encyclopedia of Agrophysics. Springer, Netherlands, pp. 717–722.
22. Javaux, M., Schröder, T., Vanderborght, J., Vereecken, H., 2008. Use of a threedimensional detailed modeling approach for predicting root water uptake. Vadose Zone J. 7, 1079–1088.
23. Kashiwagi, J., Krishnamurthy, L., Upadhyaya, H.D., Krishna, H., Chandra, S., Vadez, V., Serraj, R., 2005. Genetic variability of drought-avoidance root traits in the minicore germplasm collection of chickpea (Cicer arietinum L.). Euphytica 146, 213– 222.
24. Kato, Y., Abe, J., Kamoshita, A., Yamagishi, J., 2006. Genotypic variation in root growth angle in rice (Oryza sativa L.) and its association with deep root development in upland fields with different water regimes. Plant Soil 287, 117–129.
25. Keating,B.A.,Carberry, P.S.,Hammer, G.L., Probert,M.E.,Robertson, M.J.,Holzworth,D., Huth,N.I.,Hargreaves,J.N.G.,Meinke,H.,Hochman, Z., Lean,Mc,Verburg,G., Snow, K.,Dimes,V., Silburn,J.P.,Wang,M.,Brown, E.,Bristow, S.,Asseng,K.L.,Chapman,S., Cown, Mc, Freebairn, R.L., Smith, D.M., C.J, 2003. An overview of APSIM, a model designed for farming systems simulation. Eur. J. Agron. 18, 267–288.
26. Kirkegaard, J.A., Lilley, J.M., Howe, G.N., Graham, J.M., 2007. Impact of subsoil water use on wheat yield. Aust. J. Agric. Res. 58, 303–315.
27. Landsberg, J.J., Fowkes, N.D., 1978. Water movement through plant roots. Ann. Bot. 42, 493–508. Leitner, D., Klepsch, S., Bodner, G., Schnepf, A., 2010a. A dynamic root system growth model based on L-systems. Plant Soil 332, 177–192.
28. Leitner, D., Klepsch, S., Ptashnyk, M., Marchant, A., Kirk, G.J.D., Schnepf, A., Roose, T., 2010b. A dynamic model of nutrient uptake by root hairs. New Phytol. 185, 792– 802.
29. Levin, A., Shaviv, A., Indelman, P., 2007. Influence of root resistivity on plant water uptake mechanism, part I: numerical solution. Transport Porous Med. 70, 63– 79.
30. Levit, J., 1980. Responses of plants to environmental stresses. Water, Radiation, Salt, and Other Stresses, vol. II. Academic Press, New York.

31. Lilley, J.M., Kirkegaard, J.A., 2007. Seasonal variation in the value of subsoil water to wheat: simulation studies in southern New South Wales. Aust. J. Agric. Res. 58, 1115–1128.

32. Lilley, J.M., Kirkegaard, J.A., 2011. Benefits of increased soil exploration by wheat roots. Field Crops Res. 122, 118–130.

33. Lobet, G., Couvreur, V., Meunier, F., Javaux, M., Draye, X., 2014. Plant water uptake in drying soils. Plant Physiol. 164, 1619–1627.

34. Lynch, J.P., 2007. Turner review no. 14. Roots of the second green revolution. Aust. J. Bot. 55, 493–512. Lynch, J.P., 2013. Steep, cheap and deep: an ideotype to optimize water and N acquisition by maize root systems. Ann. Bot. 112, 347–357.

35. Manschadi, A.M., Hammer, G.L., Christopher, J.T., DeVoil, P., 2008. Genotypic variation in seedling root architectural traits and implications for drought adaptation in wheat (Triticum aestivum L.). Plant Soil 303, 115–129.

36. Mori, M., Inagaki, M., 2012. Root development and water-uptake under water deficit stress in drought-adaptive wheat genotypes. Cereal Res. Commun. 40, 44–52.

37. Mualem, Y., 1976. New model for predicting the hydraulic conductivity of unsaturated porous media. Water Resour. Res. 12, 513–522.

38. Nakhforoosh, A., Grausgruber, H., Kaul, H.-P., Bodner, G., 2014. Wheat root diversity and root functional characterization. Plant Soil 1–19. OECD/FAO., 2012. Agricultural outlook, 2012-2021.

39. Pagès, L., Jordan, M.O., Picard, D., 1989. A simulation model of the three-dimensional architecture of the maize root system. Plant Soil 119, 147–154.

40. Passioura, J., 1977. Grain yield, harvest index and water use of wheat. J. Austr. Inst. Agric. Sci. 43, 117–120. Press, W.H., 1996. Numerical Recipes in Fortran 90: the Art of Parallel Scientific Computing. Cambridge University Press, New York Volume 2 of Fortran Numerical Recipes.

41. Preti, F., Dani, A., Laio, F., 2010. Root profile assessment by means of hydrological, pedological and above-ground vegetation information for bio-engineering purposes. Ecol. Eng. 36, 305–316.

42. Raza, A., Friedel, J., Bodner, G., 2012. Improving water use efficiency for sustainable agriculture. In: Lichtfouse, E. (Ed.), Agroecology and Strategies for Climate Change. Springer, Netherlands, pp. 167–211.

43. Richards, R.A., Passioura, J.B., 1989. A breeding program to reduce the diameter of the major xylem vessel in the seminal roots of wheat and its effect on grain yield in rain-fed environments. Aust. J. Agric. Res. 40, 943–950.

44. Rockström, J., Lannerstad, M., Falkenmark, M., 2007. Assessing the water challenge of a new green revolution in developing countries. Proc. Natl. Acad. Sci. U. S. A. 104, 6253–6260.

45. Safriel, U., Adeel, Z., Niemeijer, D., Puigdefabregas, J., White, R., Lal, R., Winslow, M., Ziedler, J., Prince, S., Archner, E., King, C., 2005. Dryland systems. In: Hassan, R., Scholes, R.J., Ash, N. (Eds.), Ecosystems Human Well-Being. Findings of the Conditions Trends Working Group of the Millennium Ecosystem Assessment, vol. 1. Island Press, Washington D.C., U.S.A, pp. 623–662.

46. Schenk, H.J., Jackson, R.B., 2002. Rooting depths, lateral root spreads and belowground/above-ground allometries of plants in water-limited ecosystems. J. Ecol. 90, 480–494.

47. Shaw, R.H., 1977. Water use and requirements of maize–a review. Agrometeorology of the Maize (corn) Crop. World Met. Organization, Geneva, Switzerland pp.119- 134 Publication 480.

48. Šimu nek, J., Huang, K., van Genuchten, M.T., 1995. The SWMS- 3D Code for Simulating Water Flow and Solute Transport in Three-Dimensional VariablySaturated Media. Research Report No. 139. U. S. Salinity Laboratory, Riverside, California. Smith, S., de Smet, I., 2012. Root system architecture: insights from Arabidopsis and cereal crops. Phil. Trans. Royal Soc. B: Biol. Sci. 367, 1441–1452.

49. Somma, F., Hopmans, J.W., Clausnitzer, V., 1998. Transient three-dimensional modeling of soil water and solute transport with simultaneous root growth, root water and nutrient uptake. Plant Soil 202, 281–293.

50. Sperry, J.S., Adler, F.R., Campbell, G.S., Comstock, J.P., 1998. Limitation of plant water use by rhizosphere and xylem conductance: results from a model. Plant Cell Environ. 21, 347–359.

51. Sposito, G., 2013. Green water and global food security. Vadose Zone J. 12, https:// www. soils. org/publications/ vzj/articles/12/4/vzj2013.02.0041.

52. Tardieu, F., 2012. Any trait or trait-related allele can confer drought tolerance: Just design the right drought scenario. J. Exp. Bot. 63, 25–31.

53. Thomson, B.D., Siddique, K.H.M., 1997. Grain legume species in low rainfall Mediterranean-type environments II. Canopy development, radiation interception, and dry-matter production. Field Crops Res. 54, 189–199.

54. Tollenaar, M., Lee, E.A., 2002. Yield potential, yield stability and stress tolerance in maize. Field Crops Res. 75, 161–169.

55. Trachsel, S., Kaeppler, S.M., Brown, K.M., Lynch, J.P., 2013. Maize root growth angles become steeper under low N conditions. Field Crops Res. 140, 18–31.

56. Vadez, V., Rao, S., Kholova, J., Krishnamurthy, L., Kashiwagi, J., Ratnakumar, P., Sharma, K.K., Bhatnagar-Mathur, P., Basu, P.S., 2008. Root research for drought tolerance in legumes: quo vadis. J. Food Legumes 21, 77–85.

57. Vadez, V., 2014. Root hydraulics: the forgotten side of roots in drought adaptation. Field Crops Res. http:// dx.doi.org/10.1016/j.fcr.2014.03.017.

58. van Genuchten, M.T., 1980. Closed-form equation for predicting hydraulic conducvtivity of unsaturated soils. Soil Sci. Soc. Am. J. 44, 892–898.

59. Waines, J.G., Ehdaie, B., 2007. Domestication and crop physiology: roots of greenrevolution wheat. Ann. Bot. 100, 991–998.

60. Wasson, A.P., Richards, R.A., Chatrath, R., Misra, S.C., Prasad, S.V.S., Rebetzke, G.J., Kirkegaard, J.A., Christopher, J., Watt, M., 2012. Traits and selection strategies to improve root systems and water uptake in water-limited wheat crops. J. Exp. Bot. 63, 3485–3498.

61. Zhang, H.W., Huang, Y., Ye, X.S., Xu, F.S., 2011. Genotypic variation in phosphorus acquisition from sparingly soluble P sources is related to root morphology and root exudates in Brassica napus. Sci. China Life Sci. 54, 1134–1142.

INDEX

O

P

Q

R

S

T

U

V

W